INTEGRATED CROP PROTECTION: TOWARDS SUSTAINABILITY?

BCPC Symposium Proceedings No 63

INTEGRATED CROP PROTECTION: TOWARDS SUSTAINABILITY?

BCPC Symposium Proceedings No 63

Proceedings of a symposium organised by
The British Crop Protection Council in
association with Sustainable Farming Systems

Held at Heriot-Watt University, Edinburgh,
Scotland on 11–14 September 1995

Chaired by R G McKinlay and D Atkinson

SFS
SUSTAINABLE FARMING SYSTEMS

BCPC Registered Office:
49 Downing Street, Farnham
Surrey GU9 7PH, UK.

British Library Cataloguing in Publication Data.
A catalogue record for this book is available from the British Library.

British Crop Protection Council
Pesticide Movement to Water
(Monograph Series, ISSN 0306-3941; No 62)

ISBN 0 948404 87 6

Cover design by Major Design & Production Ltd, Nottingham
Printed in Great Britain by Major Print Ltd, Nottingham

Contents

SESSION 3:
POSTER PRESENTATIONS

SESSION 4:
MOLECULAR BIOLOGY AND GENETICS

SESSION 5:
LANDSCAPE MANAGEMENT

SESSION 6:
SYSTEMS PROJECTS

Preface

Regardless of how it is defined, we believe that there would be general agreement that sustainability is an important concept for all of the processes and practices which are essential to our survival as a dominant species on earth, now and in the future. Sustainability is a critical concept for agriculture and agricultural production systems. For agriculture to be sustained, its requirements for renewable and non-renewable inputs must be in balance with the ability to provide these resources and the outputs which can be expected from these systems. In the medium and long term, this must mean the minimal practical use of non-renewable natural resources, such as fertilisers and pesticides, manufactured with the use of oil-based products. Agricultural systems requiring minimum inputs of resources are likely to make major demands on soil biological processes as key elements in both the provision of mineral nutrients and the protection of food crops from pests, weeds and diseases.

Sustainable agricultural systems will need to be based upon a clear understanding of basic chemical and biological processes. There will not be a single agricultural system which can be described as sustainable because the requirements for resource inputs will vary with the crops being produced and the climate where that production occurs. As sustainability must involve a time factor, it is likely to be relative rather than an absolute term. In addition, the word sustainability, as applied to agriculture, questions whether efficiency of farming systems can only be measured in criteria such as final yield and quality of produce. Increasingly, environmental criteria are likely to become more important.

Sustainability is clearly a concept which raises a larger number of questions, especially in an agricultural context. Sustainability raises questions as to the practicality of maintaining agricultural production with reduced use of non-renewable natural resources, the likely future availability of non-renewable resources, the future demand for foods, our ability to manage nutrient supply to crops in the absence of artificial fertilisers and our ability to manage biological systems.

Questions such as these indicate the need for existing information on sustainable agricultural systems to be brought together as a means of further refining the above questions and defining the need for future studies. With this aim in mind, a conference was established by a group representing the interests of the British Crop Protection Council (BCPC) and the Sustainable Farming Systems (SFS) Initiative.

The British Crop Protection Council represents the interests of a wide range of organisations with interests in crop protection. Through its annual Brighton Conference and its series of Symposia, it has become one of the most influential crop protection bodies in Europe.

The Sustainable Farming Systems Initiative was established in 1992 and formally launched the following year by His Royal Highness The Prince of Wales. The SFS Initiative has based its supported research programme around the study of organic farming systems as a means of developing our understanding of soil nutrient dynamics and the interactions which occur between crop plants, weeds, pests and diseases in a range of production systems. The Initiative aims to understand the basic biological processes which control the supply of nutrients and the balance between crop plants and other organisms as both a means of developed the organic farming concept and of transferring sustainable concepts based upon biological processes into more conventional farming systems. In this way, the Initiative aims to make mainstream agriculture more sustainable.

From attempts to produce crops in the developed world, using integrated crop management techniques, to attempts to conserve the soil and suppress pest incidence through intercropping in the developing world, the issues of sustainability appear global in extent. A major problem facing man as he moves into the 21st century still appears to be the challenge of feeding an increasing global population in a manner which is environmentally sustainable. Undoubtably the challenge will require the use of an integrated approach to crop protection. Such an integrated approach must make a key contribution to our understanding of sustainability.

As a consequence, both BCPC and SFS felt that now was the appropriate time at which to hold an international conference which would address the question of whether more integrated crop protection was likely to lead to a more sustainable system of crop protection. This volume includes the text papers which were presented at the Symposium run on this subject in September 1995. The editors thank all of the contributors to the Symposium which we believe make up an authoritative reference and which we hope will be of value to researchers and many others worldwide who have an interest in sustainability as applied to agriculture, and specifically to integrated crop protection. We also wish to thank Lothian and Edinburgh Enterprise Limited and Zeneca Crop Protection for their generous sponsorship of the Symposium.

R G McKINLAY
D ATKINSON

SAC
West Mains Road
Edinburgh

Symposium Organising Committee

Programme Chairman

Dr R G McKinlay — SAC, West Mains Road, Edinburgh EH9 3JG

Organisers of Sessions

Prof D Atkinson — SAC, West Mains Road, Edinburgh EH9 3JG

Dr D H K Davies — SAC, Bush Estate, Penicuick, Midlothian EH26 0PH

Dr G Edwards-Jones — SAC, West Mains Road, Edinburgh EH9 3JG

Dr L Firbank — ITE Monkswood, Abbots Ripton, Huntingdon, Cambridgeshire PE17 2LS

Dr R Harling — SAC, West Mains Road, Edinburgh EH9 3JG

Mrs S Ogilvy — ADAS High Mowthorpe, Duggleby, Malton, North Yorkshire YO17 8BP

Mr B Sheppard — SAC, Bush Estate, Penicuick, Midlothian EH26 0PH

Mr C J Siddall — British Crop Protection Council, 49 Downing Street, Farnham, Surrey GU9 7PH

Mr G Stoddart — Zeneca Crop Protection, Leadburn House, West Linton, Peeblesshire EH46 7BE

Mr D Tyson — British Crop Protection Council, Dorf House, Widdington, Essex CB11 3SF

Mr D Younie — SAC, 581 King Street, Aberdeen AB9 1UD

Organiser of Posters

Dr R O Clements — IGER, North Wyke, Okehampton, Devon EX20 2SB

Editor-in-Chief & Press Manager

Miss F M McKim — British Crop Protection Council, Foxhill, Stanford on Soar, Loughborough, Leics LE12 5PZ

Chairmen of Sessions

Session 1

Prof D Atkinson   SAC, West Mains Road, Edinburgh EH9 3JG

Session 2

Dr D H K Davies   SAC, Bush Estate, Penicuick, Midlothian EH26 0PH

Session 4

Dr W Spoor   SAC, West Mains Road, Edinburgh EH9 3JG

Session 5

Dr L Firbank   ITE Monkswood, Abbots Ripton, Huntingdon, Cambridgeshire PE17 2LS

Session 6

Mrs S Ogilvy   ADAS High Mowthorpe, Duggleby, Malton, North Yorkshire YO17 8BP

Session 7

Mr M Talbot   Biomathematics and Statistics Scotland, University of Edinburgh, James Clerk Maxwell Building, The King's Buildings, Mayfield Road, Edinburgh EH9 3JZ

Prof J B Dent   Institute of Ecology and Resource Management, University of Edinburgh, School of Agriculture Building, West Mains Road, Edinburgh EH9 3JG

Mr C Mackie   SAC, 581 King Street, Aberdeen AB9 1UD

Session 8

Mr G Stoddart   Zeneca Crop Protection, Leadburn House, West Linton, Peeblesshire EH46 7BE

Session 9

Dr G Edwards-Jones SAC, West Mains Road, Edinburgh EH9 3JG

Session 10

Dr R G McKinlay   SAC, West Mains Road, Edinburgh EH9 3JG

Acknowledgement

The Symposium Organising Committee wish to thank Lothian and Edinburgh Enterprise Limited and Zeneca Crop Protection for their generous sponsorship.

Lothian and Edinburgh Enterprise Limited

ZENECA
Crop Protection

Abbreviations

Where abbreviations are necessary the following are permitted without definition

acceptable daily intake	ADI
acid equivalent	a.e.
active ingredient	a.i.
approximately	c.
body weight	b.w.
boiling point	b.p.
British Standards Institution	BSI
by the author last mentioned	idem.
centimetre(s)	cm
Chemical Abstracts Services Registry Number	CAS RN
compare	cf.
concentration x time product	ct
concentration required to kill 50% of test organisms	LC_{50}
correlation coefficient	r
cultivar	cv.
cultivars	cvs.
day(s)	d
days after treatment	DAT
degrees Celsius (centigrade)	°C
dose required to kill 50% of test organisums	LD_{50}
dry matter	d.m.
Edition	Edn
editor	ed.
editors	eds.
emulsifiable concentrate	EC
freezing point	f.p.
for example	e.g.
gas chromatography-mass spectrometry	gc-ms
gas-liquid chromatography	glc
gram(s)	g
growth stage	GS
hectare(s)	ha
high performance (or pressure) liquid chromatography	hplc
hour	h
infrared	i.r.
integrated crop management	ICM
integrated pest management	IPM
International Standardisation Organisation	ISO
in the journal last mentioned	ibid.
Joules	J
Kelvin	K
kilogram(s)	kg
least significant difference	LSD
litre(s)	litre
litres per hectare	l/ha
mass	m
mass per mass	m/m
mass per volume	m/V
mass spectrometry	ms
maximum	max.
melting point	m.p.
metre(s)	m
milligram(s)	mg
milligrams per litre	mg/litre
milligrams per kg	mg/kg
millilitre(s)	ml
millimetre(s)	mm
minimum	min.

Ministry of Agriculture Fisheries and Food (England & Wales)	MAFF
minute (time unit)	min
molar concentration	M
no observed adverse effect level	NOAEL
no observed effect concentration	NOEC
no observed effect level	NOEL
nuclear magnetic resonance	nmr
number average diameter	n.a.d.
number median diameter	n.m.d.
organic matter	o.m.
page	p.
pages	pp.
parts per million	ppm
pascal	Pa
percentage	%
post-emergence	post-em.
power take off	p.t.o.
pre-emergence	pre-em.
pre-plant incorporated	ppi
probability (statistical)	P
relative humidity	r.h.
revolutions per minute	rev/min
second (time unit)	s
standard error	SE
standard error of means	SEM
soluble powder	SP
species (singular)	sp.
species (plural)	spp.
square metre	m^2
subspecies	ssp.
surface mean diameter	s.m.d.
suspension concentrate	SC
technical grade	tech.
temperature	temp.
that is	i.e.
thin-layer chromatography	tlc
time for 50% loss; half life	DT_{50}
tonne(s)	t
ultraviolet	u.v.
United Kingdom	UK
United States Department of Agriculture	USDA
vapour pressure	v.p.
variety (wild plant use)	var.
volume	V
weight	wt
weight by volume (mass by volume is more correct)	wt/v (m/V)
weight by weight (mass by mass is more correct)	wt/wt (m/m)
wettable powder	WP

less than	<
more than	>
not less than	≮
not more than	≯
Multiplying symbols-	Prefixes
mega (x 10^6)	M
kilo (x 10^3)	k
milli (x 10^{-3})	m
micro (x 10^{-6})	μ
nano (x 10^{-9})	n
pico (x 10^{-12})	p

Session 1

Socio-Economic and Political Framework in the European Union for Sustainable Farming Systems

Chairman   Professor D ATKINSON

Session Organiser   Dr R G McKINLAY

ANALYSING THE ECONOMIC CONSEQUENCES OF SUSTAINABILITY IN FARMING SYSTEMS

J.P.G. WEBSTER

Farm Business Unit, Wye College (University of London), Ashford, Kent. TN25 5AH

ABSTRACT

In many economies where a supported agriculture contributes but a small percentage of national income and where the mainly non-agricultural population has interests in the management of the rural environment, there is increasing interest in the concept of sustainability. Public pressure manifests itself in many ways, including adjustments to agricultural support systems which may include elements of cross-compliance or modifications to the economics of farming systems, which in turn lead managers to reduce inputs. Better knowledge on the part of such managers in relation to the complexities of the environmental management of their crops may allow them to maintain gross margins at reduced levels of inputs and yields. Such changes, whilst maintaining farm incomes, may reduce the demand for inputs, including labour. Research results tend to be location-specific so, with highly variable ecologies across regions, it is very difficult to forecast the regional economic impacts of the adoption of sustainable systems. At the national level, aggregate reductions in the production of some commodities would assist in the achievement of the GATT agreements to reduce subsidised exports.

INTRODUCTION

The case for a move from existing farming systems to "sustainable" farming systems is based on the argument that, in the long run, current systems are leading to an undesirable rural environment in social, economic and ecological senses and thus a decline in the utility of a representative member of society (Turner, 1993). An undesirable social environment because fewer people are employed in agriculture, more people living in rural areas have less to do with agriculture and those who do remain are socially isolated and suffer a decline in rural social services (Pretty and Howes, 1993). The economic problems arise from the supported nature of agriculture with declining farm incomes relative to other sections in society, but where the CAP is still of major significance to the EU budget. The ecological problems arise from the contribution of pollution from heavy use of pesticides and fertilisers, and from the specialised nature of much of animal agriculture leading to problems of disposal of natural wastes and the loss of diversified habitat for natural predation (Hodge and Dunn, 1992).

An implied theme in the sustainability literature appears to be the distinction between closed systems versus open systems in agricultural production. Sustainability appears to involve a move towards a more closed system whereby byproducts are consumed on the farm and the market intrudes only at the final product stage. This could entail a diversified, mixed livestock and cropping system whose archetype might be the subsistence farmer occasionally selling food products in order to gain money to pay for tools or schooling for his children. At the other extreme, there is the specialist producer selling only a single product, all of whose inputs are bought in the market. Modern technology now permits economies of scale both in production and in transport between regions. Thus the comparative advantage of particular localities is enhanced and specialisation takes place.

But it is necessary to be more specific about the characteristics of sustainable systems before discussing the impacts of their adoption. Pretty and Howes (1993) reviewed much of the sustainability literature and defined it in terms of five goals. These included (i) a more thorough incorporation of natural processes into the agricultural production processes, (ii) a reduction in the use of off-farm inputs, (iii) a greater use of the biological and genetic potential of plant and animal species, (iv) an improvement in the match between cropping patterns and physical limitations to ensure long-term sustainability of current production levels and (v) profitable, whole-farm management to conserve soil, waste, energy and biological resources. They also emphasised that they regarded sustainable systems as a "loosely defined" middle ground between "organic agriculture and high input industrialised agriculture" (Pretty and Howes, op.cit. p8).

MAKING THE CONCEPTS OPERATIONAL

Table 1. Some Suggested Changes to Farming Systems and Practices to Enhance Sustainability (after Pretty and Howes, 1993).

Changes to Husbandry Practices:-

Reduce pesticide usage by being more selective; adopt IPM

Patch spraying, beetle banks, wild flower strips

Use of natural predators; crop mixtures; resistant varities; multiline varieties

Improve fertilizer efficiency; timing and placement; adopt global positioning systems

Incorporate legumes and catch crops for maintenance of soil stability and fertility

Maintain hedges, coppices as windbreaks, wildlife corridors and reservoirs

Changes to Farming Systems:-

Adopt diversified farming systems involving both crops and livestock

Balance intensity of livestock production with arable area for disposal and use of wastes

Lengthen and diversify rotations to improve pest management and soil fertility and thus reduce requirements for external inputs.

Pretty and Howes also made a number of suggestions as to what changes might take place in farming systems and practices in order to further sustainability. Husbandry changes include reductions in pesticide and fertiliser use by improved ecological management of the social and non-cropped areas. Systems changes include a diversification of crops within arable rotations plus a return to more mixed livestock and cropping systems. Some of their suggestions are listed in Table 1.

A difficulty with suggestions of this type is that it is easy to confuse means and ends. Are these practices to be regarded as ends in themselves? If so, how does one know when a "sustainable" system is achieved? Who makes the judgement? What indicators are to be used and what tradeoffs between them are acceptable? It does seem that the subject area is suitable for a Logical Framework Analysis (Coleman, 1987) in which ends are clarified, means are identified and targets set. Without agreement on indicators, it will surely be difficult to make the desired progress.

AN ECONOMIC FRAMEWORK FOR FARM SYSTEMS CHANGES

It is possible to categorise these changes towards sustainability in terms of their economic characteristics (White et al., 1993).

Efficiency changes

These include the adoption of practices which involve a change in the use of an input which leads to sustainability gains. Reductions in nitrogen applications thus give sustainability gains if nitrate leaching into groundwater or runoff into watercourses is reduced. If maintenance of profitability is the criterion, then a sustainability practice will be adopted by farmers so long as the value of the reduction in yield is less than the value of the fertiliser saved.

The same principle applies to the application of pesticides. Whilst the relationship between pesticide inputs and yield is very much more complex, it is still true that farmers will adopt the practice so long as gross margins are maintained.

Substitution changes

These involve the replacement of one input by another with the aim of improving the sustainability of the farming system. Examples might include the use of nitrogen generated by legumes within a rotation as a substitute for inorganic nitrogen, or the substitution of mechanical weeding for the use of herbicides in intensive vegetable production.

In order to estimate the economic impact of such a change, we need to know the rate of substitution between the two inputs for a given level of output. For example in the mechanical weeding case, it is clear that initial reductions in herbicides might relatively easily be substituted for mechanical weeders. But further reductions would become increasingly difficult. The optimal point is where the extra cost of more mechanisation is just equal to the costs foregone by reducing the use of herbicides.

Redesign of cropping and livestock systems

The redesign of systems involves the addition or, less likely in the sustainability debate, the deletion of products, or changes in enterprise size within the farming system. The optimal product mix for a farm is based on the available resources plus the relationships between the various products. Sustainability clearly implies a reduction in external inputs and an increased reliance on the interrelationships between the production systems of a number of commodities. Possibilities include the introduction of legumes into the rotation, the use of diversified crop and livestock farming systems, and diversification of cropping systems to reduce pest problems.

For the farmer, the point of maximum profit is where the rate of substitution of one product for the other is equal to the ratio of the product prices. In other words, where the increase in total gross margin arising from increasing the area of one crop just extinguishes the consequent reduction in total gross margin from the other crop (Barnard and Nix, 1979, p38).

This kind of analysis can be used to explain the reasons for specialisation where there are returns to scale in production. Given the fixed cost structure involved in the production of milk (labour, fixed equipment) and in the production of cereals (labour, machinery), it is clear that a farm of, say, 50 hectares would find it difficult to operate a diversified system profitably. The relationship between the two enterprises is such that the requirement of resources for even a small dairy enterprise would seriously impede the ability to carry on an efficient cereals enterprise. In such a case, the diversified farmer would probably have to employ contractors to farm the cereal area, thus increasing his exposure to external inputs.

PROGRESS IN GENERATING THE REQUIRED INFORMATION

Data Requirements

The above three-way categorisation also enables us to specify the types of data needed. In each case, we need to know how the suggested shift will impact upon both the economics and the sustainability indicators following, of course, their adequate definition. Where efficiency changes are being considered, we need to know how the gross margin will change, and what will be the effect on the sustainability indicators. When substitution changes are being considered, we need to know how the costs of production will change and the impact on sustainability of the simultaneous change of both factors. A less harmful input may be substituted in place of a more harmful one. Knowledge will be needed about how the change affects the particular ecological surroundings in which the change will take place. Clearly this will vary from locality to locality, depending on soil type, topography and other factors. Finally, when the redesign of systems is being considered, the mechanisms of the interactions between the crops in terms of the sustainability indicators has also to be known. This will include knowledge of mineral balances of N, P and K between the different crops, as well as pest/predator relationships under the various possible cropping combinations. It seems likely that obtaining the data will be neither straightforward nor cheap, even given the definition off appropriate sustainability indicators.

Some Experimental Results in Low Input Agriculture

Over the past 15 years MAFF, together with a number of collaborating agencies and companies, has supported a major set of programmes to develop farming systems which, while remaining profitable, safeguard the environment as far as possible. Since much of the debate has concerned the reduction of potentially harmful external inputs, it is not surprising that this is an area on which emphasis has been placed.

Table 2 lists five of these programmes which have often been aimed at elucidating the 'ecological' impacts of modified input levels as much as their profitability. As time has passed, and as scientists have increasingly recognised the potential for system changes rather than simple input reduction, the move has been towards system redesign in the form of the manipulation of rotations or the introduction of new crops.

Table 2. Some Experiments to Investigate Changes in Arable Agriculture

	date	Treatment	Treatment Levels	
BOXWORTH	1983-9	Pesticide levels	"minimum"	"supervised"
			"full insurance"	
SCARAB	1990 -	Pesticide levels	"current"	"low"
TALISMAN	1990 -	Rotations	"standard"	"green"
LIFE	1990 -	Rotations	"conventional"	"integrated"
		Input levels	"standard"	"low"
Link - IFS	1992 -	Rotations	"conventional"	"integrated"

Sources: Grieg-Smith et al(1992), Cooper(1990), Jordan and Hutcheon(1993), Prew(1992)

But there is still a long way to go. It has to be recognised that, as far as efficiency changes are concerned, we are dealing with a continuous relationship between input levels, yields and sustainability indicators. Whilst two levels, such as "current" and "low", will give some idea as to where the curve might be, it will not be much help in locating an optimum trade-off between sustainability indicators and yield. Likewise, ideas about feasible rotations are continuously changing (Jordan and Hutcheon,1993) so, by the time the results are available, new knowledge may have outdated the experimental rotations. Nevertheless, such experiments are essential because they can provide the basic science which underlies the complex interactions between crops and which is essential in the development of sustainability indicators. Once this knowledge is available, progress in the redesign of systems can be speeded up by adopting a modelling approach.

CONSEQUENCES OF THE ADOPTION OF SUSTAINABLE FARMING SYSTEMS

We now address the likely economic impact of the adoption of sustainable systems at the individual farm level, the regional level and the national or supranational level. For various reasons it is difficult to make any quantitative estimates, but it may be possible to identify the general directions of change and to indicate those factors which appear likely to be important in determining the magnitude of change.

The Farm Level

The major determinants of change at the farm level will be the existing farm system; the soil type, climate and topography; and the surrounding natural ecology. Where farming systems already depend on a diversified crop and livestock complex, and where there is integration of natural processes, there will be less need for change. However, where a high input system involving relatively few enterprises has developed, major change may well be needed if the objectives listed earlier are to be achieved. It is difficult to see, for example, large scale arable units in the Eastern Counties of the UK making major alterations including the incorporation of livestock into the system without a considerable impact on farm income. Likewise, changes would have to occur in the structure of pig production if limits were to be placed on the ratio of animals to hectares of arable land as in the Netherlands or Denmark. Whilst dairying in the west of England might be regarded as relatively sustainable, stocking rates would have to decline if inorganic fertiliser applications were to be reduced. An initial conclusion therefore must be that there will be considerable variability in impact, depending upon the existing conditions in a particular locality.

It is important to remember that, other things being equal, the adoption of these changes will usually lead to some loss of income for farmers. This is because, if it were otherwise, farmers would already have adopted the recommended practices and there would be less need for change. The *ceteris paribus* assumption is important, though. In practice there are two major variables which need not remain constant.

The first is the level of knowledge of the farmer and the second is the set of policies under which farmers operate. If farmers are shown that sustainability can actually increase their incomes albeit at the cost of improved managerial skills, then they are likely to adopt its practices. But if there remains a cost to the adoption of these practices, even though there are public benefits, we can expect farmers to be less enthusiastic. This applies particularly to those farmers who are already under financial pressures from, for example, the small size of their operation or the lack of other opportunities for generating family income (Gasson, 1988).

Table 3. Pesticide Application on Wheat 1982 - 1992, kg ai per ha

Year	Fungicide	Herbicide	Insecticide	Molluscicide	Seed treatment	Total
1982	1.37	5.80	0.23	N/A	0.06	7.46
1988	1.64	4.16	0.15	0.06	0.00	6.01
1990	1.38	2.77	0.11	0.02	0.02	4.30
1992	1.33	2.20	0.08	0.02	0.05	3.66

Source: MAFF, Pesticide Usage Surveys

The second variable is the set of support policies adopted by government. The reform of the CAP in 1992 involved a move from product price support to area payments, with the aim of maintaining incomes whilst reducing production via the set-aside provisions. A clear implication has been that, with reduced product prices, the marginal revenue from the last unit of variable input will be reduced, and so farmers are likely to reduce the level of inputs. It is too early to quantify this particular effect, since aggregate use of inputs of fertiliser and pesticides has been decreasing since around 1988. Furthermore, table 3 shows how pesticide use on wheat has been declining over the past decade, possibly also in response to declining real product price. Finally, the devaluation of the pound against the ecu in 1992, resulting in higher than anticipated product prices and area payments, has confounded this effect. But there is some evidence that one impact on arable systems of the reforms may be an increase in the amount of pulses and legumes in crop rotations (Donaldson et al., 1995).

We conclude, therefore, that under a "no policy change" scenario, and without some investment in the agronomic managerial skills of farmers, the impact of sustainability upon farm incomes is likely to be quite variable but generally negative. Cain et al. (1995), investigating the loss of profits consequent upon the mandatory adoption of specified practices, confirm this. A question which immediately arises is: what sort of policies would lead to the adoption of sustainable systems? Does it have to be mandatory practices or management agreements, or are there alternative market-based policies? This aspect must clearly be part of the agenda of any publicly funded research programme into sustainability.

The Local and Regional Level

Similar factors also affect the degree and direction of the impacts at regional level. Since much of the research currently being carried out is location-specific, it is difficult to know whether or not a system which is sustainable in one part of the country is necessarily sustainable in another. We can, therefore, not be specific about the regional impacts but we can speculate about the effects on the supply industries and their markets, the commodity processing industries and their markets, and finally the consumption impacts caused by changes in agricultural income.

One of the principles of sustainability involves decoupling the farming systems from purchased inputs. These will generally include fertiliser and pesticides but may include other goods and services such as machinery. This seems likely to lead to further contraction in the markets which have seen the wholesale elimination of family-owned local merchanting businesses over the past decade or so. There is some suggestion that demand for labour (for weeding and hedgelaying; Pretty and Howes, 1993) would increase, but these highly seasonal operations seem unlikely to do more than provide casual jobs at certain times of the year. If yields are reduced then demand for harvest and postharvest labour will also be reduced.

If the adoption of sustainability involves reductions in the amount of product, there are likely to be consequences in the commodity processing sector. With a reduction in inputs and hence yields, there is likely to be under-utilised capacity in the processing sector, much of which is located regionally. In addition, transport costs per tonne would increase, leading to increased concentration in the processing sector. This sector is already subject to rapidly changing technology and significant economies of scale. Sustainability would probably exacerbate the changes currently taking place. McCorriston (1995) has developed a methodology for investigating these downstream effects of changes in environmental policy. Furthermore, it would be important to maintain quality levels since, if sustainability involved an increase in variability of quality of product (Fenemore and Norton, 1985), there might be a loss in competitiveness with respect to high quality product imported into the region. Making a rather different point, Hanf and Verreet (1994) remind us that the substitution rate between a decrease in pesticides residues in food and drinking water and an increase in mycotoxins in food products is as yet unknown.

There is an argument that regional sustainability might include local processing and the development of niche products in national or international food markets. Whilst there are many successful examples of such products (cheeses, wines, etc.), it is difficult to see them becoming a major generator of income in every locality.

It is important also to recognise the potential consumption effects of sustainability. We have seen that without changes in support policies, a mandatory move towards sustainability would likely reduce farm incomes. Such a move would have impacts on the local community because the spending of farmers and others would be reduced. Rural shops, already under pressure from supermarket development, garages, repair facilities and other rural enterprises, would thus suffer. Whilst there may be benefits in terms of environmental externalities to the rural non-agricultural population and to urban-based visitors to areas where sustainability is practised, it is difficult to see anything other than a widening of the gap between farm and non-farm incomes if specific support measures are not adapted. There is much regional variation in the importance of non-farm and off-farm income to the rural economy (Gasson, 1988). But is seems clear that where farm income is important - and this will include much of the more remote areas of the UK - the decline of the rural economy would not be halted.

Aggregating to the UK and European Level

As yet there is little basis for informed speculation as to the impacts of sustainability at the national or European level. Most of the experimental results are highly location-specific and may not necessarily apply across the very variable ecologies which are found both within the UK and across Europe. It thus seems clear that both the income and environmental effects of adopting sustainable systems, however they are defined, will be very variable since the levels of external input use across Europe vary greatly (Brouwer et al., 1993). The Mediterranean ecologies and farm structures may be expected to react very differently as compared with those found in northern Europe. However if policies were adopted which did stimulate a move in the direction of sustainability then the relative comparative advantage of different regions might change, leading to significant adjustments of location of production and of cropping patterns within particular locations. Forage-based animal production might become more concentrated in the wetter north-western areas of the continent, whilst fruit and vegetable production would be further concentrated in regions where solar energy and natural water supplies were plentiful.

Predicting the aggregate economic consequences is even more hazardous. If, as some of the UK data appears to suggest, it is possible to reduce yields and input levels without compromising gross margin levels, then production-related support costs would be reduced. Likewise, such production adjustments would make the GATT agreements on levels of aggregate support more easily achievable. Whilst this applies to supported crops, there might well be significant effects on the price of unsupported crops such as intensively grown fruit and vegetables.

CONCLUSION

Sustainability is a concept which has yet to be made operational in many agricultural situations. Whilst the aspirations involved are relatively clear, it is the case that a full range of indicators for different sets of ecological circumstances remain to be developed.

A major component is the reduction in the use of external and possibly harmful inputs in the agricultural system. Experimental results suggest that in some cases gross margins can be maintained using fewer inputs, but the identification and measurement of indicators of sustainability is still under development. Much of the work has related to comparisons between current practice and organic systems, whereas the sustainable system appears usually to be regarded as something in between, and so extrapolation from organic systems may be inappropriate.

A concern of the proponents of sustainability appears to be a move to mixed livestock and arable systems. What little low-input livestock experimental work exists has not generally involved a mixture of crops and livestock. The economies of scale in each of these branches, together with the existing farm size structure in UK agriculture would suggest that wholesale change in this direction is unlikely.

Extension of the experimental arable results would seem not to compromise farm income for some arable farmers. But it is not known what proportion of farmers fit in this category. Beyond these cases, it seems likely that in the absence of policy changes to support such a move the effect on farm incomes would be negative. We need to know what changes in policy would foster sustainability on the majority of our farms. The alternative seems to be payments for mandatory practices in the form of management agreements.

Given that the general ethos of sustainability involves a reduction in external inputs and a move towards internal self sufficiency, it is hard to avoid the conclusion that the immediate effects on the existing local rural economy would not be positive. It seems likely that there would be further concentration in both the upstream and the downstream sectors. Whilst it is sometimes suggested that sustainability might increase the demand for labour, that labour is likely to be unpaid, seasonal, or casual.

Finally, it is not yet possible to say what the effects of a move towards sustainability across the EU might be. Whilst the EU is the policy-making unit for European agriculture as a whole, the effects of its policies have differing consequences across the member states. Under present policies, if yields were reduced and income maintained then support costs would be reduced and the GATT agreements made easier to achieve.

REFERENCES

Brouwer, F.M.; Terluin, I.J.; Godeschalk, F.E. (1994). *Pesticides in the EC*, The Hague: LEI-DLO, pp 159.
Barnard, C.S.; Nix, J.S. (1979). *Farm Planning and control*, Cambridge: University Press, pp600

Cabinet Office (1995). *Forward Look of Government-funded Science, Engineering and Technology*. London: HMSO, p. 49.
Cooper, D.A. (1990). Development of an Experimental Programme to Pursue the Results of the Boxworth Project. *Brighton Crop Protection Conference - Pests and Diseases*, 1, pp 153-162.
Coleman, G. (1987). Logical Framework Approach to the Monitoring and Evaluation of Agricultural and Rural Development Projects. *Project Appraisal*, 2, 4, pp251--259
Donaldson, A.B.; Flichman, G.; Webster, J.P.G. (1995). Integrating Agronomic and Economic Models for Policy Analysis at the Farm Level: The Impact of CAP Reform in Two European Regions. *Agricultural Systems*, 48, pp 163-178.
Fenemore, P.G.; Norton, G.A. (1985) Problems in Implementing Improvements in Pest Control; A Case Study of Apples in the UK. *Crop Protection*, 4, 1, pp51--70
Greig Smith, P.; Frampton, G.; Hardy, A.R. (1992) *Pesticides, Cereals and the Environment; The Boxworth Project*. London: HMSO, pp 282.
Hanf, C-H.; Verreet, J-A. (1994). Consequences of a Total Ban on Fungicide Application on Agriculture and Agribusiness. In: Michalek, J.; Hanf, C-H. *op. cit.*, pp 205-226.
Hodge, I.; Dunn, J. (1994) *Rural Change and Sustainablity; A Review of Research*. Swindon: Economic and Social Research Council, pp139.

Jones, M.J. (1993). Sustainable Agriculture: an Explanation of a Concept. In: *Crop Protection and Sustainable Agriculture*, Chadwick, D.J. and Marsh, J. (Eds.), Chichester: Willey (CIBA Foundation Symposium 177), pp 30-37.

Jordan, V.W.L.; Hutcheon, J.A. (1993). *Less Intensive Integrated Farming Systems for Arable Crop Production and Environmental Protection.* London: The Fertiliser Society, Proceedings no. 346, pp 32.

Gasson, R. (1988) Farm Diversification and Rural Development. *Journal of Agricultural Economics*,**39**, 2, pp175-182.

McCorriston, S. (1994) *Environmental Policy in Vertically Related Markets, in* Michalek, J.; Hanf, C-H. (1994). (Eds) *The Economic Consequences of a Drastic Reduction in Pesticide Use in the EU.* Kiel: Wissenschaftsverlag Vauk K.G., pp109-122

Michalek, J.; Hanf, C-H. (1994). (Eds) *The Economic Consequences of a Drastic Reduction in Pesticide Use in the EU.* Kiel: Wissenschaftsverlag Vauk K.G., pp 352.

Pretty, J.N.; Howes, R. (1993). *Sustainable Agriculture in Britain: Recent Achievements and New Policy Challenges.* London: International Institute for Environment and Development, Research Series, **2**, (1), pp 74.

Prew, R. (1992). Development of Integrated Arable Farming Systems for the UK. *Proceedings, HGCA Conference on Cereals Research and Development*, pp 242-254.

Turner, R.K., (1993). *Sustainable Environmental Economics and Management; Principles and Practice.* London; Bellhaven Press, pp386.

White, D.C.; Braden, J.B; Hornbaker, R.H. (1993). Economics of Sustainable Agriculture. In: *Sustainable Agriculture Systems*, Hatfield, J.L. and Karlen, D.L. (Eds.), London: Lewis Publishers, pp 230-258.

Wijnands, F.G.; Vereijken, P. (1992). Regionwise Development of Prototypes of Integrated Arable Farming and Outdoor Horticulture. *Netherlands Journal of Agricultural Science*, **40**, pp 225-238.

THE POLICY APPROACH TO SUSTAINABLE FARMING SYSTEMS IN THE EU

J S MARSH

Centre for Agricultural Strategy, University of Reading, Berks.

Thomas Malthus, an early economist, took an essentially pessimistic view of the earth's ability to support people, based on his understanding of the conflict between trends in population size and the available resources. However, in the 197 years since his essay on population was published, the world's population has increased by 4.5 billion. For many people living standards are higher than ever before and, possibly for the first time, we now have the ability, even if we lack the organisation, to provide a nutritious diet for all those people currently alive. This sort of experience has led people to discount Malthusian gloom in favour of a technologically based optimism.

Current concerns about sustainability question that optimism. From the work of the Club of Rome in the early 1970's, the publication of The Global 2000 Report to the President(1982), the report of the Bruntland Commission (1987) and successive international conferences, including the UN Conference on Environment and Development, (1992) have come increasingly informed analyses of the fragility of the economic system and the excessive demands it makes on global natural resources. Agriculture is deeply implicated in these concerns. Not only is the provision of food a sine qua non of continued human survival but the industry is the largest single user of the world's land resources.

It is, however, not just the scale and necessity of agriculture which makes it crucial in debates about sustainability, it is also the fact that to a degree unknown in other consumer industries, it is deeply affected by government policies. This provides both a challenge and an opportunity. The challenge is to devise policies which properly reflect to consumers and producers the longer run values of the resources used in production. The opportunity arises because there exists a wealth of knowledge and a diversity of policy instruments.

This paper is concerned with how sustainability has become an issue for farm policy, with the options open to the industry and to governments to devise sustainable farming systems and with the role of the European Union in implementing such policies.

1. The imperatives for sustainable farming systems.

(i) What is at risk of not being sustained?

Farming systems are the meeting point of natural, economic and social systems, each of which has its own dynamics. For farming systems to survive, they have to be simultaneously sustainable in each of these dimensions. If they are not then the farming system will have to change. Adjustments to reach a lasting system in any one dimension may be incompatible with its sustainability in the others. As a result changes which originate, for example, in the natural world, are likely eventually to result in changes in industry and the social structures which it supports.

The object of farming is to modify natural resources in ways which add to their usefulness to man. Productive plants and animals are developed, protected from competition and disease and multiplied to the extent that markets or government policies justify. The whole of this process is concerned with changing rather than sustaining the existing natural environment. If biological sustainability is conceived in terms of retaining all existing species, farming is an incompatible activity. At least three sorts of concern about modern farming have been voiced in recent years. First, that the system is self-destructive; its dependence on a small range of biological and other natural resources means that it will ultimately be incapable of continuation. It may simply exhaust the set of finite resources upon which it depends or lead to a catastrophe if pests or diseases attack the favoured, 'economic' species.[1] Second, the power of modern farming to suppress competitive biota threatens a loss of biodiversity which will impoverish future generations. Losses on this score may be economic, they may lessen our chances of eliminating some human disease and they may reduce the quality of life for all people.[2] Third, the extraordinary increase in the recent ability of farming to manipulate the natural environment through mechanisation, the use of agro-chemicals and biotechnology, raises ethical questions. For example, how far is it acceptable for humanity to suppress other species, can one generation justify actions which diminish the alternatives open to its successors?[3]

Farming is a business. To survive it has to be able to reward the resources it employs at rates which will retain them in their current activities. Farming systems represent the response of farmers to continuously changing economic circumstances Productivity raising technologies tend to lower prices, those who cannot compete at these prices ultimately cease to farm. Higher levels of income in a growing economy mean that the cost of labour, whether hired or provided by the farmer himself, is rising in real terms. If that cost is to be covered the farm business has to generate additional revenues over other costs. Farming systems have adjusted by increasing farm size, increasing productivity and reducing employment. Within the EU the agricultural labour force declined from around 25% in the mid 1950's to less than 7% by 1992. Yields of crops and animals have increased as a result of breeding and the use of fertilisers and supplementary feeding.

This has made heavier demands on some natural resources. If their rate of use is to be restricted in order to protect the interests of future generations, some existing farming systems will not be sustainable. However, responding to environmental demand may create new sources of revenue for some farmers. Governments, in richer societies may choose to pay for some farming activities which currently do not generate adequate revenue, for example those which have shaped treasured landscapes. If these policies accurately capture the values of the population, they will lead to a more economically efficient use of resources even if many existing farm and other businesses disappear. Sustainability from this perspective is concerned with the continued value of the industry's outputs not just with the preservation of particular inputs or specific farm businesses. The challenge to the industry is to devise farming systems which respond to the new set of values.

Changing farming systems have profound social and political consequences. In many parts of Europe farmers are still major players in rural communities. There are strong pressures to preserve such communities. Culturally they represent distinctive traditions and provide much of the diversity which makes life in Europe more interesting. Politically they are often represented by relatively cohesive groups which can exercise power within national government structures. Where these groups are strong, it is often assumed that rural and agricultural policies are identical. However, for most of Europe this has become a serious oversimplification. In many 'rural areas' the bulk of income and jobs is no longer on farms or even related to agriculture. For many city dwellers the existing countryside represents a means of escape and recreation. Where governments in an attempt

---

1   Such concerns led the US Board of Agriculture in 1984 to appoint a committee to study the science and policies that influenced the adoption of alternative productive systems designed to combat pollution from nitrates, pesticides and anti biotic residues, soil erosion and the depletion of aquifers. The committee's report, Alternative Agriculture, – National Research Council National Academy Press Washington 1989 recommended changes in US Federal Agricultural Policy, in Research and Development designed to support and encourage the application of alternative farming systems.

2   See for example, Goodman D and Redclift M, – Refashioning Nature, Food Ecology and Culture Routledge London and New York 1991 notably Chapter 2 'The passing of rural society' and Chapter 6 The Food System and the Environment.

3   Such a view is implicit in the Bruntland report's definition of sustainability which requires each generation to meet its own needs without lessening the opportunities for future generations to meet theirs – Our Common Future: the report of the World Commission on Environment and Development – Oxford University Press 1987

to achieve social sustainability, have sought to protect traditional types of farming[4] it has proved necessary to do more than preserve the status quo. If people are to remain in farming, their incomes have to keep pace with the growth of income elsewhere in the economy. One response has been to enable farmers to sell more. This has led to surplus production, high budget costs and international trade conflicts. It has also increased the pressure of farming on the natural environment.

These underlying conflicts have led commentators to question the concept of sustainable development.[5] Viewed in purely economic terms, where the measure can be in terms of a flow of output which is valued more highly in successive years, continuous development is conceptually possible. Values may grow as much because of developments in understanding and appreciation as in physical quantities consumed. In natural resource terms, the contradiction between development and sustainability is more difficult to avoid but even here greater efficiency in the use of resources, stemming from improved understanding and methods may mean that although ultimately a ceiling must be reached, in practical terms this does not prevent continued real economic growth. The issue boils down to a race between the rate of increase in the use of natural resources and the rate of growth in demands upon them. In contrast, there appears to be no equivalent practical or conceptual escape route from the conflict between economic development and existing social systems. Here the most useful concept is not so much sustainability as acceptable rates of change. Given that change may be inescapable for both natural resource and economic reasons, social implications become a central concern of policy makers.

(ii) Evidence of unsustainability.

Technologies have greatly enhanced crop yields and livestock productivity in Europe Land has been used more intensively, more than offsetting the decline in its area. Demands for land for other uses, including housing, road building and recreation have expanded with the growth of the EU economy. A number of indicators suggest that current systems may not be sustainable. These are not evenly distributed across the Union and in some instances are a cause for anxiety only in rather limited areas.

In countries where farming is most intensive pollution of both ground and surface waters has resulted from highly intensive livestock systems and arable farming practices. The escape of fertilisers or farmyard manure into water systems may lead to algal blooms or eutrophication. Raised levels of nitrates in drinking water have led to the imposition of EU wide standards. Pesticides which escape into the environment may damage non-target plants and animals leading to anxiety about their impact on human health.

The range of commercial varieties now used on farms has greatly narrowed. Dependence upon too small a genetic base raises anxieties about the vulnerability of species to diseases to which the variety concerned may have little resistance This loss of biodiversity within agriculture reduces the options open to plant and animal breeders and may make the industry less able to cope with future demands, whether from the market or as a result of environmental pressures. Combined with the continuing loss of wild varieties[6] this has led to increased demands for policies to preserve biodiversity.

---

4 For example much of the debate about the reform of the CAP has been concerned with the future of the small family farm. The Commission in describing the goals of the Reform of the Cap in its annual report for 1992, identifies as two of its six goals: a certain redistribution of support to the benefit of more vulnerable enterprises, and: continued employment for a sufficiently high number of farmers, while encouraging a certain mobility as regards production factors, notably land, in order to create more efficient production structures. The Agricultural Situation in the Community 1992 Report CEC Brussels and Luxembourg 1993

5 See for example the report of the Club of Rome Donella H and Dennis L Meadows et al., The limits to Growth Washington: Potomac Associates 1972

In some places soil erosion has become a matter of concern. Mechanisation enables soils to be tilled which would have been beyond traditional methods. The use of manufactured fertilisers can enable yields to be maintained in the short run, even if the underlying soil structure is damaged or its organic contents depleted. Erosion problems are site specific and in most of the UK and the northern countries of the EU are not yet perceived as critical. In some regions in the South of the Community, however, there is evidence of erosion both from the distant past and more recently.

Throughout Europe farming populations have been in decline in the final decades of this century. In regions with good transport or where modern industries have established themselves, total rural populations have not declined, indeed the number of residents has increased. In the remoter regions, in hill and mountain areas, however, the decline of farming has been associated with a process of depopulation. The tendency has been for the younger and more able adult population to seek employment elsewhere, leaving villages with an increasing proportion of older people and small children. Such villages become decreasingly able to support the overhead costs of transport, medical services, schools and shops, social structures which make life more tolerable. Such communities become increasingly disadvantaged. The ultimate result is often the abandonment of such settlements.

The departure of the farming population need not automatically lead to the abandonment of land nor the death of rural communities. Farms can be amalgamated and the larger units provide a basis for the livelihood of a new farming household. New economic activities can provide fresh employment and support a population which sustains the social infra-structure. However, where there is no land market, where farms are badly fragmented or where there are tax advantages in remaining nominally a farmer although being resident in a city, structural change within farming may not take place. In this situation land which is potentially productive may fall out of use. In some parts of the Community, for example in Greece and in the Massif Central in France this loss of land to agriculture is regarded as a serious problem. Similarly where communications are poor, planning restrictions obstruct and the social infra-structure is already inadequate, it is difficult to attract new enterprise or to prevent the outward flow of people.

(iii) Who demands sustainability?

The loss of land to farming is not universally regarded as a problem. It provides an opportunity for other types of ecology, managed or natural to develop. It frees resources for other industries such as forestry or recreation. It may diminish the aggregate supply reaching the markets for farm goods and so strengthen prices. Such observations illustrate that while the idea of sustainability is generally regarded as important, the content of what different groups demand differs quite widely.

Scientists studying how natural phenomena develop have drawn attention to anthropogenic sources of global change. Their primary concern is to explain the relationship between human actions and the natural world. Where changes in important variables are seen to result, their responsibility is to explore these, to examine their longer run consequences and to alert the rest of society should damaging outcomes seem to be threatened. They may also help in identifying possible modifications in human behaviour which would ameliorate any undesired outcomes. Recent work on global warming and the potential this may have for climate change provides a good example[7]. Agriculture, through its use of fossil fuels, through the destruction of forests to extend the farmed area and because of release of green-house gases associated with ruminant livestock and paddy rice, is part of the cause. It may also be part of the solution as a means of capturing renewable energy from the sun and of recycling rather than increasing the carbon dioxide content of the atmosphere by replacing fossil fuels[8]. Science has to be concerned with the balance of these effects and how it may be managed to promote human welfare.

---

6   See Missing species never to be seen again Times London 4th March 1995.

Within Europe green politics have become a matter of widespread interest. Green spokesmen make much of the need to change economic systems, including farming, in order to attain a greater degree of sustainability.[9] Their critique goes much further than farming. It questions the basis of a consumer driven society. In this it has something in common with traditional puritan values which have formed part of the development of thought within Europe. Although it's overall message may not command a majority in any country it has influenced the attitudes of many other parties, particularly where it focuses on one or two high profile issues. It is not politically safe to be seen as 'non' green.

Consumers, too, have conflicting interests in sustainability. On the one hand they are anxious to ensure that the goods and services which they want are available at affordable prices, and will continue to be available for their children. On the other hand, they feel threatened by demands to change lifestyles. Owners of second homes, commuters and car users all feel targeted. Consumers, in general, seek technological solutions which will enable the lifestyles which they prefer and to which they aspire to be sustained.

Two classes of consumer demand are especially important for farmers. There is evidence of interest in the way food is produced. For some this stems from concerns for health. For others it is an expression of a belief that 'modern' foods have sacrificed taste for price and appearance. An important group are critical of contemporary farming because it uses fertilisers and pesticides and seeks to source supplies from production systems which do not make use of such aids. Farmers have to assess how far they should adjust their systems in the light of such consumer attitudes. They have to determine the cost of alternative systems, the additional value which products might reasonably be expected to receive, the scale of the market and their ability to compete with other farmers in supplying such a market. In most of the EU, despite much discussion, these goods remain a small proportion of the total food market.

Consumer demands for rural resources are also important for farmers. As affluence increases more land is needed for recreation, transport and housing. The appearance of the countryside and the relative abundance of wild plants and animals becomes a matter of importance to new rural residents who have no link to farming. Farming practices which have been accepted as normal by traditional farming based societies may be questioned.[10] The outcome of such concerns is likely to be an increasingly regulated environment within which farming systems will have to operate.

Whilst demands for sustainability figure high on the political agenda it is clear that this does not lead to a consistent programme of action. Expert opinion is far from unanimous. Pressure groups may use the term as a weapon but be unprepared to accept limitations which affect their own interests. Policies which impose costs on consumers or taxpayers rapidly encounter the gulf between verbal assent and the willingness to pay. Meantime, autonomous changes in real incomes and consumer aspirations may make the task of attaining any form of sustainability, in natural resource use, in economic terms or in social arrangements increasingly difficult.

---

7 See Climate Change The UK Programme, CM2427 HMSO London

8 For a discussion of the situation in the UK see Silsoe Research Institute & ADAS Towards a UK Research Strategy for Alternative Crops – Silsoe Research Institute 1994 and Carruthers SP, Miller FA and Vaughan CMA- Crops for Energy and Industry – Centre for Agricultural Strategy 1994.

9 For a discussion of the development of environmental ideas see: Economics of Natural Resources and the Environment Pearce D W and Turner R K – Harvester Wheatsheaf New York and London 1990 For an earlier American view see, for example, Erlich P R and Erlich A H – Population Resources Environment Issues in Human Ecology W H Freeman and Company San Francisco 1970, who concluded – pp322 " The basic solutions involve dramatic and rapid changes in human attitudes especially those relating to reproductive behaviour, economic growth, technology, the environment and conflict resolution".

10 A number of cases have reached the courts in which neighbours have complained about the disturbance caused by cocks crowing. See for example the Times 18.8.94

2. Options for policy

How far any system is sustainable depends upon the rate at which it uses up non-renewable resources. In so far as policy sets a priority on more sustainable farming systems it implies a change in the way in which resources are currently used in agriculture. In discussing what might be options for changing current practice it is therefore essential to understand what forces have led to and underpin the present resource disposition. These include economic, political and technological considerations.

Within the EU both market forces and political decisions play a major part in determining farming systems. The Common Agricultural Policy, (CAP) has regulated the price level within the EU for most agricultural products. The actual prices received by individual farmers have been affected by exchange rates, by the quality of their output and by the efficiency of agricultural marketing systems. Some farming costs are also directly influenced by the CAP, the prices of feedingstuffs or of store livestock, for example. Others are influenced by the CAP, but only indirectly and in the longer term, for example the price of land. Within each member country the performance of the economy in relation to interest rates, inflation, exchange rates and the price of labour is of critical importance.

It is within this framework, which he cannot control, that the individual farmer has to generate sufficient profit to remain in business. His ability to do so will be affected by the physical characteristics of his farm, by the climate within which he works and by his own skills. If resource uses are to be changed to achieve greater sustainability, it is through this framework that the appropriate signals will have to be given. Where consumers take a serious view of the relationship between farming practice and sustainability they may change their purchases to reflect this. The market provides a measure of the extent of this demand within any particular society. Such evidence as we have suggests that whilst environmental benefits are regarded as a selling feature for many products, it is unlikely that this alone would result in changes on a scale which decisively altered current farming systems.[11]

This increases the responsibility of the policy makers. This has been recognised. It is notable, for example, that the 1992 package of changes in the CAP, generally known as the 'MacSharry reforms' include an agri-environmental package. An important development has been the emergence of a substantial area of consensus between some green lobbyists and some farmer pressure groups. The implicit goal of the farmers has been to ensure the continued availability of funds currently received in the form of price support. As price support becomes vulnerable both because of its budget cost and because of the commitments of the EU in GATT, so payments in return for introducing more sustainable farming systems become increasingly attractive. For those seeking to promote sustainability and other 'green' issues, this affords an opportunity to capture substantial public funding. Even where governments question the validity of the claims made the existence of such large pressure groups, supported by expert presentation in the media, represents a political force which is difficult to resist.

In the longer term both the political and the economic framework are profoundly affected by technological developments. Technology both adds to and changes the relative productivity of inputs. As a result those who can apply new methods earn higher rewards at current prices. Two long term consequences result. First, they tend to bid up the price of those inputs which are fixed in supply, such as land. Second, as additional output reaches the market, product prices fall or support costs rise. In time the squeeze between prices and costs tends to restore the return on capital and

---

11 For example the UK's most profitable retail group, Marks and Spencer have ceased to carry organic vegetable, and the largest farming organisation the CWS farms, have abandoned production of organic products in favour of integrated farming systems which seek to minimise but not eliminate the use of pesticides and other farm chemical inputs.

labour to its previous level. However, for the individual farmer there is no choice, only those who use the new methods or find new markets will be able to survive. Those who cannot, or who choose not to do so, will suffer reduced income and eventually be unable to retain the resources used within their business.

Technological progress is likely to change rather than sustain existing farming systems. It may increase overall sustainability, simply by reducing the quantity of inputs needed to produce a given level of output. It may lead to more fragile systems if it depends upon increased use of an input which is scarce. If benign effects are to prevail, then policy needs to adapt so as to devise a framework within which it is profitable for farmers to use technologies which reduce resource consumption rather than those which may exacerbate the risks of unsustainability. To do so is far from easy. For example, where technology leads to production in excess of the level of consumption for which consumers willingly pay, then the implied policy requirement is a withdrawal of resources from the industry. Far from promoting this many politicians prefer to talk about maintaining agricultural employment and the family farm. At its worst this attitude may discourage research and farming innovations which increase productivity.[12]

In seeking to devise a framework that will encourage sustainable farming systems ministers have at their disposal a wide variety of policy instruments. The more important include:

Price policies:

These may affect the prices received by farmers or the prices they pay for inputs. Final product prices may be increased by restricting overall supplies to a market. In this case farmers gain. Prices may also be increased by taxes on all supplies or, if it is intended to benefit domestic production, simply on those which are imported. Prices in the market place may be allowed to find their own level but, if these are judged to be too low, the receipts of farmers can be increased by subsidies. In this case the burden of support falls on the taxpayer rather than the consumer. Subsidies or taxes can also be used to influence the prices paid for inputs.

Such manipulation of costs and returns changes the framework within which farm businesses have to operate. They affect calculations about the most profitable scale and method of operation. They affect the ability of farmers to compete and, within the EU have been limited to decisions by the Community rather than by individual governments. Such policies are often relatively easy to administer, they have direct appeal to those who seem to gain and, especially where the cost is borne by consumers, may not produce strongly negative responses from those who pay. Their impact on sustainability is less easy to assess. At most only some of the relevant parts of the framework are susceptible to manipulation by price policy. Economic models can establish the most profitable combination of inputs and outputs under given price assumptions. However, the underlying assumptions built into such models, about farmers goals, their levels of skill and the mobility of resources within the farm business are relatively crude. Furthermore over time conditions in the rest of the economy may make assumptions about factor and product prices decreasingly reliable. Where policy makers are concerned with the very long run consequences of farming systems, such models cannot provide much certainty about the impact of price manipulation.

Structural policies:

Structural policies seek, characteristically, to encourage farm amalgamation or enlargement, to provide physical or marketing infrastructure or to help farmers to retire early or find new, non-farming employment. They influence directly the longer term pattern of resource use in farming.

---

12 The UK Strategy for Sustainable Development stresses both the need to make further reductions in support levels and to research ways of reducing agriculture's adverse environmental impacts, of developing new crops including energy crops and of finding new uses for established crops. Ch 15 Sustainable Development The UK Strategy CM2426 HMSO 1994

They can be regarded as part of agricultural policy or, more usefully, as part of an overall process of regional economic adjustment. Policies of this sort may fund education and advisory services which increase the personal choices open to the agricultural population. They can subsidise changes on the farm which form part of new farming systems, for example, irrigation, power supplies and the construction of new farm buildings. Structural funds may also be employed to ease pressure on farming systems by promoting the development of diversified activities, for example, forestry, bed and breakfast, craft industries and food processing and farm shops. Such policies have an opportunity to reward changes in farming systems which are judged to promote a more sustainable use of resources. Their effectiveness depends on their uptake. Where existing businesses yield a satisfactory return to their owners, it will require substantially more funding through structural support to persuade them that the work and risks involved in changing the farming system are justified. In terms of the sustainability of natural resources, however, it is not clear that the greatest need is where farm incomes are lowest. Changes may be more urgent among some farms whose present businesses are robust and who will find little to attract them in modest support for structural change.

Market information and public education:

There are good reasons to believe that many people are prepared to modify their purchasing habits to take account of issues such as sustainability if they can do so at little or no personal cost. Providing information about the environmental costs of alternative products enables people to make such choices. There is a danger that such claims may be used to promote sales without being based on independent evidence. Governments may require manufacturers or food distributors to include accurate information on product labels. They may also ensure that producers, particularly those who operate on a small scale are informed about market opportunities for products which are genuinely 'green'. This is an extension of the traditional role of government in establishing and enforcing standards of measurement and purity. In the UK, for example the government has supported the development of a UK Register of Organic Producers, with clearly defined indicators about what this implies. This enables those who believe that organic production provides a more sustainable food supply, to adjust their buying behaviour appropriately.

Formal and informal education play a major role in determining behaviour. An informed public is better placed to judge what regulatory interventions may be needed to ensure sustainability and is also more likely to moderate its own buying habits in ways which encourage the development of sustainable farming systems. In a largely urban community people have little first hand knowledge of agriculture and are vulnerable to propagandist and romantic notions about how it actually operates. Education which includes both the underlying sciences involved in food production and the role of the agricultural and food industries in food supply and in shaping the countryside is especially important.[13]

Research and Development Policies:

Technologies which enable resources to be used more productively may originate within businesses or as 'bright ideas' by outsiders. Fundamentally, they depend upon understanding both the resources and the business concerned. In seeking to promote more sustainable farming systems governments can encourage improved understanding by their support for research and development.

Whilst such a general principle may give rise to little argument, the allocation of research and development expenditure is a much more contentious matter. There is debate about whether such activities should be wholly funded by the industry which, it is assumed, will benefit from the

---

13 The Council for Environmental Education – sponsored a national programme on Education and Training for Business and the Environment. This included a number of studies such as one on the Rural Environment which is to be published by the Pluto Press London. Teachers are often offered teaching aids by pressure groups who wish to interest children in environmental matters.

changes made possible. There are arguments about the distribution of research and development funds between areas of science. Many critical questions remain unresolved. For example, is it more sensible to direct efforts at solving problems caused by current practice or should 'bolt on solutions' be eschewed in favour of more radical changes in the entire farming system? How far should the choice be left to scientists or given to industry? Should we concentrate more on research and development which would foster the early application of present best practice – or put our efforts whole heartedly into finding better methods of production?

None of these questions is susceptible of easy answers. Within the UK we have recently conducted our first Technology Foresight exercise[14]. The goal is to improve just these sorts of decision. The process which involved the academic, the business and the official world, has demonstrated how important such discussion is and how difficult to achieve.

Government may not only fund research and development itself but also create conditions conducive to it by other policies. Regulations relating to patents and intellectual property must influence business decisions about research and development. Taxation systems may give relief to expenditures associated with research and the early stages of development. Regulations about the licensing of new products or the use of bio-technological inputs may encourage or deter investment.

Research and development are readily recognised as important[15] in seeking more sustainable farming systems. The results, of what may prove to be considerable expenditures, are less easy to forecast. Not only is the outcome of research necessarily uncertain but it is impossible to know precisely what changes will have occurred in the economy, the time at which research results and changes in society will take place and how these events will influence the ability of UK businesses to compete in the market. Retrospectively, however, we know that it is advances in technology which have led to higher living standards and have determined the success of businesses throughout the world. The risks of neglecting research are far greater than those of choosing the wrong topic.

Environmental policies:

Virtually every policy has an environmental impact, however, a number of environmental concerns cross the boundaries of several policy areas and it is here that specific environmental policies are needed. Concerns about global warming, which affect all energy using activities, land use questions, housing, transport, water, forestry and tourism are also relevant to agriculture. Environmental policies have to balance conflicting claims, including goals relating to sustainability. Since the Rio Summit a number of these environmental goals have become international commitments. In response governments publish reports on progress which are monitored by the United Nations Commission on Sustainable Development.

Environmental policies although designed as part of an overall strategy operate through changing the activity of individual industries. For agriculture they are likely to include an increasing range of regulatory requirements concerned with water, with maintaining access to the countryside and with the development of recreational activities[16]. In essence they change the framework within which the farmer has to reach his decisions.

3. The European Dimension.

The extent to which the Common Agricultural Policy has dominated the development of farming systems in the EU makes it central to any consideration of prospects for changing farming methods. Until 1992 the policy relied almost wholly on manipulating prices to achieve its goals. A

---

14 Progress through Partnership – Report from the Steering Groups of the Technology Foresight Programme 1995 OST HMSO 1995

15 See for example para 15.18 in Sustainable Development The UK Strategy op cit

16 See for example OECD Agricultural and Environmental Policies OECD Paris 1989

clear conflict emerged. Whilst the incomes of many farmers remained relatively low, higher prices led to production at costs greatly in excess of the market value of the output. This not only stressed the budget and disrupted world trade it also meant that significant quantities of natural resources were being systematically wasted. Given this outcome the most important thing that the CAP could do for sustainability was to cut surplus output.

Within the EU there exist great differences both of circumstance and of attitude, which grow as new members join. The problems of the Mediterranean countries differ from those of the countries of north west Europe for which the CAP was initially designed. Since 1995 the Community has to come to terms with the challenges posed by sparsely populated, afforested areas in Finland as well as those of densely populated countries such as the Netherlands. Even within countries the requirements for sustainable farming systems differ among regions. In Germany the new Lander have thrown into sharp relief the inappropriate nature of policies based on the interests of Bavarian farmers. In the UK, systems which are wholly sustainable in East Anglia would be disastrous in the Scottish Highlands or in Northern Ireland.

Differences exist not only in circumstance but in aspiration. We have differing perceptions of what constitutes an attractive landscape. Our attitudes to animal welfare are not the same. In some countries landlord tenant relationships are well established and accepted. In others small owner occupiers characterise the farming population. Inevitably these lead to different priorities for policies about sustainability. Social problems of rural communities dominate much of the South of the EU. Among the potential members in Central and Eastern Europe the legacy of communism has created a need to build new economic institutions and, in some places, to deal with problems of pollution on a scale more worrying than in the West. In North West Europe natural resource concerns, the preservation of natural habitats, animal welfare and the maintenance of landscape tend to dominate debate.

Such differences mean that the same policies have unequal effects. This makes it difficult for the Council to reach decisions. Even where agreement is reached there may be marked differences in application. In 1992 the Council of Ministers accepted a package of measures which reduced production by a combination of lower prices and supply control and provided compensation for farmers whose incomes were expected to fall. The particular method of supply control chosen for the arable sector was 'set aside'. In order to qualify for compensation all but the smallest farmers had to take part of their arable area out of production. Complicated rules were established about how this 'set aside' area was to be used. From an early stage permission was given for 'industrial' crops to be planted. There was also concern that land set aside should be managed taking account of environmental criteria – such rules affected the dates at which land could be cultivated, the use of fertilisers and sprays and the requirements relating to cutting weeds. In order to ensure that supply was limited, the area set aside had to be rotated around the farm. This was intended to prevent farmers taking out only their least productive fields. However, many environmental interests wished land to be left uncultivated for much longer periods. As a result the rules were adapted to allow farmers to qualify for compensation by choosing to set aside a rather larger proportion of their arable area on a permanent basis so that longer run projects of environmental value could be undertaken.

In economic terms the 1992 package of set aside measures is wasteful. Not only does it keep in production relatively unproductive land whilst forcing some of the better land into idleness but it involves a very substantial administrative exercise and creates possibilities for fraud. Its impact on the sustainability of farming systems is less clear. One object was to provide continued support for the small family farm. An economically more rational solution , to allow market prices to fall until supply matched demand, would have meant that many family farms would have suffered much larger reductions in income. Where land is idled, the pressure of farming on the natural environment has been lessened. However, where industrial crops are grown this is unlikely to be the case. Had economic forces been allowed to cut output by lowering prices, it is arguable that more resources would have been released for environmental purposes. Certainly the capital value of farm land would have fallen and the process of facilitating the introduction of viable but more extensive farming systems made less costly.

The 1992 measures also changed the beef and sheep regimes. An important feature of the new regimes is the stocking density limits imposed on the number of animals for which farmers can claim compensation. It introduces an element of 'cross compliance', where the receipt of a specific benefit is made conditional upon action to improve the environment. At first sight this seems to be a move towards more sustainable farming systems. However, it also illustrates the complexity of such an approach. The stocking capacity of land varies. To impose a single limit means that while some land may be under-utilised other land may be over-stretched. At the same time, the environmental 'good', – lower stocking rates – may well evaporate if market prices rise to a point at which the loss of compensation would be more than offset by returns from the market. Thus as a tool to encourage sustainable systems 'cross compliance' is both blunt and fragile.

Such considerations illustrate the difficulties which exist where the CAP moves from setting a price/cost framework in order to attain more complicated goals. There is a clear distinction between policies concerned with the overall volume of production and those which seek to steer agriculture in ways which require monitoring at the level of the farm business. The overall level of output and the application of price or other policies which bring about the output required in the Community as a whole must be a matter for central decision. Social and environmental considerations, in contrast, lead to attempts to influence the activity of individual farms or regions. It is not practicable to do this from the centre. As a result the EU, when introducing policies such as Environmentally Sensitive Areas(ESAs), had allowed considerable local autonomy in devising management plans for participating farmers. The agri-environmental package within the 1992 reforms similarly, allows for differences in implementation within the member countries.

A degree of local freedom may not be too serious provided that the methods used are genuinely de-coupled from production . Provided the EU has overall authority to accept or reject such measures, within a general framework of subsidiarity, so that they do not undermine the single market, policies designed in this way offer a way forward. Two sorts of difficulty are likely to arise. First, complete decoupling may not be easy to achieve. If support for sustainable systems is given in ways which affect the quantity produced, 'playing fields will not be level'. Farming and environmental groups who do not benefit are likely to press national governments for equivalent support. If they succeed it will be impossible for the EU to control the level of its output. Second, the share of farming in the economy of member states varies considerably. The need for change may be greatest in countries with the largest proportion of their employment in farming and with the lowest levels of per capita income. They are likely to demand common funding to promote sustainable farming. This would involve considerable inter member transfers. These will be visible and, as a result, are likely to be politically contentious. Countries which seem likely to lose from such policies will legitimately point out that farming affects resource use in only 2.8% of the overall economy. To divert resources in this direction at the expense of other sectors, particularly energy and transport, may make it more difficult to achieve an acceptable and sustainable standard of living for most of the EU's citizens.

Considerations of this type suggest that although the adjustment of agricultural policies to encourage sustainable farming systems may be approved in principle, it will face strong competition for public resources with other sectors. Success may require simultaneous adjustments in areas well beyond agricultural policy. If this were the case the CAP might lose part of its dominance of EU budgets, allowing space for a growth in regional and environmental policies which were applied not just to one sector but to the economic areas within which they made the greatest contribution to improving the long run capacity of the Community to raise the standard of living of its people. The farm lobby and some environmental groups might see this as a defeat; for the Community it could prove to be an important victory.

Session 2
Biological Control

Chairman and
Session Organiser   Dr D H K DAVIES

INTERCROPPING AND BIOLOGICAL CONTROL OF PESTS AND WEEDS.

J.THEUNISSEN

Research Institute for Plant Protection (IPO-DLO)
Binnenhaven 5, 6700 GW Wageningen, The Netherlands.

ABSTRACT

The potential and limitations of intercropping and biological control of pests and weeds are discussed. In perhaps most intercropping systems biological control is one of the effects which are responsible for the observed suppression of pest populations. Intercropping also has a direct controlling effect on weed populations in the field. The relationship between weed suppression and intercrop competition is an essential element in the economic viability of intercropping based production systems. Some technical, economical and psychological constraints of intercropping are discussed.

INTRODUCTION

Due to increasing human populations and related environmental problems, discussions on how to achieve sustainable development in socio-economic, industrial and agricultural aspects of society are taking place. The precarious prospects for food production for a growing world population elicits opposing arguments on the correct approach. There are experts who argue that all scientific, technical and chemical means at our disposal should be directed primarily to higher food production to meet the expected demand. According to others this would mean playing "va banque" and would end in disaster because of the eventual destruction of once productive agricultural land. They advocate the development of agricultural production methods which keep the agroecosystems concerned in such a condition that productivity is at least stabilised, if not increased, on a long term basis. Such development would safeguard future, long term production capacity and healthy, good quality products. Sustainable production systems are increasingly demanded by the market and consumers.

Integrated crop protection is gradually being developed to minimize the inputs of pesticides in the production of agricultural commodities. Elements of integrated control such as biological control, microbial control, breeding for resistance, modifying the cropping system have to be combined in managing the crop-pest(s) combinations, and the technical and economical possibilities.

INTERCROPPING

Intercropping is: "the cultivation of two or more species of crop in such a way that they interact agronomically" (Vandermeer, 1989). During the last decade numerous studies have been carried out to evaluate intercropping effects in terms of yield and pest control (Vandermeer, 1989; Andow, 1991). Attention has been focussed on field vegetable crops (Andow et al. 1986; Theunissen, 1994; Theunissen et al., 1995).

A general effect of intercropping in vegetables is suppression of the population of most insect pests. This has been found in cabbage crops (e.g. Ryan et al., 1980; Kenny and Chapman, 1988; Kloen & Altieri, 1990; Hofsvang, 1991; Theunissen & Schelling, 1992), in carrots and onions (Uvah & Coaker, 1984), in fennel (Theunissen, 1994b), in leek (Theunissen & Schelling, 1993), and in field beans (van Rheenen et al., 1981; Tingey & Lamont, 1988).

One of the effects of intercropping is the mutual interference between the plant species concerned. This is expressed as competition between main crop and the undersown intercrop, or between mixed crops, and may involve a degree of weed suppression. We found a considerable degree of weed suppression by taller clovers such as Trifolium repens cultivars, used as an undersown intercrop. Lower growing clovers such as many cultivars of Trifolium subterraneum accordingly showed less weed suppression. In a sense the total or partial control of weeds by an intercrop may be considered as a form of biological weed control. Competition with the main crop and weed suppressing ability seem to be parallel processes in an intercropping situation. Both will have to be considered in the technical and economical evaluation of the utility of intercropping as a pest management principle.

Loss of harvestable yield and acceptable weed control management are the main technical constraints for practical implementation of intercropping. We found that mowing intercrops between rows of leek minimized the inter-crop competition. This resulted in the same plant size and weight of leek undersown with *T. subterraneum* when compared to a monocrop. An additional effect was the timely removal of most weeds before they seeded. This contributed greatly to a check on weed populations in an ecologically acceptable way. Yield analysis at harvest showed for white cabbage and leek that, in terms of financial results, the total harvestable yield is less important than the marketable yield. In fresh vegetables quality is important and marketable weight determines the economic cropping result. In comparisons between unsprayed monocrops and intercrops, the quality distribution of the harvested plants from intercrops showed a significant shift to the best quality classes in white cabbage (Theunissen *et al.*, 1995) and leek (Theunissen and Schelling, unpublished data) (fig.1). Translated into monetary values this means a better revenue per hectare by application of intercropping. Detailed balances of inputs/output must show the economic consistency of intercropping based cropping systems.

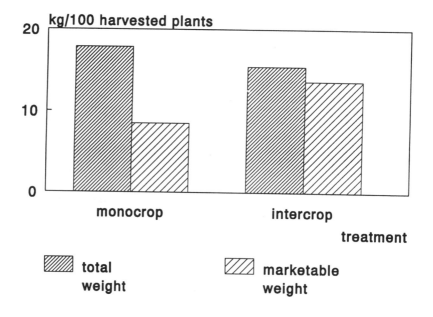

Fig.1. Yield data from an experiment in leek using *T. subterraneum* cv. Geraldton as an undersown intercrop. The total harvestable weight of market trimmed leek plants was larger in the unsprayed monocrop, but the saleable, marketable yield was higher in the intercrop owing to better quality.

BIOLOGICAL CONTROL

Our original reason to try undersowing in vegetable crops was an attempt to create a habitat for natural enemies in intensively cropped field vegetables. In this way we hoped to enhance the chances of natural biological control of pests in these crops. In monocrops, biological control was found to be always too little and too late, partly due to climatic conditions and the biological uniformity in intensive commercial cropping. In a landscape with a very scarce natural vegetation (host plants), and regularly sprayed with pesticides, natural enemies had little chance to survive in large numbers. Field trials with Brussels sprouts, being host plants for many pest species, intercropped with *Spergula arvensis*, showed increased parasitation of cabbage aphids (*Brevicoryne brassicae*) (Theunissen and den Ouden, 1980). Prior and subsequent research on intercropping by other workers and our group showed that stimulation of biological control is one of a number of possible mechanisms which lead to pest population suppressing effects (fig.2).

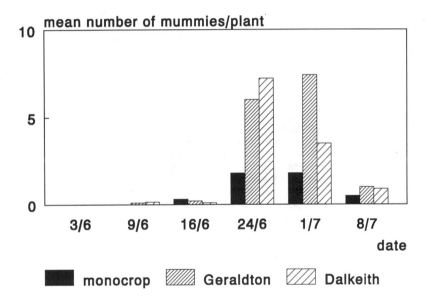

Fig.2. Parasitation of aphids in iceberg lettuce in monocrop and *T. subterraneum* undersown treatments. During the cropping period the numbers of aphid mummies were counted in plants sampled in the treatments. No pesticide applications were used. The weights of harvested plants did not differ between the intercropped and monocropped treatments.

The "enemies hypothesis" (Root, 1973; Risch, 1981) directly refers to biological control as a major intercropping effect. Direct evidence of increased levels of predators was found in clover undersown cabbage (Theunissen *et al.* 1995) and in relay intercropping in cantaloupe (Bugg *et al.*, 1991). While diversification of a cropping ecosystem might stimulate the presence and survival of natural enemies, it is not an automatic process that they indeed interfere with the pests of the main crop. For instance, no effect of natural enemies on *Thrips tabaci* populations was found in intercropped leek, despite drastic reduction of thrips populations (Theunissen, unpublished data). Immigrating pests may react on contrasts between (Smith, 1976) or spectral reflectance intensity (Costello, 1995) of the crop and its background. Caterpillars become more easily infected with virus diseases in the changed microclimate of the crop canopy in intercropped fields. Migrating catepillars of *Mamestra brassicae* get lost in the intercrop and starve, thus causing an extra 50% mortality of these caterpillars in a young cabbage crop where plants do not yet touch (van de Fliert, unpubl. data). Leek plants change in host plant quality for thrips when undersown with clover. These are a few examples of mechanisms apart from direct biological control which can have similar effects on the pest populations concerned.

IMPLEMENTATION

Polycultures are very common all over the tropics and are used for a variety of reasons (Vandermeer, 1989). However, in the intensive temperate regions, cropping practises including mixed cropping is considered economically inefficient, given the ample availability of cheap pesticides to counter the disadvantages of growing monocrops. The costs of human labour and the mechanization of cropping in many countries are also incentives to monocropping. The situation changes drastically when the pesticides become ineffective or the consumer starts to worry about the safety of eating fresh products, or the way in which these products have been grown.

The magnitude of the problems with *Thrips tabaci* in leek crops all over western Europe are a classical example of results obtained by a cropping system which relies totally on the use of pesticides. The quickly decreasing sensitivity of thrips for most, if not all, accepted insecticides resulted in uncontrollable populations, in spite of numerous and still increasing applications. Additional causes may be an increasing area of leek cropping, continuous cropping during the entire year and a very narrow crop rotation scheme. The outcome is a low quality vegetable, producable only with excessive pesticide use and at a high cost.

Undersowing, mixed cropping and other forms of intercropping do not have a place in the present commercial cropping methods and in the considerations of growers. For that reason alone implementation of intercropping based production methods will be difficult. Apart from these psychological hurdles intercropping must be an economically viable approach. Even when the reduction in the pest population (and sometimes disease) is acknowledged, growers have to accept a change from total yield towards marketable yield as criterion for economic success. Increasing consumers' awareness on food safety and methods of production will stimulate, more than scientific arguments can do, a market demand for healthy and clean food products. Integrated production, that is a combination of methods which each influence the agro-ecosystem to realize economically and ecologically acceptable production methods, may provide solutions.

REFERENCES

Andow, D.A. (1991) Vegetational diversity and arthropod population response. *Annual Review of Entomology* **36**, 561-586.

Bugg, R.L.; Wäckers, F.L.; Brunson, K.E.; Dutcher, J.D.; Phatak, S.C. (1991) Cool-season cover crops relay intercropped with cantaloupe: influence on a generalist predator, *Geocoris puncticeps* (Hemiptera: Lygaeidae). *Journal of Economic Entomology* **84**, (2), 408-416.

Costello, M.J. (1995) Spectral reflectance from a broccoli crop with vegetation or soil as background: influence on immigration by *Brevicoryne brassicae* and *Myzus persicae*. *Entomologia Experimentalis et Applicata* **75**, 109-117.

Hofsvang, T. (1991) The influence of intercropping and weeds on the oviposition of the brassica root flies (*Delia radicum* and *D. floralis*). *Norwegian Journal of Agricultural Sciences* **5**, 349-356.

Kenny, C.J.; Chapman, R.B. (1988) Effect of an intercrop on the insect pests, yields, and quality of cabbage. *New Zealand Journal of Experimental Agriculture* **16**, 67-72.

Kloen, H.; Altieri, M.A. (1990) Effect of mustard (*Brassica hirta*) as a non-crop plant on competition and insect pests in broccoli (*Brassica oleracea*). *Crop Protection* **9**, 90-96.

Rheenen, H.A. van; Hasselbach, O.E.; Muigai, S.G.S. (1981) The effects of growing beans together with maize on the incidence of bean diseases and pests. *Netherlands Journal of Plant Pathology* **87**, 193-199.

Risch, S.J. (1981) Insect herbivore abundance in tropical monocultures and polycultures: an experimental test of two hypotheses. *Ecology* **62**, 1325-1340.

Root, R.B. (1973) Organization of a plant-arthropod association in simple and diverse habitats: the fauna of collards (*Brassica oleraceae*). *Ecological Monographs* **43**, 95-124.

Ryan, J.; Ryan, M.F.; McNaeidhe, F. (1980) The effect of interrow plant cover on populations of the cabbage root fly, *Delia brassicae* (Wiedemann). *Journal of Applied Ecology* **17**, 31-40.

Smith, J.G. (1976) Influence of crop background on aphids and other phytophagous insects of Brussels sprouts. *Annals of Applied Biology* **83**, 1-13.

Theunissen, J. (1994a) Intercropping in field vegetable crops: pest management by agrosystem diversification - an overview. *Pesticide Science* **42**, 65-68.

Theunissen, J. (1994b) Effects of intercropping on pest populations in vegetable crops. *"Integrated control in field vegetable crops"*, *IOBC Bulletin* **17** (8), 153-158.

Theunissen, J.; den Ouden, H. (1980) Effects of intercropping with Spergula arvensis op pests in Brussels sprouts. *Entomologia Experimentalis et Applicata* **27**, 260-268.

Theunissen, J.; Schelling, G. (1992) Cabbage-clover intercropping: oviposition of *Delia radicum*. *Proceedings of the Section Experimental and Applied Entomology of the Netherlands Entomological Society* **3**, 191-196.

Theunissen, J.; Schelling, G. (1993) Suppression of *Thrips tabaci* populations in intercropped leek. *Mededelingen van de Faculteit Landbouwwetenschappen Rijksuniversiteit Gent* **58**/2a, 383-390.

Theunissen, J.; Booij, C.J.H.; Lotz, L.A.P. (1995) Effects of intercropping white cabbage with clovers on pest infestation and yield. *Entomologia Experimentalis et Applicata* **74**, 7-16.

Tingey, W.M.; Lamont, W.J. (1988) Insect abundance in field beans altered by intercropping. *Bulletin of Entomological Research* **78**, 527-535.

Uvah, I.I.I.; Coaker, T.H. (1984) Effect of mixed cropping on some pests of carrots and onions. *Entomologia Experimentalis et Applicata* **36**, 159-167.

Vandermeer, J. (1989) *The ecology of intercropping*. Cambridge University Press, 1-237.

DEVELOPING STRATEGIES FOR THE NEMATODE, *PHASMARHABDITIS HERMAPHRODITA,* AS A BIOLOGICAL CONTROL AGENT FOR SLUGS IN INTEGRATED CROP MANAGEMENT SYSTEMS

M. J. WILSON, L.A. HUGHES, D.M. GLEN.

IACR-Long Ashton Research Station, Department of Agricultural Sciences, University of Bristol, Long Ashton, Bristol, BS18 9AF, UK

ABSTRACT

Slugs are likely to increase in importance as pests in integrated crop management systems. The nematode parasite, *Phasmarhabditis hermaphrodita*, has the necessary attributes to be used as a successful biological control agent against slugs and is already on sale in the UK as a molluscicide for use by domestic gardeners. However, for this nematode to be successfully used in arable agriculture, strategies for using the nematode must be optimised so as to provide protection against slugs at low doses, and protection from slug damage must be reliable under a wide range of environmental conditions at and after the time of application. Research to develop strategies for cost-effective use of this nematode is described. Application of the nematode one or two weeks before drilling a susceptible crop (oilseed rape) appeared to have no advantages over application at the time of drilling. However, shallow incorporation of nematodes applied to dry soil was advantageous in protecting a wheat crop from slug damage. In laboratory tests, slugs avoided feeding and resting on soil treated with nematodes, suggesting that it may be feasible to apply nematodes in bands to protect crops from slug damage.

INTRODUCTION

Slugs are important pests of many agricultural and horticultural crops throughout Europe, North and Central America, Asia and Australasia (South, 1992). In the UK, the most important crops damaged by slugs in economic terms are winter wheat, oilseed rape and maincrop potatoes (Port & Port, 1986). These pests are generally controlled using molluscicides formulated as bait pellets. However, the chemicals available are sometimes ineffective and may adversely affect non-target organisms. Furthermore, many of the features associated with sustainable, integrated farming systems are likely to lead to an increase in slug problems (Glen *et al.*, 1994b). These features include more varied crop rotations, non-inversion tillage, incorporation of crop residues and the use of cover crops to reduce nitrogen leaching.

The bacterial feeding nematode, *Phasmarhabditis hermaphrodita*, is a parasite of slugs which kills many pest species (Wilson *et al.*, 1993a). It was discovered parasitising slugs at IACR-Long Ashton and developed as a biological molluscicide in collaboration with the Agricultural Genetics Company Ltd.. *Phasmarhabditis hermaphrodita* forms a developmentally arrested non-feeding larva (dauer larva) which is the infective stage. Dauer larvae infect slugs by entering their shell sac above the mantle, develop into adults and reproduce. Eventually, the slug dies and the nematodes spread over the slug cadaver reproducing until the food supply is depleted. The juveniles then fail to develop into adults and form new dauer larvae which disperse in order to infect new slugs. *Phasmarhabditis hermaphrodita* can be mass-reared *in vitro* in rich media containing specific bacteria (Wilson *et al.*, 1993b; 1995a). It is currently being produced in fermenters by MicroBio Ltd (a subsidiary of the Agricultural Genetics Company), and is sold for use as a molluscicide to domestic gardeners in the UK (Glen *et al.*, 1994a). The ability of this nematode to protect crops from slug damage has been demonstrated in a series of field experiments in different crops (Wilson *et al.*, 1994a,b; 1995c).

Research on the use of *P. hermaphrodita* as a biological molluscicide is complemented by a much larger volume of research on the use of entomopathogenic nematodes (families Heterorhabditidae and Steinernematidae) as biological insecticides. Much of this work gives insight into the potential problems which have been, or might be encountered, with the use of *P. hermaphrodita* for slug control. Nevertheless, the use of entomopathogenic nematodes is largely restricted to high value horticultural crops and the problems of using these nematodes as biocontrol agents in arable crops have, in the main, still to be addressed. The paper describes recent research as part of a project in the LINK Programme "Technologies for Sustainable Farming Systems", which aims to establish the principles for cost-effective use of *P. hermaphrodita* in arable crops, especially in integrated and less intensive crop management systems. Areas where future strategic research may help the exploitation of *P. hermaphrodita* and related species as biocontrol agents for slugs are highlighted.

NEMATODE STRAINS

All research and development on *P. hermaphrodita* as a biological molluscicide was done using a single strain of the nematode (UK1) isolated at IACR-Long Ashton. It is possible that by collecting new strains of this nematode and the related nematode *P. neopapillosa,* which also parasitises slugs (Wilson *et al.*, 1993a), nematode strains may be found that are more virulent, easier to produce in fermenters or able to withstand a greater range of environmental conditions than the current strain. Development of such strains would make the product more cost-effective. Furthermore, it may be possible to select better strains of bacteria for rearing the nematode. Different species of bacteria can have profound effects on yields of nematodes in culture (Wilson *et al.*, 1995a) and on the ability of nematodes to infect and kill slugs (Wilson *et al.*, 1995b). *Phasmarhabditis hermaprhodita* is mass produced for commercial use in fermenters where it is grown in monoxenic culture with the bacterium, *Moraxella osloensis*. *Moraxella osloensis* was found to be the most suitable bacterium out of 13 bacterial isolates tested. However, it is possible that there may be other bacteria which would produce greater yields of more pathogenic nematodes. This approach to improving the efficacy or *P. hermaphrodita* as a biocontrol agent for slugs has considerable potential for future investigation.

REDUCING APPLICATION RATE

The currently recommended application rate of *P. hermaphrodita* as a molluscicide for garden use is 3×10^9 dauer larvae ha^{-1}. This rate was recommended following three field trials, one in mini-plots with Chinese cabbage (Wilson *et al.*, 1994a), one in field plots of winter wheat (Wilson *et al.*, 1994b) and one in protected lettuce (Wilson *et al.*, 1995c). In all these trials, a range of nematode doses were applied evenly to the soil surface of replicated experimental plots at the time of crop planting or sowing and slug damage assessed and numbers recorded (Wilson *et al.*, 1994a,b;1995c). In all three trials 3×10^9 nematodes ha^{-1} or a lower dose was found to give protection equivalent to methiocarb pellets (Draza) applied at the recommended rate of 5.5 kg ha^{-1}. The relationship between nematode dose and reduction in slug damage differed between trials. In the Chinese cabbage trial, protection improved with increasing nematode dose between 1×10^8 and 8×10^8 ha^{-1}, but showed little or no further improvement at higher doses of up to 2×10^{10} ha^{-1}. In the winter wheat and lettuce trials, there was a linear increase in plant protection with increasing nematode dose for all doses between 1×10^8 and 1×10^{10} ha^{-1} (the entire range of doses tested). Even though a dose lower than 3×10^9 ha^{-1} provided good protection in the Chinese cabbage trial, it is unlikely that application rates could be much reduced without the risk of compromising efficacy using current application strategies. It is interesting to note that the recommended application rate for *P. hermaphrodita* is similar to that recommended for application of entomopathogenic nematodes to the soil surface. These doses are extremely high, bearing in mind typical host densities (e.g. for slugs usually less than 500 m^{-2}) and the low numbers of nematodes needed to kill slugs. It is therefore possible that research into the fate and behaviour of applied nematodes may permit lower doses than those currently recommended to be used in the future.

TIMING OF APPLICATION.

In two field experiments with Chinese cabbage (Wilson *et al.*, 1994a), weekly or bi-weekly assessments of slug damage were made in each plot. This enabled nematode efficacy to be studied in relation to time after application. In both trials, the protection given by the nematode increased over the first two weeks, before stabilising. In the previously described wheat trial (Wilson *et al.*, 1994b), nematodes were added immediately after the crop was drilled. Since the worst damage done by slugs to wheat is the hollowing of the seeds shortly after sowing, and the worst damage to oilseed rape is done by feeding on the seedlings immediately after emergence, it was hypothesised that *P. hermaphrodita* might perform better if the nematodes were applied one or two weeks before drilling, rather than immediately after.

An experiment in August-September 1992 investigated the effect of different timings of nematode application on slug damage to a crop of autumn-drilled oilseed rape. Individual plots (12 x 12 m) were marked out before oilseed rape was drilled and three replicate plots were either left untreated or treated with methiocarb pellets at the recommended rate immediately after drilling, or treated with *P. hermaphrodita* 2 weeks or 1 week before drilling or immediately after drilling. Nematodes were applied at the recommended field rate (3×10^9 ha^{-1}) to the soil surface in 555 l ha^{-1} of water using a knapsack sprayer fitted with a coarse anvil nozzle. Numbers of oilseed rape seedlings were assessed three weeks later

(Table 1).

Nematode application significantly reduced slug feeding and, thus, increased numbers of rape plants established, irrespective of the timing of application. There appeared to be no benefit from application up to 2 wks before drilling compared with application immediately after drilling. In this experiment, the protection given by methiocarb pellets was superior to that given by the nematodes. The moisture content of the top 1 cm of soil at all application dates was below 10 % w/w; it is likely that this led to nematode death through desiccation. It may also have reduced the ability of nematodes to move down through the soil to avoid exposure to ultra violet light.

Table 1. Mean square root numbers of oilseed rape seedlings 0.25 m^{-2} in untreated plots, plots treated with methiocarb immediately after drilling and plots treated with 3 x 10^9 P. *hermaphrodita* ha^{-1}. Soil moisture content in the top 1 cm of soil at the time of nematode application is also shown.

	Unt-reated	Nematodes applied at drilling	Nematodes 1 wk before drilling	Nematodes 2 wk before drilling	Methiocarb applied at drilling	S.E.D. 24 d.f.
Plants m^{-2}	2.393	3.709	3.31	3.905	4.731	0.3675
% soil moisture	*	7.0	7.1	9.1	*	*

INCORPORATION OF NEMATODES INTO SOIL

It is known that the persistence of entomopathogenic nematodes when applied to the soil surface is poor and this is generally considered to result from mortality induced by desiccation and solar radiation (Gaugler, 1988). Certainly, we have obtained the best results in field trials, in both wheat and oilseed rape, when nematodes have been applied to moist soil. In many cases, application of entomopathogenic nematodes is followed by irrigation; Georgis & Gaugler (1991) recommended application of 1-2 cm of water after nematode application. The volumes of water required for this would preclude the use of this method in arable agriculture in the UK, with the exception of the potato crop. Other attempts to increase survival of soil-applied entomopathogenic nematodes have used machinery to inject the nematodes below the soil surface. While this technique has been successful in Australia (Berg et al., 1987), it has proved less effective in the U.S.A (Klein & Georgis, 1994), and also requires specialised machinery not available to most farmers.

A simple technique we have used to reduce nematode mortality on the soil surface is to incorporate the nematodes into the soil immediately after application. This was done in a field experiment in winter wheat in autumn 1994. The experiment was a split-plot design with plots of 12 x 16 m. One half of each plot was treated with P. *hermaphrodita* at the recommended field rate one day after the plots were drilled and the remaining half was left untreated. Nematodes were applied using a knapsack sprayer in 520 l ha^{-1} of water per plot. Plots were either left uncultivated after nematode application or cultivated using tractor-mounted spring tines immediately after application. The depth of cultivation was adjusted so

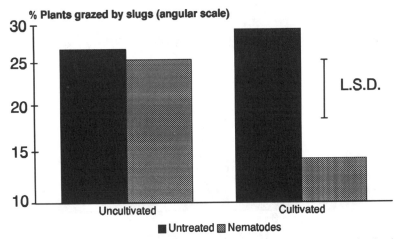

Figure 1 Percent of wheat plants grazed by slugs in untreated sub-plots and sub-plots treated with *P. hermaphrodita* which were cultivated to 2, 5 or 10 cm with spring tines or left uncultivated

that the tines worked the soil to a depth of approximately 2, 5 or 10 cm. Slug damage was assessed when the wheat had reached Zadoks' growth stage 11. There was no evidence of significant differences between the three cultivation depths in their effects on nematode efficacy; therefore, data for all three cultivation depths were combined for the analysis (Fig 1).

The top 1 cm layer of soil contained only 4.9 % (w/w) water at the time nematodes were applied, at approximately mid-day with strong sunlight. Thus, it is not surprising that nematodes left on the soil surface had no significant impact on slug damage. In spite of these harsh conditions, there was a significant, 50% reduction in slug damage on plots that were cultivated after nematode application. This suggests that shallow incorporation of nematodes into soil protected them from desiccation.

PARTIAL TREATMENT OF SOIL

Results from a field trial in wheat in autumn 1991 suggested that slug behaviour was altered in plots where nematodes were applied to soil (Glen *et al.*, 1994a; Wilson *et al.*, 1994b). Laboratory experiments were done to investigate whether slugs can detect the presence of *P. hermaphrodita* in soil and avoid areas of soil treated with the nematode, and whether this response could contribute to a reduction in feeding damage. An experiment was done using five plastic boxes, 26 x 13 x 9 cm high, lined with a 2 cm deep layer of soil aggregates. Half the soil surface of each box was treated with approximately 100 *P. hermaphrodita* larvae cm^{-2} and the remaining half was untreated. Five adult *Deroceras reticulatum* were added to each box. The insides of the boxes were lined with 0.8 mm woven copper mesh, over which slugs do not crawl, thus confining slugs to the soil surface. Ten Chinese cabbage leaf discs, each 3 cm in diameter were placed in each box, five in the nematode-treated half and five in the untreated half, at opposite ends of the box.

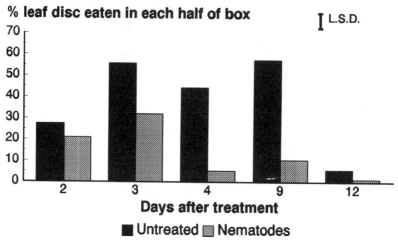

Figure 2. Mean percentage area of Chinese cabbage leaf consumed by slugs in untreated and nematode-treated halves of soil surface in boxes.

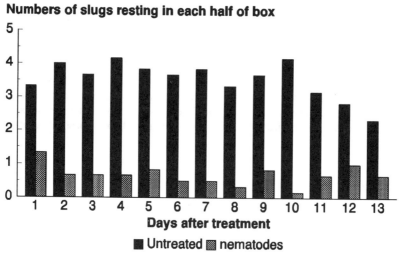

Figure 3. Numbers of slugs found resting in untreated and nematode-treated halves of soil surface in boxes (n=5).

The percentage of each leaf disc eaten by slugs was recorded the next day and the leaf discs were removed. This procedure was repeated at intervals over 12 days (Fig. 2). Every day, numbers of slugs resting on untreated and nematode-treated halves of boxes were recorded (Fig. 3). Slugs consumed significantly more in the untreated half than in the nematode-treated half at all assessments except the first (day 2) and the last (day 12) (Fig. 2), by which time most slugs had stopped feeding as a result of nematode infection. On all days, more slugs were found resting on the untreated than the nematode-treated half. It is hoped that this repellent effect of *P. hermaphrodita* can be exploited, in addition to its effects in killing slugs, to reduce nematode doses by applying the nematodes in narrow bands around susceptible crops. This strategy may be particularly useful in crops grown in distinct rows some distance apart, such as sugar beet and many vegetable crops.

INOCULATIVE RELEASE

Certain cropping systems lead to an increase in slug population size. This is particularly true of oilseed rape, where the dense crop canopy provides conditions ideal for slugs. Winter wheat crops following rape are often severely damaged by slugs. The majority of field trials with *P. hermaphrodita* have used an inundative release of large numbers of nematodes to give a relatively rapid reduction in slug damage to crops. It may be possible to treat oilseed rape crops with a low dose of nematodes in autumn or spring which will stop the slug population building up and, thus, protect a following crop. The damp humid conditions which favour slugs should also favour nematode persistence and activity. However, initial tests of inoculative release have not given encouraging results.

CONCLUSIONS

Phasmarhabditis hermaphrodita is an effective biological molluscicide. It is more selective than available chemical molluscicides and, thus, could form part of integrated crop management systems. There is much scope for strategic and applied research to improve the efficacy of *P. hermaphrodita* and related species as biocontrol agents for slugs, since the genus *Phasmarhabditis* has been studied little. By considering the factors most likely to limit efficacy of *P. hermaphrodita*, and by developing appropriate application techniques, it should be possible to develop cost-effective, reliable strategies for using the nematode in arable crops. Different strategies, which take into account application method, timing, incorporation and selective placement, will probably be needed for the different types of crops damaged by slugs. Research on the biology and ecology of *P. hermaphrodita*, could eventually lead to the ability to manipulate and exploit natural epizootics of this nematode in integrated crop management systems.

ACKNOWLEDGEMENTS

This work was jointly funded by the Agricultural Genetics Company and the UK Ministry of Agriculture, Fisheries and Food under the Link Programme, "Technologies for Sustainable Farming Systems". IACR receives grant-aided support from the Biotechnology and Biological Sciences Research Council of the United Kingdom.

REFERENCES

Berg, G.N.; Williams, P.; Bedding, R.A.; Akhurst, R.J. (1987). A commercial method of application of entomopathogenic nematodes to pasture for controlling subterranean insect pests. *Plant Protection Quarterly*, **2**, 174-177.

Gaugler, R. (1988). Ecological consideration in the biological control of soil-inhabiting insects with entomopathogenic nematodes. *Agriculture, Ecosystems and Environment*, **24**, 351-360.

Georgis, R.; Gaugler, R. (1991). Predictability in biological control using entomopathogenic nematodes. *Journal of Economic Entomology*, **84**, 713-720.

Glen, D.M.; Wilson, M.J.; Pearce, J.D.; Rodgers, P.B. (1994a). Discovery and investigation of a novel nematode parasite for biological control of slugs. *Proceedings of the Brighton Crop Protection Conference - Pests and Diseases*, 617-624.

Glen, D.M.; Wiltshire, C.W.; Wilson, M.J.; Kendall, D.A.; Symondson, W.O.C. (1994b). Slugs in arable crops: key pests under CAP reform, in *Aspects of Applied Biology*, **40**, *Arable Farming Under CAP Reform*. 199-206.

Klein, M.G.; Georgis, R. (1994). Application techniques for entomopathogenic nematodes *Proceedings of the VIth International Colloquium on Invertebrate Pathology and Microbial Control*, Montpellier, France, 483-484.

Port, C.M.; Port, G.R. (1986). The biology and behaviour of slugs in relation to crop damage and control. *Agricultural Zoology Reviews*, **1**, 253-297.

South, A. (1992). *Terrestrial Slugs, Biology, Ecology and Control*. London: Chapman & Hall, 428 pp.

Wilson, M.J.; Glen, D.M.; George, S.K. (1993a) The rhabditid nematode *Phasmarhabditis hermaphrodita*, as a potential biological control agent for slugs. *Biocontrol Science and Technology*, **3**, 503-511,

Wilson, M.J.; Glen, D.M.; George, S.K.; Butler, R.C. (1993b) Mass cultivation and storage of the rhabditid nematode, *Phasmarhabditis hermaphrodita*, a biocontrol agent for slugs. *Biocontrol Science and Technology*, **3**, 513-521.

Wilson, M.J.; Glen, D.M.; George, S.K.; Wiltshire, C.W. (1994a) Mini-plot field experiments using the rhabditid nematode, *Phasmarhabditis hermaphrodita*, for biocontrol of slugs. *Biocontrol Science and Technology*, **4**, 103-113.

Wilson, M.J.; Glen, D.M.; George, S.K.; Pearce, J.D.; Wiltshire, C.W. (1994b). Biological control of slugs in winter wheat using the rhabditid nematode, *Phasmarhabditis hermaphrodita*. *Annals of Applied Biology*, **125**, 377-390.

Wilson, M.J.; Glen, D.M.; Pearce, J.D.; Rodgers, P.B. (1995a) Monoxenic culture of the slug parasite, *Phasmarhabditis hermaphrodita* (Nematoda: Rhabditidae) with different bacteria in liquid and solid phase. *Fundamental and Applied Nematology*, **18**, 159-166.

Wilson, M.J.; Glen, D.M.; George, S.K.; Pearce, J.D. (1995b). Selection of a bacterium for the mass production of *Phasmarhabditis hermaphrodita* (Nematoda : Rhabditidae) as a biocontrol agent for slugs. *Fundamental and Applied Nematology*, (in press)

Wilson, M.J.; Glen, D.M.; George, S.K.; Hughes, L.A. (1995c). Biocontrol of slugs in protected lettuce using the rhabditid nematode *Phasmarhabditis hermaphrodita*. *Biocontrol Science and Technology*, **5**, (in press).

THE POTENTIAL FOR CONTROLLING THE CABBAGE ROOT FLY [*DELIA RADICUM*] BY RELEASING LABORATORY-REARED PARASITOIDS

S. FINCH

Horticulture Research International, Wellesbourne, Warwick CV35 9EF, UK

ABSTRACT

The life-cycles of the three parasitoids that attack the immature stages of the cabbage root fly are described together with possible ways of using them in the field. Ways of arresting parasitoids in the vicinity of brassica crops by growing flowering plants as feeding sites, by introducing grassy banks as overwintering refuges and/or by undersowing crops to make the environment more diverse, are discussed. Most of the paper is concerned with the problems inherent in employing parasitoids in the field. These include, how to produce sufficient parasitoids economically and more importantly how to minimize competition not only from other parasitoids but also from the other natural control agents, such as predatory ground beetles. The possibilities for using parasitoids on a large field scale are also discussed.

INTRODUCTION

During the last 75 years, many authors (see Tomlin *et al.*, 1992) have suggested that it might be possible to control field populations of the cabbage root fly (*Delia radicum* L.) using parasitoids. However, no one has yet used parasitoids successfully in this way.

The aim of this review is to determine from the published work which of the methods suggested to date seems most feasible. As I intend this review to be critical rather than encyclopaedic, there are likely to be more questions than answers. Nevertheless, by raising the contentious matters, I hope to identify both a profitable future research programme and those areas where additional research is required.

LIFE-CYCLE OF THE PARASITOIDS

Although five species of Braconidae, three of Eucoilidae and four of Ichneumonidae have been reared from cabbage root fly pupae (for authors see Coaker & Finch, 1971), the eucoilid *Trybliographa rapae* (Westw.) is the only hymenopterous parasitoid of major importance. This insect lays its eggs in all three larval instars of the fly (Wishart & Monteith, 1954) and has been recorded from 60% of the individuals in some samples of overwintering pupae (Wishart *et al.*, 1957). Parasitism of pupae by two beetle species of the genus *Aleochara* is also common. *Aleochara bilineata* (Gyll) is usually more common than *A. bipustulata* (L), possibly because the larvae of *A. bipustulata* find difficulty in entering the puparium of the cabbage root fly, which is thicker than the puparium of its preferred host, the bean seed fly *Delia platura* (Mg.) (Wishart *et al.*, 1957). The two species of *Aleochara* regularly parasitize 20-30% of cabbage root fly pupae (Finch & Collier, 1984) and occasionally 60% (Wishart et al., 1957). The two parasitoid beetles have life-cycles that are out of phase with each other. *A. bipustalata* overwinters as the adult whereas *A. bilineata* overwinters as the first-instar larva within the puparium of its host. Although a few authors have done detailed studies of the life-cycles of the wasp *T. rapae* (Wishart & Monteith, 1954) and the beetle *A. bilineata* (Colhoun, 1953), most references describe the parasitoids merely as mortality factors in the population dynamics of one, or more, pest species of *Delia*. Therefore, a major aim of the future programme will be to do detailed studies of the life-cycles of the parasitoids to determine how the various species interact, particularly under field conditions.

PROPOSED WAYS OF USING PARASITOIDS IN THE FIELD

The work to date has centred on *A. bilineata*, mainly because it prefers cabbage root fly pupae to bean seed fly pupae, and because, unlike the wasp *T. rapae*, it is carnivorous throughout its life-cycle. Hence, researchers believe that *A. bilineata* will act both as a predator and parasitoid providing it enters an infested brassica crop sufficiently early in the life-cycle of the pest. Two ways have been suggested for increasing the impact of the two parasitoid beetles. The first is to release *A. bilineata* inundatively at the time the pest fly starts to oviposit. The beetles would then be active much earlier than normal in brassica crops and could first eat the eggs of the fly, to lower the overall pest infestation, and then lay for their progeny to parasitize the remaining pest insects. The second method proposed is more complex (Ahlstrom-Olsson & Jonasson, 1992) and is based on using the overwintering adults of the beetle *A. bipustulata* to eat the eggs and early-instars of the cabbage root fly larvae, and the second beetle *A. bilineata* to eat the later-instars and also parasitize the pupae. This method involves placing considerable amounts of mustard meal around the base of brassica plants to attract and stimulate females of the saprophagous bean seed fly to lay in the mustard meal. The chemicals associated with bean seed fly larvae feeding within this meal then attract the overwintering adults of *A. bipustulata*, which lay in the meal so that their progeny can parasitize the bean seed fly pupae. As such parasitoid beetles are then at the sites where the later-emerging cabbage root fly lay, it is hoped that the beetle adults will feed on cabbage root fly eggs to mature their own eggs, and in this way lower the overall cabbage root fly infestation. The volatile chemicals associated with the feeding of those cabbage root fly larvae that manage to establish on the brassica plants will then attract first the parasitoid wasp *T. rapae*, and later the second beetle parasitoid, *A. bilineata*.

CONSERVING AND "ENHANCING" THE NUMBERS OF PARASITOIDS IN FIELD CROPS

Under current farming practices, the major impact of naturally-occurring polyphagous predators and parasitoids is to maintain certain soil-pest populations at more or less constant levels from year to year. As insecticides all act in a density-independent manner, large fluctuations in pest populations cannot be tolerated if reduced application rates of insecticide are to remain consistently effective (Suett & Thompson, 1985). Therefore, our principle aim at present is to "conserve" the existing levels of natural pest control. Unfortunately, in most cases, there is little quantitive information on the contribution made to overall pest control by the various biological agents. What is known, is that parasitism can be either high or low in pest populations. What is not known, is whether we can manipulate the field environment to ensure that parasitism is high in all instances. Many researchers within the IOBC Working Group "Integrated Farming Systems" (see Vereijken & Royle, 1989) are now attempting to increase populations of predators and parasitoids by the introduction of additional feeding sites and refuges. The problem with this approach is in assuming that the more beneficial insects there are in the area, the better will be the control of pest species. Pest control may not improve within such systems, particularly if the increase in the numbers of predators/parasitoids reflects merely the increase in the numbers of alternative sources of prey/hosts within the new habitats. Although it might seem like semantics, "increasing" and "enhancing" parasitoid numbers are not synonymous. The word enhance means "to add to the effect". In the field, it is relatively easy to "increase parasitoid numbers" whereas "enhancing their effects" is much more difficult. For example, parasitoid numbers can be "increased" by releasing parasitoids into fly-infested crops. However, if such parasitoids disperse before laying, then obviously they will not "enhance" the levels of pest control locally.

CHANGING THE ENVIRONMENT - ITS EFFECTS ON PARASITOIDS OF THE CABBAGE ROOT FLY

Flowering plants and grassy banks

Improving the general environment for predatory insects and parasitoids is attempted usually by growing flowering plants, in or around monocultures, to provide the beneficial insects with additional sources of food in the form of nectar and/or pollen (Finch,1988). This approach might help to increase the fecundity of *T. rapae*, once the types of flowers visited have been identified. This approach will not directly improve the environment for either species of *Aleochara*, however, as they feed almost entirely on animal protein.

Attempts are being made to introduce refuges, such as grassy banks, into crop fields, to increase the numbers of predators and parasitoids that overwinter successfully (Thomas *et al.*, 1991). This approach will not be effective with *T. rapae* or *A. bilineata* as both overwinter within cultivated fields inside the puparia of their host flies. Refuges might be effective, however, for *A. bipustulata*, particularly if brassica crops are grown after hay (*Lolium perenne* L.), as the plant material left once hay crops have been harvested can support large numbers of bean seed flies and could leave large numbers of *A. bipustulata* overwintering in such fields. The main problem with this approach is that there is no information on whether the refuges increase overall survival or simply alter the distribution of the overwintering beetles.

Undersown crops

The final way to make the environment more diverse within cultivated crops is to undersow the main crop with a forage crop, such as clover (*Trifolium repens* L.). Such systems have an adverse effect on pest insects largely because the pest insects are adapted to plants growing in bare soil (Kostal & Finch, 1994). Unfortunately, the beetle parasitoids of the cabbage root fly are also strongly adapted to bare soil situations and so undersowing with clover reduces the effectiveness of *A. bilineata*, though it appears to improve slightly the levels of parasitization by *T. rapae* (Langer, 1992).

PROBLEMS TO BE CONSIDERED WHEN RELEASING PARASITOIDS OF THE CABBAGE ROOT FLY

Short-season crops

Three factors operate against the use of "classical" biological control in brassica crops: the short growing season of many crops, the transience of the crops due to rotation, and the demand for increasingly high-quality produce. These three factors are complementary as, in the short-term relationship between pest and crop, the natural enemies of the pest have insufficient time to establish their superiority before the crop is damaged and no longer of high quality. This is particularly true for the cabbage root fly, as this fly generally enters brassica crops shortly after they have been transplanted and before the transplants have had time to establish adequate root systems. Therefore, as most brassica crops need to be protected more or less as soon as they are planted, the parasitoids will have to be released shortly after the crop is planted if they are to be effective as predators.

Rearing sufficient parasitoids

The main difficulty in mass-rearing parasitoids of the cabbage root fly is that an artificial diet has not yet been developed and so the host insect still has to be reared on swedes (*Brassica napus* var. *napobrassica*), a method which is both labour-intensive and physically-demanding. There is always the possibility of rearing the parasitoids on the closely-related onion fly (*Delia antiqua*), for which there is an artificial diet (Ticheler *et al.*, 1980) but getting the parasitoids to "switch" back to preferring the cabbage root fly might then become a problem. Perhaps the percentage of beetles that can "switch" from developing on onion fly pupae to developing on cabbage root fly pupae can be maintained at a high level by regularly transferring the insect culture back onto cabbage root fly pupae after having reared them on onion fly pupae for a fixed number of generations? However, it is possible that the beetles may have to be reared solely on the cabbage root fly if this is the species against which they are going to be used in the field.

Cost of parasitoid production

This will depend upon the number of staphylinid beetles required. The only two estimates made at present vary between 20,000 (Bromand, 1980) and 650,000 (Hertveldt *et al.*, 1984) beetles per hectare. Using the rearing technique described by Whistlecraft *et al.* (1985), 10 hours of labour would be needed to produce 20,000 *A. bilineata*. Although the work required for this approach might seem daunting, the costs appear to compare favourably with insecticidal control. For example to treat 20,000 plants (1ha) with chlorfenvinphos granules requires 11.2kg of product at a cost of £4.63/kg, or about £52/ha. Therefore, providing the hourly wage, plus overheads, of the workers producing the beetles does not exceed £5, the cost of control using these parasitoid beetles could be similar to that of using insecticide. Application costs might vary, but as chlorfenvinphos granules are applied generally as a sub-surface band at a cost of about £30/ha, there appears to be sufficient flexibility to keep within this cost even if the beetles have to be released manually.

Distributing the parasitoids within the crop to be protected

According to Esbjerg & Bromand (1977), *A. bilineata* released into brassica crops disperse at the rate of about 6.5m per day. From studies with beetles marked with radioisotopes, they concluded that for the control of cabbage root fly, batches of several hundred beetles should be placed at each release point, which should be spaced no more than 20m apart to ensure that the beetles spread throughout the crop as quickly as possible. Based on such data, releasing beetles from 16-20 points/ha should not create problems. To be effective, it is likely that the beetles will have to be distributed in this way, as female cabbage root flies are distributed more or less evenly through brassica crops (Finch & Skinner, 1973).

Competition with other parasitoids and with other natural enemies

Although many people believe that an array of parasitoids is preferable to using just one species, when the parasitoids compete for the same resource, the presence of more than one species can prevent the build-up of large parasitoid populations. Reader & Jones (1990) showed that there were marked differences in both the species emerging and in overall mortality when cabbage root fly pupae were attacked by larvae of both the wasp and the beetle. The data of Reader & Jones (1990) were used in an earlier paper (Finch, 1995) to develop a method, based on the timing of when the wasp and beetle larvae entered the host insect, to determine whether the wasp or the beetle survived or whether both perished. To be successful the beetle larvae must enter the fly pupae early in their development. Hence, the competition between the two parasitoids is biased heavily in favour of the wasp, as even if the beetle larva finds a cabbage root fly larva shortly after it has pupated, the beetle larva still needs between 12 and 36 hours to chew its way through the wall of the puparium before it can start feeding on the fly pupa (Colhoun, 1953).

One of the problems of releasing any biological agent into the field is that it has to compete with the established natural enemies. At worst, the release may simply upset the overall local balance so that the existing predators feed in a density-dependent manner on the released parasitoids until the balance is re-established. At such times, it is questionable whether polyphagous predatory beetles can be regarded as "beneficial" insects.

EFFECTS OF CULTURAL PRACTICES ON PARASITOID NUMBERS

Changes in the level of parasitism at a particular site can occur either gradually or rapidly. For example, at one site in Denmark, Bromand (1980) noted a gradual decline from 1971 to 1975 in the numbers of overwintering cabbage root fly pupae parasitized by *A. bilineata*. He associated this decline with a decrease in the area of swedes being grown, which resulted in a larger proportion of the beetles failing to find brassica crops. In contrast, rapid reductions in the levels of parasitism can be caused by some of the soil insecticides applied regularly to control the cabbage root fly (Coaker, 1966: Bromand, 1980), the parasitoids being more susceptible to these insecticides than the host insect (El Titi, 1980;

Finch & Skinner, 1980).

Apart from the adverse effects of pesticides on parasitoids, studies are needed on why ploughing has a greater effect on parasitoids than on pest populations (Finch & Skinner, 1980) and why parasitized pupae are rarely recovered from crops grown in highly-organic soils (Finch, unpublished data).

DISCUSSION

It is clear from this review that trying to enhance the activity of any one parasitoid within a group complex is not easy, as changes to "improve" the environment for one species invariably have the opposite effect on one of the other species. For example, growing flowering plants alongside crop boundaries could provide additional feeding sites for the wasp. However, such plants would undoubtedly support additional prey species that could lower the impact of the beetles as predators of the cabbage root fly (Finch, 1988).

The main problem with all of the systems used until now to "enhance" the numbers of polyphagous predators and parasitoids, is that the various treatments have only added to the numbers of beneficial insects present rather than "enhancing" their effects as pest control agents. This raises the question of whether it will ever be possible to improve pest control by making crop boundaries more diverse, as the associations between the numbers of predators and their prey and the numbers of parasitoids and their hosts, is finely balanced in such systems. In general, insects only attain pest status in the types of unbalanced systems that occur in agriculture and, in particular, in large monocultures. Therefore, perhaps the only way to resolve this problem will be to treat the unbalanced systems with an unbalanced control measure, such as the release of higher than normal numbers of laboratory-reared predators/parasitoids. At present we use specific insecticides to control the cabbage root fly, so by analogy we may also need to use specific, rather than general, biological agents for the types of pest control needed in ephemeral cultivated crops.

Many authors have indicated that the beetle A. bilineata is probably the most appropriate parasitoid to rear and release against the cabbage root fly. However, the published data indicate that this would be true only in localities where the wasp T. rapae does not occur. Although such localities can be found in Canada (Wishart & Monteith, 1954), the wasp was found to be the dominant species in 10 countries in northern Europe (Finch et al., 1985). A second drawback to releasing the beetle A. bilineata as a predator, is that it would have to be released at the start of each and every fly generation as, being a pupal parasitoid, no matter how early it is released its offspring will always emerge 2-3 weeks later than the pest fly.

The second alternative is to release the wasp, but as this is not predatory, the benefits of high levels of parasitism would accrue only in subsequent generations. There does, however, seem to be considerable scope for increasing levels of parasitization by the wasp, as no beetles emerged from samples of cabbage root fly pupae collected from Belgium, Denmark, Eire, Germany, Northern Ireland and The Netherlands, and in many of these countries the levels of parasitzation by the wasp rarely exceeded 5% (Finch, unpublished data). Therefore, in these countries, competition between the two major parasitoids appeared to be of little importance. Presumably the best way to reduce the competition from the beetle

A. *bilineata* in a specific locality in the UK would be to release wasps early into a crop so they parasitize a higher proportion of the early-instars of the fly and hence reduce considerably those instances of multiparasitism where the beetle larva was likely to survive. In addition, if *T. rapae* is to be reared in the laboratory, it would help if the fly larvae could be reared on an artificial diet to make them more accessible to the wasps. Alternatively, high levels of parasitism may be easy to achieve if the wasps will attack first-instar larvae shortly after they emerge from the eggs.

The third parasitoid species, *A. bipustulata*, has the advantage that it overwinters as the adult and hence is active when the fly is laying in the early spring. Whether this beetle can be used in conjunction with the wasp to give adequate levels of control requires testing. What is certain, however, is that if attempts are made to arrest this beetle around the base of brassica plants by adding organic material, something other than mustard meal (Ahlstrom-Olsson & Jonasson, 1992) should be used, as when Ahlstrom-Olsson tested this material at Horticulture Research International Wellesbourne in 1992, the mustard meal attracted preferentially the cabbage root fly and hence plants surrounded by mustard meal were damaged more severely than plants without mustard meal.

ACKNOWLEDGEMENT

I thank the Ministry of Agriculture, Fisheries and Food (contact: Dr A.R. Thompson) for supporting this work as part of Project HH1815SFV.

REFERENCES

Ahlström-Olsson, M.; Jonasson, T. (1992) Mustard meal mulch - a possible cultural method for attracting natural enemies of brassica root flies into brassica crops. *OILB/SROP Bulletin* **15/4** : 171-175.

Bromand, B. (1980) Investigations on the biological control of the cabbage root fly (*Hylemya brassicae*) with *Aleochara bilineata*. *Bulletin OILB/SROP* **3/1**: 49-62.

Coaker, T.H. (1966) The effect of soil insecticides on the predators and parasites of the cabbage root fly (*Erioischia brassicae* (Bouché)) and on the subsequent damage caused by the pest. *Annals of Applied Biology* **57** : 397-407.

Coaker, T.H.; Finch, S. (1971) The cabbage root fly, *Erioischia brassicae* (Bouché). *Report of the National Vegetable Research Station for 1970:* pp 23-42.

Colhoun, E.H. (1953) Notes on the stages and the biology of *Baryodma ontarionis* Casey (Coleoptera: Staphylinidae), a parasite of the cabbage maggot, *Hylemya brassicae* Bouché (Diptera: Anthomyiidae). *Canadian Entomologist* **85** : 1-8.

El Titi, A. (1980) Die Veränderung der Kohlfliegenmortalitat als Folge der chemischen Bekämpfung von anderen Kohlschadlingen. *Zeitschrift fur angewandte Entomologie* **90: 401-412.**

Esbjerg, P.; Bromand, B. (1977) Labelling with radioisotopes, release and dispersal of the rove beetle, *Aleochara bilineata* Gyll. (Coleoptera: Staphylinidae) in a Danish cauliflower field. *Tidsskrift for Planteavl* **81** : 457-468.

Finch, S. (1988) Entomology of crucifers and agriculture. Diversification of the agroecosystem in relation to cruciferous crops. In: *The Entomology of Indigenous and Naturalized Systems in Agriculture.* Eds. M.K. Harris & C.E. Rogers. Westview Press: Boulder Colorado, pp. 39-71.

Finch, S. (1995) A review of the progress made to control the cabbage root fly (*Delia radicum*) using parasitoids. *Acta Jutlandica* (In press)

Finch, S.; Collier, R.H. (1984) Parasitism of overwintering pupae of cabbage root fly, *Delia radicum* (L.) (Diptera: Anthomyiidae), in England and Wales. *Bulletin of Entomological Research* **74** : 79-86.

Finch, S.; Skinner, G. (1973) Distribution of cabbage root flies in brassica crops. *Annals of Applied Biology* **75**: 1-14.

Finch, S.; Skinner, G. (1980) Mortality of overwintering pupae of the cabbage root fly [*Delia brassicae*]. *Journal of Applied Ecology* **17**: 657-665.

Finch, S.; Bromand, B.; Brunel, E.; Bues, M.; Collier, R.H.; Foster, G.; Freuler, J.; Hommes, M.; Van Keymeulen, M.; Mowat, D.J.; Pelerents, C.; Skinner, G.; Stadler, E.;Theunissen, J. (1985) Emergence of cabbage root flies from puparia collected throughout northern Europe. In: *Progress on Pest Management in field Vegetables.* R. Cavalloro and C. Pelerents (Eds). C. P.P. Rotondo - D.G. XIII - Luxembourg No. EUR 10514. Rotterdam: Balkema, pp. 33-36.

Hertveldt, L.; Van Keymeulen, M.; Pelerents, C. (1984) Large scale rearing of the entomophagous rove beetle *Aleochara bilineata* (Coleoptera: Staphylinidae). *Mitteilungen aus der Biologischen Bundesanstalt für Land- und Forstwirtschaft* **218**: 70-75.

Kostal, V.; Finch, S. (1994) Influence of background on host-plant selection and subsequent oviposition by the cabbage root fly (*Delia radicum*). *Entomologia experimentalis et applicata* **70**: 153-163.

Langer, V. (1992) The use of a living mulch of white clover on the control of the cabbage root fly (*Delia radicum*) in white cabbage. *IOBC/WPRS Bulletin* **15/4** : 102-103.

Reader, P. M.; Jones, T.H. (1990) Interactions between an eucoilid [Hymenoptera] and a staphylinid [Coleoptera] parasitoid of the cabbage root fly. *Entomophaga* **35**: 241-246.

Suett, D. L.; Thompson, A.R. (1985) The development of localised insecticide placement methods in soil. *British Crop Protection Council Monograph* **28** : 65-74.

Thomas, M.B.; Wratten, S.D.; Sotherton, N.W. (1991). Creation of 'island' habitats in farmland to manipulate populations of beneficial anthropods. *Journal of Applied Ecology* **28**: 906-917.

Ticheler, J.: Loosjes, M.; Noorlander, J. (1980) Sterile-insect technique for control of the onion maggot, *Delia antiqua*. In: *Integrated Control of Insect Pests in the Netherlands*, pp. 93-97, Pudoc, Wageningen.

Tomlin, A. D.: McLeod, D.G.R.; Moore, L.V.; Whistlecraft, J.W.; Miller, J.J.; Tolman., J.H. (1992) Dispersal of *Aleochara bilineata* [Col.: Staphylinidae] following inundative releases in urban gardens. *Entomophaga* **37**: 55-63.

Vereijken, P.; Royle, D.J. (1989) Editors of: Current Status of Integrated Farming Systems Research in Western Europe. *IOBC/WPRS Bulletin* **12/5** : 76pp.

Whistlecraft, J.W.; Harris, C.R.; Tolman, J.H.; Tomlin, A.D. (1985) Mass-rearing technique for *Aleochara bilineata* (Coleoptera: Staphylinidae). *Journal of Economic Entomology* **78**: 995-997.

Wishart, G.; Monteith, E. (1954) *Trybliographa rapae* (Westw.) (Hymenoptera: Cynipidae), a parasite of *Hylemya* spp. (Diptera: Anthomyiidae). *Canadian Entomologist* **86** : 145-154.

Wishart, G.: Colhoun, E.H.; Monteith, E. (1957) Parasites of *Hylemya* sp. (Diptera: Anthomyiidae) that attack cruciferous crops in Europe. *Canadian Entomologist* **89**: 510-517.

THE POTENTIAL OF PARASITOID STRAINS IN BIOLOGICAL CONTROL; OBSERVATIONS TO DATE ON *MICROCTONUS* SPP. INTRASPECIFIC VARIATION IN NEW ZEALAND

S.L. GOLDSON, C.B. PHILLIPS, M.R. MCNEILL, N.D. BARLOW

AgResearch, Canterbury Agriculture and Science Centre, PO Box 60, Lincoln, Canterbury, New Zealand

ABSTRACT

Over the last 15 years New Zealand researchers have been actively involved in the classical biological control of forage weevil pests. In the 1980s the lucerne pest *Sitona discoideus* Gyllenhal (Coleoptera: Curculionidae) was successfully suppressed by the parasitoid *Microctonus aethiopoides* Loan (Hymenoptera: Braconidae). Since 1990 an additional programme has been developed to examine the impact of the recently introduced parasitoid *Microctonus hyperodae* (Hymenoptera: Braconidae) on New Zealand's worst ryegrass pest, the Argentine stem weevil *Listronotus bonariensis* (Kuschel)). The imported *M. hyperodae* founder populations were collected from a wide range of ecoclimatic zones.

During both of the programmes the importance of ecotypes has become increasingly apparent. The New Zealand *M. aethiopoides* population revealed phenological behavioural patterns quite different from those observed in Mediterranean Europe, North America and Australia. As a result of these observations, part of the current research into *M. hyperodae* has been developed to specifically explore any inherent ecotypic differences and their implications.

This contribution reviews progress to date in researching ecotypic differences and comments on the potential of ecotypes in classical biological control. This is with particular reference to recently developed DNA-based techniques that permit differentiation between populations at the subspecific level.

INTRODUCTION

Biological control has been sporadically subjected to criticisms such as those of Krebs (1972) who considered it to be akin to gambling. This may well be true when benefits to agricultural production are expected to be provided as quickly as possible and at minimum expense. This low cost approach however reduces the potential to gain scientific understanding from biological control initiatives. Scientific investigation into the principles of biological control is the only way to impart certainty into the approach's decision-making and optimisation (Greathead, 1986) and thus reduce the gambling element.

Biological control research is closely linked to a range of science areas and has for some time been recognised as a branch of applied ecology more than agriculture (Krebs, 1972). Agroecosystems cannot be treated in isolation from neighbouring natural areas partly because introduced organisms are likely to invade all suitable habitats, agricultural or otherwise (Howarth, 1991; Ferguson *et al.*, in press).

New Zealand agriculture provides simplified habitats that are well-suited to the study of biological control. The New Zealand indigenous arthropod fauna comprises c. 20 000 spp. that have been geographically isolated for over 100M years and is therefore unique (Emberson, 1994). In contrast, pest species and their control agents are often exotic and therefore ecologically remote from the native fauna. Many northern hemisphere plants have also been introduced into New Zealand and are similarly differentiated from the native fauna. Because of these circumstances many of the confounding effects such as have been discussed by Waage (1990) that arise from interactions between numerous closely related species can be avoided. Two New Zealand studies based on exotic *Microctonus* spp. parasitoids illustrate how this situation has allowed the beginning of analysis into the role and impact of intraspecific variation in biological control. Some of the approaches, findings and opportunities are presented in this contribution.

THE IMPORTANCE OF INTRASPECIFIC VARIATION IN BIOLOGICAL CONTROL

Since the turn of the century, species have ceased to be regarded as something fixed and uniform but are now considered to be polytypic comprising many subspecies and local populations (Mayr *et al.*, 1953). Arising from this, there has been a proliferation of terms to describe these entities such as subspecies, race, microspecies, ecotype, variant, clone, line, strain and biotype (Steiner, 1994). In addition to such diverse terminology, the situation has been confounded further by the lack of obvious fixed criteria for discrimination between subspecific groups (Gonzalez *et al.*, 1979). For the purpose of this contribution, subspecific groups will hereafter be referred to as 'strains' and when discussed the general assumption will be that their genotypes differ from those of other groups within the species being considered. These issues and the associated genetics have recently been reviewed in relative depth by Narang *et al.* (1994).

It may be argued that in recent years there has been increasing interest in strains in biological control. Such a development has probably arisen in part from the improving quality and availability of biological control theory (e.g. Mackauer *et al.*, 1990; Narang *et al.*, 1994; Godfray, 1994). The growth in interest has probably also been related to increasing awareness that exotic biological control agents can threaten non-target native species; particularly those that are insular in their distributions (Howarth, 1991). Ironically, a wide host range was seen to be desirable as it allowed a biological control agent to exist on other species when the target species were scarce (e.g. DeBach & Bartlett, 1964; Watt, 1965). A direct consequence of the increasing concern about the environmental impacts of biological control agents has been a reduction in the number of acceptable biological control species. The ability to make useful multiple introductions as discussed by Huffaker *et al.* (1971) is therefore now considerably reduced and there is a greater imperative than ever for suitable species to be used to their maximum potential. Fortuitously, this need for improved precision in recognition, re-examination, selection and exploitation of suitable agents has coincided with the advent of DNA-based discriminatory methods. Analyses of some sequences of ribosomal RNA and transfer RNA can be particularly useful in distinguishing between parasitoid strains (Narang *et al.*, 1994).

Roush (1990) examined in a general way the importance of intraspecific variation to biological control. He suggested that, to collect a founder population of a biological control agent that comprises a reasonable expression of a species' allelic diversity, 20 individuals may form an absolute minimum while populations of over 100 probably offer little additional advantage. The need to maintain genetic variation in the importation and colonisation processes was also emphasised. This included examination of how to best release different sexually reproducing strains collected from separate sites. The extent of understanding remains inconclusive and it is still debatable as to whether different strains should be released separately at different sites, separately at the same site or hybridised prior to release (Roush, 1990). Roush (1990) also pointed out that there are few examples in the literature where genetic variation has clearly influenced the success of biological control agents; efforts to demonstrate the importance of genetic variation are complicated by the need to discern between phenotypic and genotypic variation in biological control success.

The probable importance of strains can be illustrated using two biological control programmes involving *Microctonus* Wesmael spp. parasitoids in New Zealand. While neither of these programmes necessarily represents uniquely inspired planning or execution, both have had a notable measure of success with respect to pest suppression.

Research with *Microctonus aethiopoides*

This is an example of a parasitoid species which has a very variable biological control performance in different regions and/or with different host species.

The lucerne (*Medicago sativa* L.) pest *Sitona discoideus* Gyllenhal (Coleoptera: Curculionidae) was first discovered in New Zealand in 1974. It rapidly became a severe pest in this country, particularly in light stony soils that preclude any build up of mineralised nitrogen because of leaching (Goldson *et al.*, 1985). Massive spring *S. discoideus* larval populations of up to 4000 m^{-2} frequently destroyed all of the plant's nitrogen-fixing rhizobial root nodules (Goldson *et al.*, 1984; Goldson *et al.*, 1985), often resulting in mid-season production losses of up to 50% (Goldson *et al.*, 1985). This was generally manifest as an abrupt cessation of growth once a threshold of 1100-2000 larvae m^{-2} had been exceeded (Goldson *et al.*,

1985). It was postulated that these thresholds occurred due to loss of the crop's photosynthetic ability when demand for nitrogen exerted by re-establishing nodules exceeded that contributed by any remaining intact nodules (Goldson *et al.*, 1988a). The monocultural nature of lucerne stands made these damage thresholds very easy to establish.

The detailed nature of the study on *S. discoideus* and its impacts subsequently permitted the construction of full population dynamics analyses using several years' data (Goldson *et al.*, 1988b), which allowed the impact of the biological control agent *Microctonus aethiopoides* Loan (Hymenoptera: Braconidae) to be measured. *M. aethiopoides* was introduced into New Zealand in 1982 from Morocco and Greece via South Australia (Aeschlimann, 1983). While this parasitoid had earlier shown indifferent pest suppression elsewhere (e.g. Hopkins, 1985), it was successful in New Zealand (Goldson *et al.*, 1990; Barlow & Goldson, 1993). This occurred because 3% of the Canterbury *M. aethiopoides* population did not enter sympathetic aestivation with their hosts as has been consistently reported in other countries (e.g. Abu & Ellis, 1976; Cullen & Hopkins 1982), but continued to develop in pre-aestivatory weevils. Due to this atypical development, infected weevils became stranded in the lucerne fields unable to migrate out to the aestivation sites after their spring emergence (Goldson *et al.*, 1984). This resulted in adult wasps eclosing and remaining in the lucerne ready to attack further weevils as soon as they emerged. Later, after their autumnal return, unparasitised post-aestivatory populations were also immediately 'ambushed' by the pre-existing field populations of atypically-developed *M. aethiopoides* adults (Goldson *et al.*, 1990). The combined effect of such patterns of attack resulted in an average 60% *S. discoideus* parasitism after aestivation but prior to the bulk of weevil egg-laying (Goldson *et al.*, 1990). This is far higher than has been found elsewhere (e.g. Hopkins, 1985) and as such has been shown to be essential to the success of the programme (Barlow & Goldson, 1993) in spite of compensatory effects of density-dependent survival by the larvae (Goldson *et al.*, 1988b). The reason for this difference in *M. aethiopoides* behaviour compared to that observed elsewhere remains unknown. It is notable however that parasitoids from the same source did not show the same characteristics in Australia (Cullen & Hopkins, 1982).

The atypical New Zealand *M. aethiopoides* population behaviour is paralleled by variations in the species' biology noted elsewhere. Loan and Holdaway (1961) observed significantly different levels of survival amongst French and Moroccan strains of *M. aethiopoides* on *S. cylindricollis* Fahraeus. Studies in the United States have similarly found *M. aethiopoides* to be highly active against alfalfa weevil (*Hypera postica* (Gyllenhal) (Coleoptera: Curculionidae))(Dysart & Day, 1976) but not against *S. hispidulus* Fabricius or *S. cylindricollis* (Day *et al.*, 1971).

Primitively, *M. aethiopoides* is distributed throughout Europe including Sweden, France, Croatia, Rumania, Russia, Ukraine and Uzbekistan (Loan, 1975) as well as the Mediterranean including Morocco, Greece, Algeria, Tunisia, Portugal, Spain and Italy (Aeschlimann, 1980). Given such a wide distribution, it may be argued that at least some of the observed variation may be based on genetic differences and to this effect, considerable work based on morphology and behaviour has already been done (Adler & Kim, 1985; Sundaralingam, 1986). However, at least some of the variation that has been attributed to strains could be phenotypic effects arising from varied host associations (Phillips *et al.*, 1993), as have been found in other parasitoid species (e.g. Janzon, 1986; Johnson *et al.*, 1987).

No published DNA-based analyses have been conducted to analyse the genetic diversity of *M. aethiopoides*. Such an approach would assist in understanding the basis for behavioural and morphological differences in populations. It may be argued that, had there been a better understanding of the implications of *M. aethiopoides* strains at the outset, more appropriate release strategies could have been developed. Notwithstanding this however, the application of new techniques for subspecific differentiation could still greatly improve the potential of *M. aethiopoides*. In the first instance this would probably involve the collection of different European and Mediterranean populations of *Sitona* spp. and *Microctonus* spp. for DNA analysis. In this way, the New Zealand, Australian and American *Microctonus* spp. and *Sitona* spp. populations would be able to be compared to the original geographical populations, and an understanding of strainal basis for the variable performance of *M. aethiopoides* developed. Such a study could also reveal whether the relatively broad host range of the New Zealand *M. aethiopoides* population (Ferguson *et al.*, 1994) is genetically based. It can be speculated that these parasitoids may be showing some kind of subspecific 'hybrid vigour' and insight into this would help resolve some of the issues raised by Roush (1990).

Research with *Microctonus hyperodae*

This programme represents an experimental analysis of the significance of intraspecific variation in the context of a practical biological control effort.

Study into the biological control of Argentine stem weevil (*Listronotus bonariensis* (Kuschel)) (Coleoptera: Curculionidae) using a virtually unknown parthenogenetic parasitoid *Microctonus hyperodae* Loan (Hymenoptera: Braconidae, Euphorinae) (Loan & Lloyd, 1974) permitted some of the theoretical questions raised during the *M. aethiopoides* programme to be tackled experimentally from the start. It was intended that the study involving *M. hyperodae* should be in part a response to the challenge of workers like Waage (1990) and Roush (1990) who have lamented the lack of systematic field experimentation into biological control. Waage (1990) pointed out how little study there has been using replicated introductions of different agents in different regions without appropriate controls, while Roush (1990) noted that with a little creativity it should be possible to find situations where experimental releases could be arranged without sacrificing the urgency for solving pest problems.

Background
L. bonariensis apparently established in New Zealand without its suite of natural enemies at the turn of the century and is now recognised as New Zealand's worst ryegrass pest. In particular, the tiller-mining larval stages cause insidious damage that is often characterised by lack of recovery of pasture after summer drought (e.g. Whatman, 1959) and/or change in composition to clover or weed dominance (e.g. May, 1961; Goldson & Trought, 1980). Pesticides have not been cost-effective against *L. bonariensis* and much of the current management practice of this species depends on the use of plant resistance based largely on ryegrass infected with the endophyte fungus *Acremonium lolii* Latch, Christensen and Samuels (e.g. Prestidge *et al.*, 1982; Prestidge & Ball, 1991). *A. lolii* produces toxins in the grass which deters feeding by *L. bonariensis*, but the fungus has also been implicated in stock health problems (e.g. Fletcher, 1993).

It was against this background that the potential of classical biological control was recognised as an ideal complement to plant resistance; moreover, it had been noted elsewhere that there can be synergy between plant resistance and biological control (e.g. van Emden, 1982). In the late 1980s attention was therefore directed towards the South American parasitoid *M. hyperodae*. During an initial visit to South America in 1988 (Goldson *et al.*, 1990), *M. hyperodae* was far more abundant and widely distributed than was expected based on Loan and Lloyd (1974) and Lloyd's unpublished CIBC reports. The opportunity therefore presented itself to 'design' a biological control programme from the outset.

The widespread distribution of *M. hyperodae* (Goldson *et al.*, 1990) permitted the collection of seven populations from contrasting habitats in Brazil, Uruguay, Argentina and Chile. After quarantine and host range testing permission was granted to release the species (Goldson *et al.*, 1992). To date, this work has produced very promising results with levels of parasitism of over 80% recorded within three years of the species' release and clear indications of pest suppression (Goldson *et al.*, 1994).

Ecological experimentation and assessment of the importance of strains to the success of *M. hyperodae*
The various *M. hyperodae* strains and the range of New Zealand climate zones into which the parasitoid was released offered a good opportunity for ecological experimentation. Since 1991 there have been 24 releases throughout New Zealand, the majority of which comprised over 10 000 infected weevils each (Goldson & Barker, 1995). Considerable effort was made in each release to ensure that equal numbers of each strain were released at all sites. In this way it was hoped that analysis of spatial and temporal variation in the strains' establishment patterns could be made. This objective has already been achieved in part using a morphometrical approach whereby the South American east and west coast *M. hyperodae* strains can be differentiated (Phillips & Baird, 1995). From this work there is clear evidence that in the three years immediately following the species' release the east coast strains have been more successful than those from the west coast (Phillips *et al.*, 1994; Phillips, unpublished data). Work is currently being carried out on DNA-based methods of strain differentiation to take this study further. *M. hyperodae* collected from the field in New Zealand are being stored at -80°C pending the development of suitable methods.

At its most rewarding, this study should yield information on questions such as whether the same strain established throughout New Zealand, whether the rapidly establishing strains are usurped by others later and whether some strains migrate more rapidly than others. The DNA-based techniques being developed for this study could also shed new light on the nature and genetic stability of *M. hyperodae* parthenogenecity.

Analysis of factors that pre-dispose a biological control agent to success

Many authors (e.g. Messenger & van den Bosch, 1971; Caltagirone, 1985) have indicated the importance of identifying the correct strain of a biological control agent for biological control purposes although exactly how to do this remains uncertain. Ehler (1990) summarised eight factors seen by a number of workers to be important attributes of natural enemies (e.g. fitness and adaptability, high searching capacity, sufficient power of increase etc), although it has been recognised that this approach has limitations such as the lack of coincidence of all factors within one species (Waage 1990). Roush (1990) pointed out that of all the purported factors that do contribute to the success of biological control agents, in general only their adaptability to local climate conditions and ability to avoid host defence seem to be critical.

In the *L. bonariensis* biological control programme, the seven South American strains of *M. hyperodae* have been maintained separately in laboratory culture. This has permitted laboratory-based analysis of biological variation between the *M. hyperodae* strains to be conducted concurrently with the parasitoid release programme. It is hoped that identification of each strain's key biological characteristics, combined with the experimental basis of the parasitoid release programme, will eventually allow some definition of what confers adaptive advantage to the strains in New Zealand's different climate zones. Some progress has been made in this respect already. It has been found for example that the Brazilian strain of *M. hyperodae* does not enter photoperiodically-induced diapause whereas the others do (Goldson *et al.*, 1993); such an observation could provide insight into the adaptive value of *M. hyperodae* diapause in different parts of New Zealand. Recent work has also shown that by laying a mean of 62 eggs per parasitoid, the fecundity of the Uruguayan *M. hyperodae* strain is almost twice that of the four other strains analysed to date (unpublished data).

DNA-based investigation into the geographical origin of the New Zealand *L. bonariensis* population

During collection of the different South American strains of *M. hyperodae* populations, 12 South American populations of *L. bonariensis* were also collected. These were imported live into New Zealand and preserved at -80°C with a view to developing DNA-based techniques to determine the geographical origin of the New Zealand weevil population. Using a method based on PCR-RAPDS, Williams *et al.* (1994) demonstrated that the genome of New Zealand population is very similar to that of the Australian population and that there has apparently been very little change in the genome of the New Zealand population since it was accidentally introduced. Williams *et al.* (1994) and Williams (1994) were also able to show that there is very little genetic variation in the New Zealand populations from one region to the next and there is also weaker evidence that the founder population was probably genetically quite limited. Williams *et al.* (1994) achieved their original aim by demonstrating that the New Zealand/Australian populations of *L. bonariensis* probably originated in the River Plate area near Uruguay.

The use of DNA-based methods in this way helps to answer long-asked questions. For example, the genetic uniformity of the New Zealand population means that differences in the patterns of establishment of the different *M. hyperodae* strains are not due to varying host genetics. Furthermore, knowledge of the likely area of origin of the pest species could allow the importance of *L. bonariensis-M. hyperodae* co-evolution to be compared with the importance of collecting *M. hyperodae* from an area with a similar climate to the intended release region. To this effect, the preliminary observations that the 'east coast' parasitoids appear to be the most successful (Phillips *et al.*, 1994) and that the Uruguayan strain has a comparatively higher fecundity, perhaps points to co-evolution rather than climate matching as a critical factor

Finally, if it is possible to define the geographical origin of a pest species, this would indicate to biological control practitioners where to look for other co-evolved control agents such as pathogens.

BIOLOGICAL CONTROL THEORY AND MODELLING

The role of theory and models in biological control have been reviewed by Waage (1990), Karieva (1990), Barlow and Goldson (1990,1993) and Barlow (1993). In spite of the abundance of theory on parasite/host and predator/prey interactions, little use has been made of this in practical biological control, let alone in the specific area addressing the effectiveness of different strains of the same agent. This may be because theory has addressed different questions to those asked by the practitioners, or because any answers obtained are at too strategic a level to have a meaningful impact on tactics. Certainly there are difficulties in applying the simple models which underpin much of this theory to complex real-world problems (Barlow, 1993). Potentially, however, models of intermediate complexity offer a number of practical benefits (Barlow & Goldson, 1990, 1993; Barlow, 1993) which include: 1) predicting the outcome and success of a specific introduction; 2) aiding in the selection of the most appropriate agent(s) or strain(s); 3) predicting the impact of exotic agents on ecosystems and non-target species; 4) increasing understanding of the processes involved; 5) aiding in the identification and interpretation of critical field data; and 6) optimising management of existing and introduced biocontrol agents.

Modelling was used in both case studies described here. For the *S. discoideus* programme it was necessarily retrospective, but it added to the understanding of the parasitoid's success, particularly by demonstrating that *M. aethiopoides* did account for the full extent of observed pest suppression. The model also confirmed the hypothesis that observed levels of parasitism over summer could be obtained by atypical development of those parasitoids oviposited in newly-emerged adult weevils (Goldson *et al.*, 1990).

In the case of the *L. bonariensis* programme, modelling continues to be carried out concurrently with ecological analysis. Prior to the parasitoid's release, a model of its phenology based on development rate data for the different strains obtained under quarantine, showed the extent of host/parasitoid synchrony and the expected number of parasitoid generations in various regions of New Zealand. It suggested that small differences in development rates between strains were unlikely to be significant, but conversely the absence of diapause in the Brazilian strain could be disadvantageous by exposing larvae and pupae to additional winter mortality. Possibly for the first time, the model also provided advance predictions of anticipated parasitoid behaviour and impact, based on field experiments on parasitoid attack rates soon after release (Barlow *et al.*, 1993; Barlow *et al.*, 1994). However, it became apparent that knowledge of the pest was insufficient, nevertheless this had the useful consequence of guiding further ecological analysis in order to refine predictions of parasitoid impact and final steady states (if any). These predictions include the effects on pasture damage and the likely consequences of combining biological control with resistant, high-endophyte grasses, all of which will require continuing and intensive data collection. Such work will be tested against the actual outcomes in order to help understand the importance of ecotypic pre-adaptation and climate matching.

CONCLUSION

There is growing international recognition of the need to maintain biodiversity and to avoid the dangers that introduced polyphagous biological control agents may present to non-target species. This has reduced the range of suitable agents available for biological control and could be one reason for increasing interest in subspecific variation. Such variation may be used to optimise the selection of appropriate strains within a beneficial species. These developments have coincided with the development of new DNA-based methods that allow differentiation between strains.

The New Zealand work described in this contribution suggests that intraspecific variation could be important to the success of two biological control programmes based on *M. aethiopoides* and *M. hyperodae*. It is hoped that some of the approaches taken in the *M. hyperodae* programme have started to address the challenges issued by Roush (1990) and Waage (1990) who have called for a more experimental approach to biological control.

Finally, the success and nature of the impacts of *M. aethiopoides* and *M. hyperodae* indicated that common factors may be involved. In both cases the weevil species built up to enormous numbers, as often happens in the absence of naturally occurring regulatory factors. New Zealand has numerous other examples of weeds and pests (both invertebrate and vertebrate species) that have done this. With respect to the programmes discussed in this contribution, Goldson *et al.* (1994) suggested that the parasitoids were able to progress in a way very similar to their respective pest host species and presumably for similar reasons. They contended that relatively unfilled niches have offered little constraint to rapid population growth. Finally, both parasitoid species have high searching efficiencies (Barlow & Goldson, 1993; Goldson & McNeill, 1994) and *M. hyperodae*, if not both species, originated in regions where hosts are relatively rare in their unmodified habitats. Consequently the parasitoids' host-finding abilities are probably well in excess of what is needed in New Zealand's pastoral habitat.

REFERENCES

Abu, J.F.; Ellis, C.C. (1976) Biology of *Microctonus aethiopoides*, a parasitoid of the alfalfa weevil, *Hypera postica* in Ontario. *Environmental Entomology, 5,* 1040-1042.

Adler, P.H.; Kim, K.C. (1985) Morphological and morphometric analyses of European and Moroccan biotypes of *Microctonus aethiopoides* (Hymenoptera: Braconidae). *Annals of the Entomological Society of America, 78,* 279 -283.

Aeschlimann, J.-P. (1980) The *Sitona* (Col. Curculionidae) species occurring on *Medicago* and their natural enemies in the Mediterranean region. *Entomophaga, 25,* 139-153.

Aeschlimann, J.-P. (1983) Sources of importation, establishment and spread in Australia, of *Microctonus aethiopoides* Loan (Hymenoptera: Braconidae), a parasitoid of Sitona discoideus Gyllenhal (Coleoptera: Curculionidae). *Journal of the Australian Entomological Society, 22,* 325-331.

Barlow, N.D. (1993) The role of models in an analytical approach to biological control. *Proceedings of the 6th Australasian Conference on Grassland Invertebrate Ecology,* 318-325.

Barlow, N.D.; Goldson, S.L. (1990) Modelling the impact of biological control agents. *Proceedings of the 43rd New Zealand Weed and Pest Control Conference,* 282-283.

Barlow, N.D.; Goldson, S.L. (1993) A modelling analysis of the successful biological control of *Sitona discoideus* by *Microctonus aethiopoides* in New Zealand. *Journal of Applied Ecology, 30,* 165-178.

Barlow, N.D.; Goldson, S.L.; McNeill, M.R. (1994) A prospective model for the phenology of *Microctonus hyperodae* (Hymenoptera: Braconidae), a potential biological control agent of Argentine stem weevil in New Zealand. *Biocontrol Science and Technology, 4,* 375-386.

Barlow, N.D.; Goldson, S.L.; McNeill, M.R.; Proffitt, J.R. (1993) Measurement of the attack behaviour of *Microctonus hyperodae* as a classical biological control agent of Argentine stem weevil *Listronotus bonariensis. Proceedings of the 6th Australasian Conference on Grassland Invertebrate Ecology,* 326-330.

Barratt, B.I.P.; Ferguson, C.M.; Evans, A.A. (1995) Assessing the fate of introduced biological control agents in natural and agricultural ecosystems. *Proceedings of the Asia Pacific Agri-industry Community Conference,* in press.

Caltagirone, L.E. (1985) Identifying and discriminating among biotypes of parasites and predators. *Biological Control in Agricultural IPM Systems Academic Press Inc.,* 189-200.

Cullen, J.M.; Hopkins, D.C. (1982) Rearing, release and recovery of *Microctonus aethiopoides* Loan (Hymenoptera: Braconidae) imported for the control of *Sitona discoideus* Gyllenhal (Coleoptera: Curculionidae) in South Eastern Australia. *Journal of the Australian Entomological Society, 21,* 279-284.

Day, W.H. (1971) Reproductive status and survival of alfalfa weevil adults: Effects of certain foods and temperatures. *Annals of the Entomological Society of America, 264,* 208-212.

DeBach, P.; Bartlett, B.R. (1964) Methods of colonisation, recovery and evaluation. In: *Biological Control of Insect Pests and Weeds,* O. DeBach, (Ed), London: Chapman and Hall, pp. 402-426.

Dysart, R.J.; Day, W.H. (1976) Release and recovery of introduced parasites of the alfalfa weevil in eastern North America. *Production Research Report No 167,* United States Department of Agriculture, 61p.

Ehler, L.E. (1990) Introduction strategies in biological control of insects. In: *Critical Issues in Biological Control,* M. Mackauer, L.E. Ehler and J. Roland (Eds), Andover, Hants: Intercept, pp. 111-134.

Emberson, R.M. (1994) Taxonomic impediments to the development of sustainable practices in conservation and production. *Proceedings of the 43rd Annual Conference of the Entomological Society of New Zealand,* 71-78.

Ferguson, C.M.; Roberts, C.M.; Barratt, B.I.P.; Evans, A.A. (1994) Distribution of the parasitoid *Microctonus aethiopoides* Loan (Hymenoptera: Braconidae) in southern South Island *Sitona discoideus* Gyllenhal (Coleoptera: Curculionidae) populations. *Proceedings of the 47th New Zealand Plant Protection Conference,* 261-265.

Fletcher, L.R. (1993) Grazing ryegrass/endophyte associations and their effect on animal health and performance. In: *Proceedings 2nd International Symposium Acremonium/Grass Interact,* Plenary Papers, D.E. Hume, G.C.M. Latch, H.S. Easton (Eds), pp. 115-120.

Godfray, H.J.C. (1994) *Parasitoid Behavioural and Evolutionary Ecology,* Princetown, New Jersey: Princetown University Press, 473p.

Goldson, S.L.; Barker, G.M. (1995) Progress to date with Argentine stem weevil biological control. *Proceedings of 47th Ruakura Farmers' Conference,* 42-46.

Goldson, S.L.; McNeill, M.R. (1994) Preliminary investigation into the longevity and fecundity of *Microctonus hyperodae,* a parasitoid of Argentine stem weevil. *Proceedings of the 47th New Zealand Plant Protection Conference,* 215-218.

Goldson, S.L.; Trought, T.E.T. (1980) The effects of Argentine stem weevil on pasture composition in Canterbury. *Proceedings of the 33rd New Zealand Weed and Pest Control Conference,* 46-48.

Goldson, S.L.; Frampton, E.R.; Proffitt, J.R. (1988b) Population dynamics and larval establishment of *Sitona discoideus* (Coleoptera: Curculionidae) in New Zealand lucerne. *Journal of Applied Ecology,* **25,** 177-195.

Goldson, S.L.; Frampton, E.R.; Barratt, B.I.P.; Ferguson, C.M. (1984) The seasonal biology of *Sitona discoideus* Gyllenhal (Coleoptera:Curculionidae). *Bulletin of Entomological Research,* **74,** 249-259.

Goldson, S.L.; Jamieson, P.D.; Bourdôt, G.W. (1988a) The response of field-grown lucerne to a manipulated range of insect-induced nitrogen stresses. *Annals of Applied Biology,* **113,** 189-96.

Goldson, S.L.; Proffitt, J.R.; McNeill, M.R. (1990) Seasonal biology and ecology of *Microctonus aethiopoides* (Hymenoptera: Braconidae), a parasitoid of *Sitona* spp. (Coleoptera: Curculionidae), with special emphasis on atypical behaviour. *Journal of Applied Ecology,* **27,** 703-722.

Goldson, S.L., Proffitt, J.R. and McNeill, M.R. (1993) The effect of photoperiod on the diapause behaviour of the Argentine stem weevil parasitoid *Microctonus hyperodae* and its adaptive implications. *Proceedings of the 6th Australasian Conference on Grassland Invertebrate Ecology,* 363-368.

Goldson, S.L.; Barker, G.M.; Barratt, B.I.P.; Barlow, N.D. (1994) Progress in the biological control of Argentine stem weevil and comment on its potential. *Proceedings of the New Zealand Grasslands Association,* **46** 39-42.

Goldson, S.L.; McNeill, M.R.; Phillips, C.B.; Proffitt, J.R. (1992) Host specificity testing and suitability of the parasitoid *Microctonus hyperodae* Loan (Hymenoptera: Braconidae, Euphorinae) as a biological control agent of *Listronotus bonariensis* (Kuschel) (Coleoptera: Curculionidae) in New Zealand. *Entomophaga,* **37,** 438-498.

Goldson, S.L.; Dyson, C.B.; Proffitt, J.R.; Frampton, E.R.; Logan, J.A. (1985) The effect of *Sitona discoideus* Gyllenhal (Coleoptera:Curculionidae) on lucerne yields in New Zealand. *Bulletin of Entomological Research,* **75,** 429-442.

Gonzalez, D.; Gorlh, G.; Thompson, S.N.; Adler, J. (1979) Biotype discrimination and its importance to biological control. In: *Genetics in Relation to Insect Management*, M.A. Hoy and J.J. McKelvey Jr (Eds), New York: Rockefeller Foundation Publications, 129p.

Greathead, D.G. (1986) Parasitoids in classical biological control. In: *Insect Parasitoids*, J. Waage and D. Greathead (Eds), London: Academic Press, pp. 289-318.

Hopkins, D.C. (1985) Controlling sitona weevil with insecticides. *Proceedings of the 4th Australasian Conference on Grassland Invertebrate Ecology*, 94-98.

Howarth, F.G. (1991) Environmental aspects of biological control. *Annual Review of Entomology*, **36**, 485-509.

Huffaker, C.B.; Messenger, P.S.; DeBach, P. (1971) The natural enemy component. In: *Biological Control*, C.B Huffaker (Ed), Plenum Press, pp. 16-61.

Janzon, L. (1986) Morphometric studies of some *Pteromalus* Swederus species (Hymenoptera: Chalidoidea) with emphasis on allometric relationship, or: are ratios reliable in chalcid taxonomy? *Systematic Entomology*, **11**, 75- 82.

Johnson, N.F.; Rawlins, J.E.; Pavuk, D.M. (1987) Host-related antennal variation in the polyphagous egg parasitoid *Telenomus alsophilae* (Hymenoptera: Scelioidae). *Systematic Entomology*, **12**, 437-447.

Karieva, P. (1990) Establishing a foothold for theory in biological control practice: using models to guide experimental design and release protocols. In: *New Directions in Biological Control*, R.R. Baker, P.E. Dunn (Eds), USA: Alan R. Liss, New York, pp. 65-81.

Krebs, C.J. (1972) *Ecology: the experimental analysis of distribution and abundance*, New York: Harper and Row, 694p.

Loan, C.C. (1975) A review of the Haliday species of *Microctonus* (Hymenoptera: Braconidae, Euphorinae). *Entomophaga*, **20**, 31-41.

Loan, C.; Holdaway, F.G. (1961) *Microctonus aethiops* (Nees) auctt. and *Perilitus rutilus* (Nees) (Hymenoptera: Braconidae), European parasites of *Sitona* weevils (Coleoptera: Curculionidae). *Canadian Entomologist*, **93**, 1057-1079.

Loan, C.C.; Lloyd, D.C. (1974) Description and field biology of *Microctonus hyperodae* Loan n. sp. (Hymenoptera: Braconidae, Euphorinae) a parasite of *Hyperodes bonariensis* in South America (Coleoptera: Curculionidae). *Entomophaga*, **19**, 7-12.

Mackauer, M.; Ehler, L.E.; Roland, J. (1990) *Critical Issues in Biological Control*, M. Mackauer (Ed), Andover, Hants: Intercept, 330p.

May, B.M. (1961) The Argentine stem weevil, *Hyperodes bonariensis* Kuschel on pasture in Auckland. *New Zealand Journal of Agricultural Research*, **4**, 289-297.

Mayr, E.; Linsley, E.G.; Usinger, R.L. (1953) *Methods and Principles of Systematic Zoology*, New York: McGraw-Hill.

Messenger, P.S.; van den Bosch, R. (1971) The adaptability of introduced biological control agents. In: *Biological Control*, C.B Huffaker (Ed), New York: Plenum Press, pp. 68-92.

Narang, S.K.; Bartlett, A.C.; Faust, R.M. (1994) *Applications of Genetics to Arthropods of Biological Control Significance*, Boca Raton, Florida: CRC Press, 199p.

Phillips, C.B.; Baird, D.B. (1995) A morphometric method to assist in defining the South American origins of *Microctonus hyperodae* Loan (Hymenoptera: Braconidae) established in New Zealand. *Biocontrol Science and Technology*, in press.

Phillips, C.B.; Baird, D.B.; Goldson, S.L. (1994) The South American origins of New Zealand

Microctonus hyperodae parasitoids as indicated by morphometric analysis. *Proceedings of the 47th New Zealand Plant Protection Conference,* 220-226.

Phillips, C.B.; Goldson, S.L.; Emberson, R.M. (1993) Host-associated morphological variation in Canterbury (New Zealand) populations of *Microctonus aethiopoides* Loan (Hymenoptera: Braconidae, Euphorinae). *Proceedings of the 6th Australasian Conference on Grassland Invertebrate Ecology,* 405-414.

Prestidge, R.A.; Ball, O.J-P. (1991) The role of endophytes in alleviating plant bioic stress in New Zealand. In: *Proceedings of the Second International Symposium of Acremonium/grass Interactions,* Plenary Papers, D.E. Hume, G.C.M. Latch and H.S. Easton (Eds), AgResearch, New Zealand: pp. 141-151.

Prestidge, R.A.; Pottinger, R.P.; Barker, G.M. (1982) An association of *Lolium* endophyte with ryegrass resistance to Argentine stem weevil. *Proceedings of the 35th New Zealand Weed and Pest Control Conference,* 119-122.

Roush, R.T. (1990) Genetic variation in natural enemies: critical issues for colonisation in biological control. In: *Critical Issues in Biological Control,* M. Mackauer, L.E. Ehler and J. Roland (Eds), Andover, Hants: Intercept, pp. 263-288.

Steiner, W.M. (1994) Genetics and insect biotypes: evolutionary and practical implications. In: *Applications of Genetics to Arthropods of Biological Control Significance,* S.K. Narang, A.C. Bartlett and R. M. Faust (Eds), Boca Raton, Florida: CRC Press, pp. 1-17.

Sundaralingam, S. (1986) Biological morphological and morphometric analyses of populations of *Microctonus aethiopoides* Loan (Hymenoptera: Braconidae). Unpublished PhD thesis, Pennsylvania State University, USA, 80p.

van Emden, H.F. (1982) Principles of implementation of IPM. In: *Proceedings of Australasian Workshop on Development and Implementation of IPM,* P.J. Cameron, C.H. Wearing and W.M. Kain (Eds), Auckland: Government Printer.

Waage, J. (1990) Ecological theory and the selection of biological control agents. In: *Critical Issues in Biological Control,* M. Mackauer, L.E. Ehler and J. Roland (Eds), Andover, Hants: Intercept, pp. 135-157.

Watt, K.E.F. (1965) Community stability and the strategy of biological control. *Canadian Entomologist,* **97**, 887-895.

Whatman, C.P. (1959) Damage to pasture by wheat stem weevil. *New Zealand Journal of Agriculture,* **98**, 551-552.

Williams, L.C. (1994) Genetic diversity in *Listronotus bonariensis* (Kuschel) and geographical origin of the New Zealand population. Unpublished PhD thesis, Lincoln University, Canterbury, New Zealand, 131p.

Williams, L.C.; Goldson, S.L.; Baird, D.B.; Bullock, D.W. (1994) Geographical origin of an introduced insect pest, *Listronotus bonariensis* (Kuschel), determined by RAPD analysis. *Heredity,* **72**, 412-419.

PROMOTING NATURAL BIOLOGICAL CONTROL OF SOIL-BORNE PLANT PATHOGENS

MARK P. McQUILKEN

Plant Science Department, The Scottish Agricultural College, Auchincruive, Ayr, Scotland, KA6 5HW, UK

ABSTRACT

Despite a considerable research effort, there are very few commercially-available biological control agents of soil-borne plant pathogens. An alternative approach to biological control is to manipulate the existing population of microbial antagonists. Examples of both agricultural and horticultural practices which manipulate existing microbial antagonists and promote natural biological control of soil-borne plant pathogens are reviewed. The potential for the future use of these practices in different crop production systems is discussed.

INTRODUCTION

Soil-borne plant pathogens have been with us ever since agriculture began, and include about 50 genera of fungi as well as a few bacteria and viruses. They infect foliage, stems, seeds and actively growing roots. Effective control of fungal soil-borne plant pathogens has usually been achieved by the intensive application of fungicides. In recent years, however, there have been considerable changes in attitude towards their widespread use in disease control programmes. A major problem has been the development of pathogen resistance. Currently recommended programmes, involving the alternate use of fungicides with different modes of action, are not always effective. Moreover, increasing public awareness concerning the levels of fungicides, as well as other agrochemical residues in plants and the environment, has led to more stringent regulations on their use. It is also likely that a number of fungicides, which are currently on the market, will be withdrawn in the future. Consequently, the need for safer and more effective disease control methods that can be used as alternatives or supplements to conventional fungicides has become urgent.

Biological control, using introduced microbial inocula, is one strategy available. However, despite a considerable research effort by both academia and industry, very few biological control agents have been developed commercially (Rhodes, 1992). For example, in the UK, there are only three examples and these are confined to use in horticulture. Nevertheless, there are many forms of biological control that do not involve the direct application of commercially-produced inocula, but rely on manipulating the existing population of microbial antagonists. These include cultural practices such as crop rotations, tillage, the incorporation of organic amendments and composts, and the ploughing-in of green manures (Palti, 1981; Campbell, 1989; 1994). There is also the possibility of promoting biological control by the application of fertilisers, periodic flooding and solar heating the soil (Cook and Baker, 1983).

This paper reviews examples of both agricultural and horticultural practices which manipulate the existing population of microbial antagonists and promote natural biological control of soil-borne plant pathogens. It also discusses the potential for their future use in different crop production systems.

TILLAGE

Control of soil-borne plant pathogens can be achieved by certain tillage practices (Cook & Baker, 1983). For example, ploughing buries diseased crop residues and pathogen propagules such as sclerotia, and also leads to a more rapid breakdown of the pathogen's food base. The propagules eventually die and the inoculum potential decreases. Microbial antagonists in the soil are likely to be involved in decreasing the viability of the pathogen propagules.

ROTATIONS

Rotations have been used for many years to reduce the inoculum potential of plant pathogens (Campbell, 1989; 1994). Besides providing plant nutritional and other agricultural benefits, rotations deprive pathogens of their hosts so that they have to survive for long periods in the soil. During this survival period, the pathogens may die of starvation, or be parasitised and lysed by antagonistic microorganisms.

SUPPRESSIVE SOILS

There are some soils in which diseases fail to develop even though the pathogen is present. These disease-suppressive soils may be associated with abiotic factors, such as the pH or the clay and mineral content. However, there are soils in which suppression is caused by microorganisms. For example, Lumsden et al. (1987) described the suppressive soils of the traditional Mexican chinampa agroecosystem, which involves the incorporation of high levels of organic materials, including manures and crop wastes, and also mineral nutrients from aquatic sediments. The incidence of *Pythium* damping-off within these soils is low, and they are suppressive to introduced *Pythium*. This phenomenon is associated with high levels of microbial activity, particularly fluorescent pseudomonads and saprophytic *Fusarium* spp.. Similarly, soils in Hawaii suppressive to damping-off caused by *P. splendens* contain high calcium levels and a high population of soil microorganisms (Kao & Ko, 1986).

Soils suppressive to *Fusarium* wilt of muskmelon occur in the Chateaurenard region of France (Alabouvette, 1986). Suppression is due to competition between the natural saprophytic *Fusarium* spp. and the pathogen *Fusarium oxysporum* f. sp *melonis*. Transferring small amounts of this soil to pasteurised conducive soil will control *Fusarium* wilt from introduced inocula, but sterilisation of the suppressive soil removes this ability. This feature is characteristic of suppression caused by microorganisms. Transferring samples of suppressive soil to container media and disease-conducive soils may be of use in the future in controlling diseases in horticulture.

Soils may develop suppressive characteristics during prolonged monoculture. The classic example is take-all decline. This phenomenon has been extensively researched for many decades, but it is still not fully understood (Hornby, 1979; Rouxel, 1991). It is likely that disease suppression following continued cropping of wheat is caused by a change in the microbial population of the rhizosphere to one which is antagonistic to the take-all fungus, *Gaeumannomyces graminis* var. *tritici* (Hornby, 1979).

ORGANIC AMENDMENTS

The addition of organic amendments to soil is known to stimulate the activity of antagonistic microorganisms and to control a wide range of pathogens. For example, various partially degraded crop residues and manures have been shown to control *Fusarium* spp., *Rhizoctonia solani*, *Thielaviopsis basicola* and *Sclerotium* spp. (Lumsden *et al.*, 1983). Control of such pathogens is mainly due to a reduction in inoculum potential, or suppression of germination and growth. This may involve the production and release of antibiotics, competition for nutrients or parasitism by resident antagonistic microorganisms.

The addition of chitin to soil is perhaps one of the best documented examples of using organic substances to control soil-borne plant pathogens. For example, wilt of peas, caused by *Fusarium oxysporum* f. sp. *pisi*, was reduced by up to 82% following the incorporation of chitin several weeks before planting (Khalifa, 1965). A general increase in the actinomycete population was correlated with a fall in the inoculum potential of *F. oxysporum* f. sp. *pisi* together with a reduction in pea wilt. Amendment of soil with chitin has also been shown to control *R. solani* and *Sclerotium rolfsii* (Sneh *et al.*, 1971).

COMPOSTS

In China, Japan and other countries in Asia, composted organic waste materials have been used for many years as organic fertilisers and to control soil-borne plant pathogens. However, the potential of these materials to control plant pathogens in the West has only been recognised during the last three decades (Hoitink & Fahey, 1986). Composts prepared from sewage sludge, municipal solid wastes, tree barks as well as other materials have given control of a number of plant pathogens both in the field and under glass.

Amendment of soil with composted sewage sludge over a four year period reduced the incidence of lettuce drop caused by *Sclerotinia minor* (Lumsden *et al.*, 1986). Disease suppression was correlated with increased microbial activity, but also changes in N, P, Mg, Ca and total organic matter content of the soil. Similarly, incorporation of composted organic household waste gave a reduction in root rot of peas, beans and beetroot, caused by *P. ultimum* and *R. solani* (Schüler *et al.*, 1989). Unfortunately, these two types of composted materials have the potential of introducing heavy metal contamination. This factor must be carefully considered before widespread application is considered.

Composted tree barks incorporated into container growing media have been shown to control damping-off pathogens during seedling production (Stephens & Stebbins, 1985). Other composted materials which have shown potential for the control of damping-off in container media include hemlock (*Tsuga heterophila*) bark (Kai *et al.*, 1990), liquorice (*Glycyrrhiza glabra*) roots (Hadar & Mandelbaum, 1986) and grape marc (grape skin, seeds and stalks left over after wine processing) (Gorodecki & Hadar, 1990). Disease control has been suggested to result from the populations of antagonistic bacteria and fungi which colonise the composted materials.

Recently, foliar sprays of compost extracts produced from composted organic materials have shown some potential for controlling a number of diseases, including potato blight and Botrytis grey mould (Weltzien, 1992; McQuilken *et al.*, 1994). Antagonistic microorganisms within the extracts probably suppress the pathogens by direct inhibition of germination and growth. Before compost extracts can be widely used to suppress diseases on edible crops, it will be necessary to determine the possible environmental and toxicological hazards of extracts based on municipal waste or sewage sludge.

GREEN MANURES

Green manuring involves the incorporation into soil of fresh organic material, other than just plant residues, which has been grown either *in situ* or elsewhere (Campbell, 1989). The process is known to encourage general microbial activity in soil, and there are several examples where plant pathogens have been effectively controlled. One of the best examples is the control of common scab of potatoes caused by the actinomycete, *Streptomyces scabies*. Early work by Millard & Taylor (1927) indicated that incorporating green manures in the planting trench increased microbial activity which in turn anatagonised *S. scabies*. Ploughing-in *Medicago lupulina* as a green manure was shown to reduce take-all in wheat and to significantly increase the rhizosphere population of bacteria, particularly fluorescent pseodomonads (Lennartsson, 1988). Similar work has shown that actinomycetes antagonistic to the take-all pathogen increase in response to various green manure treatments (Campbell, 1994). Ploughing-in of green manures in cotton production in the USA has been shown to control the root pathogen, *Phymatotrichum omnivorum* (Cook & Baker 1993) An increase in microbial activity, particularly *Trichoderma* spp., was correlated with colonisisation and destruction of sclerotia of the pathogen.

Caution must be taken in using green manures as some can encourage disease development. For example, using *Sesbania* spp. (tropical woody legumes) as a green manure increased the incidence of damping-off caused by *Pythium* and *Rhizoctonia*. (R. Campbell, personal communication).

FERTILISERS

Fertilisers have been implicated in stimulating the indigenous soil population of microbial antagonists. For example, in Australia, application of sulphur to soil in order to maintain a low pH reduced root rot and heart rot of pineapple caused by *Phytophthora cinnamomi* (Cook & Baker, 1983). Control was attributed to a decrease in zoosporangium formation of *P. cinnamomi* and an increase in the antagonist *T. viride*. Take-all of cereals can be suppressed when crops are provided with ammonium rather than nitrate nitrogen (Cook & Baker, 1983). The ammonium nitrogen lowers the rhizosphere pH, which increases the availability of trace nutrients and stimulates the activity of antagonists.

FLOODING

Flooding has been shown to provide highly effective biological control of soil-borne pathogens. For example, Leggett & Rahe (1985) demonstrated that flooding weakened the sclerotia of the white rot pathogen (*Sclerotium cepivorum*) of onions in soils of British Columbia, thereby enabling microbial degradation of the sclerotia to occur. Similarly, flooding soil in Florida has been found to be effective in eliminating sclerotia of *Sclerotinia sclerotiorum*. Again, microbial colonisation and degradation of previously weakened sclerotia was thought to occur.

SOIL SOLARISATION

Heating soil can be accomplished by a process known as solarisation, in which heat from the sun penetrates a clear plastic sheeting placed on top of moist soil (Katan, 1981). Solarisation raises the soil temperature to kill pathogens by direct destruction, or it weakens the pathogens to such an extent that they are attacked by resident antagonists which survive the heating process. The process has been used successfully to control *Sclerotium rolfsii*, *Sclerotinia sclerotiorum* and *Fusarium* spp. in soil. In all cases, control was mainly due to microbial colonisation and degradation of pathogen propagules previously weakened by the sublethal temperatures produced by solarisation. Unfortunately, the use of solarisation for disease control is restricted to countries with a high insolation.

CONCLUSIONS & FUTURE PROSPECTS

Manipulating the existing population of microbial antagonists using a range of cultural practices is a form of biological disease control, which has obvious environmental benefits. Many of these practices which involve the application of large quantities of organic materials are labour intensive and expensive. Consequently, they are better suited to low technology agricultural systems, especially in developing countries where labour costs are relatively inexpensive and fungicides are either too expensive to buy or commercially unobtainable. In labour intensive China, the widespread and intensive use of organic materials has led to the absence of important root diseases (Kelman & Cook, 1977).

As labour costs are very expensive in Western agriculture, there are limits to the kinds of practices that can be used. However, as many of these practices as possible should be adopted and their increased use in integrated disease management systems, in combination with reduced fungicide applications, is likely to help in preventing or delaying the onset of pathogen resistance. Rotations and green manuring have been used to some extent in conventional, and more fully in organic agricultural systems. These merit further investigations as a means of reducing disease problems and over use of fungicides.

REFERENCES

Alabouvette, C. (1986) *Fusarium* wilt-suppressive soils from the Chateaurenard region: review of a 10-year study. *Agronomie* **6**, 273-284.

Campbell, R. (1989) *Biological control of microbial plant pathogens*, Cambridge: Cambridge University Press, pp. 218.

Campbell, R. (1994) Biological control of soil-borne diseases: some present problems and different approaches. *Crop Protection* **13**, 4-13.

Cook, R. J.; Baker, K. F. (1983) *The Nature and Practice of Biological Control of Plant Pathogens*, St Paul, Minnesota: American Phytopathological Society, 539 pp.

Gorodecki, B.; Hadar, Y. (1990) Suppression of *Rhizoctonia solani* and *Sclerotium rolfsii* diseases in container media containing composted separated cattle manure and composted grape marc. *Crop Protection* **9**, 271-274.

Hadar, Y.; Mandlebaum, R. (1986) Suppression of *Pythium aphanidermatum* damping-off in container media containing composted liquorice roots. *Crop Protection* **5**, 88-92.

Hoitink, H.A.; Fahy, P.C. (1986) Basis for the control of soil-borne plant pathogens with composts. *Annual Review of Phytopathology* **24**, 93-114.

Hornby, D. (!979) Take-all decline: a theorist's paradise. In: *Soil-borne Plant Pathogens*, B. Schippers & W. Gams (Eds), London: Academic Press, pp. 133-156.

Kai, H.; Veda, T.; Sakaguchi, M. (1990). Antimicrobial activity of bark-compost extracts. *Soil Biology and Biochemistry* **22**, 983-986.

Kao, C.W.; Ko, W.H. (1986) The role of calcium and microorganisms in suppression of cucumber damping-off caused by *Pythium splendens* in a Hawaiian soil. *Phytopathology* **76**, 221-225.

Katan, J. (1981) Solar heating (solarisation) of soil for control of soilborne pests. *Annual Review of Phytopathology* **19**, 211-236.

Kelman, A; Cook, R.J. (1977) Plant pathology in the People's Republic of China. *Annual Review of Phytopathology* **15**, 409-429.

Khalifa, O. (1965) Biological control of *Fusarium* wilt of peas by organic soil amendments. *Annals of Applied Biology* **56**, 129-137.

Leggett, M.E; Rahe, J.E. (1985) Factors affecting the survival of sclerotia of *Sclerotium cepivorum* in the Fraser Valley of British Columbia. *Annals of Applied Biology* **106**, 255-263.

Lennartsson, M. (1988) Effects of organic soil amendments and mixed species cropping on take-all disease of wheat. In: *Global Perspectives on Agroecology and Sustainable Agricultural Systems Vol. 2*, P. Allen & D. Van Drusen (Eds), Santa Cruz, California: University of California, pp.575-580a.

Lumsden, R.D.; Garcia-E.,R; Lewis, J.A.; Frias-T, G.A.. (1987) Suppression of damping-off caused by *Pythium* spp. in soil from the indigenous Mexican Chinampa agricultural system. *Soil Biology & Biochemistry* **19**, 501-508.

Lumsden, R.D.; Lewis, J.A.; Papavizas, G.C. (1983) Effect of organic matter on soilborne plant diseases and pathogen antagonists. In: *Environmentally Sound Agriculture*, W. Lockeretz (Ed), New York: Praeger Press, pp.51-70.

Lumsden, R.D.; Millner, P.D.; Lewis J.A. (1986) Suppression of lettuce drop caused by *Sclerotinia minor* with composted sewage sludge. *Plant Disease* **70**, 197-201.

McQuilken, M.P.; Whipps, J.M.; Lynch J.M. (1994) Effects of water extracts of a composted manure-straw mixture on the plant pathogen *Botrytis cinerea*. *World Journal of Microbiology & Biotechnology* **10**, 20-26.

Millard, A.W.; Taylor, C.B. (1927) Antagonism of microorganisms as the controlling factor in the inhibition of scab by green manuring. *Annals of Applied Biology* **14**, 202-216.

Palti, J. (1981) *Cultural practice and infectious crop diseases*. New York: Springer-Verlag.

Rhodes, D.J. (1992) Microbial control of plant diseases. In: *Disease Management in Relation to Changing Agricultural Practice*, A. R. McCracken and P.C. Mercer (Eds), Belfast: SIPP/BSPP, pp. 102-108.

Rouxel, F. (1991) Natural suppressiveness of soils to plant diseases. In: *Biotic Interactions and Soil-borne Diseases*, A.B.R. Beemster, G.L. Bollen, M. Gerlagh, M.A. Ruissen, B. Schippers and A. Tempel (Eds), Amsterdam: Elsevier, pp.287-296.

Schüler, C.; Biala, J.; Bruns, C.; Gottschall, R.; Ahlers, S; Vogtmann, H. (1989) Suppression of root rot on peas, beans and beetroots caused by *Pythium ultimum* and *Rhizoctonia solani* through amendment of growing media with composted household waste. *Journal of Phytopathology* **127**, 227-238.

Sneh, B.; Katan, J.; Henis, Y. (1971) Mode of inhibition of *Rhizoctonia solani* in chitin-amended soil. *Phytopathology* **61**, 1113-1117.

Stephens, C.T.; Stebbins, T.C. (1985) Control of damping-off pathogens in soilless container media. *Plant Disease* **69**, 494-496.

Weltzien, H.C. (1992) Biocontrol of foliar fungal diseases with compost extracts. In: *Microbial Ecology of Leaves*, J.H. Andrews and S.S. Hirano (Eds), New York: Springer-Verlag, pp.430-450.

A BRIEF REVIEW OF THE STATUS AND PROBLEMS OF BIOLOGICAL CONTROL OF WEEDS

D.H.K. DAVIES

Crop Systems, SAC, Bush Estate, Penicuik, EH26 0PH, Scotland, UK

ABSTRACT

Approaches to control of weeds by use of biological organisms are described. Rust fungi which have well developed and stable host specificity have been successfully developed as products that can be used through conventional farm equipment products. However, commercial and registration pressures are hindering their development. Arthropods have occasionally been used successfully, but maintaining populations has presented problems. There have been no useful breakthroughs in the development of specific prokaryotic and viral-based products, nor in the use of nematodes. If suitable mostly specific products can be produced that can be used through farm equipment, then there is no marketing barrier so long as the product is perceived as useful. However, it is considered that only highly valuable niche markets will be commercially worthwhile for the present. Genetic manipulation of organisms may provide most the promising approaches to improving bioherbicide performance, but such developments are still open to scientific and political debates.

INTRODUCTION

The use of a living organisms to affect a weed species to such an extent that its population falls below a damaging or threshold level is called biological control. Although we associate the concept of biological control with a modern approach to integrated pest management (IPM), there is in ancient literature comment on use of grazing animals, especially geese, amongst certain crops too large or indigestible for them to tackle. The use of chickens and pigs to clean land of weeds and their propagules is a commonly understood husbandry approach in some farming systems. However, the use of modern biological methods of weed control may date back to 1902, with the release of a range of exotic organisms to control *Lantana camara* in Hawaii (Tisdell, 1990), or even earlier when a Canadian farmer suggested *Puccinia* rust could be a method of control of *Cirsium arvense* (creeping thistle) (Greaves, 1992). This approach of releasing a specific organism, which then multiples and spreads so long as the host is present in sufficient quantity, is often known as the classical approach.

CLASSICAL APPROACH

Theoretically all kinds of plant pathogens and pests that adversely affect the growth of a weed could be considered for classical biological control, including fungi, prokaryotes, viruses, nematodes and arthropods.

To develop a biological control strategy one needs to understand the distribution of the weed host and its original habitat. It is to be presumed that it is in the original habitat of the species that most of its pathogens would have co-evolved. However, some authors have argued that newer host-parasite relationships may be more effective as the host will not have developed a high level of resistance (Hokannen, 1985 vide Cullen & Hasan, 1988). Nevertheless, in practice, highly co-evolved obligate parasites have been the most effective agents (Cullen & Hasan, 1988). It is probable that this approach is most effective, however, when weed species have spread well beyond their original habitat, and there is little or no competition with the potential parasite to reduce its effectiveness. The spread of weeds is well documented and Watson (1991) suggests that 13 of the top 15 most important in the USA are imports, which should facilitate the use of highly selective pathogens from their original habitat. The parasite from the original habitat of the weed must, obviously, be able to survive in the new climate/habitat, and as well, if not more importantly be able to demonstrate satisfactory specificity so that it does not attack other neutral or beneficial organisms in the new habitat. Efforts in this area are now largely limited to groups of fungi with records of well developed and stable host specificity. One of the best examples of this approach was the introduction of the European plant, *Chondrilla juncea* into Australia, which, in the absence of grazers spread rapidly as a weed. The introduction of the associated rust, *Puccinia chondrillina*, in 1975 has resulted in a reduction in the plant equivalent to European levels (Cullen, 1988 vide Cullen and Hasan, 1990). Rust species dominate the successful introductions (Watson, 1991) because as obligate parasites they tend more easily to fit specificity requirements. However, the pathogen often spreads slowly from such introductions. This is unsuitable for rapidly growing weeds in annual crops, and is probably more suited for perennial cropping situations and grassland where slower growing weeds are a problem.

AUGMENTATIVE APPROACH

Another approach is to encourage natural pathogen populations on native weeds by augmenting the pathogen through dispersing inoculum at disease conducive periods (Charudatta, 1985). For example, *Puccinia obtegens* has been augmented in Montana, USA, to assist in the control of *Cirsium arvense* (Dyer et al, 1982 vide Charudatta, 1985). The response time tends to be slow as in the case of Classical Approaches.

BIOHERBICIDE APPROACH

Microbial herbicides

All of the microbial weed control agents developed or in use in the USA in 1985 were fungal pathogens (Charuddaltan, 1985) or mycoherbicides. This is probably still the case. In the microbial herbicide strategy, an inoculum of the pathogen is mass cultured, standardised, formulated and deficiencies in speed of spread overcome by mass inocreation by conventional farm sprayer inoculation. To the present the products developed have been natural strains, and there would probably be ideological constraints to the development of genetically altered strains (M Greaves, 1995, personal communication).

Research in the 1960s and 1970s culminated in the registration of two mycoherbicide products with the United States Environmental Protection Agency (EPA): (a) DeVine® - *Phytophthora citophthora* for the control of *Morrenia odorata* (strangler-vine) in citrus orchards in Florida, and (b) Collego® - *Colletotrichum gloeosporoides* f.sp. *aeschynomene* for *Aeschynomeme virginica* (northern joint vetch) control in rice and soyabean in south-eastern USA. Since then another six products had been used, or were near marketing by 1992 (Greaves, 1992) (Table 1). Charudattan (1991) pointed out that a ratio of six or seven commercial products out of about 130 researched attempts over the period was very good compared with commercial herbicide discoveries.

TABLE 1. Mycoherbicide products in use and reported near to market in USA, 1985-91. (Adapted from Greaves, 1992)

Product	Agent	Target weed	Status
Devine®	*Phytophthora citrophthora*	*Morrenia odorata*	In use
Collego®	*Colletotrichum gloeosporoides* f. sp. *aeschynomene*	*Aeschnyomene virginica*	In use
	Cephalosporium dispyrii	*Dispyros virginiana*	
	Fusarium oxysporum	*Orobanche spp*	
Luboa II	*Colletotrichum gloeosporoides* f. sp. *cuscutae*	*Cuscuta spp.*	In use
	Colletotrichum gloeosporoides f. sp. *clidemiae*	*Clidemia hirta*	
Casst	*Alternaria cassiae*	*Cassia spp* *Crotolaria spectabilis*	Near market
Bromal	*Colletotrichum gloeosporoides* f. sp. *malva*	*Malva pusilla*	Registered
Velgo	*Colletotrichum coccodes*	*Abutilon threophrastis*	Near market
ABG5003	*Cerospora rodmanii*	*Eichhornia crassipes*	Near market
Doctor Biosedge®	*Puccinia canaliculata*	*Cyperus exculentus*	Near market

To date, rust fungi have dominated the microbial herbicide research area. There has been very limited research on the potential for prokaryotic and viral based products. Viral mechanisms often present major problems in the requirement for a carrier to transmit the pathogen. This is also true for many bacteria. Plant pathogenic bacteria also generally lack resistant structures that readily allow packaging, storing and application (Lacy, 1991).

It may be that increasing specificity in the vector is as important as improving the pathogenicity of the pathogen. However, these relatively simple organisms are more amenable to genetic manipulation with the potential to improve both specificity and pathogenicity. I have no information that weeds are being targetted in this way, and the area remains very much one for future research.

Nematodes

There has been much less work on the use of nematodes as potential biological weed control agents. In part this is due to a lack of understanding of host ranges with a consequent concern as regards their specificity - what work there is concentrates on foliar gall-forming types (Parker, 1991). There are no nematode - based products.

Use of Arthropods

Use of insects and other arthropods presents particular problems. Although arthropods have been used in classical and augmentative approaches, with Julien *et al* (1984) suggesting they comprised 488 out of 499 species of natural enemies released, these authors also suggest that they only account for 51 successful releases. The inability of agricultural habitats, particularly annual cropping, to maintain populations is suggested as a major reason for failure by Bernays (1985) and van Emden (1990). They suggest that an intermediate host, or the widespread presence of the host outside or around the cropping system, is needed to maintain a pod of colonising individuals. The problem is reduced in perennial cropping.

Bernays (1985) also suggests that finding grazers for grass weeds presents a particular problem; they are not attacked by many insects. Dicotyledonous weeds have many more pathogens. Specificity can also be a major problem, and Bernays (1985) indicates that artificial selection pressures may be needed to induce specificity in a potential pathogen.

An approach that is being considered is the conservation biological control of weeds by encouraging changes in the environment, particularly encouraging the growth of other host plants in the neighbourhood, perhaps by sowing, which allow breaks in cropping to be survived by the insect weed pest. However, a great deal more needs to be understood about the life-cycle of potential weed pests.

Bernays (1985) and other authors confirm that insect releases are likely to be most successful in range-weed situations where the potential host is more widespread. This fact encouraged an example of recent work in the UK: an attempt to control *Pteridium aquilinum* (bracken) with introduced South African moths (*Conservula cinisigna* and *Panotima* spp.) (Lawton et al, 1988). There is no tradition in the UK for such approaches to weed control, and there have been problems with survival. Nevertheless, evaluation of the potential for such an approach to overcome an intractable problem is still considered important.

However, it is evident that much more work must be done to define strategies for use of arthropods in biological weed control. Berhays (1985) gives the following suggestions (adapted):

- a reduced input on classical biological studies, no effort on graminaceous species.

- encouragement of research on plant competition in crops so that least harmful weeds may be identified.

- an integrated research effort between weed ecologists and entomologists to establish areas where arthropods can be used to alter weed balances in favour of crops.

- co-operation with plant pathologists.

- genetic improvement of selected species of the native fauna to increase their virulence against weeds.

These suggestions would indicate an acceptance that complete control of an individual weed is not required by the farmer. The balance between what is acceptable in terms of leaving weeds and the impact on crop quality and seed return to affect other crops in the rotation is still poorly understood, and is very species specific. Nevertheless, the reduced vigour of an infested weed may make it more susceptible to very low doses of conventional herbicides, mechanical weed control, or other biological control mechanisms. Such interactions are at the heart of understanding sustainability in farming.

Selective genetic improvement of potential biological agents is still open to debate and consideration as to their ethical and environmental impact. This is probably as true for genetic modification of arthropods as much as for any other organism. However, selection rather than modification may have greater social support. Nevertheless, such genetic engineering is at an early stage of technical development (Tauber, *et al*, 1985), and the potential is unpredictable.

DEVELOPMENT AND USE PROBLEMS

Watson (1991) describes six identifiable areas of research required in the development of a product based on a pathogen for biological control purposes:

-        determination of the suitability of the target

- survey of suitable pathogens
- ecology of potential pathogens
- host specificity
- introduction and establishment into the new habitat
- evaluation of the effect on the target weed

To this should perhaps be added the impact on the crop. Complete weed control is not often needed to maximise crop yield response, so good suppression by a pathogen may be sufficient.

Once identified as a potential pathogen there are many practical barriers to development. Particularly where foreign species are imported, investigations may be required to be undertaken in quarantine, and the extent of its host specificity and impact on other natural enemies of the host, and related organisms, identified (Watson, 1991). Presenting the pathogen in a form that will maximise its effect and speed of spread is very important, as the competitive effect of weeds can occur over a comparatively short period; particularly in annual crops. Formulation of bioherbicide fungal pathogens in spore form is now well understood (Greaves, 1995, personal communication), and such materials can be applied, as in the case of Devine® and Collego®, through conventional farm sprayers. This is essential as the narrow weed control spectrum of such materials usually requires that other herbicides are also used, along with other inputs. So compatibility in tank-mixes and sequences of routine farm treatments may be required (Smith, 1991). This presents problems with the use of arthropods as biological control agents in arable crops unless the eggs can be disseminated in a similar manner to the spores of pathogens .

The narrow weed spectrum of such products may present a problem to some farmers, but single or limited weed spectra are not unknown amongst conventional herbicide products, for example fluroxypyr for control of *Galium aparine*, difenzoquat for control of *Avena* spp, and are accepted long as the need is perceived as greater than the inconvenience of tank-mixing separate treatments.

PRODUCTION, MARKETING AND THE FUTURE

From the list in Table 1, Devine® and Collego® were withdrawn in 1994 on commercial rather than technical grounds. There was considered to be a lack of market size (Greaves, 1995, personal communication), although farmers found no difficulty in using the products combined with other herbicides. Collego® may be produced by an American university. Doctor Biosedge is a *Cyperus* spp rust pathogen, but the USA EPA has asked for so much data that the market size, although substantial, may not be sufficient. These fates appear to be common to this market, and the only currently assured product may be Luboa, a Chinese government produced product for the rice market, sold very cheaply. Greaves (1995, personal communication) suggests that there may be ideological limitations to the development of the technology within the large agrochemical industry, and they also may feel that the technology investment would be high. However, he points out that a fermentation system to produce microbial and fungal spores could cost 25% of that of a novel chemical plant; and that some of the companies already have such technology available for pesticide manufacture. Nevertheless, the uncertainty of the patenting status for such products must also be a major barrier.

Funding for research is this area is only likely to be through public bodies, and this is unlikely given limits in near-market spending in most countries.

Consequently the only markets likely to succeed are highly valuable niche markets such as American lawn-care followed by urban vegetation control and aquatic weed control where legislation may increasingly restrict the use of pesticides.

Where the agrochemical companies may have a greater interest is in more clearly patentable areas, such as modifiable toxins that can be identified from pathogens. If these however are then modified to improve activity as well as ensuring patentability, they are then no different from any other organic agrochemical molecule in the testing required, and perceived potential hazards.

The genetic manipulation of organisms for specific enzymes or toxin production within the pathogen is probably the most promising approach for improving mycoherbicide performance (Greaves, *et al* 1989). This may also be true for other forms of pathogen. Such manipulation may also elucidate greater specificity. The potential for such developments will have to await clarification as to patenting rights to protect such investment, and for the scientific and philosophical debates to come to a reasonable conclusion.

REFERENCES

Bernays, E.A. (1985) Arthropods for weed control in IPM Systems In: *Biological Control in Agricultural IPM Systems*, M.A. Hay; D.C. Herzog (Eds.). Academic Press, London, 1985, 373-388.

Charudattan, R. (1985) Use of natural and genetically altered strains of pathogens for weed control. In: *Biological Control in Agricultural IPM Systems*, M.A. Hay; D.C. Herzog (Eds.). Academic Press, London, 1985, 347-372.

Charudattan, R. (1991) The mycoherbicide approach with plant pathogens. In: *Microbial Control of Weeds*. D.O. TeBeest (Ed.). Chapman and Hall, London 1991, 24-57.

Greaves, M.P. (1992) Mycoherbicides: the biological control of weeds with fungal pathogens. *Pflanzenschutz-Nachrichten Bayer* 45, 1992, **1**, 21-31.

Greaves, M.P. (1995) Personal Communication.

Greaves, M.P.; Bailey, J.A.; Hargreaves, J.A. (1989) Mycoherbicides: opportunities for genetic manipulation. *Pesticide Science*, 1989, **26**, 93-101.

Julien, M.H.; Kerr, J.D.; Chan, R.R. (1984). Biological control of weeds: an evaluation. *Protection Ecology*, 7, 3-25.

Lacey, G.H. (1991). Perspectives for biological engineering of prokaryotes for biological control of weeds. In: *Microbial Control of Weeds*. D.O. TeBeest (Ed.). Chapman and Hall, London, 1991, 135-151.

Lawton, J.H. (1988). Biological Control of bracken in Britain: constraints and opportunities. In: *Biological Control of Pests, Pathogens and Weeds: Developments and Prospects. Phil. Trans. R. Soc. Land.,* B318, 335-355.

Parker, P.E. (1991). Nematodes as biological control agents of weeds. In: *Microbial Control of Weeds*. D.O. TeBeest (Ed.). Chapman and Hall, London, 1991, 58-68.

Smith, R.J. (1991) Integration of biological control agents with chemical pesticides. In: *Microbial Control of weeds*. D.O. TeBeest (Ed.). Chapman and Hall, London, 1991, 189-208.

Tauber, M.J.; Hoy, M.A.; Herzog, D.C. (1985). Biological control in agricultural IPM systems: a brief overview of the current status and future prospects. In: *Biological Control in Agricultural IMP Systems*. M.A. Hoys; D.C.Herzog (Eds.). Academic Press, London, 1985, 3-12.

Tisdell, C.A. (1990) Economic importance of biological control of weeds and insects. In: *Critical Issues in Biological Control*. M. Mackauer, L.E. Ehler, J. Roland (Eds.). Intercept, Andover, UK, 1990, 301-316.

Watson, A.K. (1991). The classical approach with plant pathogens. In: *Microbial Control of Weeds*. D.O. TeBeest (Ed.). Chapman and Hall, London, 1991, 3-23.

Session 3
Poster Presentations

Poster Organiser   Dr R O CLEMENTS

CLOVER: CEREAL BI-CROPPING

R.O. CLEMENTS

Institute of Grassland and Environmental Research, North Wyke Research Station, Okehampton, Devon, EX20 2SB, UK

D.A. KENDALL

IACR - Long Ashton Research, Dept. of Agricultural Sciences, University of Bristol, Long Ashton, Bristol, BS18 9AF, UK

G. PURVIS

UC Dublin, Faculty of General Agriculture, Belfield, Dublin 4,IE

T. THOMAS

Teagasc, Oak Park Research Centre, Carlow, IE

N. KOEFOED

Danish Institute of Plant and Soil Science, Research Centre Foulum, PO Box 21, DK 8830 Tjele, DK

ABSTRACT

A system for growing cereals, especially winter wheat, that requires substantially reduced inputs of N fertilizer, agrochemicals and management time is being developed in joint work between five research stations in Europe. The system relies on an understorey of white clover, (*Trifolium repens)* which becomes permanent and perennial to fix and supply N for cereal crops. The clover is defoliated in autumn and the cereal direct-drilled into it. Following cereal harvest, for silage or grain, the clover understorey is allowed to recover, defoliated in autumn and a second cereal crop direct-drilled into it to repeat the cycle. Altered physical conditions in the crop environment mean that weed, pest and disease problems are usually greatly reduced, with consequent savings in agrochemicals.

INTRODUCTION

Various attempts have been made to reduce over-production by limiting the agricultural area used for growing cereals. Chief among them are the various set-aside options, and although these largely achieve their aim of reducing cereals output they suffer serious disadvantages, not least of which are the socio-economic effects and the public perception that farmers are being paid for doing nothing. The impact of set-aside on wildlife, pests, diseases and weeds also leads to criticism of set-aside, as does the cost to the taxpayer of its administration and support.

Probably an alternative approach is to develop systems of cereal farming which produce lower yields, but that remain economically viable because they require greatly reduced inputs of fertilisers and agrochemicals. This in turn could obviate many of the environmental problems encountered with modern high input cereal production. A clover: cereal bi-cropping system is one such possible approach.

DESCRIPTION OF CEREAL: CLOVER SYSTEM

The system, which we are continuing to develop is simple and straightforward. It relies initially on establishing a sward of pure white clover (*Trifolium repens*) which becomes permanent and perennial. There are several different ways of establishing the clover which include i) producing a seed-bed and sowing clover seed into it (with no grass or other companion crop), (ii) under-sowing clover with a 'nurse crop' of spring cereal which is cut for whole-crop silage, (iii) sowing a mixture of grass and clover seed, allowing a clover rich sward to develop, (which can of course be used for livestock production) and then killing the grass with a herbicide - a low dose of paraquat or glyphosate works well.

In autumn the clover is grazed-off with sheep, or preferably cut and ensiled by mid-September, and then mown-short. Cereal seeds are direct-drilled into the defoliated clover. The cereal is allowed to develop and given a small amount of nitrogen (N) fertilizer (50 kg N/ha is suggested) in early spring. Subsequently the cereal relies on the N fixed by the clover. The cereal crop can be either cut for whole-crop silage, in which case the clover component provides a valuable protein-rich supplement, or else the cereal is left to grow to maturity and harvested for grain. The clover understorey survives, is left to recover and is then grazed or preferably cut for silage again and a second cereal crop is drilled into it to repeat the cycle. Work to date shows that two successive cereal crops can be grown in this way. Currently, an EU funded group are working on a project to develop this system for practical use and to investigate a wide range of specific aspects of its use.

WEED, PEST & DISEASE INCIDENCE

Weeds

In the several experiments completed (e.g. Jones & Clements, 1993) or in progress (Clements *et al.*, in press) broad leaved weeds have generally not posed a significant problem; they are easily smothered by the clover canopy, which becomes very dense. However at some sites grass weeds, especially annual meadow grass (*Poa annua*), have been very troublesome. The situation appears to be exacerbated if the clover is grazed, rather than cut for silage. Grazing tends to be uneven, certainly compared with cutting, and leaves bare areas which are easily colonised by *Poa*. Clearly the number of *Poa* seeds in the seed-bank has a profound effect on the likely development of a weed problem. The *Poa* seed-bank may be especially large in ex-grass compared with ex-arable fields.

However, recent evidence suggests that although it can be severe the problem of annual meadow grass appears to be easily avoided by spraying the area prior to drilling the cereal in autumn with an appropriate herbicide e.g. low-dose of paraquat, which does not harm the clover permanently.

Pests

Aphids are the major pest of cereal crops, but in early work and some current experiments their numbers are greatly reduced in bi-cropped areas. There may be a number of factors contributing to this e.g. (i) the relatively low N status of the cereal which provides a less than ideal diet for aphids, (ii) the presence of large numbers of predatory beetles and spiders in the clover understorey and (iii) the ever-present green mantle of clover may interfere with aphids ability to locate cereal crops.

In at least one series of recent experiments aphid numbers have been extremely high, but this may be as a consequence of *Poa* infestations on the plot.

Diseases

The major virus diseases of cereals are aphid borne. As result of aphid numbers usually being reduced in our bi-cropping system so too are virus problems. Fungal diseases are also greatly reduced. Deadman & Soleimani (1994) showed that this is because many of the major fungal diseases are splash-borne and the clover canopy virtually prevents disease progression. Also, trash at the base of bi-cropped areas decomposes very rapidly, thus reducing the carry-over of some diseases including *Fusarium*, but this requires further investigation (M.L. Deadman personal communication).

FERTILIZER, AGROCHEMICAL & MANAGEMENT INPUT

There is a need to supply adequate levels of P and K fertilizer to support vigorous clover and cereal growth. However, N fertilizer levels are greatly reduced, as alluded to above, (50 cf. 150-200 kg N/ha for a conventional crop). The level of agrochemical usage for weed, pest and disease control are also greatly reduced and clearly there are fewer management decisions to be made regarding their choice and timing.

CEREAL YIELDS

In the first year of use yields of cereal within the bi-cropping system are about 60% of those of conventional crops (Table 1), but because of the greatly reduced inputs required the gross margin remains at about 90% of that for conventional crops. The clover's contribution of fixed N is expected to increase in subsequent years and lead to improved cereal yields with repeated cycles of the system.

TABLE 1.   Whole crop silage and grain yields on bi-cropping experiment at Long Ashton 1994

| | Silage yield | | | Grain yield | | |
	t/ha	s.e.m.	(df)	t/ha	s.e.m.	(df)
Bicropped	9.8	0.32	(19)	4.9	0.25	(19)
Conventional	15.9	0.33	(19)	8.3	0.26	(11)

OTHER ADVANTAGES OF THE BI-CROPPING SYSTEM

Casual observations show that the cereal: clover bi-cropping system greatly reduces soil erosion - presumably because the perennial and dense crop cover impedes sheet and rill water flow. The perennial nature of the crop with its complex entomofauna may also be of benefit to certain birds and wildlife species. Soil organic matter increases markedly under bi-cropping, which may be a further advantage.

LIMITATIONS

One limitation may be that livestock are needed to utilise the clover, but many ostensibly arable farms have a sheep enterprise and clearly these would be able to make full use of the legume component. Agro-climatological limitations are being explored in our current programme of work.

ACKNOWLEDGEMENTS

This work is funded by MAFF (UK), Teagasc (IE), DIPSS (DK) and the EU (AIR 93-0392).

REFERENCES

Clements, R.O.; Kendall, D.A.; Asteraki, E.J.; George, S. (In press) Clover: cereal bi-cropping, a progress report. Proceedings British Grassland Society, 50th Anniversary Conference, Harrogate, December 1996.

Deadman, M.; Soleimani, J. (1994) Cereal: clover bi-cropping - implications for crop disease incidence. Proceedings Fourth Research Meeting, British Grassland Society, Reading 1994, 79-80.

Jones, L.; Clements, R.O. (1993) Development of a low input system for growing wheat (Triticum vulgare) in a permanent understorey of white clover (Trifolium repens). Annals of Applied Biology 123, 109-119.

INVESTIGATIONS USING SEX PHEROMONE AGAINST *BEMBECIA SCOPIGERA* (LEP: SESIIDAE)

ALİ TAMER

Plant Protection Research Institute, Bağdat Caddesi, No : 250 , PK . 49 -06172 Yenimahalle, Ankara, Turkey

ABSTRACT

In this study , adult emergence and the flying period of *Bembecia scopigera* were determined in 1, 2 and 3 year- old infested sainfoin (*Onobrychis viciifolia*) fields by various methods including sweep netting , attractive bait traps , counting hollow cocoons and sex pheromone traps. In 1, 2 and 3 year - old infested fields between 1984- 1987, the first moth appeared in early and mid- July and countined to be found October or mid - September.The pest had one generation per year in Ankara.The use of sex pheromone traps were satisfactory to determine both the first moth appearance and flying period. It was concluded that a sex pheromone could be used to control the pest or alternative to pesticide use .

INTRODUCTION

Sainfoin (*Onobrychis viciifolia*) is an important forage crop in the Central Anatolia Region of Turkey (Elçi, 1975). *Bembecia scopigera* is the most harmful and difficult to control pest of sainfoin . There is no satisfactory control measures for this pest,but an appropriate method is required.

To synchronise spraying time, It is essential to observe the first adult emergence and flying period. Various methods including sweep netting , attractive bait traps, counting holow cocoons and sex pheromone traps to achieve this were tested.

Pheromones have particular advantages for pest control because they are usually highly species- specific, leave no undesirable residues in the enviroment and are effective in minute quantities. Pheromones are being used to determine when pests enter crops, when numbers have built - up sufficiently to warrant control measures being taken or to predict the correct timing for such measures (van Emden, 1989) . In this study, we compared various methods with a sex pheromone trap, to determine the first moth emergence and flying period .

MATERIALS & METHODS

Adult emergence and flying period were determined in 1, 2 and 3 year- old infested sainfoin fields using ;

a - Sweep net : To determine the first moth emergence, one hundred net sweeps at 7 day intervals in each sainfoin field.

b - Attractive bait traps : In order to find the first emergence and flying period, a bait trap including 1 part boiled grape + 5 parts water and 2-3 g of yeast extract was used.The pots of each trap which were 2 l in size were filled 3 / 4 full with prepared bait. Nine pots were set up at 25 m intervals in sainfoin fields when the first pupae were found . The prepared bait in each pot was changed every two weeks. Traps were used from June to mid- October and checked weekly.

c - Counting hollow cocoons : The first moth emergence , emergence period and percentage of hollow cocoons were found, by assassing 25 - 100 indivuals (larvae, pupae , cocoon) . The moth emergence curve was drawn for 1 year- old sainfoin fields.

d - Sex pheromone traps : The commercial clearwing moth (*Synanthedon exitiosa*) sex lure used in the study to determine the first moth emergence and flying period was (3Z, 13 Z) - 3 - 13 - Octadecadienylacetate. Pherocon 1C traps were set up in mid - field , 1- 1.5 m above ground. The capsules and traps were renewed every two weeks and three weeks , respectively. The moths in each trap were taken weekly and counted to enable a curve for flying adults to be drawn.

RESULTS & DISCUSSION

The first moth emerged on 16 July, 1984 in 1, 2 and 3 year- old sainfoin fields, using netting, attractive bait trap, hollow cocoon counting and sex pheromone. In 1985, the first moth appeared on 2 July by sweep net and hollow cocoons but on 9 July , using attractive bait traps.

The first emergence occurred on 10 and 17 July , 1986 using sex pheromone traps and counting hollow cocoons respectively in 1 year - old fields ; but on 2 July , 1986 in 2 and 3 year- old fields.

The first moth appeared on 9 and 14 July, 1987 in sex pheromone traps or by hollow cocoon counting in 1 and 2 year- old fields, respectively.

It was concluded that the use of attractive bait traps and sweep netting were unsuccessful in attractting adults and determining the flying period.

A curve of adult emergence was drawn , using hollow cocoon counting in 1 year - old sainfoin fields between 1984 - 1987 (Fig. 1). The first moth appeared on 16, 2, 17 and 14 July in 1984, 1985, 1986 and 1987, respectively. *B. scopigera* emerged from 16 July to 3 September, 1984 ; 2 July to 14 August, 1985 ; 17 July to 22 August, 1986 ; and from 14 July to 10 September, 1987. The peak was obtained on 6, 14, 7 and 20 August in 1984, 1985, 1986 and 1987, respectively.

Hollow cocoon counts showed that the first emergence occured between 16 July- 27 August, 1984 in 2 and 3 year- old infested fields ; between 2 July- 14 August, 1985 in 2 year - old fields ; between 2 July - 22 August, 1986 in 2 and 3 year- old fields ; between 14 July - 27 August , 1987 in 2 year - old field.

The moth flying period determined by sex pheromone traps in 1, 2 and 3 year- old infested fields and 1986 are shown in Fig. 2. The first moth appeared on 10 July in 1 year-old and on 2 July in 2 and 3 year - old fields in 1986. The flying period occured between 10 July- 2 September in 1 year - old field ; 2 July - 12 September in 2 and 3 year - old fields. The first peak occured on 17 and 10 July in 1, 2 year - old and 3 year- old fields , respectively ; the second and third peak s were on 31 July and 28 August, in 1, 2 and 3 year-old infested fields, respectively.

The flying period using sex pheromone traps in 1 and 2 year- old fields , 1987 is shown in Fig. 3, The first moth emerged on 9 July in both sainfoin fields. The flying period occurred from 9 July to 12 October in 1 year- old fields ; 9 July to 21 September in 2 year - old fields. The first peak was obtained on 22 and 27 July in 1 and 2 year-old felds, respectively. The second and third peak occurred on 20 August and 10 September in 1 and 2 year-old fields , respectively.

The moth flying period and emergence curve when using sex pheromones and hollow cocoon counts in 1 year- old infested fields, 1986 are shown in Fig. 4 . The first emergence occurred on 10 and 17 July in sex pheromone traps and from hollow cocoon counting , respectively. The flying period and emergence period occurred between 10 July - 2 September; 17 July- 22 August in sex pheromone traps and by hollow cocoon counting, respectively. The maximum moth numbers were trapped on 17, 31 July and 28 August in sex pheromone traps but on 7 August by counting hollow cocoon.

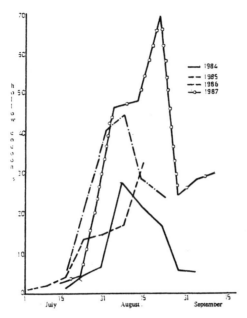

FIGURE 1. The curve of moth emergence of *B. scopigera* in 1 year-old infested sainfoin fields, using hollow cocoon counting method.

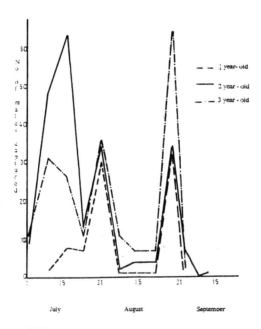

FIGURE 2. The seasonal occurrence of males of *B. scopigera* in 1986 in sex pheromone traps

82

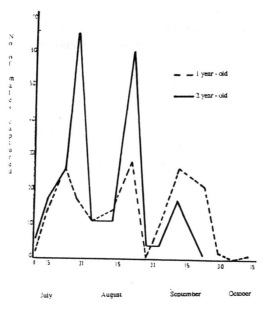

FIGURE 3. The seasonal occurrence of males of *B. scopigera* in 1987 in sex pheromone traps

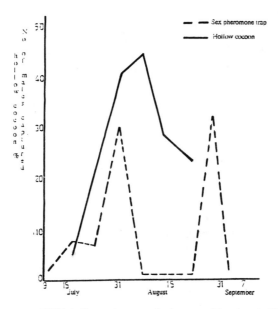

FIGURE 4. The seasonal occurrence of moth emergence of *B. scopigera* in 1986, using sex pheromone traps and a hollow cocoon counting method.

Moreover, as shown in Fig. 5 , the first emergence and catches were obtained on 9 and 14 July, 1987 in sex pheromone traps and hollow cocoon, respectively. The moths were trapped from 9 July to 12 October. Moth emergence occurred between 14 July- 10 September, using hollow cocoon counts . Moth trapping and emergence continued for 13 or 9

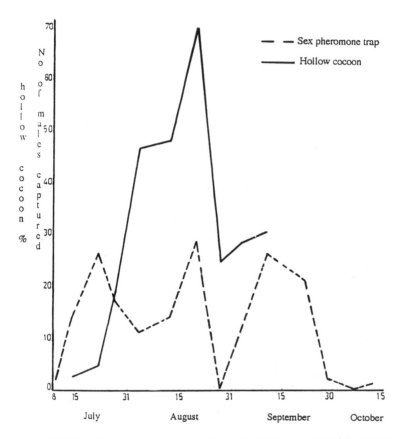

FIGURE 5 . The seasonal occurrence and the curve of moth emergence of *B. scopigera* in 1987 , using sex pheromone traps and a hollow cocoon counting method.

weeks in sex pheromone traps and hollow cocoon assesment, respectively. The peak was obtained on 22 July, 20 August, 10 September on sex pheromone, but only 20 August using hollow cocoon assesment, respectively.

It was concluded that the use of sex pheromone traps as effective in determining both the first moth appearance and flying period. As a result of the present and other studies , sex pheromones are being used to determine moth emergence and flying periods of Sesiidae , to control these pests and predict the correct timing of spraying (van Wetswinkol *et al* ., 1980 ; Neal, 1982 ; Potter & Timmons, 1984).

It is suggested that sex pheromone traps could be used to control *B. scopigera* in sainfoin fields. Attempts are being made to use pheromones in mass- trapping, confusion strategies.

REFERENCES

Elçi, Ş. (1975) Korunga. Gıda Tarım Ve Hayvancılık Bakanlığı Ziraat İşleri Genel Müdürlüğü Yayınları D - 168, 32.p

Neal, J.W.J.R. (1982)Timing Insecticide Control of Rhododendron Borer With PHeromone Trap Catches of Males, Environ. Entomol., 1981, 10, 2, 264-266, *Review of Applied Entomology*, *A* , **70**, (6), 3514 .

Potter, D. A. ; Timmons , G.M. (1984) Flight Phenology of The Dogwood Borer (Lepidoptera : Sesiidae) and Implications for Control in *Cornus florida* L. *Journal of Economic Entomol.*ogy, 1983, **76**, (5) , 1069-1074, *Review of Applied Entomology*, A, **72**, (4) , 2327 .

van Emden ; H.F. (1989) Pest Control.London, Newyork, Melbourne, Auckland.117.p

van Wetswinkel, G. ; Soenen, A ; Paternot. (1980) The Small Red- Belted Clearwing Moth *Synanthedon myopaeformis* Bork, *Review of Applied Entomology* , A, **68**, (8), 3806 .

INVESTIGATIONS ON NATURAL ENEMIES AND BIOLOGICAL CONTROL POSSIBILITIES OF *BEMBECIA SCOPIGERA* (Scopoli) (LEPIDOPTERA : SESIIDAE)

ALİ TAMER

Plant Protection Research Institute , Bağdat Caddesi No : 250 , PK . 49 -06172 , Yenimahalle , Ankara, Turkey

ABSTRACT

In field studies , two larval parasitoids *Bembecia scopigera* were found . The proportion of parasitised larvae in 1, 2 and 3 year - old infested sainfoin fields between 1984- 1987 varied from 9.7 to 42.4 %. Between years , the contribution of *Bracon crocatus* which is a new species for Turkey and *Chelonella nitens* to the rate of parasitism were about 7.5 % - 41.1 % and 0.08 % - 4.7 % , respectively. *B. crocatus* , a gregarious parasitoid of the pest , was found to have potential for biological control in the future.

Further, *Beauveria* spp. , *Fusarium* spp., *Penicillium* spp. and *Bacillus* spp. were isolated from larvae contaminated with diseases. *Bacillus thurigiensis* (16000 IU / mg a.i) was used in the field against the larvae of *B. scopigera* as biological control agent.

INTRODUCTION

Bembecia scopigera is the most important pest of sainfoin (*Onobrychis viciifolia*) fields in the Central Anatolia Region. The pest infests plants during the first year and shortens the life of the crop and results in early dislodging of the plants.Consequently control of the pest is required , by various measures including biological control, etc.

One of the oldest and most successful methods of controlling insect and related pests is by using their natural enemies , including parasitoids which attack and destroy them. Biological control agents traditionally have included predators, parasitoids and microbial pathogens (van Emden, 1989).

To find potential candidates to use as biocontrol agents, survey of possible natural enemies is often carried out in the area of the world assumed to be the centre of evolution of the pest species.The aim of the present study was to determine the natural enemies of *B. scopigera* and their effectiveness . In addition, microbial preparates were tested against the larvae .

MATERIALS & METHODS

The parasitised larvae separated from healthy larvae were counted at 7- 10 day intervals in 1, 2 and 3 year - old infested sainfoin fields between 1984- 1987. Larvae which were parasitised by gregarious and solitary parasitoids were evaluated and the percentage of them parasitised obtained . Disease larvae which have characteristics such as melting, lengthening and hardening of the body or a covering by fungus, or darkening of colour were seperated from healthy larvae.

Furthermore, a preparate including *Bacillus thurigiensis* (Bt)was used against the larvae of *B . scopigera* .

RESULTS & DISCUSSION

It was found that the preparate including Bt (16000 IU / mg a.i) had low effectiveness - 40 .7 % of the larvae killed.

In this study , two larval parasitoids were obtained from cultures as follows ; (i) *Bracon crocatus* (Hymenoptera : Braconidae) , a gregarious parasitoid . Some 4- 53 parasitoid eggs were counted in the host larvae, (ii) *Chelonella nitens* (Hymenoptera : Braconidae) , a solitary parasitoid. Diseases identified included *Beauveria* spp. , *Fusarium* spp., *Penicillium* spp. and *Bacillus* spp.

It was found that parasitism occurred between 21 May - 13 August 1984 ; 28 March- 25 September 1985 ; 16 April - 7 August 1986 and 7 May - 12 August 1987. The percentage of parasitism by the two species of parasitoid for field of various ages is shown in Table 1 .

Rhinotachina modesta , Ipobracon triangularis , Exerites roborator and Leskia aurea were a larval parasitoids of B. scopigera , but were ineffective (Bourniel & Khial , 1965 ; 1968) . Doğanlar (1982) found that the percentage of parasitism by C . nitens was less than 5 % on B. scopigera . Aubert (1978) reported that Lissonata pimplator was a larval parasitoid of B. scopigera . It was reported that parasitoids of B. scopigera were rare (Vuola & Korpela 1981).

It is concluded that B. scopigera , a gregarious parasitoid of the pest, was found to have potential for biological control in the future. Attempts are being made to rear and release this parasitoid.

REFERENCES

Aubert, J.F. (1978) Les Ichneumonides Qesut- Palearctiques Et Leurs Hötes 2 Banchinae et Suppl. aux Pimplinae.E.D.I.F.A.T.-O.P.I.D.A. Echauffour. 61370.

Bourniel . A. ; Khial , B. (1965) Note Prelimonanera Sur I a Sesie du Sainfoin. Compt. Rend. Hebd. Sean. Ces Acar Agr. , 51, (18) , 1252- 1255.

Bourniel . A. and Khial , B. (1968) Dipsosphecia scopigera Scop., La Sesie d.u Sainfoin. Annales Epiphyties, 19 , (2), 235 - 260.

Doğanlar, M. (1982)Doğu Anadolu' da bazı Lepidopter Zararlılarda Saptanan Hymenopter Parazitler, Türk.iye Bitki Koruma Dergisi , 6, (4), 197-207.

van Emden , H.F. (1989) Pest Control.London, Newyork, Melbourne, Auckland.117.p

Vuolau., M. ; Korpela, S. (1981) Suomen Lasisipisten (Sesiidae) ja Puuntuhoojen (Cossidae) elintavoista (Lepidoptera) , 7. Juurilasisiipi (Bembecia scopigera) , Mahdolliesti Maasta Löytyvat Uudet Lasisiipislajit, Korjauksia ja Taydenynksia edellisiin Osiin, Puuntuhoojo (Cossus cossus) , Haavantuhooja (Lamellocossus terebra) Joruokotuhooje (Phragmataecia castanea) , Notulae Entomologicae , 61, 103- 112.

TABLE 1 . The effectiveness of parasitoids of *Bembecia scopigera* (Scopoli.)

Year and infested sainfoin fields	Percentage of parasitoid larvae		
	Average (and range) of parasitised larvae	*B. crocatus*	*C. nitens*
1984-1 year-old	41.2 (2.7 - 63.6)	41.1 (2.7 - 63.6)	0.9 (0.0 - 0.9)
1985-2 year-old	35.8 (10.7 - 68.0)	33.4 (33.4 - 71.6)	2.5 (0.0 - 6.8)
1986-3 year-old	14.2 (1.7 - 30.7)	9.7 (1.7 - 19.8)	4.5 (0.0 - 19.9)
1984-2 year-old	26.7 (2.5 - 42.2)	25.3 (2.5 - 37.9)	1.4 (0.0 - 5.9)
1985-3 year-old	21.8 (2.0 - 47.1)	20.9 (2.0 - 45.1)	0.9 (0.0 - 5.7)
1984-3 year-old	21.4 (0.8 - 41.4)	19.7 (0.8 - 41.4)	1.7 (0.0 - 7.4)
1985-1 year-old	37.7 (13.6 - 66.3)	34.2 (13.6 - 62.3)	3.4 (0.0 - 11.4)
1986-1 year-old	12.2 (1.1 - 26.5)	7.5 (1.1 - 18.8)	4.7 (0.0 - 22.9)
1986-2 year-old	13.7 (0.0 - 31.3)	10.7 (0.0 - 23.3)	2.9 (0.0 - 9.8)
1987-2 year-old	9.7 (5.0 - 22.1)	8.9 (5.0 - 19.1)	0.7 (0.0 - 6.5)
1987-1 year-old	14.9 (1.4 - 52.5)	10.2 (1.4 - 28.8)	4.7 (0.0 - 23.8)

POTENTIAL FOR BIOLOGICAL CONTROL OF *SCLEROTINIA SCLEROTIORUM* IN WINTER OILSEED RAPE WITH *CONIOTHYRIUM MINITANS*

M. P. McQUILKEN

Plant Science Department, The Scottish Agricultural College, Auchincruive, Ayr, Scotland, KA6 5HW, UK.

S. J. MITCHELL, S. A. ARCHER

Department of Biology, Imperial College of Science Technology and Medicine, Silwood Park, Ascot, Berkshire, SL5 7PY, UK.

S. P. BUDGE, J. M. WHIPPS

Department of Microbial Biotechnology, Horticulture Research International, Wellesbourne, Warwickshire, CV35 9EF, UK.

ABSTRACT

A field trial, with winter oilseed rape, was conducted to determine the effect of soil incorporations of a maizemeal-perlite preparation of *Coniothyrium minitans* on sclerotial survival and apothecial production of *Sclerotinia sclerotiorum*. The mycoparasite infected sclerotia and decreased sclerotial survival, carpogenic germination and production of apothecia. Effects were greatest when inoculum of *C. minitans* was applied in autumn, at the time of sowing, rather than when applied in spring. *C. minitans* survived in soil for two years and spread considerable distances to infect sclerotia in control plots. Despite the inoculum potential of *S. sclerotiorum* being reduced by the *C. minitans* treatment, no disease control was obtained. The reasons for this failure of *C. minitans* to control Sclerotinia stem rot in oilseed rape, and possible strategies to improve its efficacy in the field are discussed.

INTRODUCTION

Stem rot of winter oilseed rape caused by *Sclerotinia sclerotiorum* was first recorded in England in 1973 on a crop grown in Berkshire (Anonymous, 1973, 1975) and, since then, it has occurred irregularly over a gradually increasing area as the crop has been more widely grown (Jellis *et al.*, 1984). Incidence of stem rot has generally been low in the UK and consequently crop losses have been small except in disease 'hot spots', confined mainly to the southern counties of Kent and West Sussex (Jellis *et al.*, 1984; Davies 1986). However, in more recent years, severe infections have been widespread throughout the UK, with yield losses in excess of 20% (Fitt *et al.*, 1992).

Once the pathogen is established, it is extremely difficult to control. Even though crop rotation and different cultivation methods can reduce the build-up of the pathogen in soil (Archer *et al.*, 1992), they cannot be relied upon as effective means of control because of extrinsically produced ascospores initiating infections (Williams & Stelfox, 1979). Foliar-applied fungicides, including the dicarboximides and MBCs, have been shown to give effective control provided that they are applied at the correct time just before petal fall (Bowerman & Gladders, 1993). However, no reliable disease forecasting system is currently available in the UK, and routine applications of fungicides for disease control are expensive. These problems and the environmental concerns over the use of pesticides have led to the search for biological control of stem rot.

Coniothyrium minitans has already shown potential for biological control of *S. sclerotiorum*. Solid-substrate soil incorporations of the antagonist have been shown to control *S. sclerotiorum* in sunflower (Huang, 1980), celery and lettuce (Whipps & Budge, 1992; McQuilken & Whipps, 1995), but have yet to be tested in oilseed rape. This paper reports a small-scale field trial conducted to determine the effect of solid-substrate soil incorporations of *C. minitans* on sclerotial survival and apothecial production of *S. sclerotiorum*, and Sclerotinia stem rot in oilseed rape.

MATERIALS AND METHODS

Inoculum production of fungi

Sclerotia of *S. sclerotiorum* were produced on sterilised wheat grain, cv. Armada, following inoculation and incubation at $20^{\circ}C$ for 3 weeks (Mylchreest & Wheeler, 1987). Batches of twenty sclerotia (*c.* 3-6 mm diameter) washed clean of adhering wheat were placed in Terylene net bags (*c.* 5 x 5 cm, mesh < 2 mm) for immediate use in the field trial. Maizemeal-perlite inocula of *C. minitans* were prepared using a method described previously (McQuilken *et al.*, 1995).

Field trial

The field trial was conducted at Imperial College's field station, Silwood Park, near Ascot, Berks., UK. Soil from the trial site was free of sclerotia of *S. sclerotiorum*. The trial was arranged in a randomised block design with three replicate plots for each treatment. Plot sizes were 1 m² separated by 2 m of the same crop. Five days after sowing, maizemeal-perlite inocula of *C. minitans* were evenly applied to plots (0.8 1 m^{-2}) and raked into the soil surface to a depth of *c*. 3 cm. Immediately after incorporation, ten Terylene net bags of sclerotia were buried (1-2 cm deep) individually at random positions within each 1 m^2 plot and labelled. Inocula were also incorporated into plots in the following spring just before the end of stem extension (26 March). Controls consisted of untreated plots and plots treated with inocula killed by autoclaving. To control annual grass and broad-leaved weeds, Kerb 50 W (500 g kg^{-1} propyzamide) and Dow Shield (200 g l^{-1}) were applied at recommended rates in late October and early February, respectively. A spring top dressing of nitrogen (200 kg N ha^{-1}) was applied in March. No other pesticides or fertilisers were applied.

A bag of sclerotia was removed from each replicate plot in October and then at monthly intervals thereafter. Numbers of sclerotia recovered from each bag, their viability and infection by *C. minitans* were assessed (Whipps & Budge, 1990). Survival of *C. minitans* in the soil was also monitored by soil dilution plating on Oxoid potato dextrose agar (PDA) containing Triton X-100 and Aureomycin (Whipps *et al.*, 1989). Numbers of apothecia in each 1 m^2 plot were counted at regular intervals during April, May and June. Apothecia were not removed after counting. Just before petal fall, samples of petals (30-50) were removed from each plot and plated onto PDA. Plates were incubated for 14 days at 18-20°C and scored for the presence of *S. sclerotiorum*. Disease was assessed at pod senescence (mid-July).

RESULTS

Survival and infection of sclerotia

In comparison with the control plots, consistently fewer sclerotia were recovered from plots treated with autumn soil incorporations of *C. minitans* at all monthly samplings except the first in October (Table 1). There was also a general decline with time in the numbers of sclerotia recovered from plots treated with the antagonist in autumn. Spring soil incorporations of the antagonist in March had no effect on subsequent sclerotial recovery in comparison with the corresponding controls.

The highest levels of sclerotial infection (74-100%) by *C. minitans* always occurred in plots treated with autumn soil incorporations of the antagonist (Table 2). Even by the first sampling (October; 7 weeks after soil incorporation), up to 98% of the sclerotia recovered were infected by *C. minitans*. The antagonist also spread to control plots by November (11 weeks after soil incorporation), infecting low numbers of recovered sclerotia (3-12%) throughout the trial period. Between April and August, sclerotial infection by *C. minitans* was higher in plots treated with spring soil incorporations of the antagonist than in the corresponding controls.

However, the levels of infection were considerably lower than in those sclerotia recovered from plots treated with autumn soil incorporations of the antagonist.

TABLE 1. Effect of autumn and spring soil incorporations of *Coniothyrium minitans* on percentage recovery of sclerotia of *Sclerotinia sclerotiorum* by the antagonist from plots of oilseed rape.

Sampling month	Treatment		
	Untreated control	Autumn *C. minitans*	Spring *C. minitans*
Oct	100	100	-
Nov	99±0.2[a]	75±6.6	-
Dec	97±1.4	50±6.1	-
Jan	99±0.4	35±7.8	-
Feb	99±0.6	22±7.1	-
Mar	95±2.7	17±6.6	-
Apr	97±2.1	15±6.4	85±8.3
May	98±1.3	9±3.4	98±1.7
Jun	88±2.6	18±8.9	88±6.8
Aug	65±5.1	4±1.9	78±3.6

[a]Values are means ±SE

TABLE 2. Effect of autumn and spring soil incorporations of *Coniothyrium minitans* on percentage infection of sclerotia of *Sclerotinia sclerotiorum* by the antagonist from plots of oilseed rape.

Sampling month	Treatment		
	Untreated control	Autumn *C. minitans*	Spring *C. minitans*
Oct	0	98±1.1	-
Nov	3±1.7[a]	74±10.9	-
Dec	12±4.7	85±8.9	-
Jan	7±3.1	100	-
Feb	6±2.0	97±2.9	-
Mar	8±2.9	95±2.5	-
Apr	9±2.6	97±2.2	28±8.4
May	6±2.8	94±2.5	28±6.5
Jun	7±2.8	93±4.2	38±6.1
Aug	3±1.8	98±1.4	38±9.9

[a]Values are means ±SE

Apothecial production and disease incidence

Apothecia were first observed in treatment plots on 16 April (flowering) and were present until late June (leaf senescence) (Table 3). Most apothecia were produced from late April until mid-May. No new apothecia were found after the end of May and numbers declined thereafter. During peak apothecial production (23 April - 21 May), very low numbers were produced from sclerotia in plots treated with autumn soil incorporations of *C. minitans* compared with control plots, or those treated with spring soil incorporations of the antagonist. However, spring soil incorporations of the antagonist had some effect on apothecial production later on in the growing season. On 11 June (seed development (most pods green)), fewer apothecia were present in spring *C. minitans*-treated plots compared with controls.

TABLE 3. Effect of autumn and spring soil incorporations of *Coniothyrium minitans* on numbers of apothecia produced by sclerotia of *Sclerotinia sclerotiorum* in plots of oilseed rape.

| Sampling date | Treatment | | |
	Untreated control	Autumn *C. minitans*	Spring *C. minitans*
16 Apr[a]	15±3.7[b]	1±0.8	15±5.7
30	47±7.2	2±1.6	40±13.8
7 May	30±6.1	0	26±13.0
14	60±6.4	2±1.0	52±14.7
28	24±3.1	1±0.5	15±5.1
4 Jun	24±3.1	1±0.5	15±5.1
11	14±3.2	0	1±0.5
18	2±0.9	4±3.8	0

[a]16 Apr was 31 weeks after burying sclerotia/autumn soil incorporation of *C. minitans*
[b]Values are means ±SE

Samples of petals removed from all treatment plots just before petal fall and plated onto PDA produced typical colonies of *S. sclerotiorum*. In general, numbers of petals from which the pathogen was isolated were lower in plots treated with autumn soil incorporations of *C. minitans* compared with either the controls, or plots treated with spring soil incorporations of the antagonist. At pod senescence, *S. sclerotiorum*-diseased plants were present in virtually all treatments, but not in all plots. Overall, only a low level of disease (0-20% of stems affected) was present throughout the trial and there were no significant differences between the *C. minitans*-treated plots and the control (data not shown).

Survival of *C. minitans* in soil

C. minitans exhibited a general decline in colony forming units (CFUs) cm^{-3} of soil with time in all plots treated with inocula of the antagonist. However, the antagonist could still be detected at 10^4 CFUs cm^{-3} of soil for up to two years after the last soil incorporation.

DISCUSSION

Autumn treatment with *C. minitans* reduced the inoculum potential of *S. sclerotiorum* in soil. Sclerotial recovery was reduced, and carpogenic germination and apothecial production were inhibited. These effects of *C. minitans* have been reported before in field trials against *Sclerotinia trifoliorum* in the absence of plants (Turner & Tribe, 1976), and in glasshouse trials with *S. sclerotiorum* in lettuce and celery (Whipps & Budge, 1992). However, this is the first report of such effects of *C. minitans* on *S. sclerotiorum* in the UK under field-grown oilseed rape. Spring treatments were not as effective at reducing the inoculum potential as autumn treatments. It is possible that spring treatments with *C. minitans* may not provide sufficient time for the antagonist to infect sclerotia and prevent carpogenic germination of *S. sclerotiorum*. Alternatively, it may be related to the environmental conditions prevailing following soil incorporation. Tribe (1957) correlated poor infection of sclerotia of *S. trifoliorum* by *C. minitans* in spring with low soil temperatures and greater infection in the autumn with higher soil moisture. Similar conditions may have prevailed during this trial.

Even though *C. minitans* was able to reduce the inoculum potential of *S. sclerotiorum*, no disease control was obtained. The small plot size and the low level of disease (< 20%), caused by dry weather conditions during flowering, made detection of statistically significant effects difficult. Since ascopores of *S. sclerotiorum* require water for germination, colonisation of petals and subsequent infection of oilseed rape stems, periodic irrigation of the crop during continued dry conditions at flowering and petal fall is likely to encourage disease development in future trials. In view of the possibility of long distance spread of ascospores between plots (Williams & Stelfox, 1979; Archer *et al.*, 1992), it is recommended that bigger plots separated by large guard areas of oilseed rape are also used. This will help to prevent ascosporic inocula produced in control plots from infecting oilseed rape in plots treated with *C. minitans*.

C. minitans survived in soil in this trial for up to two years and spread to infect sclerotia at a considerable distance from soil incorporation. This confirms the potential of *C. minitans* to survive and infect sclerotia of *S. sclerotiorum* in the long term (Whipps & Budge, 1992).

The reduction in inoculum potential of *S. sclerotiorum* resulting from soil incorporation of *C. minitans* indicates the potential that may exist for this biocontrol agent to control Sclerotinia disease in oilseed rape. However, further field trials are required at different sites to evaluate *C. minitans* under a wide range of soil types and conditions. It is also necessary to optimise the timing of application of the biocontrol agent to achieve practical field use. Another treatment before stem extension may also be required to attack those sclerotia which develop on plants in the rosette stage following autumn-winter infection (McQuilken *et al.*, 1994). Foliar spore sprays during flowering may also be required to prevent *S. sclerotiorum* using petals as a food base to infect plants.

ACKNOWLEDGEMENTS

We wish to thank Dr. B. E. J. Wheeler for constructive discussion throughout this work and the gardening staff at Silwood Park for field assistance. We are also grateful to Dr G. E. Jackson for assistance with data collection. Financial support for this work, provided by the Ministry of Agriculture Fisheries and Food (MAFF) and EEC, is gratefully acknowledged.

REFERENCES

Anonymous, (1973). Monthly summary of fungus and other diseases occurring in England and Wales. Harpenden: MAFF, Plant Pathology Laboratory; Nos. 5-12, May-December 1973.

Anonymous, (1975). MAFF ADAS Science Arm Annual Report, 1973. London: HMSO, 167-168.

Archer, S.A.; Mitchell, S.J.; Wheeler, B.E.J. (1992) The effects of rotation and other cultural factors on Sclerotinia in oilseed rape, peas and potatoes. *British Crop Protection Conference - Pests and Diseases 1992*, 1, 99-108.

Bowerman P.; Gladders, P. (1993) Evaluation of fungicides against *Sclerotinia sclerotiorum*. Tests of Agrochemicals and Cultivars 14. *Annals of Applied Biology* **122**, *(Supplement)*, 42-43.

Davies, J.M.L. (1986). Diseases of oilseed rape. In: *Oilseed Rape*. D.H. Scarisbrick and R.W. Daniels (Eds), London, Collins, pp. 195-236.

Fitt, B; McCartney, H.A.; Davies, J.M.L. (1992) Strategies for the control of Sclerotinia. *The Agronomist* **1**, 12-13.

Huang, H.C. (1980) Control of sclerotinia wilt of sunflower by hyperparasites. *Canadian Journal of Plant Pathology* **2**, 26-32.

Jellis, G.J; Davies, J.M.L.; Scott, E.S. (1984) Sclerotinia on oilseed rape: Implications for crop rotation. *British Crop Protection Conference - Pest and Diseases 1984*, **3**, 709-715.

McQuilken, M.P.; Mitchell, S.J.; Archer, S.A. (1994) Origin of early attacks of Sclerotinia stem rot on winter oilseed rape (*Brassica napus* sub. sp. *oleifera* var. *biensis*) in the UK. *Journal of Phytopathology* **140**, 179-186.

McQuilken, M.P.; Whipps, J.M. (1995) Production and evaluation of solid-substrate inocula of *Coniothyrium minitans* against *Sclerotinia sclerotiorum*. *European Journal of Plant Pathology* **101**, 101-110.

Mylchreest, S.J.; Wheeler, B.E.J. (1987) A method for inducing apothecia from sclerotia of *Sclerotinia sclerotiorum*. *Plant Pathology* **36**, 16-20.

Tribe, H.T. (1957) On the parasitism of *Sclerotinia trifoliorum* by *Coniothyrium minitans*. *Transactions of the British Mycological Society* **40**, 489-499.

Turner, G.J.; Tibe, H.T. (1976) On *Coniothyrium minitans* and its parasitism of *Sclerotinia* species. *Transactions of the British Mycological Society* **66**, 97-105.

Whipps, J.M.; Budge, S.P. (1990) Screening for sclerotial mycoparasites of *Sclerotinia sclerotiorum*. *Mycological Research* **94,** 607-612.

Whipps, J.M.; Budge, S.P. (1992) Biological control of *Sclerotinia sclerotiorum* in glasshouse crops. *British Crop Protection Conference - Pest and Diseases 1992*, **1**, 127-132.

Whipps, J.M.; Budge, S.P.; Ebben, M.H. (1989) Effect of *Coniothyrium minitans* and *Trichoderma harzianum* on *Sclerotinia* disease of celery and lettuce in the glasshouse at a range of humidities. In: *Integrated Pest Management in Protected Vegetable Crops*. Proceedings of CEC I0BC Joint Experts, Cabrils: 27-29 May 1987, R. Cavalloro, C. Pelerents Rotterdam: A.A. Balkema, pp. 233-243.

Williams, J.R.; Stelfox, D. (1979) Dispersal of ascospores of *Sclerotinia sclerotiorum* in relation to sclerotinia stem rot of rapeseed. *Plant Disease Reporter* **63**, 395-399.

THE POTENTIAL FOR RESISTANCE TO CYST NEMATODES IN TRANSGENIC PLANTS WHICH EXPRESS ANTIBODIES.

B.S. RAMOS, R.H.C. CURTIS*, K. EVANS*, P. BURROWS*, P.P.J. HAYDOCK

Crop and Environment Research Centre, Harper Adams University Sector College, Newport, Shropshire, TF10 8NB, UK.

*Entomology and Nematology Department, IACR - Rothamsted, Harpenden, Herts, AL5 2JQ, UK.

ABSTRACT

Potato cyst nematodes (PCN) are important pests of potato crops worldwide. Monoclonal antibodies (MAbs) are valuable tools for the identification of nematodes through their ability to recognise species-specific antigens and they also have potential for use in novel control strategies. Specific antibodies have been expressed in planta (plantibodies) that bind to plant viruses and provide some protection against viral attack. A panel of MAbs to PCN has been produced at Rothamsted and is currently being screened by indirect immunofluorescence for antibodies that recognise targets important in nematode development. The proteins recognised by the selected MAbs will be characterised. MAbs with potential to interfere in nematode development will be cloned as short-chain antibody fragments for eventual expression in plants. Protocols will be devised to analyse the efficacy of these MAbs in reducing nematode infection and development. Preliminary results on the identification and characterisation of nematode antigens recognised by Mabs with the potential to be used as 'plantibodies' are presented.

INTRODUCTION

Potato cyst nematodes (PCN) are agricultural pests of great economic importance. In the UK they are estimated to cause loss of about 10% of the value of the annual potato crop (Evans & Stone, 1977). They are root endoparasites not exceeding 1mm in length at any developmental stage. The two PCN species *Globodera rostochiensis* (Wollenweber, 1923) and *Globodera pallida* (Stone, 1973) probably originated in the Andean region of South America and were taken, sometime in the 19th century, to Europe which then became a secondary centre of distribution (Brodie, 1984). PCN have now spread throughout the world to at least 50 countries (Baldwin & Mundo-Ocampo, 1991).

PCN decrease yield directly by decreasing the size of the host root system, which in turn affects water and nutrient uptake and leaf duration (Brodie *et al.*, 1993). Further, indirect losses are incurred by the necessity to grow crops of lesser value in crop rotations and from the costs of maintaining quarantine schemes, advisory services and resistance breeding programmes. Resistant cultivars, however, cannot be used continuously because they promote the selection of virulent pathotypes. There is no commercially available cultivar with full resistance to *G.*

pallida and nematicides and crop rotation are also less effective at controlling this species of PCN.

Cyst nematodes have complex relationships with their host plants. They overwinter as juveniles in eggs contained within a cyst, formed from the female cuticle which tans on death to form a protective capsule for 200-500 eggs (Brodie *et al.*, 1993). The eggs may remain dormant in the soil for up to about 25 years. Hatching is triggered by exudates from host plant roots. Potato root diffusate (PRD) contains up to six active components, and induces hatching by a bimodal action on eggshells and juveniles. Changes brought about by PRD allow juveniles to become sufficiently active to cut through the eggshell and ultimately hatch (Doncaster & Shepherd, 1967).

The juveniles move in the soil searching for a healthy plant and invade young potato roots. Later, young potato tubers may also be subject to invasion by juveniles. Infested tubers and tubers carrying infested soil are the primary vehicles for the dispersal of the nematodes to uninfested regions. The juveniles mature into adults inside the roots or under the surface layer of the tuber. The males are active and migrate in search of young females which, on maturation, emerge through the root surface and spend their entire lives attached to the roots.

Developing juveniles absorb nourishment by the induction of a syncytial feeding site in the host roots (Wyss & Zunke, 1986). These feeding sites act as nutrient sinks and solutes are continually withdrawn by the nematode and replenished by the plant (Hussey, 1989). Nematodes fail to reach maturity if the syncytia are not formed and females often die when there is reduced food availability because of competition from adjacent nematodes. After fertilisation, an embryo develops within each egg to become a second-stage juvenile (J2). The adult female dies after completion of egg development and its cuticle tans. Development from hatching to adult at optimal temperatures takes between 38 and 45 days (Baldwin & Mundo-Ocampo, 1991).

The expression of antibodies in plants has already been demonstrated (Hiatt *et al.*, 1989; Hiatt, 1990). Plant cell culture, transformation and genetic recombination (sexual crossing of regenerated plants) are novel methods for achieving assembly of light and heavy chain pairs and immunoglobulin structures in plants. Plants expressing functional antibodies are found in the progeny of a cross between plants containing the individual heavy or light chains.

Engineering nematode resistance in crops by the production of transgenic plants expressing functional antibody fragments directed to vital nematode proteins has been proposed (Bakker *et al.*, 1993).The engineered single-chain Fv antibody (scFv) is particularly suitable for expression in plants because of its small size and the lack of assembly requirements. The variable domains (V and VL) can be amplified by polymerase chain reaction (PCR) using 'universal primers' and inserted into vectors for *Escherichia coli* expression of scFv antibody, in which the two variable domains are connected by a linker peptide. With this work, (Tavladoraki *et al.*, 1993), first reported a plant phenotype with an attenuation of viral infection derived from a constitutively expressed, virus-specific antibody. The scFv antibodies proved to be functionally stable in the cell cytoplasm. The intracellular scFvs seem particularly suitable to 'immunomodulate' selected cytoplasmic antigens, unlike whole antibody molecules which need to be targeted to the endoplasmic reticulum for correct assembly/folding and stable accumulation in plants (Tavladoraki *et al.*, 1993).

A panel of monoclonal antibodies (MAbs) to PCN antigens has been produced at Rothamsted, originally for diagnostic purposes, and from these, three MAbs have been shown to react with the nematode amphids, dorsal oesophageal glands and somatic muscles (Curtis & Evans, 1994). The whole antibody panel is now being further analysed in order to identify MAbs reacting with potential targets for antibody inhibition which could disrupt nematode development in the plant. Here we report preliminary results on the identification of MAbs reacting to PCN antigens.

MATERIALS AND METHODS

Antigen preparation

Approximately 50 cysts of *G. pallida* pathotype Pa2/3 and *G. rostochiensis* pathotype Ro1 were soaked in distilled water for 1 d and homogenised using an Eppendorf plastic homogenizer (Biomedix) in 10 μl of PBS on ice (10 cysts/ml). Homogenate of gravid white females of *G. pallida* was prepared as described above. To obtain excretory/secretory products, including stylet secretions, approximately 2000 J2 of *G. pallida* and *G. rostochiensis* were incubated with a solution of 0.2 mg/ml 5-Methoxy DMT oxalate (Research Biochemicals Incorporated) in distilled water for 4 h at room temperature. Two protease inhibitors, 1mM EDTA and 1mM PMSF, were added to the final solution. The protein concentrations were determined using a Bio-Rad assay (Bradford, 1976).

Indirect immunofluorescence

The specific reactivity of the antibodies was localised in cryostat sections of *G. rostochiensis* and G. *pallida* second stage juveniles (J2). To obtain fresh J2, cysts of *G. rostochiensis* and *G. pallida* were soaked in distilled water for 3 d and stimulated to hatch in potato root diffusate 1:3 in distilled water. Freshly hatched juveniles were collected daily and processed for microtomy by rapid freezing of a block of J2 in liquid nitrogen. The nematode block was sectioned using a cryostat and sections 7-8 μm thick were collected on poly-prep slides (Sigma), air dried and fixed in cold acetone. The post-fixed cryostat sections were soaked in 0.2% Triton X-100 in PBS for 30 min., and then incubated with goat serum diluted 1:50 in PBS for 20 min. The slides were then incubated with cell line supernatant (MAbs), overnight at room temperature. The sections were rinsed 3 times with PBS and incubated for 45 min. at room temperature with goat anti-mouse FITC Conjugate (Sigma) diluted 1:50 in PBS. After rinsing, the slides were mounted with an anti-quenching agent, Citifluor (Agar Scientific). Visualisation of bound antibodies was achieved using a microscope with an epifluorescent attachment (Olympus BH-2) fitted with a 455nm excitation filter and a 460nm secondary filter. Micrographs were obtained using an Olympus OM4 camera with Ilford XP2 400 film.
(results are shown in tables and histograms).
ELISA - Indirect ELISA was performed using the procedure described by Robinson *et al.*, 1993.

RESULTS

In the preliminary screening of 132 hybridomas raised against J2 of PCN *G. rostochiensis* and *G. pallida*, some hybridomas had shown positive reaction with more than one structure in the nematode body (Table 1). The indirect immunofluorescence test was performed more than once for the majority of hybridomas tested.

TABLE 1. Preliminary screening of 132 hybridomas by indirect immunofluorescence on cryostat sections of second stage juveniles (J2) of *G. rostochiensis* and *G. pallida*

Nematode structure	Number of hybridomas
amphids	05
anterior sense organs	01
stylet	03
stylet protractor muscle	03
stylet knob	06
procorpus membrane	04
dorsal gland amullae	07
dorsal gland extension	05
oesophageal lumen	02
metacorpus membrane	03
metacorpus pump chamber	02
subventral gland ampullae	01
nerve ring	03
dorsal gland	02
subventral glands	01
intestine canal	08
intestine granules	18
reproductive system	02
excretory system	04
nervous system	03
somatic musculature	19
cuticle	04
other structures	14
negatives	12

Thus, a variety of hybridomas binding to different nematode structures was obtained following standard immunizations and, from this panel of antibodies, six cell lines were repeatedly cloned and sub-cultured in order to obtain cell lines with a defined specificity. Indirect immunofluorescence was used to detect the range of nematode structures to which the MAbs bound (Table 2).

TABLE 2. Indirect immunofluorescence reaction of monoclonal antibodies (MAbs) raised against antigens from second-stage juveniles (J2) of *G. rostochiensis* and *G. pallida*.

Antibody	Nematode structure	Specificity of antibody
MAb GR-j5	Anterior part of the nematode head	Microvillar nerve process, dendritic and nerve processes.
	Median region of the nematode body	Nerve cord which contains neuronal process and associated neuronal cell bodies. Longitudinal nerve cords: nerve processes and cell bodies.
MAb GR-j6	Anterior part of the nematode body	Pharyngeal region, possibly related to the dorsal oesophageal gland extension and ampullae.
MAb GR-j9	Anterior part of the nematode body	Pharyngeal region, possibly related to the dorsal oesophageal gland extension or to the oesophageal lumen.
MAb GR-j7	Anterior end of nematode head	Possibly related to the style protractor muscles or to the nerve processes of the dorsal oesophageal gland ampullae.
MAb GR-j8	Cuticle	Annulations and lateral line.
MAb GR-j12	Intestine	Granules (in the gut wall).

The MAbs which reacted with the nematode nervous system were tested for reactivity with several neurotransmitters. Mabs GRj-10 and 11 reacted with adrenaline and MAb GRj-11 had an additional reactivity with GABA (gamma-amino butyric acid) and a weak reaction with acetylCoA, noradrenaline and octopamine (Table 3).

TABLE 3. Binding of MAbs to neurotransmitters in indirect ELISA (optical densities less than 3X negative control value are recorded as negative).

MAbs	GABA	5-HT	acetylCoA	Adrenaline	Noradren	Octopamine
MAb GR-j5	neg.	neg.	neg.	neg.	neg.	neg.
MAb GR-j10	neg.	neg.	neg.	0.058	neg.	neg.
MAb GR-j11	0.067	neg.	0.032	0.089	0.037	0.030

MAbs reacted with homogenates prepared from cysts containing juveniles of
G. rostochiensis and *G. pallida*, adult females of *G. pallida* and
excreted/secreted antigens produced *in vitro* by juveniles of both species
(Fig. 1). Negative controls consisted of PBS pH 7.2 and 5-Methoxy DMT 0.2
mg/ml. MAbs GRj 5,6,7 showed most reactivity with the antigens tested, and
none of the MAbs tested reacted well with *G. pallida* adult females.

Fig.1 Binding of MAbs in indirect ELISA to PCN antigens prepared from
cysts (containing juveniles) of *G. rostochiensis* and *G. pallida* (GRJ2 and
GPJ2 respectively), adult females of *G. pallida* (GPFe) and
excreted/secreted antigens collected from juveniles of both species (ES).

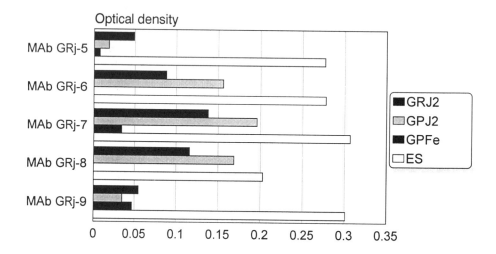

DISCUSSION

Nematode sections were screened by immunofluorescence with 132 Mabs.
Mabs were found that, in preliminary tests, appeared to react with nematode
structures including the nervous system, oesophageal glands, oesophageal
lumen, cuticle/epicuticle, stylet protractor muscles, granules in the gut.
The reactivity of five selected Mabs with PCN antigens was tested by ELISA
and binding was obtained with homogenates prepared from *G. rostochiensis*
and *G. pallida* J2 and excreted/secreted antigens. However, the reactivity
of these antibodies with a homogenate prepared from *G. pallida* gravid white
females was low.

Three Mabs reacted with the nematode nervous system, showing
immunofluorescence staining of longitudinal nerves and cell bodies. They
also seem to react with neural processes present in the nematode dorsal
oesophageal gland ampullae and amphids. The reactivity of these MAbs with
several neurotransmitters was tested in ELISA and most binding was obtained
with GABA (MAb GRj-11) and adrenaline (MAbs GRj-5 and 11). The cell lines
producing antibodies of interest were subcloned and stored in liquid
nitrogen. Further work will include characterisation of the antigens

recognised by the selected MAbs by Western blotting, isoelectric focusing, immunofluorescence using nematodes 'pieces' and immunogold labelling at the EM level. Attempts will be made to devise a system to assess antibody efficacy in interfering with the nematode life-cycle *in vitro*. The eventual aims of this programme of work are to understand more about the interactions between PCN and their potato hosts and to obtain resistance to PCN by expressing antibodies that interfere with nematode development in transgenic plants.

ACKNOWLEDGEMENTS

IACR receives grant-aided support from the Biotechnology and Biological Sciences Research Council of the United Kingdom. Bruno Ramos is in receipt of a research studentship funded by the Higher Education Funding Council for England. Financial help from Harper Adams University Sector College and MAFF is gratefully acknowledged.

REFERENCES

Bakker, J., Schots, A., Schouten, A.S., Roosien, J., van Engelen, F.J., Boer, J.M., Stiekema, W.J. & Gommers, F.J. (1993) 'Plantibodies': a versatile approach to engineer resistance against pathogens. *[Abstract] 6th International Congress of Plant Pathology*.July 28-August 6, Canada.

Baldwin, J.G. and Mundo-Ocampo, M. (1991) Heteroderinae, cyst and non-cyst forming nematodes. In: *'Manual of Agricultural Nematology'* ed. W.R. Nickle. Marcel Dekker, New York, pp. 275-362.

Bradford, M. (1976) A rapid and sensitive method for the quantitation of microgram quantities of protein utilising the principle of protein dye binding. *Analytical Biochemistry* 72: 248-254.

Brodie, B.B. (1984) Nematode parasites of potato. *In 'Plant and Insect Nematology'* ed. W.R. Nickle. Marcel Dekker, New York, pp. 167-212.

Brodie, B.B., Evans, K. and Franco, J. (1993) Nematode parasites of potatoes. In: Evans, K., Trudgill, D.L. and Webster, J.M. (eds.). Plant Parasitic Nematodes in Temperate Agriculture. *CAB International*. Wallingford, pp. 87-132.

Curtis, R.H.C. & Evans, K. (1994) Biochemical characterisation and localisation of potato cyst nematode diagnostic proteins using monoclonal antibodies. *Nematologica* 41: 283-294.

Doncaster, C.C. and Shepherd, A.M. (1967) The behaviour of second-stage *Heterodera rostochiensis* larvae leading to their emergence from the egg. *Nematologica* 13: 476-478.

Evans, K. and Stone, A.R. (1977) A review of the distribution and biology of the potato cyst nematodes *Globodera rostochiensis* and *Globodera pallida. PANS* 23(2): 178-189.

Hiatt, A. Cafferkey, R. & Bourdish, K. (1989) Production of antibodies in transgenic plants. *Nature* 342: 76-78.

Hiatt, A. (1990) Antibodies produced in plants. *Nature*, 344: 469-470.

Hussey, R.S. (1989) Disease inducing secretions of plant parasitic nematodes. *Annual Review of Phytopathology* 27: 123-141.

Robinson, M.P., Butcher, G., Curtis, R.H.C., Davies, K.D. & Evans, K. (1993) Characterisation of a 34 kD protein from potato cyst nematodes using monoclonal antibodies with potential for species diagnosis. *Annals of Applied Biology*, 123: 337-347.

Tavladoraki, P., Benvenuto, E., Trinca, S., Martins, D., Cattaneo, A. & Galeffi, P. (1993) Transgenic plants expressing a functional single-chain Fv antibody are specifically protected from viral attack. Nature, 366: 469-472.

Wyss, U. and Zunke, U. (1986) Observations on the behaviour of second-stage juveniles of *Heterodera schachtii* inside host roots. *Revue Nematologie* 9(2): 153-165.

CAN POTATO PRODUCTION BE SUSTAINED IN LAND INFESTED WITH HIGH
POPULATION DENSITIES OF THE POTATO CYST NEMATODE *GLOBODERA PALLIDA*?

S. WOODS, P.P.J.HAYDOCK

Crop and Environment Research Centre, Harper Adams University Sector College, Newport, Shropshire, TF10 8NB, UK

K. EVANS

Department of Entomology & Nematology, IACR-Rothamsted, Harpenden, Herts AL5 2JQ, UK

ABSTRACT

A field trial at Great Bolas, Shropshire, investigated the effect of oxamyl (Vydate 10G, Du Pont) on the control of the potato cyst nematode (PCN) *Globodera pallida*, and the yield of the partially resistant potato cultivar Santé. The trial was situated on a Bridgnorth series black sand soil with an initial *G. pallida* population density of 90 eggs/g of soil. Potato production would not usually be recommended on such highly infested land. In spite of the high initial population density of PCN, the trial produced acceptable ware yields of 40.8 t/ha from nematicide treated plots and 35.4 t/ha from untreated plots. Also, the overall *G. pallida* population density on the trial site decreased, which suggests that potato production that is profitable and sustainable can be obtained in highly infested land. The contribution of cultivar resistance, tolerance, nematicide and good husbandry are discussed in the context of integrated PCN management.

INTRODUCTION

The potato cyst nematodes *Globodera rostochiensis* and *G. pallida* are among the most difficult of crop pests to control in the UK. Granular nematicides such as oxamyl and aldicarb have been successful in controlling *G. rostochiensis* when used in conjunction with crop rotations and resistant cultivars. However, the integrated control of *G. pallida* has been less successful due to the lack of fully resistant cultivars and lower rates of natural population decline, making crop rotation less effective (Whitehead, 1993). These problems may have been exaggerated by the quota system which has effectively concentrated potato production on the most suitable land and thereby decreased the lengths of rotations. At present, the UK potato industry has to manage the *G. pallida* problem and a large number of producers are growing potatoes in land infested with high population densities of this species. Can this be sustained?

SUSTAINABLE POTATO PRODUCTION

Sustainable agriculture generally refers to production systems which rely on lower inputs and put more emphasis on long term and stable crop production with the least environmental damage, in contrast to focusing more on short term goals such as maximum yields (Bridge, 1995). Sustainable potato production in land infested with PCN requires the integration of control measures to achieve a constant or preferably declining pest population without causing a shift within mixed populations towards the more problematic G. pallida or to more virulent pathotypes. An additional essential component of sustainable production is that the crop is profitable to the grower.

FIELD TRIAL 1994

In 1994 a field trial at Great Bolas, Shropshire, demonstrated the effects of integrating several different control techniques on the management of the potato cyst nematode G. pallida and the yield of the potato cultivar Santé. Control techniques used were the partially resistant cultivar Santé, a granular nematicide and various agronomic practices designed to improve yield. The importance of each technique within a sustainable production system is discussed below.

Population estimation and determination of species

Accurate population estimation and determination of species is vital in order to quantify the pest burden in ground that is intended for potato production. This is important because the choice of control strategy for particular conditions may depend upon this information (Haydock & Evans, 1994). In this trial, the mean initial population densities were 80 and 96 eggs/g of soil for nematicide treated and untreated plots respectively (Table 1), and the population was determined to be pure Globodera pallida by isoelectric focusing (Fleming & Marks, 1983).

TABLE 1. Population measurements and yield

Nematicide (oxamyl) kg / ha	Pi(eggs/g of soil)	Pf (eggs/g of soil)	Pf/Pi	Ware yield (t/ha)
5.5	80	55	0.69	40.8
0	96	76	0.79	35.4

The initial population densities for the field trial were high and in most circumstances a commercial potato grower would be advised not to use such land for potato production, the upper threshold in current use being approximately 80 eggs/g of soil (T. Dawkins, pers.comm.). However, ware yields of 40.8 t/ha and 35.4 t/ha, whilst not being very high, are certainly profitable in most years. How was this achieved?

INTEGRATED CONTROL

The term integrated control in this instance refers to the combined use of several of the control methods that producers in the UK have available to manage PCN populations to provide a management package giving best possible control of the pest. The methods available in the UK are: legislative control, chemical control, plant resistance, crop rotation and possibly trap cropping. The aim is to make the most efficient use of control methods available and thereby maximise the control of the pest.

Legislative control

Legislative control of PCN is aimed at preventing their build up and spread. The legislation operates through UK Government orders, and the phytosanitary requirements of importing countries. Entry into the European community and acceptance of the PCN directive of 1969 (69/465 EEC) obliged the UK Parliament to pass an order making testing for PCN mandatory on land which is to grow potato seed for sale. The Potato Cyst Eelworm (Great Britain) Order 1973 adopted the measures required by the directive. The order established soil testing as the official method of determining whether land is free from infestation by PCN. The subsequent 1993 Plant Health (Great Britain) Order retains all the requirements of previous legislation. Movement of infested plant material between PCN sensitive countries is avoided by the phytosanitary requirements of the individual countries, which generally involve root inspection of growing potato crops and testing of soil taken from consignments ready for export. The import controls are very important because, for instance, South American populations of PCN are genetically more diverse than European populations and are able to overcome resistance conferred by genes currently incorporated in resistant cultivars in Europe (Franco & Evans, 1978); it is important that such pathotypes are not allowed to spread. Legislation, whilst not a direct method of control, may be the single most important factor in maintaining sustainable potato production in the UK. However, this legislation does not apply to home saved seed and localised spread of PCN on individual holdings is not prevented.

Chemical control

Chemical control of PCN in the UK uses two distinct types of nematicide, fumigant and non-fumigant. Fumigant nematicides, such as 1,3- dichloropropene mixture (Telone II, Dow Elanco), are used to kill eggs and juveniles in the soil. The chemical is injected into the soil in the autumn and the soil surface sealed either by rolling or by polyethylene sheeting to prevent the volatile nematicide escaping (Whitehead et al., 1975a). Use of this type of nematicide is popular in the Netherlands but new legislation, which aims to reduce the amount of agrochemicals used in that country, has meant that methods for mapping the distribution of PCN infestations are being investigated in order to make more efficient use of smaller amounts of fumigant (Schomaker & Been, 1992). Non-fumigant or granular nematicides are more widely used than fumigants for chemical control of PCN in the UK. The oximecarbamates, oxamyl or aldicarb, are applied to the seedbed prior to planting. The action of these chemicals is as acetylcholinesterase inhibitors, effectively paralysing the nematodes and preventing them from locating host roots (Nelmes et al., 1973). The extent to which a nematicide improves crop performance and limits nematode multiplication depends on the characteristics of the cultivar grown, such as its tolerance and whether or not it is resistant to the nematode population. Tolerant cultivars such as Cara show less response to nematicides in low to moderate infestations. Partially resistant cultivars such as Santé are generally less tolerant and need the protection of a nematicide to yield profitably (Gurr, 1987). Nematicides will continue to be required in sustainable potato production systems until more tolerant partially resistant cultivars are available. Fumigant nematicides do not require incorporation after injection into the soil as they diffuse through the soil pores as a gas but the non-fumigant granular nematicides require incorporation into the top 15cm of the soil to be effective (Whitehead et al., 1975b). The use of stone and clod separators for incorporating nematicides may be reducing the efficiency of granular nematicides by incorporating the nematicides too deeply or by "layering" the chemical in the soil profile (Woods et al., 1994).

<u>Resistance and tolerance</u>

Resistance to PCN in potato cultivars has been available since the 1960s, for example in the cultivar Maris Piper. The host plant is not resistant in terms defined by standard plant pathology, as it is still invaded and damaged by juveniles, and yield loss occurs. However, the capacity of the pest to reproduce successfully is reduced. Resistant cultivars limit the establishment of the feeding sites needed for female nematodes to develop, with the result that many juveniles become males. Potato cultivars with single major gene (H1) resistance to *G. rostochiensis* pathotypes Ro1 and Ro4 are available. Single major gene resistance to *G. pallida* is not available but partial resistance against *G. pallida* conferred by minor genes is available in commercial cultivars such as Santé. This cultivar was used in this field trial and successfully reduced the PCN population from 80 to 55 and 96 to 76 eggs/g of soil for oxamyl treated and untreated plots respectively (Table 1).The control provided by such cultivars is variable, depending on the virulence of the nematode populations to which they are exposed, but a decrease in the population is often obtained.

Tolerance is the ability of a plant to withstand attack by a pest without suffering undue damage (Trudgill, 1986). Resistance and tolerance are independent, but resistance may confer tolerance (Evans & Haydock, 1990). Tolerance exerts no selection pressure on the pest population but a risk with tolerant cultivars is that, without resistance, they will increase the population of the pest to levels where the tolerance is no longer able to cope with the pest burden. The tolerant cultivars Maris Piper and Cara (both of which are resistant to UK *G. rostochiensis*) have become popular over the past few years and it is possible that many fields have become heavily populated with *G. pallida* as a result. The tolerance of *G. pallida* partially resistant cultivars such as Santé could be enhanced by the use of good agronomic practice, so reducing the stress caused to the plant by factors that can be controlled. Irrigation, fertilisers, and good weed, aphid and disease control strategies may reduce stresses on the plant and allow good yields to be achieved in PCN infested land. Table 2 shows the inputs received by the field trial and explains their importance in improving the tolerance of the crop to PCN infestation.

TABLE 2. Agronomic inputs to field trial

Application	Comments
Liquid fertiliser - 1500 l/ha 10-8-11.5 preplanting. 984 l/ha 10-8-11.5 applied during planting	Application of fertiliser is important in terms of applied improving tolerance as the damaged root system of PCN infested plants is less efficient at nutrient uptake (Trudgill *et al.*, 1975b). Low soil indices of any of the major nutrients should be avoided as an infested crop with its reduced capacity for nutrient uptake will show yield reduction. Foliar applications of N, P and K may improve the tolerance of crops grown on moderately infested irrigated land.
Foliar fertiliser - four separate applications of foliar manganese at 1.25 l/ha. Four separate applications of foliar magnesium at 3.5 l/ha	Manganese and magnesium are both involved in photosynthesis. Magnesium levels in potato tissue decrease with increasing PCN infestation (Trudgill *et al.*, 1975a). The application of foliar magnesium and possibly manganese may help the potato plant to maintain its photosynthetic capability in moderately infested irrigated land.

TABLE 2. Continued

Application	Comments
Herbicides - contact and residual herbicides applied	The slower growth rates of PCN infested crops results in their reduced ability to compete with weed species.
Insecticides - dimethoate 0.62 l/ha (product rate) and cypermethrin 0.25 l/ha applied on two separate occasions	Stress from viruses transmitted by aphids may reduce the plants ability to tolerate the PCN burden, so ensuring that the plants are virus free and may improve the crops tolerance of PCN.
Fungicides - prophylactic blight spray programme	To reduce the risk of infection with potato blight (*Phytopthora infestans*).

In theory, careful choice of a resistant cultivar is important in maintaining sustainable potato production in PCN infested areas. However, in practice, market forces dictate which cultivars are grown. Several cultivars have been available over the past 10 years with varying degrees of resistance to *G. pallida,* such as Morag, Nadine & Santé, but only Nadine and Santé have been grown on any significant area. There is a need for both a concerted effort to promote cultivars with partial resistance to *G. pallida* and for more cultivars to promote. The latest, Valor, is promising but, unless the supermarkets can sell it or it has potential for processing, another useful cultivar may fail to fulfil its potential in PCN management.

Crop rotation

Crop rotation is a vital component of PCN management. PCN requires a solanaeceous host to reproduce and there are few if any weed species in the UK that act as good hosts to the pest (Whitehead, 1985). PCN declines naturally in the soil in the absence of a host as a small percentage of eggs will hatch without exposure to potato root diffusate. A decline rate of approximately 33% is expected for *G. rostochiensis* in UK soils but the rate of decline for *G. pallida* is generally slower at 15% per annum (Whitehead, 1993). This rate of decline could necessitate 10 year rotations when populations are large which is unacceptable for commercial growers. A crop rotation of 4-6 years, when used in conjunction with one or more of the other control methods, should provide a safe preplanting nematode population density (Whitehead, 1986). However, since the introduction of nematicides in the 1960s and the concentration of potato production on the most suitable land, rotations have become shorter and 1 in 3 rotations are not uncommon now.

Trap cropping

Trap cropping is a developing technique for reducing infestations on heavily infested land. By growing a vigorous cultivar such as Cara, a substantial hatch of eggs is induced. The potato crop is then lifted and destroyed when maximum hatch and invasion has occurred, but before any new cysts have matured. This technique has proved very successful, with populations of 40-465 eggs/g of soil being reduced by 75% or more in six weeks (Whitehead, 1994). However, if the crop is lifted too late an increase in PCN population density may occur. To improve chances of lifting the crop at the correct time, nematode development in the roots could be estimated by monitoring soil temperatures or observed directly by root examination. Trap cropping is still an experimental technique and should not be tried by growers without expert advice. With the introduction of set-aside in UK agriculture, the practice of trap cropping would seem a logical exploitation of the set-aside area in rotation. However, under current EC directives, the use of set-aside for trap cropping of PCN is prohibited.

ECONOMICS OF CONTROL

Sustainable potato production in the UK requires the system involved to be profitable to the grower both in monetary and agronomic terms. The cost of variable inputs and the gross margins for this trial for three price levels are shown in Tables 3 and 4 respectively.

TABLE 3. Costs of variable inputs to field trial

Input	Cost (£/ha)
Seed	600
Fertiliser	226
Foliar fertiliser	18
Nematicide	288
Sprays; Herbicides	37
Fungicides	31
Aphicides	8
Others	6
Total with oxamyl	1214
without oxamyl	926

From the gross margins it can be seen that at a market price of £150-200/t the integrated management strategy is profitable and that nematicide treatment gives an increased gross margin. At a low market price of £50/t, the cost of a nematicide treatment reduces the gross margin to the extent that untreated plots gave a higher gross margin. However, at such low market prices, the gross margins of £825.79 and £843.44 for nematicide treated and untreated plots respectively would barely cover fixed costs, leaving very little profit. Gross margin analysis does not take into account the long term effects of nematicide use on the sustainability of potato production. The reduction in mean Pf/Pi from 0.79 to 0.69 by the use of a nematicide in this trial would increase the potential profitability of the next potato crop in the rotation by reducing the nematode population in the soil.

TABLE 4. Gross margins for field trial

		Market price of potatoes		
		£50/t	£150/t	£200/t
Cost of inputs	oxamyl +	1214	1214	1214
(£/ha)	oxamyl -	926	926	926
Value of output	oxamyl +	2040	6120	8160
(£/ha)	oxamyl -	1770	5310	7080
Gross Margin	oxamyl +	826	4906	6946
(£/ha)	oxamyl -	844	4384	6154

DISCUSSION

The production of potatoes in areas of high PCN population densities can still be profitable providing good agronomic management and an integrated approach to their management is followed. Jones (1969) demonstrated the effectiveness of combining several methods for the control of *G. rostochiensis*. Various combinations of rotation, growing resistant and susceptible cultivars and using a nematicide were studied. When a resistant cultivar was grown in nematicide treated ground on a 4 year rotation, a 99.9% kill of *G. rostochiensis* was observed. The integrated control of *G. pallida* has not been as successful due to the lack of fully resistant cultivars and the apparently longer hatching period of *G. pallida*, which it is believed may reduce the effectiveness of granular nematicides. The problem of lack of resistant cultivars is further compounded by the variability of the partial resistance, which can vary in effectiveness from field to field (Whitehead *et al.*, 1987). Hancock (1994) conducted a trial to study the management of *G. pallida* populations in intensively cropped potatoes. Results indicated that continuous production of potatoes on infested ground using a nematicide, soil fumigation and the partially resistant cultivar Santé was not economically viable or sustainable. Introducing a rotation of 1 in 2 improved the economic viability of the system, but to produce a sustainable cropping system, rotations of 1 in 4 would be required. Potato production on the trial site at Great Bolas 1994, can be seen as sustainable due to a decline in the overall pest population as a result of the management implemented. If commercial considerations would allow, increasing the rotation length from one in four would undoubtedly have further benefits. However, more work is needed in the area of integrated control of *G. pallida* if the species is to be managed as successfully as *G. rostochiensis*.

ACKNOWLEDGEMENTS

Mr. S. Woods is in receipt of a Research Studentship funded by Du Pont (UK) Ltd., Dr. K. Evans is supported by MAFF. IACR receives grant-aided support from the Biotechnology and Biological Sciences Research Council of the United Kingdom. The authors would also like to thank Dr. T. Dawkins (Du Pont), Messrs B. and G. Maddocks (field trial hosts) and Mr. M. Russell (IACR-Rothamsted).

REFERENCES

Bridge, J. (1995) Sustainable and subsistence systems for nematode management. *Pesticide Science.* In press.

Evans K.; Haydock P.P.J. (1990) A review of tolerance by potato plants of cyst nematode attack, with consideration of what factors may confer tolerance and methods of assaying and improving it in crops. *Ann. Appl. Biol.*117, 703-740.

Fleming C.C.; Marks R.J.(1983) The identification of the potato cyst-nematodes *Globodera rostochiensis* and *G. pallida* by isoelectric focusing of proteins on polyacylamide gels. *Ann. Appl. Biol,* **103**, 277-281.

Franco, J.; Evans, K. (1978) Multiplication of some South American and European populations of potato cyst-nematodes on potatoes possessing the resistance genes H1, H2, H3. *Plant Pathology* **27**, 1-6.

Gurr, G.M.(1987) Testing potato varieties for resistance to and tolerance of the white potato cyst-nematode (PCN) *Globodera pallida. Journal of the National Institute of Agricultural Botany* 17,365-369.

Hancock, M. (1994) Managing potato cyst-nematode (*Globodera pallida*) in intensive potato cropping systems. *Proceedings of BCPC Pests and Diseases 1994.***2**, 899-904.

Haydock P.P.J.; Evans K. (1994) Sampling for decision making in potato cyst nematode management. *Aspects of Applied Biology* **37**, 113-120.

Jones, F.G.W.(1969) Integrated control of the potato cyst-nematode. *Proc. 5th Br. Insecticide and fungicide conference.***3**, 646-656.

Nelmes, A.J.; Trudgill, D.L.; Corbett, D.C.M. (1973) Chemotherapy in the study of plant parasitic nematodes. In *Chemotherapy of parasites* (ed A.Taylor). Oxford: Blackwell Scientific.

Schomaker C.H.; Been T.H. (1992) Sampling strategies for the detection of potato cyst nematodes: developing and evaluating a model. In *Nematology from Molecule to Ecosystem.* pp 182-194. Eds F.J. Gommers and P.W.T. Maas. European Society of Nematologists, Inc., Dundee.

Trudgill, D.L.(1986) Concepts of resistance, tolerance and susceptibility in relation to cyst nematodes. In *Cyst Nematodes*, F.Lamberti and C.E. Taylor (eds.), pp 179-189.

Trudgill D.L.; Evans K.; Parrott D.M. (1975a) Effects of potato cyst-nematodes on potato plants 1. Effects in a trial with irrigation and fumigation on the growth and nitrogen and potassium contents of a resistant and susceptible variety. *Nematologica* **21**, 169-182.

Trudgill D.L.; Evans K.; Parrott D.M. (1975b) Effects of potato cyst nematodes on potato plants II. Effects on haulm size, concentration of nutrients in haulm tissue and tuber yield of a nematode resistant and a nematode susceptible potato variety. *Nematologica* **21**, 183-191.

Whitehead, A.G.(1985) The potential value of British wild *Solanum* spp. as trap crops for potato cyst-nematodes, *Globodera rostochiensis* and *G. pallida. Plant Pathology.*34, 105-107.

Whitehead A.G. (1986) Problems in the integrated control of potato cyst-nematodes, *Globodera rostochiensis* and *G. pallida,* and their solution. *Aspects of Applied Biology.*13, 363-372.

Whitehead, A.G. (1993) Control of potato cyst-nematode. *Potato Production for Quality Markets,* Du Pont Potato Seminary February 1993.

Whitehead, A.G. (1994) Trap cropping could save valuable area. *Potato review.* 4(3), pp. 12-13.

Whitehead, A.G.; Fraser, J.E.; French , E.M.; Wright, S.M.(1975a) Chemical control of potato cyst-nematode *Heterodera pallida*, on tomatoes grown under glass. *Ann. Appl. Biol.* **80**, 75-84.

Whitehead A.G.; Tite D.J.; Fraser. J.E.; French, E.M. (1975b) Incorporating granular nematicides in soil to control potato cyst nematode, *Heterodera rostochiensis. Ann.Appl.Biol.*80, 85-92.

Whitehead, A.G.; Nichols, A.J.; Peters, C.G. (1987) Integrated control of potato cyst-nematodes. In *Report of Rothamsted Experimental Station for 1986*, Part 1, 105.

Woods, S.R.; Haydock, P.P.J.; Evans, K. (1994) Granular nematicide incorporation technique and the yield of potatoes grown in land infested with the potato cyst-nematode *Globodera pallida. Offered Papers in Nematology*, AAB Meeting, Linnean Society, December 12, 1994.

INTEGRATED PEST MANAGEMENT OF APHIDS ON OUTDOOR LETTUCE CROPS

P.R.ELLIS[1], G.M.TATCHELL[1], R.H.COLLIER[2], D.CHANDLER[1], A. MEAD[1],
P.L.JUKES[1], W.E.VICE[1], W.E.PARKER[3] & L.J.WADHAMS[4]

[1]   Horticulture Research International, Wellesbourne, Warwick, CV35 9EF, U.K.
[2]   Horticulture Research International, Willington Road, Kirton, Boston, Lincolnshire,
     PE20 1EJ, U.K.
[3]   ADAS, Woodthorne, Wergs Road, Wolverhampton, WV6 8TQ, U.K.
[4]   Institute of Arable Crops Research, Rothamsted Experimental Station, Harpenden,
     Herts, AL5 2JQ, U.K.

ABSTRACT

A series of field and laboratory experiments were done at different centres to
investigate various components of integrated pest management of aphid pests of
lettuce. The study concentrated on the four main species of aphids infesting
crops in Britain:- the currant-lettuce aphid, *Nasonovia ribisnigri*, the lettuce root
aphid, *Pemphigus bursarius*, the peach-potato aphid, *Myzus persicae*, and the
potato aphid, *Macrosiphum euphorbiae*. Immigration of alate aphids to lettuce
was monitored weekly between May and October in Warwickshire, Lancashire,
Lincolnshire and Kent with water traps and aphid population development was
monitored on lettuce crops, cv. `Saladin' planted in succession during the season.
Aphid species were identified and counted and a preliminary model was devised
for *P. bursarius* which could be used to predict aphid immigration. The
performance of several novel insecticides was determined in field experiments in
Lancashire. Three of these insecticides controlled aphids as effectively as the
approved products pirimicarb and demeton-S-methyl. The role of host plant
resistance and two semiochemicals in limiting crop colonisation were investigated
in a field experiment at HRI, Wellesbourne. Four lettuce varieties possessing
different combinations of resistance to the various aphid pests performed as
predicted, but the semiochemical treatments were not significantly different from
the untreated plots. The entomopathogenic fungus, *Metarhizium anisopliae*,
which is a pathogen of *P. bursarius,* was shown to kill aphids in a field
experiment.

INTRODUCTION

Aphids are the principal pests of lettuce grown outdoors. They have the potential to
destroy the crop or render it unmarketable. Crops of iceberg varieties, which occupy more
than 60% of the area of lettuce grown, are attacked by two key species, the lettuce root
aphid, *Pemphigus bursarius,* and the currant-lettuce aphid, *Nasonovia ribisnigri.* Two
further species which cause less direct damage but can transmit viruses to the crop are the
potato aphid, *Macrosiphum euphorbiae*, and the peach-potato aphid, *Myzus persicae*.
Lettuce crops are planted sequentially throughout the spring and summer and currently

aphid control relies on the routine and frequent use of insecticides with limited regard for the life-cycle of the pest. Individual crops may receive up to six applications of insecticide. The reduction of these inputs to limit insecticide usage and make the industry more competitive relates specifically to current MAFF policy. However, this can only be done if quality is not reduced.

Pemphigus bursarius feeds on the roots of lettuce and may kill plants, particularly in dry seasons. *Nasonovia ribisnigri* attacks the foliage, penetrating to the heart of the plant, making it unmarketable. Once crops have been colonised, these species occupy parts of the plant that are extremely difficult to reach with conventional insecticides.

Both aphid species overwinter as eggs on a woody host plant, *P. bursarius* on poplar and *N. ribisnigri* on currants. In spring, eggs hatch and after two to three generations, winged aphids fly to colonise lettuce crops. In the case of *N. ribisnigri*, further winged aphids may develop on lettuce in the summer and disperse to new lettuce crops. There is already considerable knowledge of the biology of *P. bursarius* on its winter host (Dunn, 1959) and of the timing of its migration to lettuce, but equivalent information for *N. ribisnigri* is inadequate. The timing of these key events in the biology of the pest are vital in the targeting of appropriate control strategies to limit or prevent crop colonisation.

The process of crop colonisation is a key stage at which the development of pest epidemics on lettuce can be limited. Lettuce genotypes exhibiting resistance to one or both of these aphid species are available, through a commercial breeding company, in advanced breeding lines, following identification of some of the original germplasm through MAFF funding at HRI. The use of resistant plant material limits crop colonisation. The mechanisms of resistance have been shown to be based on single genes (Einink & Dieleman, 1983; Ellis *et al.*, 1994), and so aphid strains may be able to overcome this resistance rapidly. The effectiveness of the resistant material will have a longer life if deployed for use only when necessary rather than throughout the cropping season, and when combined with other control strategies.

Aphid behaviour, particularly host finding, can be manipulated to limit crop colonisation further. It has already been shown at IACR Rothamsted that winged cereal aphids respond to host plant volatiles (semio-chemicals), and that appropriate volatiles can be used to "confuse" an aphid into mis-identifying its preferred host plant, so limiting colonisation (Pickett *et al.*, 1992).

Despite all efforts to prevent aphids landing on crops, limited colonisation will inevitably occur. Novel products, with new modes of action, have recently been released by pesticide companies, but do not yet have approval for use on lettuce. In addition, an isolate of the aphid pathogenic fungus, *Metarhizium anisopliae*, is currently under development in a MAFF funded programme at HRI for control of *P. bursarius*. A different isolate of *M. anisopliae*, that is not pathogenic to *P. bursarius*, has already been registered in the UK for the control of other pests. The appropriateness of these different products within integrated pest management (IPM) programmes has yet to be quantified.

The purpose of the project is to develop an integrated strategy for the control of aphids on outdoor lettuce through the integration of novel and existing control methods targeted at key stages in the life cycle of the pest species using accurate pest forecasting. The system developed would provide a strategy for aphid control which would lead to sustainable lettuce production and sustainable control methodologies.

MATERIALS AND METHODS

Monitoring of immigration of alate aphids and their population development

Crop colonisation and aphid population development were monitored by:-

Trapping winged aphids.
Yellow water traps fitted with vertical yellow baffles were placed in the perimeter of lettuce fields at HRI (Wellesbourne), Warwickshire, HRI (Kirton), Lincolnshire, in Kent and Burscough, Lancashire. Aphid samples were collected twice a week (May to October inclusive) and the aphids identified and counted.

Sampling `Saladin' lettuce plants.
Five sequential-sowings of the iceberg lettuce cv. `Saladin' were made at HRI (W), HRI (K), and three in Burscough. The plants were sampled twice a week throughout the season (May to October inclusive). Aphids on the foliage were identified and counted. In addition, sampled plants were lifted and their root system scored for *P.bursarius* infestation.

Novel insecticides

Field experiments were done on growers holdings in Lancashire in 1994 to compare the performance of three novel insecticides with the currently-approved products pirimicarb and demeton-S-methyl. Both seed-treatments and sprays were compared and plants sampled twice following application.

Semio-chemicals and resistant varieties

In a large field experiment at HRI, Wellesbourne in 1994 the effect of host plant resistance and of two semio-chemicals, methyl salicylate (an extract from willows) and butyl isothiocyanate (an extract from cruciferous crops) on aphid colonisation and development was compared. Untreated plots were also included in this experiment. Semio-chemicals were released from sachets which were placed in the crop on two occasions during the season. Plants were sampled on 7 occasions to record aphid abundance as well as root aphid damage.

Biological control of aphids

The fungus, *M. anisopliae*, obtained from a laboratory culture was incorporated in peat blocks at the rate of 10^7 and 10^8 spores per ml to inoculate `Saladin' lettuce plants raised in peat blocks. A wetting agent was used as a control treatment. Roots were sampled at the end of the season to determine the effects of the fungus on a severe infestation of *P. bursarius* which had colonised the crop.

RESULTS

Monitoring of immigration of alate aphids and their population development

The data from water traps and monitoring plots were analysed and graphs illustrating seasonal patterns of activity produced. Examples of two of these give some indication of aphid numbers and their phenology (Figures 1 & 2).

Thus, water trap samples indicated a single period of immigration of alate *P. bursarius* from poplar to lettuce in late June/early July. (Figure 1).

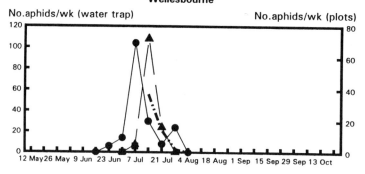

Pemphigus bursarius
Alates
Wellesbourne

FIGURE 1.   The immigration of *P bursarius* to lettuce, as indicated by water trap samples, and samples from plots of lettuce planted sequentially, at HRI Wellesbourne, 1994.

The picture for *N. ribisnigri* was more complicated (Figure 2).

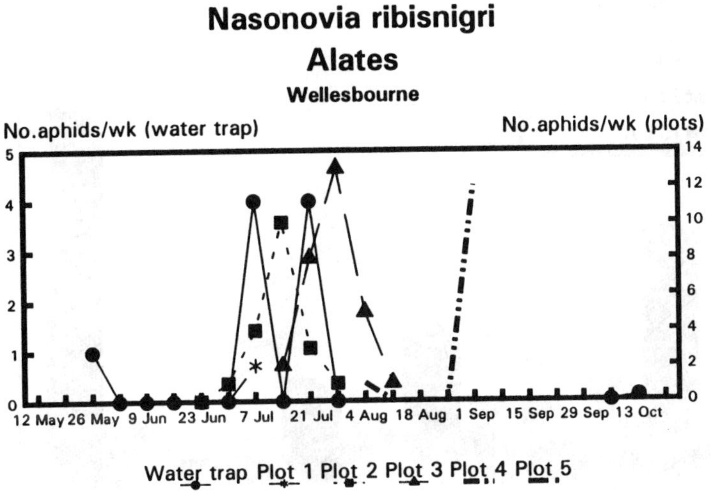

Nasonovia ribisnigri
Alates
Wellesbourne

No.aphids/wk (water trap)   No.aphids/wk (plots)

Water trap Plot 1 Plot 2 Plot 3 Plot 4 Plot 5

FIGURE 2.   The immigration of *N. ribisnigri* to lettuce as indicated by water trap
samples and samples from plots of lettuce planted sequentially, at HRI
Wellesbourne, 1994.

Immigration of *N. ribisnigri* from overwintering sites on *Ribes* species occurred at
almost exactly the same time as *P. bursarius* but further immigration was recorded in July
and again in September.

The other two aphids species, *M. euphorbiae* and *M. persicae,* colonised lettuce plants
earlier in the season but their numbers declined rapidly in late July as the numbers of
natural enemies increased in the crop.   These two species are the most important vectors
of virus diseases to the lettuce crop and so their activity must also be forecast accurately.

The results clearly showed that three of the novel insecticides have considerable potential to control the aphids on lettuce foliage (Figure 3). There is a need to collect further data, particularly that required for approval for any new compounds for use on lettuce.

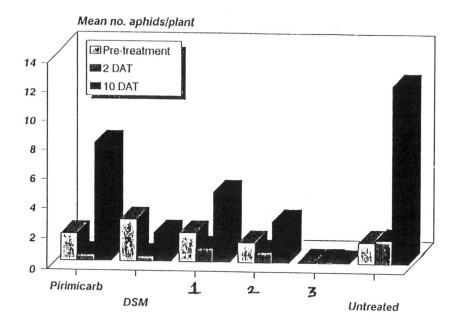

FIGURE 3. The efficacy of the commercial standard insecticides, pirimicarb and demeton-S-methyl, as compared to three coded products to control aphids on lettuce foliage at Gregson's Bridge, Lancashire. (DAT = days after treatment).

Semio-chemicals and resistant varieties

The crop was colonised by large numbers of aphids. The experiment provided valuable information on the performance of the different lettuce varieties. Thus `Saladin' was highly susceptible to all aphid species, `Beatrice' possessed total resistance to root aphid but was the most heavily colonised variety by foliage species. `Great Lakes' was susceptible to foliage species but partially resistant to root aphid whilst `Iceberg' possessed high levels of resistance to root aphid and partial resistance to foliage species. The results of releasing semio-chemicals into the plots was less clear. Overall, methyl salicylate-treated plots were the least colonised but there were interactions between variety and chemical treatment so that on `Beatrice' both semio-chemicals reduced colonisation whilst this effect was not evident on the other lettuce varieties.

Biological control of aphids

The control plants became infected as well as the inoculated lettuce and so scoring became complicated and difficult. More work is needed to develop the techniques in handling the pathogen and scoring the performance of the inocula.

DISCUSSION

It should be possible to predict accurately the timing of this migration of *P. bursarius* from poplar trees in the future. Winged *P. bursarius* leaving the lettuce crop all fly back to poplar. However, limited spread within fields can occur as young wingless individuals (crawlers) move from plant to plant or row to row and several weed species, closely related to lettuce, may serve as resevoirs of root aphid. This species is also known to overwinter in fields and colonise any lettuce crops planted out in spring or summer. The conditions governing this ability to overwinter and the degree of infestation which results need to be investigated.

N. ribisnigri activity in the middle of the season indicated movement between lettuce crops whilst the peak at the end of the season suggested the production of winged individuals which would be flying back to *Ribes*. In Europe and parts of the UK, *N.ribisnigri* is the most prevalent aphid on lettuce and so it is vitally important to be able to predict accurately its activity.

It is clearly sensible to utilise whatever sources of host resistance exist provided that the plant material is acceptable to the grower and consumer. At present no single iceberg variety of lettuce possesses resistance to all aphid species. However, seed companies and HRI(W) are developing lettuce breeding lines which possess resistance to the most important species of aphid. Varieties bred from these lines should reduce the growers' dependence on insecticides at times of the year predicted from the forecasts.

Further work is needed on both semio-chemicals and biological control agents of aphid pests before these approaches can be integrated with resistant varieties, cultural and chemical control methods.

AKNOWLEDGEMENTS

This work formed part of the LINK project No. P191 within the programme 'Sustainable farming systems' and was supported by MAFF, HDC, and Elsoms Seeds Ltd. We should like to thank these organisations for their support. We also thank Dr J.A. Blood-Smyth, M.A. Cantwell and C.Wallwork for their assistance with the experiments.

REFERENCES

Dunn, J.A. (1959) The biology of lettuce root aphid. *Annals of Applied Biology* **47**, 475-491.

Einink, A.H.; Dielemam, F.L. (1983) Inheritance of resistance to the leaf aphid *Nasonovia ribisnigri* in wild lettuce species. *Euphytica* **32**, 691-695.

Ellis, P.R.; Pink, D.A.C.; Ramsey, A.D. (1994) Inheritance of resistance to lettuce root aphid in the lettuce cultivars `Avoncrisp' and `Lakeland'. *Annals of Applied Biology* **124**, 141-151.

Pickett, J.A.; Pye, B,J,; Wadhams, L.J.; Woodcock, C.M. & Campbell, C.A.M. (1992) Potential applications of semiochemicals in aphid control. *1992 BCPC Monograph No.51 Insect Pheromones and other Behaviour-Modifying Chemicals: Application and Regulation*, 29-33.

A KNOWLEDGE-BASED SYSTEM FOR PREDICTING THE ENVIRONMENTAL IMPACTS OF SYSTEM LEVEL CHANGES TO EUROPEAN AGRICULTURAL SYSTEMS

G. EDWARDS-JONES, R. HOPKINS

SAC, West Mains Road, Edinburgh, EH9 3JG

ABSTRACT

Environmental impact assessments are routinely completed prior to initiating many industrial developments. Similar assessments are undertaken of many donor funded agricultural and rural development projects in the developing world. As increased international aid is being directed towards Eastern and Central Europe so there is increased pressure on donor agencies to assess the environmental impacts of their actions in these regions.

This paper describes a computer based system which predicts the environmental impacts which may arise from implementing change to a variety of European farming systems. The system, which was developed from a similar system aimed at assessing the impacts of change in tropical countries, identifies primary and higher order impacts associated with a range of projects via a rule-based causal network. In order to provide the user with further information on a range of topics related to the impacts identified by the rulebase, hypertext linkages are provided from the rulebase and a textual database. This database contains information on the underlying causal mechanisms of impacts, suggestions for potential mitigating activities, a glossary and a bibliography. Although the primary use of the system is in training, the potential for its use in the scoping phase of Environmental Impact Assessments is discussed.

INTRODUCTION

Although the relationships between agriculture and the environment have long been recognized, the primary causes for concern often vary with situation. For example, in most Western countries food production is adequate and agriculture is increasingly seen as a provider of non-market goods, e.g. biodiversity and landscape. In many developing countries however, food supply is insufficient and development projects are regularly initiated with the aim of increasing agricultural production. Although these projects may be successful in the short term, if they are badly designed or executed they can cause serious environmental degradation, which in turn may serve to reduce productivity.

The Food and Agriculture Organization of the United Nations (FAO) recognised that the people involved with developing and implementing development projects were generally economists and planners, who had little background in environmental science. For this reason they were generally unaware of the range of environmental impacts which may arise from development projects. FAO also recognised that planners cannot be expected to perform such tasks without adequate training in environmental science and environmental impact assessment. In part fulfillment of this training requirement FAO commissioned the production of the ECOZONE software, which predicts the environmental impacts of development projects in tropical regions. This suite of software, which is described in detail in Edwards-Jones & Gough (1994a,b) has been utilised in training over the last two years, and has proved to be a useful tool both within formal training courses / seminars and also for informal training. Further developments of the software have included translation into French and Spanish and the development of specific case-studies (e.g. Edwards-Jones & Abdel-Asiz, 1995).

Given the success of the ECOZONE concept in the developing world it was decided to develop a prototype system for use in Europe, particularly Eastern and Central Europe. Many of the countries in these regions are undergoing transition to full market economies, and the agricultural systems which have predominated over the last 40 years may undergo significant change during this transition. It was this prospect of poorly regulated, large scale change which triggered the initial investigation of training needs in the areas of agri-environment interactions and environmental impact assessment in Central and Eastern Europe. This paper describes the development, structure and potential use of software, named EurEco (European Ecozone) which was developed as a part of an FAO initiative in this area.

AIMS OF THE PROJECT

The aim of this project was to develop a computer based system that could be used for training agricultural planners and extension workers to be more aware of the environmental impacts which may arise from changing existing agricultural systems. In order to meet this aim it was decided that the system should:
   a)   Contain knowledge about a wide range of agricultural systems.
   b)   Be general enough to be suitable for training agricultural planners from all European countries.
   c)   Recognise (and represent) the complexity of environmental systems and be capable of demonstrating both the higher order and cross-sectoral effects of an impact and make explicit the interaction between environmental, economic and social systems.
   d)   Possess extensive explanation facilities.
   e)   Be a suitable for use in formal and informal training situations.

Summary of the approach adopted to the development of EurEco

As with ECOZONE, a knowledge-based approach was adopted for the development of EurEco. This approach was taken; firstly because suitable quantitative data in the domain of agri-environment interactions is scarce, but much qualitative data and experiential knowledge does exist within the knowledge-bases of individual domain experts. Secondly the large scale of the model meant that, even if suitable data had been available, a system based on numerical algorithms would have been extremely complex, large and expensive to develop. Subsequent to engineering a rulebase from the acquired knowledge, the rulebase was implemented on the computer as a causal network (Shachter & Kenley, 1989). This approach permitted accurate representation of the connectivity of environmental systems, whilst also being easily transcribed into computer code and simple to amend.

Simply predicting the likely impacts arising from any change to an agricultural system is unlikley to be adequate for training purposes, and it was a stated requirement of the sponsors that the system should contain extensive explanatory facilities. Hypertext links systems had been used in this manner in computer systems (Estep et al, 1989) hence it was decided to develop EurEco around two modules, a knowledge-base and a textual database, that would be connected via hypertext links.

Knowledge acquisition

Knowledge acquisition for EurEco was initially limited to text analysis. However, subsequent to the construction of an initial rule-base, the rules were checked and amended by domain specialists in FAO and SAC.

Software

Due to the requirement to integrate knowledge-based systems with hypertext and the necessity for the software to be easily available, inexpensive and robust, it was decided to utilise the commercially available Toolbook (Asymetrix Corporation, Washington USA) as the development environment. Toolbook is an object-oriented, card/book-based development environment which runs under Windows.

GENERAL DESCRIPTION OF THE EurEco MODEL

The users view of the model

After viewing the introductory screens users of EurEco are required to select a sector for analysis. The sectors available include arable, livestock, forestry, water resources and aquaculture (Figure 1). In a conventional analysis the user is then presented with a list of activities typical of projects within the chosen sector. The use then selects one or more activities, and the software presents the primary impacts which may occur if those activities were undertaken. The user may then select one or more of these primary impacts for further analysis. This analysis may take one of two forms, either all further impacts arising from the selected primary impact are presented as one large tree, or if the 'Next Impact' facility is selected, it is possible to follow individual impact pathways one step at a time. These two options are analogous to a full search and a directed search.

FIGURE 1. A typical user's movement through EurEco

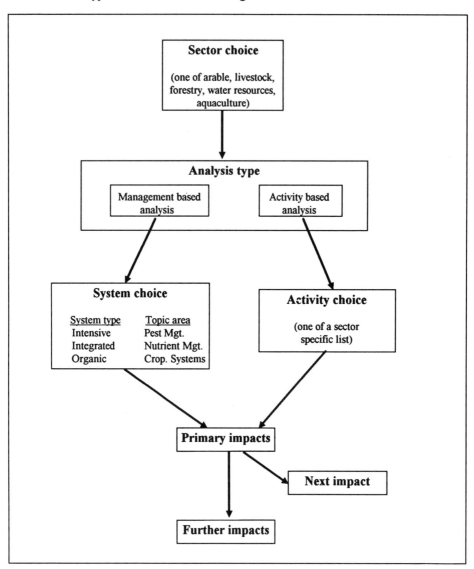

This activity based approach had proved appropriate for modelling interactions in all of the sectors in the ECOZONE software, however one of the perceived advantages of developing a similar model on a smaller scale was that a greater degree of detail could be incorporated into the model. For example it had been envisaged that EurEco would be able to detail the environmental impacts arising from management changes made to specific animal and cropping systems. It became apparent during the construction of

EurEco that while it remained appropriate to consider the environmental impacts arising from changes to livestock systems through the existing activity-based approach, this method was not appropriate for modelling the impacts arising from changes to cropping systems. This was because many of the practices undertaken within arable systems are common across crops. For example, many possible changes in pest and soil management and in the management of field margins all lead to broadly similar impacts regardless of the crop being grown. This overlap lead to a cluttered and confused user-interface and considerable repetition and redundancy within the rule base. For these reasons an alternative approach to the activity-based analysis was developed for the analysis of the environmental impacts arising from changes to cropping systems.

In this so-called, management based analysis, users specify whether they are which to analyze conventional, organic or integrated cropping systems. Having made this selection users are then select a topic area for analysis. This may be one of pest management, soil and nutrient management or cropping systems. Having completed this selection the user is presented with a list of activities which may be undertaken in that topic area under that farming system. Some examples of the different activities available for pest management in the three farming systems are given in Table 1. After selecting one or more of these activities the user is able to identify the primary and further impacts of the activity in the manner described above. It is possible to enter the textual database from any stage of the impact identification process, and having found the relevant information, to return to the appropriate impact prediction screen.

Representation of primary and higher order impacts in EurEco

Primary and higher order impacts are presented to the user through a combination of numerical notation and paragraph indents. In this system the number signifies the level of impact, i.e. '1' for a primary impact caused directly by the project, and '2' for secondary impacts caused by a primary impact. See Table 2 for an example.

In reality the number of higher-order impacts is potentially vast and the knowledge within the system is structured in order to permit realistic simulation of this process, however when the system is in normal use no more than 5 levels of impacts are presented to the user. This limit was implemented in order to provide a balance between demonstrating the real complexity of environmental systems and the need to keep search times short.

The hypertext information system (HTIS)

In order to render the HTIS more amenable to search by inexperienced users it was partitioned in to several sections, and upon initial entry into the system the user may chose which section to enter. The main sections include a glossary which contains a brief definition/description of terms and phrases, a text encyclopaedia which contains textual information on sectors, activities and impacts as would be found in a normal book, and a section entitled "Mitigation" which discusses possible methods of avoiding or mitigating impacts of activities.

TABLE 1. Activities listed under three farming systems for the topic area of 'Pest Management'.

Farming system		
Intensive	Integrated	Organic
use of herbicides and pesticides	use of pest/disease resistant crop varieties	no use of synthetic pesticides
improved aerial pesticide application	use of biological controls	use of pest/disease resistant crop varieties
increased pesticide usage	reduced dose applications	use of biological controls
	decreased pesticide usage	
	improved pesticide timing	
	improved pesticide placement	
	improved vehicle pesticide application	
	use of integrated pest management	
	use of biodegradable pesticides	
	use of encapsulated pesticides	
	use of herbicides and pesticides	
	use of systemic pesticides	

DISCUSSION

Generally mathematical models are useful for modelling systems for which we have a good understanding and sufficient data to quantify relationships. Conversely knowledge-based systems are well suited for modelling systems of which we have a good understanding but little available data (Stone, 1992). For this reason knowledge-based systems are increasingly being used to model environmental systems at a relatively large spatial scale (e.g. Fedra et al, 1991). In this situation they utilise knowledge to make some general predictions which may be accurate, but are unlikely to be precise. For example compare the output of a model for predicting soil erosion on a certain study area with that of a knowledge-based systems, such as EurEco. The former will give a precise prediction which is only applicable to the defined study area, while the latter will give predictions which may be imprecise for any one situation, but which will be valid over many situations. While the lack of quantification may be a disadvantage in some situations, this disadvantage must be weighed against the difficulty of developing quantitative models which are equally applicable in all situations.

Given our current state of knowledge, it is almost impossible to imagine the development of generic models which would be able to predict the multi-dimensional impacts which typically arise from any development project. Until this becomes possible then knowledge-based systems probably have a role to play in training personnel about the likely environmental impacts arising from any change, and also perhaps in the so-called scoping stage of environmental impact assessment. In the scoping stage all possible impacts are identified and the important ones are selected for further study (Glasson et al, 1994). Knowledge-based systems, similar to, but probably slightly more sophisticated than, EurEco and ECOZONE, may play a useful role in this process.

TABLE 2. An example of the output of EurEco. Here the activity is "reduced dose applications" which may lead to five primary impacts. All five of these could be analysed further, but for the purposes of clarity only the further impacts arising from "decreased pesticide residues in soil" are shown.

Activity: reduced dose applications

Primary impacts
    1,decreased pesticide in surface water
    1,decreased pesticide in groundwater
    1,decreased pesticide residues in soil
    1,risk of poor control of pests
    1,decreased production costs

Further impacts of: decreased pesticide residues in soil
      2,improved wildlife habitat
      2,decreased pesticide leaching
        3,decreased pesticide in drinking water
          4,improved human health
        3,decreased pesticide in surface water
          4,improved wildlife habitat,
        3,decreased pesticide in groundwater
          4,decreased pesticide in drinking water

Regardless of their use, whether it be for training or in scoping, the imprecise nature of the predictions of knowledge-based systems must be recognised, and human expertise will nearly always be required to interpret the output and put it in its local context. It was partly for this reason that the textual database was included within EurEco. The idea being that the knowledge-based systems would suggest all possible impacts, but with the aid of the information in the database, local experts could identify the more and less probable impacts for their situation. In this way a degree of precision could be brought into the predictions.

Although EurEco is clearly subject to some important limitations, such as a lack of quantification of impacts, both in terms of importance and magnitude, and the assumption that interactions within environmental systems may be modelled in a purely deterministic manner, the potential of such systems for training has been demonstrated with the

ECOZONE software. However it must be noted, that to date official training activities utilising EurEco have been limited. The view of an international workshop which considered the immediate agricultural training needs of Eastern and Central Europe was that the requirement for training in extension and basic production techniques was far more important than that in agri-environment interactions. It appears however, as though this attitude is starting to change and the EurEco software is scheduled to be used in an FAO training initiative in Slovakia in July 1995. Despite this recent development, the attitudes in Eastern and Central Europe to agri-environment interactions provides an interesting contrast to that in many African and Asian countries. The latter are regularly faced with the immediacy of environmental degradation, and are keen to develop more environmentally benign agricultural systems.

ACKNOWLEDGMENTS

We gratefully acknowledge the funding and help provided by FAO during this project.

REFERENCES

Edwards-Jones, G.; Abdel-Asiz, I. (1995) Environmental impacts of agricultural development projects: an Egyptian case-study. *Proceedings of the 2nd International symposium on 'Systems approaches for agricultural development'.* 6-8 December 1995, IRRI. Kluwer Academic Press (in press).

Edwards-Jones, G.; Gough, M. (1994a) Use of a simple knowledge-based system for training development planners about the environmental consequences of development projects. *Proceedings of the Second International Conference on Expert Systems for Development*, Bangkok 28-31 March 1994, pp. 333-337.

Edwards-Jones, G.; Gough, M. (1994b) ECOZONE: A computerised knowledge management system for sensitizing planners to the environmental impacts of development projects. *Project Appraisal* 9:37-45.

Estep, K.W.; Hasle, A.; Omli, L.; MacIntyre, F. (1989) Linnaeus: Interactive taxonomy using the Macintosh computer and Hypercard. *Bioscience* 39: 635-638.

Fedra, K.; Winkelbauer, L.; Pantulu, V.R. (1991) *Expert systems for environmental screening: An application in the Lower Mekong Basin.* IIASA, Laxenberg, Austria.

Glasson, J.; Therivel, A.; Chadwick, A. (1994) *Introduction to Environmental Impact Assessment* UCL Press, London.

Stone, N.D. (1992) Artificial intelligence approaches to modeling insect systems. In: *'Basics of Insect Modelling'* (Eds. J L Goodenough and J M McKinion). ASAE, St. Joseph. MI. pp. 37-52.

Shachter, R.D.; Kenley, R.C. (1989) Gaussian Influence Diagrams. *Management Science* 35:527-550.

THE IMPLICATIONS OF IMPROVING THE CONSERVATION VALUE OF FIELD MARGINS
ON CROP PRODUCTION.

N.H. JONES, K. CHANEY, A. WILCOX

Crop and Environment Research Centre, Harper Adams, Newport, Shropshire, TF10 8NB.

N.D. BOATMAN

Allerton Research and Educational Trust, Loddington House, Loddington, East Norton, Leicestersire, LE7 9XE.

ABSTRACT

Two field studies were conducted in Shropshire and Leicestershire during 1993/94 to quantify the effects of field margin management on cereal production. In the first, it was demonstrated experimentally that growing the crop up to the field margin gave a greater overall yield. Crops adjacent to a wildflower/grass strip yielded the next highest, whilst the poorest yield was obtained from a conservation headland adjacent to a sterile strip.

In the second study, a survey of winter wheat headlands revealed that grain yields were significantly less at the crop edge compared to 12 m into the crop, whilst weed biomass was significantly greater near to the field margin and decreased on moving towards the centre of the field.

INTRODUCTION

Field margins are a prominent feature of farm landscapes in Britain. However, the increase in agricultural productivity over the last thirty years has had a dramatic effect on these semi-natural areas, particularly in terms of hedgerow removal. Many thousands of kilometres of hedgerows have been removed to facilitate the operation of larger machinery (Barr et al., 1993). The mis-application of fertiliser and the application of herbicides, either deliberately, or accidentally through spray drift, have seriously reduced the botanical diversity found both at the base of remaining hedgerows and within arable fields. The loss of certain primary producers has been shown to have severe implications on important food chains and has resulted in a serious reduction in the number of species, for example gamebirds (Sotherton & Rands, 1987). However, the requirements for agriculture and wildlife may be complimentary, since the maintenance of a diverse, perennial ground flora will also discourage weed populations within the boundary, as well as supporting a wider variety of birds and beneficial insects (Marshall, 1988; Lakhani, 1994; Morris & Webb, 1987).

Crop yields from the headland area are often lower than that of the midfield (Boatman & Sotherton, 1988; Speller et al., 1992; Sparkes et al., 1994). The headland is used for turning agricultural machinery during cultivation, drilling, spraying and harvesting operations, which may directly lead to crop damage, soil compaction, double application of seed, fertilisers and pesticides. Shading by tall boundary vegetation and competition for water from tree and shrub roots may also cause additional yield losses (Fielder, 1987). However, in some cases the crop may benefit from the shelter effect of hedges which may in turn increase yields (Marshall, 1967).

Various methods of field margin management have been proposed, but have focused mainly on wildlife conservation, and limited efforts have been made at quantifying the effects of management strategies on crop production. This paper describes preliminary results from the first year of two experiments which aims to redress this balance. Results are also presented for a survey of winter wheat headland grain yields and weed amounts.

MATERIALS AND METHODS

Field margin management experiment

A replicated field experiment was conducted within winter wheat headlands (cultivar Hunter) at two locations, the Harper Adams College Farm, Shropshire and the Loddington Estate, Leicestershire. The aims of the experiment were to investigate the effects of field margin management practices on crop production. The experimental treatments were: (i) Cropping up to the field margin with a fully sprayed headland. (ii) Cropping up to the field margin with a conservation headland. (iii) Leaving a 1 m wide strip next to the field margin to naturally regenerate. (iv) A 1 m wide sterile strip with a fully sprayed headland. (v) A 1 m wide sterile strip with a conservation headland. (vi) A 1 m wide strip planted with a mixture of perennial grasses and wildflowers.

Plots were marked out in the headland areas in a randomised block design, with three blocks of six treatments at each site. Plots measured 14 m x 12 m at the Shropshire site, and 10 m x 12 m at the Leicestershire site. Permanent and destructive quadrats (0.25 m²) were established in the plots at 0, 1, 2, 3, 4 and 11.5 m from the field margin. The plots were assessed at Zadoks growth stage 31 and 59 (Tottmann, 1987) and at harvest. Estimates of percentage ground cover by each species present were recorded within the permanent quadrats at each assessment date. All vegetation within the destructive quadrats was cut by hand at ground level and the crop and the weeds separated and weighed at each assessment date for the Leicestershire trial. At the Shropshire site quadrats were cut by hand at GS31 and GS59, at harvest the plots were harvested with a plot combine and subsamples of grain were collected.

Survey

A detailed survey of winter wheat headlands was conducted during August 1994. Sixteen headlands were sampled, nine in Shropshire and seven in Leicestershire. A series of four transects were set out at each site, running at right angles to the field boundary, from the crop edge to 12 m into the field. Quadrats (0.25 m²) were placed along the transects at 0, 1, 2, 3, 4, and 11.5 m from the crop edge. All vegetation within the quadrats was harvested and separated into crop or weeds. It was noted whether the headland was a turning or non-turning headland, and the aspect of the site was recorded.

RESULTS

Field margin management experiments

The experiment was analysed using ANOVA at each site.

GS31 and GS59

At GS31 and GS59 treatment has a significant effect on total crop dry weight (GS31 Shropshire $F_{5,50}=6.904$, P<0.001 & GS31 Leicestershire $F_{5,50}=5.839$, P<0.001, GS59 Shropshire $F_{5,50}=12.341$, P<0.001 & GS59 Leicestershire $F_{5,50}=3.163$, P<0.05). On both occasions the crop to the edge sprayed and the crop to the edge conservation treatments yielded higher than the other treatments (Table 1). Quadrat position was highly significant (GS31 Shropshire $F_{5,50}=88.490$, P<0.001 & GS31 Leicestershire $F_{5,50}=6.523$, P<0.001, GS59 Shropshire $F_{5,50}=85.650$, P<0.001 & GS59 Leicestershire $F_{5,50}=35.414$, P<0.001), with crop dry weights generally increasing with distance from the crop edge (Table 2).

There were significant differences between treatments for weed dry weight at GS31 and GS59 (GS31 Shropshire $F_{5,50}=4.972$, P<0.001 & GS31 Leicestershire $F_{5,50}=4.043$, P<0.01, GS59 Shropshire $F_{5,50}=4.176$, P<0.01 & GS59 Leicestershire $F_{5,50}=3.117$, P<0.05) (Table 1). Quadrat position was significant (GS31 Shropshire $F_{5,50}=3.025$, P<0.05 & GS31 Leicestershire $F_{5,50}=9.453$, P<0.001, GS59 Shropshire $F_{5,50}=15.889$, P<0.001 & GS59 Leicestershire $F_{5,50}=5.167$, P<0.001) and weed dry weights were generally greater from the headland area than from the quadrats positioned at 11.5-12 m from the crop edge (Table 2).

TABLE 1. Mean crop (whole plant) and weed dry weights (g/m²) for each treatment at GS31 and GS59

		Treatment						
		Crop to edge sprayed	Crop to edge conserv.	Natural regen.	Sterile strip sprayed	Sterile strip conserv.	Wildflower /grass	LSD*
GS31								
Shrops.	Crop	98.15	106.00	78.27	80.13	78.39	74.69	3.33
	Weed	12.04	10.03	14.95	7.75	5.94	9.54	4.11
Leics.	Crop	52.43	56.20	38.72	39.10	28.55	40.16	12.08
	Weed	8.98	6.92	5.06	4.23	7.93	3.07	3.27
GS59								
Shrops.	Crop	672.33	614.54	464.44	481.00	440.82	513.93	75.54
	Weed	32.53	41.40	73.53	41.47	40.61	48.93	20.14
Leics.	Crop	594.03	587.68	479.00	541.91	495.13	484.14	87.78
	Weed	20.56	41.37	18.92	15.03	37.43	22.67	17.57

(* Least significant difference between treatment means at P<0.05, 34d.f.)

TABLE 2. Effect of distance from crop edge on mean crop and weed dry weights (g/m²)

		Distance from crop edge (m)						
		0-0.5	1-1.5	2-2.5	3-3.5	4-4.5	11.5-12	LSD*
GS31								
Shrops.	Crop	18.02	44.10	88.65	109.36	117.87	137.64	47.15
	Weed	8.90	13.22	9.01	9.23	6.98	12.90	13.72
Leics.	Crop	28.01	51.32	48.07	34.82	37.46	55.50	40.34
	Weed	7.57	4.30	5.25	5.99	11.75	1.33	10.92
GS59								
Shrops.	Crop	178.34	273.70	633.74	647.63	693.59	760.06	252.16
	Weed	81.67	81.97	32.96	33.78	27.48	20.59	67.24
Leics.	Crop	208.62	498.57	602.85	542.16	616.78	718.88	283.01
	Weed	36.64	23.81	28.72	23.38	41.58	1.85	58.65

(* Least significant difference between distance means at P<0.05, 4 d.f.)

Harvest

At harvest, treatment had a significant effect on grain yield at the Leicestershire site ($F_{5,50}$=7.053, P<0.001). The crop to the edge fully sprayed and the crop to the edge conservation treatments yielded higher than the other treatments. The conservation headland marginally outyielded the conventional crop to the edge headland. Of the remaining treatments, the headland next to the grass / wildflower strip yielded the highest, whilst the conservation headland next to the sterile strip produced the poorest yield (Table 3). Similar results were obtained at the Shropshire site, though differences between treatments were not significant. The Shropshire site was harvested using a plot combine and so detailed quadrat yields are not available. Combine yields are not presented here. Quadrat position was highly significant at the Leicestershire site ($F_{5,50}$=50.358, P<0.001), with grain yield increasing on moving away from the crop edge for all treatments (Table 3).

Assessments of weed dry weight were only made for the Leicestershire site. The amount of weed material differed between treatments ($F_{5,50}$=19.471, P<0.001). The two conservation headland treatments contained the most weed material, whilst the remaining treatments contained relatively few weeds. Quadrat position was significant ($F_{5,50}$=3.186, P<0.05), weed dry weights once again decreased on moving away from the crop edge for all treatments (Table 3).

TABLE 3. Mean grain yield and weed dry weights at harvest for the Leicestershire site

| | | Distance from crop edge (m) | | | | | | |
		0-0.5	1-1.5	2-2.5	3-3.5	4-4.5	11.5-12	treatment mean
Crop to edge sprayed	yield t/ha	4.57	6.02	5.77	6.55	7.19	6.55	6.11
	weeds g/m²	33.31	6.33	0.80	1.68	1.17	1.81	7.52
Crop to edge conserv.	yield t/ha	5.54	5.84	6.64	5.97	6.87	8.13	6.50
	weeds g/m²	36.91	43.81	32.09	50.29	46.71	1.00	35.14
Natural regen.	yield t/ha	-	5.38	5.35	6.20	6.78	6.77	5.08
	weeds g/m²	16.28	5.53	65.69	2.23	23.08	5.21	9.67
Sterile strip sprayed	yield t/ha	-	6.02	5.84	6.13	6.82	7.43	5.37
	weeds g/m²	40.32	0.39	10.08	3.76	38.01	1.87	15.74
Sterile strip conserv.	yield t/ha	-	5.92	4.63	4.99	5.25	6.35	4.52
	weeds g/m²	26.76	109.19	123.55	101.60	147.35	11.60	86.67
Wildflower /grass	yield t/ha	-	5.23	6.11	5.50	7.77	8.39	5.50
	weeds g/m²	18.51	7.99	2.15	2.12	13.12	14.05	9.66

(Least significant difference between treatment means : yield = 0.76, P<0.05, 34 d.f.)
(Least significant difference between treatment means : weeds = 20.17, P<0.05, 34 d.f.)

Both grain yield and weed dry weights varied significantly between sites (Grain yield $F_{15,225}$=23.674, P<0.001, weed dry weight $F_{15,225}$=18.832, P<0.001), grain yields ranged from 5.13 t/ha to 9.56 t/ha, whilst weed dry weights varied from 4.09 g/m^2 to 113.19 g/m^2. Quadrat position was highly significant for grain yield ($F_{5,225}$=54.293, P<0.001) and weed dry weight ($F_{5,225}$=19.766, P<0.001). Grain yield became significantly greater as distance from the field edge increased, while weed dry weights decreased on moving away from the crop edge (Figure 1). There was a significant interaction between site and quadrat position for both grain yield and weed dry weight. Differences in yield were recorded between headland and main field (11.5-12 m) quadrats (Table 4). Differences ranged from a reduction in yield of 47 % on the headland to an increase of 13 %, though usually the headland yielded less compared to the main field. Weed dry weights were greater from the headland area compared to the main field quadrats, especially for site 16 which was a conservation headland, and so would be expected to contain a greater amount of weed material.

There was no significant difference between turning and non-turning headlands. Aspect had a significant effect, with north facing headlands yielding slightly higher than south facing ones ($F_{1,228}$=30.287, P<0.001).

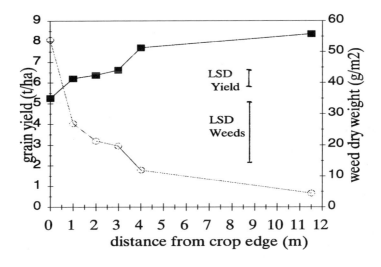

Figure 1 Survey mean grain yields and mean weed dry weights for sixteen sites

TABLE 4. Difference between headland and main field quadrat yields for 16 sites

Site	Mean Headland Yield (t/ha)	Mean Field Yield (t/ha)	% Difference
1	5.80	5.13	+13.06
2	4.60	5.70	-19.23
3	4.62	8.77	-47.32
4	7.18	8.15	-11.90
5	5.03	6.03	-16.58
6	7.07	7.51	- 5.86
7	6.04	8.42	-28.27
8	7.43	9.31	-20.24
9	6.33	9.68	-34.57
10	7.54	7.92	- 4.80
11	7.45	9.26	-19.57
12	6.25	8.34	-25.01
13	7.30	9.07	-19.56
14	6.04	8.61	-29.85
15	9.25	11.10	-16.70
16	4.98	8.96	-44.42
Mean	6.43	8.37	-23.12

Sites 1-9 were in Shropshire, sites 10-16 were in Leicestershire.

DISCUSSION

Results for the first year show that although the average yield at the field edge is low, taking 1 m out of production, either by creating a sterile strip of bare ground, or by sowing or leaving to natural regeneration significantly reduced overall yields. May et al. (1994) recorded similar findings where the lowest wheat yields occurred where the crop was grown with a sterile strip and the highest where wheat was grown up to the field edge. However, the amount of income that is lost due to a lower yield when the outer 1 m of the field is taken out of crop production may be outweighed by other benefits of creating what is essentially an extended field margin. For example, the expansion of the perennial ground flora at field edges, either using sown species or natural regeneration, can help to control annual weed species of hedgerows such as *Galium aparine* and *Bromus sterilis*, which may invade adjacent crops (Marshall, 1989), and also enhance populations of beneficial insects by providing suitable overwintering sites (Thomas et al., 1991).

Generally the headland areas tended to yield less than the main field. This suggests that losses incurred by reducing inputs into this area would be proportionally less than if inputs were reduced on another part of the field.

The survey showed that there was no difference between turning and non-turning headlands. Aspect had a significant effect, with north facing headlands yielding slightly higher than south facing ones. This may have been because north facing headlands received more shelter, this agrees with Marshall (1967) who found that the sheltering effects of hedges can lead to increased yields in some circumstances.

The field margin management experiments are being repeated at the same sites during 1994/95. As well as recording crop yield and weed dry weights, measurements will also be made of soil compaction and the fertiliser spread pattern over the headland area to attempt to find out what factors affect yield differences over the headland area.

A second survey of winter wheat headlands will be conducted in 1995. This time the sampling distance will be extended to 30 m into the field. The survey area will also be widened to cover calcareous soils, which make up a large amount of the cereal area in the U.K.

From these observations it is aimed to produce recommendations for a more integrated approach to the management of field margins to improve their conservation value, whilst still meeting the agricultural objective of economic crop production.

ACKNOWLEDGEMENTS

Thanks are due to the staff at Harper Adams and the Allerton Research and Educational Trust for their co-operation in running the experiments and the survey. N.H. Jones is funded by the Higher Education Funding Council for England (H.E.F.C.E.).

REFERENCES

Barr, C.J.; Bunce, R.G.H.; Clarke, R.T.; Fuller, R.M.; Furse, M.T.; Gillespie, M.K.; Groom, G.B.; Hallam, C.J.; Hornung, M.; Howard, D.C.; Ness, M.J. (1993) *Countryside Survey 1990*. Main Report. London : Department of the Environment.

Boatman, N.D.; Sotherton, N.W. (1988) The agronomic consequences and costs of managing field margins for game and wildlife conservation. *Aspects of Applied Biology* **17**, *Environmental Aspects of Applied Biology Part 1*, 47-56

Fielder, A.G. (1987) Management options for field margins - an agricultural advisor's view. In : *Field Margins*, J.M. Way and P.W. Greig-Smith (Eds.) BCPC Monograph No. **35**, Thornton Heath : BCPC publications, 85-94

Lakhani, K.H. (1994) The importance of field margin attributes to birds. In : *Field Margins : Integrating Agriculture and Conservation*, N.D. Boatman (Ed.) BCPC Monograph **58**, Farnham : BCPC publications, 77-84

Marshall, E.J.P. (1988) The Ecology and Management of Field Margin Floras in England. *Outlook on Agriculture*, **17(4)**, 178-182

Marshall, E.J.P. (1989) Distribution patterns of plants associated with arable field edges. *Journal of Applied Ecology*, **26**, 247-257

Marshall, K.J. (1967) The effect of shelter in the productivity of grasslands and field crops. *Field Crop Abstracts* **20**, 1-14

May, M.J.; Ewin, C.; Mott, J.; Pack, R.; Russell, C. (1994) Comparison of five different boundary strips - interim report of first two years study. In : *Field Margins : Integrating Agriculture and Conservation*, N.D. Boatman (Ed.) BCPC Monograph **58**, Farnham : BCPC publications, 259-264

Morris, M.G.; Webb, N.R. (1987) The importance of field margins for the conservation of insects. In : *Field Margins*, J.M. Way and P.W. Greig-Smith (Eds.) BCPC Monograph No. **35**, Thornton Heath : BCPC publications, 53-65

Sotherton, N.W.; Rands, M.R.W. (1987) The environmental interest of field margins to game and other wildlife

: a Game Conservancy view. In : *Field Margins*, J.M. Way and P.W. Greig-Smith (Eds.) BCPC Monograph No. **35**, Thornton Heath : BCPC publications, 67-75

Sparkes, D.L.; Scott, R.K.; Jaggard, K.W. (1994) The case for headland set-aside. In : *Field Margins : Integrating Agriculture and Conservation*, N.D. Boatman (Ed.) BCPC Monograph **58**, Farnham : BCPC publications, 265-270

Speller, C.S.; Cleal, R.A.E.; Runham, S.R. (1992) A comparison of winter wheat yields from headlands with other positions in five fen peat fields. In : *Set-Aside*, J. Clarke (Ed.) BCPC Monograph No. **50**, Farnham : BCPC publications, 47-50

Thomas, M.B.; Wratten, S.D.; Sotherton, N.W. (1991) Creation of "island" habitats in farmland to manipulate populations of beneficial arthropods : predator densities and emigration. *Journal of Applied Ecology*, **28**, 906-917

Tottman, D.R. (1987) The decimal code for the growth stages of cereals, with illustrations. *Annals of Applied Biology* **110**, 441-454.

SLUG DAMAGE TO CLOVER AND WHEAT GROWN SINGLY AND IN MIXTURES

S.K. GEORGE, D.A. KENDALL

IACR-Long Ashton Research Station, Department of Agricultural Sciences, University of Bristol, Long Ashton, Bristol, BS18 9AF, UK

R.O. CLEMENTS, E.J. ASTERAKI

Institute of Grassland & Environmental Research, North Wyke Research Station, Okehampton, Devon, EX20 2SB, UK

ABSTRACT

Feeding damage by the grey field slug, *Derocerus reticulatum*, was compared on two cultivars of white clover (*Trifolium repens*) and on three growth stages of wheat (*Triticum vulgare*), grown singly or as mixtures in controlled environment conditions. Slug damage to both clover and wheat, irrespective of clover cultivar or wheat growth stage, was significantly less in mixtures than in single (monoculture) plantings. The greatest reduction in damage to wheat at all growth stages occurred in wheat-clover mixtures with clover cv.Milkanova (50-60% less than in wheat monocultures). This clover cultivar was slightly more susceptible to slug damage than the other clover, cv.Donna, used in the experiment.

INTRODUCTION

Growing successive crops of winter wheat with a permanent companion crop of white clover may have considerable potential as a continuous low-input, dual purpose (bicropping) production system for grain and animal fodder or silage. Once established, the wheat component of the system is harvested for grain in the usual way and the wheat straw baled and removed. The flush of clover growth after harvest is then grazed or cut for silage in autumn and the next crop of winter wheat direct-drilled into this close-cropped sward to repeat the cycle. Nitrogen fixation by the permanent ground cover of clover provides most of the nitrogen required by the developing wheat crop and allows substantial reduction in the use of inorganic fertiliser (Jones & Clements, 1993).

Evidence from recent field experiments suggests that bicropped wheat is also less prone to damage by some invertebrate pests (e.g. slugs and aphids) and fungal pathogens (e.g. *Septoria* and *Fusarium*) and, thus, may require fewer pesticide inputs than wheat grown in conventional monoculture (Deadman & Soleimani, 1994; Clements & Kendall, 1995). However, further research is needed on the epidemiology of pests and pathogens in crop mixtures, in order to fully exploit and maximise the potential benefits of bicropping on pest and disease control.

Slugs can destroy large areas of winter wheat, especially in less intensive farming systems, by feeding on seeds and seedlings at crop establishment (Glen et al., 1994). Thus, crop management practices that tend to reduce slug damage are of considerable interest. A controlled environment experiment was done at Long Ashton in 1994 to investigate and compare slug grazing on wheat at different growth stages when planted in monoculture or with slug-susceptible and slug-resistant white clovers.

MATERIALS & METHODS

Five replicates of eleven clover, wheat and clover-wheat mixtures (Table 1) were grown in covered propagator trays (22 x 15 cm) in John Innes compost (loam, peat, grit mixture, 6:4:2 m/V) with VitaxQ4 nutrients (3.3 g/l at pH 6.5). Initially the trays were kept in a controlled environment room at 18°C day (8h light) and 15°C night (16h dark) in order to establish the various plant populations. The two white clover cultivars, cv. Milkanova (slug- susceptible) and cv. Donna (slug-resistant), were sown first. After 14 days, the clover seedlings in each tray were thinned to three rows of five plants, corresponding to a field seed-rate of 10kg/ha. The three growth stages of wheat (cv. Hereward) were then established in their respective trays by sowing seed 10 days, 5 days and 1 day before the start of the experiment.

TABLE 1. Summary of multifactorial treatments of two cultivars of white clover (cv. Donna and cv. Milkanova) and three growth stages of winter wheat (cv. Hereward).

Treatment No.	Clover cultivar	Wheat growth stage (Zadocks)	
1	Donna	(no wheat)	
2	Donna	Seed	(GS03)
3	Donna	1 leaf	(GS10)
4	Donna	3 leaf	(GS13)
5	Milkanova	(no wheat)	
6	Milkanova	Seed	(GS03)
7	Milkanova	1 leaf	(GS10)
8	Milkanova	3 leaf	(GS13)
9	(no clover)	Seed	(GS03)
10	(no clover)	1 leaf	(GS10)
11	(no clover)	3 leaf	(GS13)

Grey field slugs (Derocerus reticulatum) were collected from bran bait traps in fields of winter wheat at Long Ashton and kept for 1-2 weeks at 10°C in covered plastic sandwich boxes (16 x 28 x 9 cm) lined with moist cotton wool (about 30 slugs/box) until the start of the experiment. Every three days the slugs were fed a fresh mixture of the two clover cultivars used in the experiment and, at the same time, any uneaten leaves and dead slugs were removed.

The experiment was done in a controlled environment chamber at 10°C with a 10h light, 14h dark cycle and high relative humidity. One adult slug of known weight was put into each propagator tray and allowed to feed for 14 days. After this period, the slug was removed and re-weighed. The amount of feeding damage to each of the clover and/or wheat plants in each tray was estimated visually to obtain the percent leaf or seed tissue grazed. Treatments were compared by analysis of variance in Genstat 5, using a logit transformation of the percent tissue damage/plant.

RESULTS

Changes in slug weight during the experiment did not differ significantly between the experimental treatments.

Slug grazing/plant was consistently greater on clover cv. Milkanova than on cv. Donna irrespective of the presence or absence of wheat or the growth stage of wheat, although the overall difference in susceptibility to damage of the two clover cultivars was not significant (Table 2A). Both cultivars had significantly more tissue damage/plant in the absence of wheat than in the presence of wheat, with least damage where wheat seeds were present (Table 2B). The percentage of clover plants damaged in each treatment followed a similar pattern (Table 2).

Wheat seed was always the most susceptible growth stage to slugs with progressively less damage/plant at GS10 and GS13, irrespective of the presence or absence of clover (Table 3A). All growth stages of wheat suffered most slug grazing (tissue damage/plant) in the absence of clover (i.e. in monoculture) and least in the wheat-Milkanova mixtures, with the wheat-Donna mixtures intermediate (Table 3B). As for clover, the percentage of wheat plants damaged by slugs in each treatment corresponded with the amount of tissue damage/plant (Table 3).

TABLE 2. Mean percent area of clover leaf tissue grazed by slugs (expressed as logits with back-transformed percentages in brackets) and the percentage of clover plants damaged.

Treatment (see Table 1)		Tissue damage/plant		% Plants damaged
A. 1,2,3,4	Donna	-3.995	(1.3)	9.3
5,6,7,8	Milkanova	-3.852	(1.6)	10.2
	s.e.d. (df=98)	0.1017		0.599
B. 1,5	Clover (no wheat)	-3.552	(2.3)	12.0
4,8	Clover + wheat GS13	-3.839	(1.6)	10.3
3,7	Clover + wheat GS10	-3.951	(1.4)	9.6
2,6	Clover + wheat seed	-4.353	(0.8)	7.8
	s.e.d. (df=98)	0.144		0.85

TABLE 3. Mean percent area of wheat seed or leaf tissue grazed by slugs (expressed as logits with back-transformed percentages in brackets) and the percentage of wheat plants damaged.

Treatment (see Table 1)		Tissue damage/plant		% Plants damaged
A. 4,8,11	Wheat GS13	-4.729	(0.4)	27.7
3,7,10	Wheat GS10	-4.532	(0.6)	21.1
2,6,9	Wheat seed	-4.107	(1.1)	31.4
	s.e.d. (df=112)	0.1838		4.3
B. 9,10,11	Wheat (no clover)	-4.217	(1.0)	35.0
2,3,4	Wheat + Donna	-4.348	(0.8)	28.4
5,6,7	Wheat + Milkanova	-4.804	(0.3)	16.8
	s.e.d. (df=112)	0.1838		4.3

DISCUSSION

In our experiments, slug damage to winter wheat (cv. Hereward) seed and seedlings was significantly reduced to 30-80% of that in wheat monoculture by the presence of a companion crop of white clover. The greatest reductions in damage to wheat (30-50%) occurred in wheat-clover mixtures with a slug susceptible clover, cv. Milkanova. Hence, choice of clover cultivar may be important for the integrated control of slug damage in less-intensive (low-input) wheat-clover bicrops. In view of these laboratory findings, several wheat-clover mixtures are currently being tested in small plot field experiments to determine their relative susceptibility to slugs and other pests.

Our results provide evidence that crop mixtures (i.e. increased plant diversity) in arable cropping systems could be beneficial for integrated pest management, at least for a generalist herbivore like the grey field slug (*Derocerus reticulatum*). In this case, the reduced levels of damage to both wheat and clover when grown in mixtures can be explained as a dilution effect whereby the total amount of herbivory (or damage) is spread between two plant species; this probably reduces the specific damage to each component of the cropping mixture. This conclusion is supported by the measurements of slug weight before and after the experiment. These data indicate that the overall level of herbivory in each of the experimental treatments was not significantly different.

ACKNOWLEDGEMENTS

This work was funded by MAFF commission CE0403. IACR receives grant-aided support from the Biotechnology and Biological Sciences Research Council of the United Kingdom.

REFERENCES

Clements, R.O.; Kendall, D.A. (1995) Bi-croppers do it in   pairs. *New Farmer   and Grower, Spring 1995*, pp. 28-29.

Deadman, M.L.; Soleimani, M.J. (1994) Cereal-clover   bicropping: implications   for crop disease   incidence. *Proceedings Fourth Research   Meeting,   British   Grassland Society, Reading 1994*, 79-80.

Glen, D.M.; Witshire, C.W.; Wilson, M.J.; Kendall, D.A.;   Symondson, W.O.C.   (1994) Slugs in arable crops: key   pests under CAP reform? *Aspects of   Applied Biology* **40**,   *Arable farming under CAP reform*, 199-206.

Jones, L.; Clements, R.O. (1993) Development of a low input   system for   growing wheat (*Triticum vulgare*) in a   permanent understorey of white   clover   (*Trifolium repens*). *Annals of Applied Biology* **123**, 109-119.

EVALUATION OF ENERGY USAGE FOR MACHINERY OPERATIONS IN THE DEVELOPMENT OF CEREAL CLOVER BICROPPING SYSTEMS.

J V G DONALDSON, J HUGHES, S D DIXON

IACR Long Ashton Research Station, Department of Agricultural Sciences, University of Bristol, Long Ashton, Bristol, BS18 9AF. UK.

R O CLEMENTS

IGER, North Wyke Research Station, Okehampton, Devon, EX20 2SB. UK.

ABSTRACT

The adoption of a Wheat Clover bicropping system in its initial establishment year shows increased energy usage (and a significant loss in crop production) compared with conventional grown wheat, large savings in energy usage are made in later years particularly when second and subsequent (continuous wheat) crops are grown; in these crops the savings in labour and machinery costs alone will virtually offset the loss of crop value due to lower output each year whilst the bicropping system remains viable.

INTRODUCTION

A joint project (AIR 3 CT 93-0893) between five European Contractors began in the Spring of 1994 to study the exploitation of a sustainable low-input and reduced-output system for arable crops. Four of the sites (including IACR - Long Ashton) would study large-scale whole system testing to compare winter wheat grown over a 3 year period; the wheat would be grown with or without a clover understorey.

In the Spring of 1993, two fields in grass leys of approximately 4 ha each were designated for the project. After the first silage cut in May 1993 one field would be glyphosated, ploughed and then sown to a clover ley, subsequently this field would be direct seeded, using a Hunter Rotaseeder, to winter wheat in October 1993. The comparison field would receive fertiliser for a further cut for silage during the summer and then would be glyphosated, ploughed and sown to winter wheat using conventional farm cultivators and drills. This field would then be managed according to local best farming practice to maximise output.

This paper ascribes an Energy Value Factor for each mechanical operation in each field to allow comparisons to be made between the two systems, to ensure that the concept of "low-input" is true in terms of "energy" as well as the financial cost of variable inputs of fertiliser, pesticides etc.

MATERIALS AND METHODS

The Energy Value Factor (Donaldson *et al*, 1994) is calculated according to the formula

$$\frac{Energy\ Factor}{(kW\ hr^{-1}\ ha^{-1})} = \frac{Tractor\ (kW)\ power\ required \times 10}{\frac{Forward}{Speed}\ (km\ hr^{-1}) \times \frac{Implement}{Width}\ (m) \times \frac{Field\ Efficiency\ (\%)}{100}}$$

This formula has already been successfully applied to measure the farming systems comparison in the Long Ashton Research Station LIFE experiment (Jordan & Hutcheon, 1994 and Donaldson *et al*, 1994). A tractor of suitable size (kw) is chosen to match the working requirements of the implement, the forward speed (km hr^{-1}) is measured mid run, and a figure for field efficiency (Witney, 1988) is calculated but with reference to local practice and experience. The working width of the implement (m) will be the effective working width, for example the working width of a baler is not the width of the balers pick-up reel but the width of crop cut from which the swath was produced.

All field operations for both fields were recorded in diary form, with all the relevant information required to calculate the Energy Factor for each mechanised operation.

After harvest of the grain, the straw is baled and removed in the bi-cropping field, the clover is allowed to grow and is then grazed tightly by sheep to achieve minimum impedance to the Rotaseeder when drilling the winter wheat. (Depending on the growth of the clover, it could be cut and baled for silage in early September.)

RESULTS

Using the above guidelines energy usage figures were calculated for the two fields for the establishment year, the first wheat crop and where necessary predicted for the second wheat crop.

As 1995 figures are as yet not to hand it is assumed that a conventionally grown second wheat will yield 12.5% less than a first wheat (Nix 1994). It is further assumed that the yield of the bicropped wheat in the second year is likely to be comparable with the yield in the first year.

TABLE 1. Energy factors for each operation during the establishment year, March 1993 - September 1993

Operation	Energy Factor (kw hr^{-1} m ha^{-1})	
	Conventional	Bicropping
Top dressing	14	14
Round-up (glyphosate) spray	-	10
First cut silage	149	149
Mow	(33)	(33)
Ted	(31)	(31)
Row up	(43)	(43)
Big bale	(36)	(36)
Remove bales	(6)	(6)
Top dressing	14	-
Plough	-	114
Springtine	-	32
Speedkult	-	26
Ring roll	-	21
Drill clover	-	31
Ring roll	-	28
Herbicide spray	-	10
Second cut silage	149	-
Round-up spray	10	-
Top clover	-	36
Top clover	-	36
TOTAL	**336**	**507**

( ) - Included in total above - First cut silage

145

TABLE 2. Energy factors for each operation when growing first wheat crop, September 1993 - August 1994.

Operation	Energy Factor (kw hr^{-1} ha^{-1})	
	Conventional	Bicropping
Plough	114	-
Springtine	32	-
Speedkult	26	-
Drill winter wheat	33	-
Direct drill winter wheat	-	57
Chain harrow	-	12
Basal fertiliser	14	14
Herbicide Spray	-	10
Top Dressing	14	-
Herbicide spray	10	-
Top dressing	14	14
Herbicide/fungicide spray	10	-
Growth regulator spray	10	-
Fungicide spray	10	-
Aphicide spray	10	10
Combine	54	39
Big bale straw	36	30
Remove bales	6	4
TOTAL	**393**	**190**

TABLE 3. Energy factors for each operation when growing of second wheat crop, September 1994 - August 1995.

Operation	Energy Factor (kw hr^{-1} ha^{-1})	
	Conventional	Bicropping
Sub soil	77	-
Ring roll	21	-
Plough	98	-
Springtine	32	-
Speedkult	26	-
Drill winter wheat	33	-
Direct drill winter wheat	-	57
Herbicide spray	-	10
Herbicide & aphicide spray	10	-
Slug pellets	-	4
Basal fertiliser	14	14
Top dressing	14	-
Herbicide spray	-	10
Growth regulator spray	10	-
Top dressing	14	14
Fungicide spray GS 35	10	-
Herbicide spray	10	-
Top dressing	14	-
Fungicide spray GS 39-45	10	10
Fungicide spray GS 59	10	-
Combine	54	39
Big bale straw	36	30
Remove straw	6	4
TOTAL	**499**	**192**

GS - Growth Stage

TABLE 4. Total Energy factors for each season

Year	Energy Factor (kw hr^{-1} ha^{-1})	
	Conventional	Bicropping
1. Clover establishment/ silage production	336	507
2. First wheat crop	393	190
3. Second wheat crop	499	192
TOTAL	1228	889
Percentage %	100	72

TABLE 5. Crop Production/Value for each system

Year/Crop	Conventional		Bicropping	
	t/ha	Value £/ha	t/ha	Value £/ha
1. First cut silage	21.3	479.25	21.3	479.25
Second cut silage	14.4	324.00	-	-
Autumn grazing (Sheep)	-	-		37.50
Sub Total		803.25		516.75
2. First wheat crop				
Grain	8.27	909.70	4.86	534.60
Straw	4.70	94.00	1.65	33.00
Autumn grazing (Sheep)	-	-	-	37.50
Sub Total		1003.70		605.10
3. Second wheat crop (Estimated figures[1])				
Grain	7.24	796.40	4.86	534.60
Straw	3.56	71.20	1.65	33.00
Autumn grazing (Sheep)	-	-	-	37.50
Sub Total		867.60		605.10
TOTAL		2674.55		1726.95
Percentage %		100		65

DISCUSSION

The summary of energy usage figure for the first three years (Table 4) show that overall considerable energy savings are being made on mechanical operations in the bicropping system, some 28% less than in the conventional system.

In the initial establishment year (Table 1), the bicropping has a higher total energy factor due to the mechanical operations for establishing the clover; this also coincides with the loss of crop production (second cut silage) that is obtainable in the conventional system.

In the second season when first wheat crops are grown, savings in energy usage are seen in the bicropping system (Table 2); in this system the wheat is established by non-inversion tillage using the Rotaseeder, further savings are made through less interventions to apply pesticides, plant growth regulators and the number of Nitrogen top dressings.

The third season of cropping, with second wheat also shows the same savings as the second year for the bicropping system, although the conventional system does show a somewhat higher energy use due to the need to sub soil. However the saving of energy usage of over 60%, 192 compared to 449 kW hr^{-1} ha^{-1} has to be set against a loss of crop production value of about 30%, £605 compared to £867 for the conventional - a smaller reduction than with the first wheat when the conventional crop produced over £1,000 ha^{-1} output.

If the energy factor figures are used as an indicator of fixed costs (excluding rents) then according to Nix (1994) an arable farm of 100-200 ha would have labour, power and machinery costs in the order of £370 ha^{-1}, a 60% saving on which would represent £220 ha^{-1} thus virtually negating the crop production value loss on the bi-cropping system when growing second and subsequent wheat crops.

REFERENCES

Donaldson, J. V. G; Hutcheon, J. A; Jordan, V. W. L; Osborne, N. J. (1994) Evaluation of energy costs for machinery operations in the developments of more environmentally benign farming systems. *Aspects of Applied Biology 40, Arable farming under CAP reform.* pp 87-92.

Jordan, V. W. L; Hutcheon, J. A. (1994) Economic viability of less-intensive farming systems designed to meet current and future policy requirement: 5 year summary of LIFE project. *Aspects of Applied Biology 40, 1994. Arable farming under CAP reform.* pp 61-68.

Nix, J. (1993) Farm Management Pocket Book (24th edition 1994). Wye College, University of London.

Nix, J. (1994) Farm Management Pocket Book (25th edition 1995). Wye College, University of London.

Witney, B. (1988) Choosing and Using Farm Machinery. *Longman Scientific & Technical.* pp 412.

HETEROPTERA DISTRIBUTION AND DIVERSITY WITHIN THE CEREAL ECOSYSTEM

S.J. MOREBY

The Farmland Ecology Unit, The Game Conservancy Trust,
Burgate Manor, Fordingbridge, Hampshire, SP6 1EF, U.K.

ABSTRACT

Heteroptera are one of the many arable insect groups that use
the crop during part of their annual life cycle. Within British
agriculture Heteroptera are not generally regarded as cereal
pests and can even be termed "beneficial" with respect to their
occurrence in the diet of many farmland bird species. Farming
practices can make the field a harsh, inhospitable environment
for non-target arthropods and this paper details the movement
and distribution of Heteroptera within cereal fields during the
summer. Possible benefits of integrated control methods for
Heteroptera are discussed.

INTRODUCTION

On farmland, all or part of the life-cycles of many arthropod
species, depend on non-cropped habitats for shelter and food at some
time of the year. However uncropped areas (excluding woodland) can
account for as little as 2% of habitats on farmland (Sotherton 1984)
Agricultural practices can make arable fields harsh, inhospitable
environments. However many non-pest species are found in, and are
often reliant on the crop habitat during one or more life stages.
Many arable insect species from a wide variety of Orders and Families
e.g. ground beetles (Coleoptera: Carabidae) use the crop during part
of the year and another, less well studied group exhibiting similar
spatial and temporal patterns are the plant bugs (Hemiptera:
Heteroptera).

The Heteroptera contain many species which overwinter in non-
cropped habitats and then disperse from the field boundary into the
field during the spring and summer when the crop and associated weed
flora can provide a temporary but favourable habitat. In Britain
Heteroptera are not generally regarded as cereal pests, and can even
be termed "beneficial" because of their occurrence in the diet of
many farmland birds (Potts, 1986). This paper uses Heteroptera as an
example of a non-pest group and details their movement and
distribution within cereal fields during the summer.

MATERIALS AND METHODS

Study area

The study was carried out on an 11 km² arable estate in northern
Hampshire between 1983-85 on which up to 65% of the area was sown to
cereals. The estate was sub-divided into three discrete areas by
natural barriers (roads and a railway embankment). Each area

contained a block of cereal fields which were sprayed in accordance with the normal farming practice and a block in which the headlands were selectively sprayed according to the guidelines laid down for Conservation Headlands (Rands, 1985; 1986; Sotherton *et al.*, 1989).

Spatial distribution

Heteroptera samples were collected 3m into cereal field headlands and 12m and 50m into the field. Samples were collected with a 0.2m² sweep net and 50 sweeps were taken at each distance. They were collected on, or around, the 23 June each year. To reduce possible variation between fields all sampling was carried out over a six-hour period (10 a.m.-4 p.m.) by the same individual. In the three years of the study a total of 36, 34, and 33 fields were sampled respectively. All Heteroptera were identified to species. Analysis was carried out on the numerically dominant species *Calocoris norvegicus* and four groups or guilds containing two or more species; grass-feeding Stenodemini, predatory Nabidae and *Anthocoris* spp., total *Other* species and the total number of Heteroptera.

All blocks contained three crop types (spring barley, winter barley and winter wheat) and each block had a minimum number of four fields of each crop. For each heteropteran group, differences between crop types, pesticide regimes and distances into the field were compared. The collected data were transformed (log n+1), and analyses of variance (ANOVA) using genstat 5 (Genstat 5 Committee) were carried on the transformed means for pesticide regimes, crop types and distance.

RESULTS

Between cereal crop types

Out of 45 comparison analyses conducted to detect significant differences in mean numbers of Heteroptera between the three crop types during 1983-85, only three such tests were significant. No differences were found in 1983. In 1984 significantly more predatory species occurred in winter wheat compared to winter barley (P<0.05); and in 1985 both spring barley and winter wheat contained significantly more *Other* species compared to winter barley (P<0.01) (Tables 1-3).

Between pesticide treatments on headlands

Higher numbers of Heteroptera were found in the headlands where selective pesticide inputs were used compared to those that were fully sprayed. However few of these differences were significant. In 1983 three groups were significantly more numerous in the selectively sprayed fields, namely *C. norvegicus* and Total Heteroptera (P<0.05), and *Other* species (P<0.01) (Table 1). In 1984, this was the case for *C. norvegicus* and Total Heteroptera (P<0.05) (Table 2). No differences were detected between treatment regimes in 1985 (Table 3).

Spatial distribution within fields

All the heteropteran groups were found in higher numbers at 3m into cereal fields compared to distances further into the crop and in most cases these densities were significantly greater than those found at 12m and 50m. Densities at 12m were also often higher than those found at 50m with many of the differences being significant (Tables 1-3).

DISCUSSION

Cereal fields are temporary habitats in arable ecosystems existing for only 6-9 months during which time destructive and disruptive events (pesticide applications, cultivations, harvesting) occur. Arable field-dwelling Heteroptera therefore are generally restricted to species that overwinter in the field boundary and move into the developing crop in spring or early summer when the crop-weed species form a suitable habitat (Moreby, 1994). Of the Heteroptera species or groups studied, only C. norvegicus successfully exploited cereal crops as a food source after the newly hatch nymphs moved out of the field boundary in the early summer. The unlimited supply of suitable food plants, (cereals and weed species) and the dilution effect with movement into the crop could explain the significant differences in densities between distances close to and far from the overwintering site. The grass-feeding Stenedomini, seemed to be dependent for food on tall grass species such as Poa trivialis, Alopecurus myosuroides and Lolium spp. and were not able to use the developing cereal as a suitable food source (Moreby, 1994). The groups of predatory Heteroptera and Other Heteroptera were also most numerous at 3m, but very low densities were found at all sites, particularly 12m and 50m. The scarcity of individuals could have resulted in most distance comparisons being non-significant. The final group, total Heteroptera, closely mirrored the distribution patterns found for C. norvegicus due to the numerical dominance of this single species over all the other groups.

While it undoubtedly failed to collect many individuals, the use of a sweep net as the method of collection did allow good comparisons to be made between sites and distances for C. norvegicus and the Stenedomini because these were predominantly found feeding on the higher, more nutritious parts of their food plants. However, while the predatory and other Heteroptera were predominantly to be found on the crop floor or low down in the cereal canopy on the weeds these two groups could have been greatly under-estimated. However similarly low densities were found in cereal field headlands at 3m (Moreby, 1994) using a D-Vac suction sampler (Dietrick, 1961).

The different cereal varieties did not seem to affect numbers of Heteroptera to any significant degree. However, favourable weather conditions resulting in an early ripening crop such as winter barley could result in the cereal quickly declining in suitability in July-August. Growth/development trials conducted throughout the summer found significantly greater survival of C. norvegicus on cereals, ranked in the order of spring barley - winter wheat - winter barley (Moreby S.J., unpub. data).

Direct or indirect decreases in numbers of many beneficial and other non-target arthropods caused by pesticide use are well documented (Potts & Vickerman, 1974; Vickerman & Sunderland, 1977; Coombes & Sotherton, 1986; Potts, 1986; Sotherton et al., 1987; Inglesfield, 1989; Somerville & Walker, 1990; Chiverton & Sotherton, 1991; Davis et al., 1991). Toxic effects of pesticides against Heteroptera are less well studied. Direct effects of fungicides and herbicides lead to low levels of mortality. However herbicides can cause significant indirect mortality via their effect on food plants (Moreby, 1991; 1994). As a result, the lower numbers of Heteroptera in the fully sprayed fields were not unexpected. Insecticides commonly used to control cereal aphids have also been shown to be toxic to Heteroptera (Moreby, 1991).

The use of integrated control methods has the potential of increasing numbers of cereal-dwelling Heteroptera, and reduced pesticide rates and/or the use of more selective compounds, particularly insecticides, would benefit Heteroptera directly. Enhancement of natural enemies to reduce insecticide application against Aphididae would have benefits on heteropteran survival and the resulting biocontrol would only have a limited impact on Heteroptera as many aphid-specific predators, such as the larvae of Syrphidae, Coccinellidae and Neuroptera prey on sedentary colonies of aphids rather than individual, relatively fast moving heteropteran nymphs. Planting of flower-rich strips to attract winged, beneficial species could also provide a supplementary food source particularly if planted next to early ripening cereals such as winter barley. However the use of less nitrogen could reduce the nutritional quality of the cereal and as for cereal aphids, Heteroptera could find cereals in such farming systems employing low levels of nitrogen fertiliser less suitable as host plants for shorter periods. Later sowing of crops could have benefits to Heteroptera following reduced pesticide use and perhaps a better synchrony between more favourable cereal growth stages and dispersing nymphs. While minimal tillage is unlikely to have any effect on Heteroptera since most species overwinter in the field boundary, mechanical weeding or organic production systems could have similar effects to herbicides and remove potential food plants which the use of Conservation Headlands help to preserve.

ACKNOWLEDGEMENTS

The author would like to thank the Manydown Company for permission to work on their land.

REFERENCES

Chiverton, P.A; Sotherton, N.W. (1991) The effects on beneficial arthropods of the exclusion of herbicides from cereal crop edges. Journal of Applied Ecology 28, 1027-1040.
Coombes, D.S.; Sotherton, N.W. (1986) The dispersion of polyphagous predators from their overwintering sites into cereal fields and factors affecting their distribution in the spring and summer. Annals of Applied Biology 108, 461-474.

Davis, B.N.K.; Lakhani, K.Y.; Yates, T.J. (1991) The hazards of insecticides to butterflies of field margins. *Agriculture, Ecosystems & Environment* **36**, 151-161.

Dietrick, E.J. (1961) An improved backpack motorised fan for suction sampling of insects. *Journal of Economic Entomology* **54**, 394-395.

Inglesfield, C. (1989) Pyrethoids and terrestrial non-target organisms. *Pesticide Science* **27**, 387-428.

Moreby, S.J. (1991) Laboratory screening of pesticides against *Calocoris norvegicus*, a non-target heteropteran in cereals. *Tests of Agrochemicals and Cultivars* No. 12, (*Annals of Applied Biology* **118**, Supplement) 8-9.

Moreby, S.J. (1994) The influence of agricultural practices on Heteroptera in arable field margins. *Unpublished M.Phil Thesis, University of Southampton.*

Potts, G.R. (1986) *The Partridge: Pesticides, Predation and Conservation.* Collins, London, pp. 274.

Potts, G.R.; Vickerman, G.P. (1974) Studies of the cereal ecosystem. *Advances in Ecological Research* **8**, 107-197.

Rands, M.R.W. (1985) Pesticide use on cereals and the survival of grey partridge chicks: a field experiment. *Journal of Applied Ecology* **22**, 49-54.

Rands, M.R.W. (1986) The survival of gamebirds (Galliformes) chicks in relation to pesticide use on cereals. *Ibis* **128**, 57-64.

Somerville, L.; Walker, C. Eds (1990) *Pesticide Effects on Terrestrial Wildlife.* London: Taylor & Francis. 404pp.

Sotherton (1984) The distribution and abundance of predatory arthropods overwintering on farmland. *Annals of Applied Biology* **105**, 423-429.

Sotherton, N.W.; Moreby, S.J.; Langley, M.G. (1987) The effects of the foliar fungicide pyrazophos on beneficial arthropods in barley fields. *Annals of Applied Biology* **111**, 75-87.

Sotherton, N.W.; Boatman, N.D.; Rands, M.R.W. (1989) The 'Conservation Headland' experiment in cereal ecosystems. *The Entomologist* **108**(1&2), 135-143.

Vickerman, G.P.; Sunderland, K.D. (1977) Some effects of dimethoate on arthropods in winter wheat. *Journal of Applied Ecology* **14**, 767-777.

TABLE 1. Mean number of Heteroptera (± S.E.) collected by 50 sweeps in cereal fields taken on 23rd June 1983 and significant differences between crops, treatments and distances.

Treatments	Distance	Calocoris norvegicus	p	Predatory species	p	Stenodemini	p	Other Heteroptera	p	Total Heteroptera	p
Selectively sprayed cereal fields											
n=20	3m	123.65 ± 38.82	1. *** 2. *** 3. ***	0.25 ± 0.16	1. * 2. NS 3. NS	1.05 ± 0.46	1. ** 2. ** 3. NS	0.90 ± 0.29	1. *** 2. NS 3. NS	125.80 ± 38.82	1. *** 2. *** 3. ***
	12m	10.15 ± 4.53		0.05 ± 0.05		0.05 ± 0.05		0.10 ± 0.07		10.50 ± 4.52	
	50m	0.10 ± 0.07		0.10 ± 0.07		0.30 ± 0.13		0.45 ± 0.25		0.95 ± 0.32	
Fully sprayed cereal fields											
n=16	3m	54.63 ± 23.31	1. *** 2. *** 3. ***	0.19 ± 0.10	1. * 2. NS 3. NS	2.56 ± 0.96	1. *** 2. *** 3. ***	0.19 ± 0.10	1. NS 2. NS 3. NS	57.56 ± 23.37	1. *** 2. *** 3. ***
	12m	3.81 ± 1.67		0.00 ± 0.00		0.31 ± 0.25		0.06 ± 0.06		4.19 ± 1.88	
	50m	0.00 ± 0.00		0.06 ± 0.06		0.44 ± 0.16		0.13 ± 0.13		0.63 ± 0.22	
Crop differences		NS		NS		NS		NS		NS	
Treatment differences		*		NS		NS		**		*	

* = P ≤ 0.05   ** = P ≤ 0.01   *** = P ≤ 0.001   NS = Not Significant

1. = 3m vs. 12m   2. = 3m vs. 50m   3. = 12m vs. 50m   sb = spring barley   wb = winter barley   ww = winter wheat

TABLE 2. Mean number of Heteroptera (± S.E.) collected by 50 sweeps in cereal fields taken on 26th June 1984 and significant differences between crops, treatments and distances.

Treatments	Distance	Calocoris norvegicus	P	Predatory species	P	Stenodemini	P	Other Heteroptera	P	Total Heteroptera	P
Selectively sprayed cereal fields											
n=17	3m	68.00 ± 25.06	1. *** 2. *** 3. **	1.18 ± 0.64	1. NS 2. NS 3. NS	11.18 ± 4.52	1. *** 2. *** 3. NS	2.88 ± 0.24	1. NS 2. NS 3. NS	83.12 ± 26.98	1. *** 2. *** 3. **
	12m	8.53 ± 3.07		0.47 ± 0.31		0.82 ± 0.26		1.88 ± 0.58		11.71 ± 3.63	
	50m	0.35 ± 0.15		0.24 ± 0.24		0.65 ± 0.26		1.12 ± 0.45		2.35 ± 0.72	
Fully sprayed cereal fields											
n=17	3m	31.12 ± 10.14	1. *** 2. *** 3. NS	1.29 ± 0.89	1. NS 2. * 3. NS	3.29 ± 2.00	1. ** 2. ** 3. NS	2.18 ± 0.73	1. ** 2. ** 3. NS	37.88 ± 11.20	1. *** 2. *** 3. ***
	12m	1.18 ± 0.54		0.29 ± 0.19		0.47 ± 0.19		0.53 ± 0.33		2.47 ± 0.66	
	50m	0.41 ± 0.19		0.06 ± 0.06		0.65 ± 0.30		0.47 ± 0.26		1.59 ± 0.40	
Crop differences		NS		* wb/ww		NS		NS		NS	
Treatment differences		*		NS		NS		NS		*	

$*$ = P ≤ 0.05 $**$ = P ≤ 0.01 $***$ = P ≤ 0.001   NS = Not Significant

1 = 3m vs. 12m   2 = 3m vs. 50m   3 = 12m vs. 50m   sb = spring barley   wb = winter barley   ww = winter wheat

TABLE 3. Mean number of Heteroptera (± S.E.) collected by 50 sweeps in cereal fields taken on 1st July 1985 and significant differences between crops, treatments and distances.

Treatments	Distance	Calocoris norvegicus	P	Predatory species	P	Stenodemini	P	Other Heteroptera	P	Total Heteroptera	P
Selectively sprayed cereal fields											
n=17	3m	66.24 ± 15.62	1. *** 2. *** 3. **	0.82 ± 0.37	1. ** 2. NS 3. NS	0.71 ± 0.25	1. ** 2. NS 3. NS	1.77 ± 0.41	1. ** 2. ** 3. NS	69.53 ± 15.48	1. *** 2. *** 3. *
	12m	3.35 ± 1.19		0.00 ± 0.00		0.24 ± 0.11		0.29 ± 0.17		3.94 ± 1.25	
	50m	0.53 ± 0.23		0.24 ± 0.11		0.24 ± 0.18		0.29 ± 0.29		1.35 ± 0.56	
Fully sprayed cereal fields											
n=17	3m	44.63 ± 19.68	1. *** 2. *** 3. **	0.38 ± 0.16	1. NS 2. NS 3. NS	1.44 ± 0.99	1. NS 2. NS 3. NS	1.06 ± 0.55	1. ** 2. ** 3. NS	47.50 ± 20.00	1. *** 2. *** 3. **
	12m	8.56 ± 4.22		0.13 ± 0.09		0.19 ± 0.14		0.69 ± 0.30		9.56 ± 4.37	
	50m	1.44 ± 0.79		0.31 ± 0.15		0.19 ± 0.10		0.50 ± 0.20		2.44 ± 0.94	
Crop differences		NS		NS		NS		** sb/wb ** ww/wb		NS	
Treatment differences		NS		NS		NS		NS		NS	

* = P ≤ 0.05   ** = P ≤ 0.01   *** = P ≤ 0.001   NS = Not Significant

1 = 3m vs. 12m   2 = 3m vs. 50m   3 = 12m vs. 50m   sb = spring barley   wb = winter barley   ww = winter wheat

PCR-BASED DETECTION OF *PHYTOPHTHORA* SPECIES IN HORTICULTURAL CROPS

D. E. L. COOKE, J. M. DUNCAN, I. LACOURT

Department of Mycology and Bacteriology, Scottish Crop Research Institute, Invergowrie, Dundee, DD2 5DA

ABSTRACT

Accurate and rapid detection and diagnosis of Phytophthora diseases in horticultural crops is of key importance in the management of the disease. The minimising of disease spread via infected rootstock and a move away from prophylactic fungicide application are two major benefits of such a procedure. The Polymerase Chain Reaction (PCR) offers great potential as a highly specific tool to achieve this. This paper reports on the design and optimisation of such a PCR-based detection system based on ribosomal DNA sequences of several important *Phytophthora* species of horticultural crops. Specific detection of the fungus in infected root material has now been achieved.

INTRODUCTION

The genus *Phytophthora* represents an important group of plant pathogenic fungi which are responsible for large scale losses of tropical and temperate crops. They are a particular problem in vegetatively propagated horticultural crops, being spread on infected planting material.

The sustainability of both the propagation and cultivation stages of many horticultural crops is severely threatened by *Phytophthora* spp. In order to minimise disease problems in nurseries, plants are often treated prophylactically with high doses of fungicides which rather than solving the problem may actually exacerbate it. Many fungicides have been shown to suppress rather than kill the *Phytophthora* spp. (Duncan, 1985) so apparently healthy material may be widely distributed, furthermore *Phytophthora* which has been exposed to strong selection pressure increases the potential for the development of fungicide resistance. In order to minimise this threat it is important that a scheme for the accurate and specific detection and diagnosis of *Phytophthora* is established.

An example of this which represents an important problem to raspberry and strawberry growers is *Phytophthora fragariae* which is a major limiting factor in crop growth, requiring the application of fungicides which are expensive and potentially damaging to the environment. It is now recognised that worldwide spread of the strawberry and raspberry varieties of this pathogen has been brought about via the movement of infected rootstock (Duncan, 1993). In an effort to stem this it has been declared a quarantine organism in many countries, which means stocks must be guaranteed "disease free" before importation.

There are however many difficulties in detection and diagnosis of *Phytophthora* spp. The non-specific symptoms on the root, crown or stem base makes visual confirmation of the presence of *Phytophthora* difficult. Currently, detection and diagnosis of *Phytophthora* relies on visual inspection, bait testing (Duncan *et al.*, 1993), isolation of the fungus on selective media or diagnostic kits based on polyclonal antibodies, each of these has its disadvantages compared to a PCR-based system.

THE USE OF RIBOSOMAL DNA SEQUENCES AS THE BASIS FOR PHYTOPHTHORA DETECTION.

The fundamental importance of the ribosomes in protein synthesis means DNA sequence homology can be seen in almost all forms of life. Despite the highly conserved "core sequences", there is some variation in other rDNA which allows phylogenetic separation at many levels from kingdoms through to genera (Bruns et al., 1991). The advantages of using rDNA sequences as target sites for PCR primers is now widely recognised. There is an abundance of publications on rDNA sequence variation, sequences are rich in informative regions, mutation rates are known and many copies are present in each nucleus thus increasing the sensitivity of detection. Since one is looking at a very tightly defined region, species can be added to the analysis at any time resulting in an expanding sequence database.

Many 18S and 28S ribosomal subunit sequences have been published and variation between genera noted. Spacer regions of the ribosomal repeat unit which are not thought to play a functional role, are less conserved and have been reported to show interspecific variation in plants (Sun et al., 1994) and fungi (Lee & Taylor, 1992; Zambino & Szabo; 1993, Sherriff et al., 1994) suggesting their suitability for the purposes of molecular detection.

The following is brief description of recent progress at SCRI on the molecular variation of ITS1 and ITS2 regions of *Phytophthora* spp. and their utility in the diagnosis and detection of *P. fragariae* and other species. Specific details of the procedures used and primers sequences will appear in subsequent papers.

PRIMER DESIGN

Using a set of PCR primers designed for the amplification of fungal spacer regions (White et al., 1990) ITS 1 and ITS 2 regions from many *Phytophthora* spp. were amplified and found to be approximately 220 and 400 bp long, respectively. The species sequenced so far are *P. fragariae* var. *fragariae*, *P. fragariae* var. *rubi*, *P. cambivora*, *P. cinnamoni*, *P. megasperma*, *P. nicotianae*, *P. cryptogea*, *P. citricola*, *P. drechsleri*, *P. infestans*, *P. idaei*, *P. pseudotsugae*, and *P. cactorum*. The double stranded PCR products were manually sequenced and aligned using multiple sequence alignment software on Seqnet (Daresbury Laboratory). Both ITS1 and ITS2 regions were sufficiently conserved to allow an accurate alignment, some regions were identical in all species tested and others showed considerable variation.

Two distinct types of sequence were noted, those from non-papillate species sharing regions of homology not seen in papillate and semi-papillate species which formed a separate sub-group. This rich source of defined sequence allowed the design of PCR primers specific for a particular species. To date, primers have been designed for *P. fragariae*, *P. cambivora*, *P. cinnamomi* and *P. nicotianae*. There is however sufficient sequence variation to design primers for almost any given species.

PRIMER TESTING

Each of the specific primers was tested against pure DNA from the 13 *Phytophthora* species detailed above and three showed excellent specificity, resulting in amplification products from many isolates of that species but no amplification products from other species. The *P. nicotianae* primer also amplified a product from *P. infestans*, which as a papillate species, was shown to be closely related to *P. nicotianae* (unpublished results based on ITS sequence homology). Preliminary studies on the sensitivity of the procedure on pure DNA showed the lower limit of detection to be 100 fg, which may be improved with nested PCR procedures.

Improvements in DNA extraction procedures and PCR efficiency have resulted in the successful amplification of *P. fragariae* DNA from both raspberry and strawberry roots which represents a major step forward in the diagnosis of the disease from a procedure taking weeks to one which may be completed in a matter of hours.

DISCUSSION

We now have a system for the detection of specific species of *Phytophthora* in infected material which has great potential as a diagnostic tool. More rapid disease diagnosis forms an integral part of any integrated control package and should result in more timely application of fungicides and reduce numbers of unnecessary applications. A system of careful monitoring of propagation material and attention to improved phytosanitary conditions should result in an overall reduction in the incidence of *Phytophthora* in propagation stocks, the benefits of which would be passed on to growers.

In order to carry out larger scale detection programmes further work on the development of protocols for efficient amplification from zoospores will be necessary. Such work is now underway at SCRI and should result in a system of monitoring plant stocks via the trapping and detection of zoospores in drainage water from pots, water from recirculating irrigation systems or environmental monitoring of pathogen populations in streams or soil.

ACKNOWLEDGEMENTS

The authors thank the Horticultural Development Council and the Scottish Office, Agriculture and Fisheries Department for funding this project.

REFERENCES

Bruns T. D.; White T. J. & Taylor J. W. (1991) Fungal molecular systematics. *Annual Review of Ecology and Systematics*, **22**, 525-564.

Duncan J. M. (1985) Effect of fungicides on survival, infectivity and germination of *P. fragariae* oospores. *Transactions of the British Mycological Society*, **85**, 585-593.

Duncan J. M. (1993) Data sheets on Quarantine Organisms: *Phytophthora fragariae.* In: *Quarantine Pests for Europe,* I. M. Smith; D. G. McNamara; P. R. Scott & K. M. Harris (Eds.), Wallingford: CAB International, pp. 599-608

Duncan J. M.; Kennedy D. M.; Chard J.; Ali A. & Rankin P. A. (1993) Control of *Phytophthora fragariae* on strawberry and raspberry in Scotland by bait tests. *Plant Health and the European Single Market, BCPC Monograph No.* **54**, 305-308.

Lee S. B. & Taylor J. W. (1992) Phylogeny of five fungus-like protoctistan *Phytophthora* species, inferred from the internal transcribed spacers of ribosomal DNA. *Journal of Molecular Biology and Evolution,* **9**, 636-653.

Sherriff C.; Mitzi J. W.; Arnold G. M.; Lafay J.; Brygoo Y. & Bailey J. A. (1994) Ribosomal DNA sequence analysis reveals new species groupings in the genus *Colletotrichum. Experimental Mycology,* **18,** 121-138.

Sun Y.; Skinner D. Z.; Liang G. H. & Hulbert S. H. (1994) Phylogenetic analysis of *Sorghum* and related taxa using internal transcribed spacers of nuclear ribosomal DNA. *Theoretical and Applied Genetics,* **89,** 26-32.

White T. J.; Bruns T.; Lee S. & Taylor J. (1990). Amplification and direct sequencing of fungal ribosomal RNA genes for phylogenetics. In: *PCR protocols: A guide to Methods and Applications,* M. A. Ignis; D. H. Gelfand; J. J. Sninsky & T. J. White (Eds.) San Diego: Academic Press, pp. 315-322.

Zambino P. J & Szabo L. J. (1993) Phylogenetic relationships of selected cereal and grass rusts based on rDNA sequence analysis. *Mycologia,* **85**, 401-414.

THE EFFECTS OF ARABLE FIELD MARGIN MANAGEMENT ON THE ABUNDANCE OF BENEFICIAL ARTHROPODS

R.E. FEBER, P.J. JOHNSON, H. SMITH, M. BAINES & D.W. MACDONALD

Wildlife Conservation Research Unit, Department of Zoology, University of Oxford, South Parks Road, Oxford OX1 3PS

ABSTRACT

We examine the extent to which the restoration and management of arable field margins can enhance their potential as habitat for Linyphiidae (Araneae) and for Staphylinidae (Coleoptera). In 1987, extended-width field margins were established around six arable fields at Wytham, Oxford. The field margins were subject to ten contrasting management treatments. Arthropods were sampled from the field margin swards using vacuum suction sampling, between 1987 and 1991. Linyphiid abundance was reduced both by mowing and by spraying with glyphosate herbicide. Staphylinid abundance was also reduced by mowing, but there was no significant effect of spraying. Effects of mowing on both groups differed with the timing of mowing and the date of sampling, with spring and summer mowing having the most severe effects on abundance. There was a tendency for staphylinid abundance in the crop to be greater adjacent to uncut field margin treatments. Neither linyphiid nor staphylinid abundance was significantly affected by establishing field margin swards with a grass and wild flower seed mixture, rather than by natural regeneration.

INTRODUCTION

Modern farming methods do not provide high quality habitats for the great majority of invertebrates. Many species of agricultural weeds, which formerly supported a rich and varied insect fauna, are now rare and associated primarily with field edges. Heavy use of pesticides within the crop has contributed to this loss of invertebrate interest by causing direct mortality. Field margins formerly provided more permanent refuges and reservoirs for invertebrates on farmland, but these too have suffered from agricultural intensification, with hedgerow removal, and deliberate and accidental pesticide and fertiliser applications, impoverishing the plant and animal assemblages.

In recent years, studies have shown that the manipulation of crop edge (Chiverton & Sotherton, 1991) and linear 'island' (Thomas et al., 1992) habitats can increase polyphagous predator densities in arable systems. We have shown elsewhere that the restoration and management of field margins around arable fields can potentially enhance agricultural habitats for butterflies (Feber & Smith, 1995; Feber et al., 1994). As well as having value for nature conservation, field boundaries can also harbour invertebrate predators of crop pests. In this paper we describe the effects of contrasting methods of field margin management on the abundances of two groups of invertebrates, the Linyphiidae (money spiders) and Staphylinidae (rove beetles) whose predatory lifestyles make them of potential benefit to farmers.

Linyphiidae are the dominant spiders in arable crops in Europe, and an important component of the polyphagous predator complex. They make horizontal webs both on and above the ground, depending on the species, and have been shown to have a significant impact on aphid populations early in the season (Sunderland *et al.*, 1986). Staphylinidae are common in grassland and agricultural habitats and also have significant impacts on aphid populations (Dennis & Wratten, 1991). Both groups have ecological characteristics which made them likely to respond to our manipulations of field margin sward structure and composition. We present results which show the effects on Linyphiid and Staphylinid abundance of mowing, sowing with a grass and wild flower seed mixture, and spraying with glyphosate herbicide. We consider the implications of our results for the biocontrol of invertebrate pests in arable systems.

METHODS

In autumn 1987 we created 2 m wide field margins around arable fields at the University of Oxford's farm at Wytham. These comprised the original margin, about 0.5 m wide (the "old" margin), and a fallowed extension of about 1.5 m on to cultivated land (the "new" margin). Swards were established on the fallowed strips either by allowing natural regeneration ("unsown" swards) or by sowing a mixture of wild grasses and forbs ("sown" swards). Plots 50 m long were established on both sward types and subjected to the following management regimes: unmown, or mown (with cuttings removed) in (a) summer only (b) spring and summer or (c) spring and autumn. Two further treatments were imposed on unsown plots only: (a) mown in spring and summer with hay left lying and (b) unmown, but sprayed with glyphosate in late June or early July. The plots were mown in the last weeks of April, June and September ("spring, "summer" and "autumn") respectively. Glyphosate (3 l/ha Roundup in 175 litres water) was first sprayed in 1989. The treatments were randomised in eight complete blocks, each block occupying a single field. From the time that they were fallowed, the field margins were protected from fertiliser and spray drift.

We used D-vac suction sampling to quantify the abundance of our target groups in six blocks of the experiment. The D-vac samples comprised five, 30-second "sucks" taken at 10m intervals along both old and new margins, within each plot. This approximated to a total sample area of 0.5 m^2 per plot in each of the old and new margins. In May and July of 1989 samples were also taken at distances of 2 m and 10 m into the crop, adjacent to the field margins. Each sample was transferred to a polythene bag and immediately cooled to reduce activity. The live invertebrates were extracted by pootering within a few hours of collection and stored in 70% alcohol until they were sorted. Sampling took approximately two days.

In 1987 and 1988 samples were taken in September only. In 1989, 1990 and 1991 samples were taken in May, July and September, after the spring and summer cuts, but before the autumn cut. In this paper we present data for Linyphiids and Staphylinids from the samples taken from the new margins between 1987 and 1991.

Analyses

All analyses of the effects of the experimental treatments on the measured variables were performed on appropriately transformed data by analyses of variance. Specific hypotheses, implicit in the design of the experiment, concerning the relative effects of particular treatments, were tested using planned comparisons. We then split the treatment effect into the main effects of sowing and cutting, by excluding the treatment in which the mown hay was left lying and the sprayed treatment. We performed a three-way analysis of variance on these remaining treatments which formed a 2 x 4 factorial structure (i.e. sown or unsown x 4 types of cut x 6 blocks).

Univariate repeated measures ANOVA was used for this analysis since measures of the dependent variables made on different dates on the same plot could not be treated as independent. Significance levels for individual samplings are presented as asterisks on the figures.

RESULTS

Effects of sowing

Sowing with a wildflower seed mixture had no significant effects on the abundance of either the Linyphiidae ($F_{(1,83)}=0.00$, $P=0.944$) or the Staphylinidae ($F_{(1,34)}=0.10$, $P=0.757$).

Effects of mowing

Mowing resulted in an overall significant reduction in the abundances of both linyphiids $F_{(3,83)}=13.6$, $P<0.001$) and staphylinids ($F_{(3,83)}=38.7$, $P<0.001$). The effect differed with the timing of sampling for both linyphiids (cut x round, $F_{(3,83)}=13.6$, $P<0.001$) and staphylinids (cut x round, $F_{(18,498)}=7.4$, $P<0.001$). The effects of the mowing regimes on each group are described below and mean abundances are shown in Table 1.

Mowing in summer only

Significantly fewer linyphiids were recorded from treatments which were mown in the summer, relative to those which were left unmown ($F_{(1,105)}=23.8$, $P<0.001$), over the three years. There was, however, a significant interaction between year and mowing, with the most substantial effects recorded in 1990 and 1991. In each year, linyphiid abundance was severely reduced following the summer mowing. Under this regime, numbers tended to recover by the autumn and increased further by the following spring (Figure 1). Mowing in summer only resulted in consistently higher abundances of linyphiids in spring than the other mowing regimes (Figure 1).

Staphylinids also performed poorly on the plots mown in summer only ($F_{(1,43)}=22.3$, $P<0.001$). Abundances were significantly lower on the mown than the unmown treatments in the summer samples in both 1990 and 1991, and the autumn samples of 1989 and 1990 (Figure 1).

Mowing in spring and summer

This mowing regime resulted in a more persistent and severe reduction in linyphiid abundance than mowing in summer only ($F_{(1,105)}=31.6$, $P<0.001$). Abundances in both spring

TABLE 1. Mean abundances of Staphylinidae and Linyphiidae per m² on plots under different mowing regimes throughout the experiment. Su = mown in summer only, Sp + Su = mown in spring and summer, Sp + Au = mown in spring and autumn. See figures for significances.

	Not mown	Mown Su	Mown Sp + Su	Mown Sp + Au
		Staphylinidae		
Sep 87	13.6	–	–	–
Sep 88	9.4	18.6	8.0	13.4
May 89	17.0	–	20.6	–
Jul 89	19.6	–	20.0	–
Sep 89	19.6	18.6	26.4	8.0
May 90	37.0	27.0	31.4	26.0
Jul 90	15.6	7.2	15.4	20.0
Sep 90	22.6	14.6	11.4	21.4
May 91	72.0	50.6	36.0	38.0
Jul 91	15.6	6.4	4.6	10.4
Sep 91	30.6	24.0	30.4	31.4
		Linyphiidae		
Sep 87	12.4	–	–	–
Sep 88	60.0	67.0	68.0	53.4
May 89	22.6	–	9.4	–
Jul 89	89.4	–	24.5	–
Sep 89	50.0	45.5	46.4	66.4
May 90	24.0	10.4	9.0	8.0
Jul 90	26.0	7.0	7.6	24.0
Sep 90	41.4	44.6	25.4	34.4
May 91	25.6	19.0	5.0	7.6
Jul 91	18.6	4.6	5.4	18.6
Sep 91	82.2	83.5	96.5	193.0

and summer in 1989 and 1990 were significantly lower on mown than unmown treatments (Figure 1). Although numbers on the mown treatments tended to recover in the autumns of 1990 and 1991, in all years they were substantially reduced again by mowing in the following spring (Figure 1).

Spring and summer mowing also significantly decreased staphylinid abundance ($F_{(1,43)}$=21.1, $P<0.001$) although, in contrast to the linyphiids, its effects on this group were no more severe than those of mowing in summer only. In 1989, for example, there was no significant difference in staphylinid abundance between mown and unmown plots in any sampling round under this regime (Figure 1). In 1991, staphylinid abundance on mown plots had recovered sufficiently by September to be indistinguishable from unmown plots.

Mowing in spring and autumn

Mowing in spring and autumn significantly lowered linyphiid abundance ($F_{(1,105)}$=9.9, P=0.003), although, of all the mowing regimes, this generally had the smallest and least persistent effects (Figure 1). However, the low linyphiid abundance after the spring in this, and the spring and summer, mowing regime, in all years, has unfavourable implications for biocontrol (see Discussion).

Staphylinid abundance was also lowered by a spring and autumn mowing regime ($F_{(1,43)}$=7.1, P=0.012).

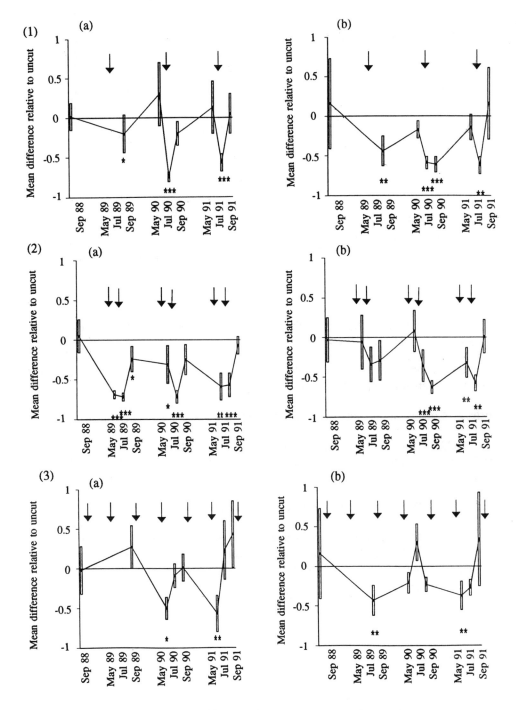

FIGURE 1. Mean proportionate differences between unmown plots and plots mown in (1) summer only (2) spring and summer and (3) spring and autumn for (a) Linyphiidae and (b) Staphylinidae. Arrows indicate time of mowing.

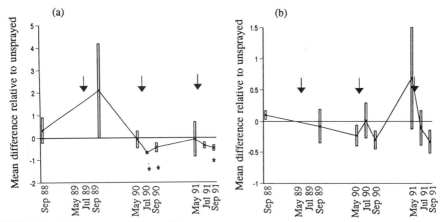

FIGURE 2. Mean proportionate differences between unmown plots and plots sprayed once-annually with glyphosate herbicide for (a) Linyphiidae and (b) Staphylinidae. Arrows indicate time of spraying.

Effects of spraying

Spraying once-annually with glyphosate significantly lowered linyphiid abundance overall ($F_{(1,105)}$=18.1, $P<0.001$; Figure 2).

By contrast to the Linyphiidae, staphylinid abundance was not significantly affected by glyphosate spraying ($F_{(1,43)}$=0.352, P=0.88).

Effects of treatment on abundances of Linyphiids and Staphylinids in the adjacent crop

Field margin treatment had no significant effect on the abundances of linyphiids sampled at 2 m or 10 m into the crop in either May ($F_{(4,20)}$=2.01, P=0.132 and $F_{(4,20)}$=1.60, P=0.213 respectively) or July $F_{(4,8)}$=0.98, P=0.471 and $F_{(4,8)}$=0.80, P=0.556 respectively) in 1989.

In May 1989, fewer staphylinids were associated with the plots cut in spring and summer than with those left uncut, at 2 m into the crop ($F_{(1,20)}$=4.05, P=0.058). The abundance of staphylinids associated with mown plots was also lower at a distance of 10 m into the crop ($F_{(1,15)}$=4.35, P=0.054). In July, there was a significant interaction between mowing and sowing, with sowing increasing abundance in the crop adjacent to mown plots ($F_{(1,6)}$=7.37, P=0.035). Abundance of staphylinids 2 m into the crop was also significantly lower adjacent to spring and summer cut plots than to unmown plots ($F_{(1,8)}$=7.38, P=0.026). There was no main treatment effect on staphylinid abundance at 10 m into the crop ($F_{(4,6)}$=0.94, P=0.499).

DISCUSSION

The ways in which the field margins were managed affected the abundances of both the Linyphiidae and the Staphylinidae. The dominant influences were those which altered the structural diversity of the sward. In general, factors which reduced the structural diversity of the sward also reduced the abundance of both groups. Thus, both were more abundant in the absence of regular annual cutting. In each of the four years after establishment of the experiment, abundance of linyphiids and staphylinids was much higher on field margin plots which were left uncut than on those which were cut. However, there was no evidence for a cumulative deleterious effect of management, although this may not be apparent over our relatively short sampling period.

All of the mowing regimes lowered the abundances of the Linyphiidae and the Staphylinidae. Mowing in mid-summer (whether alone, or in addition to mowing in spring) had larger negative effects on the linyphiids than mowing in spring or autumn. However, although the effects of cutting in spring were short-lived under the spring and autumn regime, they may have had a disproportionate influence on the potential effectiveness of both the linyphiids and the staphylinids for suppressing aphids in the crop. Winder (1990) demonstrated that polyphagous predators were most important during the early stages of aphid population development.

Although unmown swards supported consistently higher numbers of linyphiids and staphylinids than mown swards, leaving swards unmanaged over the longer term decreases their plant species richness, with possible consequences for the invertebrate assemblage. In our experiment, by 1991, plant species richness was significantly lower on unmown pots than on those mown in spring and autumn (Smith *et al.*, 1993). This may eventually reduce the abundance of phytophagous species on which the predators depend, although such effects were not apparent over our sampling period.

Spraying once annually with glyphosate had deleterious effects on the abundance of linyphiids. Spraying affected the physical complexity of the vegetation as dead stems collapsed (Smith *et al.*, 1993), and may also have had indirect effects on the Linyphiidae through its influence on phytophagous prey species which depend on the plant species composition. There was no significant effect of spraying on the abundance of staphylinids, which are less dependent on the structure of the vegetation.

We were unable to detect any effects of sowing on the abundance of linyphiids or staphylinids on the new margin extensions. Differences in sward composition caused by sowing, rather than natural regeneration, such as an increase in plant species richness (Smith *et al.*, 1994), did not appear to be exploited by either of the two groups. The benefits to invertebrates of sown swards are likely to depend on their precise species composition. Sowing can have negative effects on some invertebrates by excluding plant species on which they depend; the common stinging nettle (*Urtica dioica*), for example, is an important host of polyphagous predators (Perrin, 1975) and its abundance was significantly reduced by sowing (Smith *et al.*, 1994). Conversely, plant species with specific characteristics might be included in a seed mixture to encourage other groups. Thus, for example, Thomas *et al.* (1992) showed that grass species which form dense tussocky growth provide a high density of overwintering sites for carabid beetles.

We found little evidence that different field margin management regimes affected the abundance of Staphylinidae or Linyphiidae in the crop adjacent to the margin in summer, although there was a tendency for staphylinids to be more abundant opposite unmown rather than mown plots. However, the length of the plots in relation to the dispersal abilities of the groups (Thomas *et al.*, 1991; Curry, 1994) may have partially masked any treatment effects in the crop. Nonetheless, the magnitude of the differences between treatments provides strong evidence that the ways in which field margins are managed on a farm scale could potentially make a significant contribution to an integrated approach to crop protection.

ACKNOWLEDGEMENTS

This work was funded by English Nature, with additional support from the Ernest Cook Trust, the People's Trust for Endangered Species, the Co-op Bank and the Whitley Animal Protection Trust.

REFERENCES

Chiverton, P.A.; Sotherton, N.W. (1991). The effects on beneficial arthropods of the exclusion of herbicides from cereal crop edges. *Journal of Applied Ecology*, **28**, 1027-1039.

Curry, J.P. (1994) *Grassland Invertebrates*. London: Chapman & Hall.

Dennis, P; Wratten, S.D. (1991) Field manipulation of populations of individual staphylinid species in cereals and their impact on aphid populations. *Ecological Entomology*, **16**, 17-24.

Feber, R.E.; Smith, H. (1995) Butterfly conservation on arable farmland. In: *Ecology and Conservation of Butterflies*, A.S. Pullin (Ed.). London: Chapman & Hall.

Feber, R.E.; Smith, H.; Macdonald, D.W (1994). The effects of the restoration of boundary strip vegetation on the conservation of the meadow brown butterfly (*Maniola jurtina*). In: *Field Margins - Integrating Agriculture and Conservation*, N.D. Boatman (Ed.), *BCPC Monograph No. 58*, Farnham: BCPC Publications, pp. 295-300.

Perrin, R.M. (1975) The role of the perennial stinging nettle, *Urtica dioica*, as a reservoir of beneficial natural enemies. *Annals of Applied Biology*, **81**, 289-297.

Smith, H; Feber, R.E.; Macdonald, D.W (1994) The role of wild flower seed mixtures in field margin restoration. *Field Margins - Integrating Agriculture and Conservation*, N.D. Boatman (Ed.), *BCPC Monograph No. 58*, Farnham: BCPC Publications, pp. 289-294.

Smith, H.; Feber, R.E.; Johnson, P.; McCallum, K.; Plesner Jensen, S.; Younes, M.; Macdonald, D.W. (1993). *The conservation management of arable field margins. English Nature Science No. 18*. Peterborough: English Nature.

Sunderland, K.D.; Fraser, A.M.; Dixon, F.G. (1986). Field and laboratory studies on money spiders (Linyphiidae) as predators of cereal aphids. *Journal of Applied Ecology*, **23**, 433-447.

Thomas, M.B.; Wratten, S.D.; Sotherton, N.W. (1991) Creation of 'island' habitats in farmland to manipulate populations of beneficial arthropods: predator densities and emigration. *Journal of Applied Ecology*, **28**, 906-917.

Thomas, M.B., Wratten, S.D.; Sotherton, N.W. (1992) Creation of 'island' habitats in farmland to manipulate populations of beneficial arthropods: predator densities and species composition. *Journal of Applied Ecology*, **29**, 524-531.

Winder, L. (1990) Predation of the cereal aphid *Sitobion avenae* by polyphagous predators on the ground. *Ecological Entomology*, **15**, 105-110.

FOCUS ON FARMING PRACTICE - AN INTEGRATED APPROACH TO SOLVING CROP PROTECTION PROBLEMS IN CONVENTIONAL AND ORGANIC AGRICULTURE.

A. R. LEAKE

CWS Agriculture, Stoughton, Leicestershire, LE2 2FL, U.K.

ABSTRACT

Five years evaluation of farm scale organic trials have demonstrated the importance of crop rotation, cultural and biological control in suppressing crop antagonists. Such a systems approach appears to provide opportunities for conventional agriculture to use inputs in a more controlled and precise manner.

In order to investigate an integrated approach to crop protection on a commercial basis a farm scale trial was established, using technical and financial support from a crop protection specialist and fertiliser producer, along with expertise and land provided by a major farming company. The 150 acre farm was established in 1993 using a 7 course integrated rotation. The trial will evaluate the economics, technical feasibility and environmental impact of an Integrated Farming System and compare this with conventional and organic practice at the same site.

INTRODUCTION

CWS Agriculture, the farming division of the Co-operative Wholesale Society, farms 20,000 ha of land at 27 locations around Great Britain using agrochemicals and fertilisers characteristic of West European agriculture. In the late 1980's British retailers reported a rapid increase in demand for organic produce leading to predictions that this market could account for 20% of produce by the year 2000. Literature searches revealed that both the technical viability and financial performance of this farming system was poorly documented. In order to ascertain this information as well as to gain practical experience, 105 ha were put into organic conversion in the spring of 1989, achieving full approval to Soil Association standards by 1991 (Anon, 1992). This experiment continues and full results will be reported when the 7 course rotations are completed.

The experimental acreage was increased by a further 60 ha in 1993 to examine the practical management and economic consequences of an Integrated Farming System. The principle requirements of Integrated Farming are defined by the International Organisation for Biological Control (El Titi et al., 1993).

Practical and financial support is provided by fertiliser manufacturer Hydro Agri and crop protection specialists Profarma. This collaboration was considered necessary due to the high technical requirement which integrated farming demands, particularly in relation to the optimal and precise use of fertilisers and crop protection products.

METHODS

The organic and integrated farms were both set up to investigate the systems approach to suppressing crop antagonists. This involves using cultural, biological and mechanical techniques. The organic system is limited to these alone while the integrated system permits targeted intervention with crop protection products and fertilisers. The use of these products is only permitted where yield loss is likely to be significant. Diagnostics, early warning systems and thresholds form an important part of the decision making process with patch spraying and low dose treatment to remove only the damaging portion of the pest population commonly practised. These techniques are compared with conventional farm practice at the same site.

The three farming systems rely on contrasting techniques to achieve a common aim: maximum profitability of the whole rotation whilst sustaining or improving the land for future cropping.

All operations are logged and costed using the Central Association of Agricultural Valuers Guide To Costings (Anon, 1994) adjusted to take account of local conditions. Inputs are costed at purchase price plus application costs. The key aspects of each system are set out as follows:

Crop Rotation

								Area (ha)
Organic (mixed)	Ley	Ley	Ley	Wheat	Oats	Beans	Wheat	20
Organic (stockless)	G/M	G/M	Wheat	Oats	Beans	Wheat	SAS	25
Integrated	Grass/ Clover	Grass/ Clover	Wheat	SAS	Wheat	Beans	Wheat	28
Conventional (integrated rotation)	Grass	Grass	Wheat	SAS	Wheat	Beans	Wheat	28
Conventional (conventional rotation)	OSR	Wheat	Beans	Wheat	SAS	OSR	Wheat	15

(OSR - oilseed rape, SAS - set-aside, G/M - green manure)

Crop Varieties

Organic: Disease resistance and competitive ability (height and ground cover) to suppress weeds are of primary importance. Milling varieties of cereals to achieve organic premiums.

Integrated: High yielding varieties with good disease resistance, standing power and vigour to compete with weeds. Feed varieties only.

Conventional: High yielding varieties which respond to crop protection and fertiliser inputs. Feed varieties only.

Cultivations

Organic: Ploughing essential to destroy weeds and bury weed seeds below the germination zone.

Integrated: Light cultivation post harvest to stimulate volunteer and weed seed germination, and incorporate trash. Crop established using one pass Rau Rotosem, which loosens, cultivates, drills and packs. Ploughing once per rotation if weed seed burden becomes high in the top 10cm soil.

Conventional: Plough 1 year in 3 with heavy discing in the intervening years.

Crop Establishment

Organic: Drilling seldom before first week of November to reduce autumn weed strike and barley yellow dwarf virus (BYDV) infection through aphid colonisation. The lower yielding potential of organic crops means that little loss is likely to be incurred by such a delayed drilling. Higher seed rate to compensate for poorer germination conditions, slug damage and loss of plants through mechanical weeding. Good plant populations required to compete with weeds and compensate for lower tiller numbers due to limited N availability.

Integrated:	Crops following grass/clover: Drilling usually the last week of September to take up mineral N released by sward destruction. Weed competition is potentially lowest at this point in the rotation.
	Other crops: Drilling date targeted at mid to late October with higher seed rates as for the organic system.
Conventional:	Mid-September to mid-October to ensure crop establishment while soil and weather conditions are good.

Crop Protection - Weeds

Organic:	Mechanical only where conditions are suitable, preferably autumn and spring. Larger weeds require 2 passes at right angles to each other. Tap rooted plants are not susceptible. Lower N availability restricts growth of aggressive species, and reduces potential seed shed.
Integrated:	Threshold based systems using weed numbers per m², or crop equivalent indexes while considering the risk of seed with long viability being shed causing problems later in the rotation. Volunteers and weeds germinated by post harvest cultivation are sprayed off with low rate glyphosate. A range of strategies are being examined including use of low dose contact only materials, spring applied to suppress weeds; combined techniques of low dose herbicide in conjunction with mechanical weeding. Products which are selective, of low leachability, volatility, mammalian toxicity and persistence, are preferred. The protection of water quality is cited as a major consideration on this site since the soil is prone to cracking and the fields drain directly into the River Sence.
Conventional:	Post emergence autumn applied residual materials, at reduced rates where appropriate, followed by treatment of specific problems with spring applied contact products.

Crop Protection - Diseases

Organic:
No treatments used but very little disease generally experienced due to less 'soft' growth and more open canopies.

Integrated:
Resistant varieties and precise N use. Diagnostics and early warning systems used and where disease pressure is likely to be high a low dose is applied. Crop monitoring between growth stages 32 and 39 determine the requirement for flag leaf sprays using between half and full rate.

Conventional:
Multi low dose approach using protectant and curative fungicide mixtures. First applications in the autumn and at regular intervals through to flag leaf. Ear wash sprays where crop yield potential is high and disease pressure evident.

Crop Protection - Pests

Organic:
Biological control only. No crops have been lost to pest attack to date. Pest populations appear to build up in large numbers on individual plants, but the crop as a whole remains relatively clean. Slugs have caused crop failure on heavy and wet areas which could not be rolled after sowing. Cereal crops tend to mature earlier than conventional crops and consequently aphid numbers seldom reach yield threatening levels.

Integrated:
Threshold levels observed for aphids, bulb fly, orange blossom midge and slugs. Populations are monitored by counting individuals on plants (aphids), extracting eggs from soil samples (wheat bulb fly) by catching adults using attractive sticky cards (orange blossom midge) or using unbaited activity monitors prior to sowing (slugs).

Conventional:
Treatments applied at first sign of pest attack. Where fungicide or herbicide applications are scheduled low rate insecticide is incorporated in the mixture.

Crop Nutrition

Organic (mixed): Grass/clover leys, composted farmyard manure (FYM) from the suckler herd and leguminous break crops. Green covers during the winter period.

Organic (stockless): Grass/clover leys cut and mulched. Cereal crop prior to set-aside period undersown with grass and clover to achieve maximum N fixation and crop biomass. Leguminous break crops. Permitted products of low solubility e.g. rock phosphate to maintain P and K levels.

Integrated: Nitrogen targeted to expected crop yield using a computer based field specific prediction system called "Extran Plan" (Anon, 1991-92). First application is made in mid March using NPK compound fertiliser followed by a second application mid April of ammonium nitrate. The third and final application is fine tuned using a chlorophyll meter and tissue tests. Mineral N measurements taken post harvest to ascertain accuracy of strategy and mid February to establish soil levels prior to fertilisation. Experiments are underway to measure the amount of mineralisation which occurs after different cultivation techniques to be included in the calculation and atmospheric deposition is being recorded. P + K levels are maintained using the Field Nutrient Balance Scheme which replaces losses through crop off-take.

Conventional: N fertilisation for optimum yield. Standard farm practice based on historical experience. Product choice is based on price, usually Urea for early dressings and applied in 2 splits. Ammonium nitrate for applications after mid April. Rotational P + K fertilisation.

RESULTS AND DISCUSSION

It is inappropriate to make judgements on the economic performance of a systems approach until at least one full rotation is completed. Similarly the condition of soil particularly in reference to fertility, % organic matter content and weed seed bank can only be measured at a fixed point preferably at the beginning and end of each rotation. The organic rotation is currently in its fourth year of full conversion with a further three years cropping to complete the rotation. The integrated and conventional comparison is currently in the second year with a further five years to complete the rotation. At the present time it would appear that under a range of conditions both systems are technically feasible and crops can be grown with limited yield penalties. Lower yields can be tolerated in the organic system since premium prices are obtainable, while the precision use of crop protection inputs and fertilisers in the integrated system should reduce production costs while maintaining yields close to conventional levels. The current relatively high commodity prices favour the conventional system since it increases the cost:benefit ratio of each input. Should the predicted decline in prices occur the integrated system is expected to show an improved margin.

Information is also being sought on a range of environmental factors to ascertain the impact of ·the farming systems as the rotations progress. Measurements are being made of flora numbers and diversity, invertebrates, small mammals, molluscs, earthworm numbers and biomass, bird census and pesticide and nitrate measurements in water courses on the site.

CONCLUSIONS

The results obtained through the Focus on Farming Practice project will provide UK agriculture with commercial and farm scale data concerning economics, technical feasibility and environmental impact of organic, integrated and conventional farming techniques. Such farm scale systems operated by an experienced farming company with technical support from commercial advisors is representative of that which is available to other UK farmers. These systems approaches deal with a range of interactions which occur in agro-ecosystems and attempt, through best farm practice and good science, to minimise the need for · intervention. Where intervention is required the scale is measured.

Information derived through the experiments will be published as results become available.

REFERENCES

Anon, (1992) The Soil Association and Organic Marketing Company Ltd. *Standards for Organic Food and Farming*.

Anon, (1994) The Central Association of Agricultural Valuers. *Guide to Costings 1994*.

Anon, (1991-92) Hydro Agri UK Ltd. *AgTec Bulletin*. pp. 12-13.

El Titi, A; Boller, E.F; Gendrier, J.P. (1993) Integrated Production - Principles and Technical Guidelines. *Bulletin OILB/4ROP Volume 16(1) 1993*.

COMPARISONS OF ENERGY OUTPUT/INPUT OF CONVENTIONAL AND ORGANIC AGRICULTURE IN SCOTLAND AND IN HUNGARY

G. BUJÁKI, P. GUZLI

Department of Crop Protection, Gödöllő Agricultural University, 2100 Gödöllô Hungary

R.G. McKINLAY

Department of Crop Protection, Scottish Agricultural College,West Meins Road, Edinburg EH9 3JG Scotland UK

ABSTRACT

A study was carried out in order to determine the differences in the energy ratio between organic and conventional farming systems in Scotland and Hungary, examining two crops, namely winter wheat and potatoes. Generally in both countries, conventional agriculture seemed to be more energy efficient particularly in the case of potatoes. In Hungary, conventional potato production was 20% more energy efficient than organic production. In Scotland this difference goes up to 100% owing to plant protection problems. On the other hand, Hungarian organic winter wheat is five times more energy efficient than conventional production. In Scotland, conventional wheat production is about 60% more energy efficient than organic producton.

INTRODUCTION

A study was carried out in order to determine the differences in the energy ratio between organic and conventional farming systems in Scotland and in Hungary.

The objective of this study was to find out whether organic farming systems could compete energetically with conventional farming systems looking at two crops, namely, winter wheat and potatoes.

There are various methods to measure energy use efficiency in agriculture. Energy ratio is the most important among them. The ratio deals with the use of energy in organic versus conventional farming systems leaving environmental aspects out of consideration. Despite its limitations the proponents of using the energy ratio technique view this method as the most satisfacotry means of measuring energy utilization efficiency.

(Pimental et.al, 1983, Singh 1993)

MATERIALS AND METHOD S

Wheat and potatoes

For this analysis, the data used came from wheat and potatoes grown under organic and conventional agricultural systems in south-east Scotland. For organic wheat production synthetic fertilisers were replaced with cattle manure, or with a combination of cattle and poultry manure. Data also came from wheat and potatoes grown under biodynamic and conventional agricultural systems in Hungary in the Gödöllő Hills.

ASSUMPTIONS AND NOTES

Soil is energy neutral.

Size of field is 1 ha in all cases.

The use of fuel is only considered on the field.

Contribution of previous crop residue to fertility of current crop   is zero.

In Scotland, organic production systems are more or less similar to biodynamic production systems in Hungary.

Energy content of 1 kg of seed - winter wheat: 18.4 MJ/kg, potato:3.18 MJ/kg

Energy content of 1 kg of synthetic fertilizer,N: 80 MJ/kg, P: 14 MJ/kg, K: 9 MJ/kg

The energy content of manure where calculated with the figures above for synthetic fertilizer.

For organic winter wheat grown in Scotland, energy ratios were calculated for the use of cattle manure and for the use of a combination of poultry and cattle manure.

For cattle and poultry manure, energy contents were calculated for both available and total nutrient contents. (Nutrients are available to the crop in the season of application, assuming no losses between store and soil.

The total energy required to supply one kg of pesticide has been estimated at about $101,3*10^6$ [Stout, 1979].

1 l of pesticide weighs 1 kg.

The energy content of preparation applied in Hungary by biodynamics is negligible.

The energy content of organic materials applied for plant protection is equal to the energy content of pesticides.

The fuel requirement of machines is the same in both countries.

The energy content of 1 l of fuel is:40 MJ/kg [Witney, 1988].

There is no energy content of by-products in this study.

The energy content of environmental pollution has been left out of consideration.

The energy content of micro elements in manure is negligible.

Scottish data are average of 1994 from south-east Scotland; hungarian data are from the area of the Gödöllö Hills during a year.

Conclusions were drawn from yield performance concerning energy efficiency, labour requirements have not been taken into account.

RESUL TS & CONCLUSIONS

Conventional agriculture is more energy efficient in both countries, particulary in the case of potatoes. In Hungary, conventional potatoes are 20% more energy efficient than organically produced potatoes. In Scotland, this difference goes up to 100% owing to plant protection problems. Hungarian organic winter wheat is five times more energy efficient than the conventionally produced wheat. In Scotland conventional wheat production is about 60% more energy efficient than organic wheat production.

Conventional winter wheat in Hungary was about four times less energy efficient than Scottish wheat owing to financial problems in Hungary, where there is a lack of nutrient supplementation in organic farming. Therefore, the inputs decreased and outputs remained the same, which resulted in a higher energy ratio. After a few years, the soil in the organic farming system in Hungary will require more nutrient inputs to obtain the same outputs which wil cause the energy ratio to decline gradually. Conventional potatoes gave the same results in both countries. Taking all factors into account, organic farming systems seem to be more energy efficient in Hungary, while conventional farming systems gave similar results in both countries.

TABLE 1.   Energy ratio in Hungary

	biodynamic	conventional
Potatoes	2.88	3.04
Winter wheat	9.00	1.74

TABLE 2.   Energy ratio in Scotland

	organic	conventional
Potatoes	1.709 (a)	3.25
	3.242 (b)	
Winter wheat	3.041 (a)	5.74
	2.932 (b)	
	5.902 (c)	
	5.504 (d)	
	2.864 (e)	
	2.767 (f)	
	4.832 (g)	
	4.562 (h)	

Notes to table 2.

Energy ratio is calculated,
potatoes (a): for the total nutrient contents.
        (b): for the available nutrient contents.

winter wheat (a): for the total nutrient contents. There is no plant protection. Only cattle manure applied.
        (b): for the total nutrient contents. There is a plant protection with Kumulus D.F.
        (c): for the available nutrient contents. There is no plant protection. Only cattle anure applied.
        (d): for the available nutrient contents. There is a plant protection with Kumulus D.F. Only cattle manure applied.
        (e): for the total nutrient contents. There is no plant protection. Combined cattle manure and poultry manure.
        (f): for the total nutrient contents. There is a plant protection . Combined cattle manure and poultry manure.
        (g): for the available nutrient contents. There is no plant protection. Cattle and poultry manure applied.
        (h): for available nutrient contents. There a is plant protection with Kumulus D.F. Applied cattle and poultry manure.

REFERENCES

Pimental, D.; Beradi, G. (1983) Energy efficiency of farming systems: organic and conventional agriculture. Amsterdam.

Singh, A. (1993) Measurement of energy use in agriculture. *Punjab Agricultural University, Punjab.*

Stout, B.A. (1979) Energy for world agriculture. Rome.

Witney, B. (1988) Choosing and using farm machines. Singapore.

EFFECT OF DENATONIUM BENZOATE ON *MYZUS PERSICAE* FEEDING ON CHINESE CABBAGE, *BRASSICA CAMPESTRIS SSP. PEKINENSIS*

M.T.M.D.R. PERERA, G. ARMSTRONG

SAC, Aberdeen, 581 King Street, Aberdeen AB9 1UD, UK

R.E.L. NAYLOR

Department of Agriculture, University of Aberdeen, 581 King Street, Aberdeen AB9 1UD, UK

ABSTRACT

Denatonium benzoate which is known to act as an antifeedant against mammals was shown to have a similar effect on *Myzus persicae*, on Chinese cabbage plants in the laboratory. Aphids on plants sprayed with distilled water lived longer and produced more nymphs per plant than aphids on plants sprayed with denatonium benzoate at concentrations varying from 0.5 to 250 ppm. Honeydew production was less on leaves and leaf discs treated with denatonium benzoate than on untreated leaves, and leaf discs. A choice test showed that the aphids tended to move on to untreated halves of leaf discs.

INTRODUCTION

Denatonium benzoate, marketed as bitrex™ (by Macfarlan Smith Ltd, Edinburgh) is an inert odourless synthetic chemical with an extremely bitter taste. It is commonly used as an additive to prevent people ingesting or chewing a wide variety of hazardous chemical products such as detergents, disinfectants, and pesticides. Denatonium benzoate has been shown to act as an antifeedant against rabbits (Menu, 1993), and it is sold for use against a variety of mammals including horses, deer, and rats. However, the effect of denatonium benzoate on insects appears to have received little attention, although there is currently a great deal of interest in other compounds such as azadirachtin, partly because of their effectiveness as antifeedants against commercially important insect pest species (Saxena, 1989). *Myzus persicae* (Homoptera : Aphididae), the peach potato aphid, is an important virus vector, and has an ability to develop resistance rapidly to successive groups of insecticides (Wege, 1994). Therefore the use of antifeedants may be a substitute for conventional insecticides. *M. persicae* is also a good subject for research, as it can be reared easily in the laboratory.

The aim of this study was to determine whether denatonium benzoate has an antifeedant effect against *M. persicae* on Chinese cabbage, *Brassica campestris ssp. pekinensis*. Experiments were carried out to determine the effect of denatonium benzoate on the survival of aphids and their production of nymphs. Honeydew production was monitored as an index of feeding rate, and a choice experiment was undertaken to determine whether aphids preferred to feed on untreated leaves than on leaves treated with denatonium benzoate.

MATERIALS AND METHODS

To determine whether denatonium benzoate affected the survival of *M. persicae* and number of nymphs produced, three-week old potted Chinese cabbage plants (cv. Nepos ez F1) were sprayed with distilled water or denatonium benzoate solution (0.5, 5, 50, 100, and 250 parts per million (ppm)). The wetting agent agral™ was added to the sprays at a rate of 0.1 ml/litre. Five plants were used for each treatment. Ten one-day old *M. persicae* nymphs were placed on each plant, and the number surviving was recorded each day until all of the original aphids were dead. Once the aphids began to produce nymphs, the number of nymphs produced was also recorded each day, and the nymphs were removed after counting.

Honeydew droplets excreted by aphids feeding on leaves and leaf discs were detected using ashless 125 mm diameter filter papers stained with bromocresol green indicator solution (Banks & Macaulay, 1964). For the experiment with leaves, five mature Chinese cabbage leaves were dipped in distilled water with agral or 250 ppm denatonium benzoate solution with agral, and three one-day old aphid nymphs were placed on the underside of each leaf. The leaves were placed in water in glass vials, so that they projected over a stained filter paper. For the experiment using leaf discs, five 20 mm diameter Chinese cabbage leaf discs were dipped in distilled water with agral or 0.5, 5, 50, 100 and 250 ppm denatonium benzoate solutions with agral and placed on hydrogel (Erin™ waterwell) in 35 mm diameter petri dishes. Three one-day old nymphs were introduced on to each leaf disc, then the petri dishes containing the leaf discs were suspended upside down over stained filter paper. The number of honeydew spots on each filter paper and the number of nymphs remaining on each leaf or leaf disc were counted after 24 hours.

For the choice experiment 35 mm diameter leaf discs were taken, with the mid rib bisecting the disc. One half of each disc was dipped in distilled water with agral and the other in denatonium benzoate (0.5, 5, 50, 100, and 250 ppm) with agral, five discs were used for each concentration. The discs were placed in 35 mm diameter petri dishes containing hydrogel. Ten one-day old peach potato aphids were introduced on the mid rib and the number of nymphs on each half of the leaf disc were counted after 24 hours.

RESULTS

The aphids on the plants treated with denatonium benzoate survived to produce nymphs, and it was possible to relate the lifespan of the original aphids, and the number of nymphs they produced, to the denatonium benzoate concentration. The aphids on plants treated with distilled water had the highest average lifespan, and this decreased as the denatonium benzoate concentration increased. The average lifespan of the aphids was significantly longer on plants sprayed with distilled water and the two lowest denatonium benzoate concentrations (0.5 and 5 ppm) than on plants sprayed with denatonium benzoate at 100 ppm and 250 ppm (Table 1). Total numbers of nymphs produced on plants followed a similar trend, in that the total number of nymphs produced per plant was significantly higher on plants sprayed with distilled water than on plants sprayed with all of the denatonium benzoate concentrations (Table 1). Significantly more nymphs were also produced on plants sprayed with the lowest concentrations of denatonium benzoate (0.5 and 5 ppm) than on plants sprayed with denatonium benzoate at 50, 100 and 250 ppm, and the total number of nymphs produced on plants treated with 50 ppm denatonium benzoate was significantly higher than on plants treated with denatonium benzoate at 100 and 250 ppm.

TABLE 1.   Average lifespan (days) of *Myzus persicae* and numbers of nymphs produced on denatonium benzoate treated Chinese cabbage plants. Values in the same row followed by the same letter are not significantly different (p>0.05). SED - Standard Error of the Difference

	Denatonium benzoate concentration (ppm)						SED
	0	0.5	5	50	100	250	
Lifespan	15.4 a	15.2 a	15.0 a	13.9 ab	10.8 b	10.4 b	1.6
Nymphs	303.6 a	253.8 b	231.2 b	186.4 c	124.8 d	123.8 d	12.9

The aphids on the whole leaves treated with distilled water produced a significantly (p<0.05) higher number of honeydew spots (8.6 ± 1.3 SEM) than the aphids on the leaves treated with 250 ppm denatonium benzoate solution (2.3 ± 1.1 SEM).   The aphids on untreated leaf discs also produced more honeydew spots than aphids on leaf discs treated with denatonium benzoate, but there   was no significant difference between any of the treatments (Table 2, p>0.05).

TABLE 2.   Average number of honeydew spots produced by *Myzus persicae* on Chinese cabbage leaf discs during two 24 h periods.   SED - Standard Error of the Difference

Day	Denatonium benzoate concentration (ppm)						SED
	0	0.5	5	50	100	250	
1	23.0	17.1	15.3	20.1	13.6	15.2	4.35
2	15.9	9.8	10.3	12.8	9.4	12.4	2.85

TABLE 3.   Total number of aphids on halves of Chinese cabbage leaf discs treated with denatonium benzoate or distilled water, after 24 hours.

Denatonium benzoate concentration (ppm)	Total Number of aphids on leaf halves		p (χ^2)
	D. benzoate treated	Distilled water treated	
0	33	27	NS
0.5	27	23	NS
5.0	17	32	<0.05
50.0	9	19	NS
100.0	10	22	<0.05
250.0	10	7	NS

In the choice experiment fewer aphids were found on the 5, 50, and 100 ppm denatonium benzoate treated halves of leaf discs than on the halves treated with distilled water, after 24 hours (Table 3). The differences were significant for the 5 and 100 ppm treatments. The aphids were able to leave the petri dishes during this experiment, and after 24 hours, less than the original ten aphids per dish remained in all of the denatonium benzoate treatments except 0.5 ppm. The greatest reduction in numbers of aphids was in the 250 ppm denatonium benzoate treatments, where only 17 out of the original total of 50 remained (although slightly more of those which remained were on the denatonium benzoate treated leaf halves). Some of the aphids moved into the control dishes.

CONCLUSION

The results of the experiment using sprayed plants suggest that treatment with denatonium benzoate would significantly reduce the rate of increase of aphid infestations, and on the basis of the measurements of honeydew production, the effect of denatonium benzoate appears to be a reduction of sap intake. The results of the choice experiment also suggest an antifeedant effect. However, the fact that the aphids would feed and produce nymphs on plants treated with denatonium benzoate indicates that it would probably not prevent them from colonising crops, and therefore would not prevent the spread of viruses. Denatonium benzoate can be taken up systemically by plants (Ondruskova *et al.,* 1992; Menu, 1993), and this may explain its apparently long-lasting effect on the aphids feeding on the sprayed plants in this study. Because of the bitter taste, denatonium benzoate could not be used on edible parts of crop plants, but if its effect was systemic, this could probably not be avoided. However, these results do suggest that if denatonium benzoate was used against mammals, for example to protect ornamental plants, it would also give some protection against insect pests.

ACKNOWLEDGEMENTS

The authors thank the Department of Agriculture, Sri Lanka for the award of a studentship through the Agricultural Research Project, Sri Lanka.

REFERENCES

Banks, C.J.; Macaulay, E.D.M. (1964) The feeding, growth and reproduction of *Aphis fabae* Scop. on *Vicia faba* under experimental conditions. *Annals of Applied Biology,* **53,** 229-242.

Menu, F. (1993) Investigation of the potential of denatonium benzoate as a spray applied rabbit repellent on grass. MSc. Thesis, University of Aberdeen.

Ondruskova, L; Spurny J; Vaclavik J. (1992) Use of bitrex (methazole and pyridate) in ornamental bulb plants. *Acta Horticulture,* **325,** 811-813.

Saxena, R.C. (1989) Insecticides from neem. In: *Insecticides of plant origin,* J.T Arnason, B.J.R Philogène,. & P. Morand (Eds), ACS symposium series 387, 110-135.

Wege, P.J. (1994) Challenges in producing resistance management strategies for *Myzus persicae, Proceedings 1994 Brighton Crop Protection Conference,* **1,** 419-426.

NEEM TISSUE CULTURE AND THE PRODUCTION OF INSECT ANTIFEEDANT AND GROWTH REGULATORY COMPOUNDS.

A.J. MORDUE (LUNTZ)

Department of Zoology, University of Aberdeen, Tillydrone Avenue, Aberdeen AB9 2TN

A. ZOUNOS, I. R. WICKRAMANANDA

Departments of Agriculture and Zoology, University of Aberdeen, AB9

E.J. ALLAN

Department of Agriculture, University of Aberdeen, 581 King Street, Aberdeen AB9 1UD

ABSTRACT

Neem tissue culture methodologies have been established for both callus and cell suspension cultures. Extraction and detection methodologies for plant metabolites with biological activity have been developed and refined. The presence of antifeedant compounds has been measured using a choice bioassay for the desert locust (*Schistocerca gregaria*), with 3-4 day old fifth instar nymphs and with azadirachtin as standard. A bioassay for insect growth regulatory effects has been established using one day old fifth instar nymphs of the milkweed bug (*Oncopeltus fasciatus*), after the method of Isman *et al.,* 1990. The antifeedant activity of a callus line, from Niger, has shown to be similar to that of another callus line derived from a different geographical location (Kearney *et al.*, 1994). Cell suspension cultures from a Ghanaian line were shown to contain antifeedant compounds in both the cells and the medium and that these increased with growth.

INTRODUCTION

The Neem tree (*Azadirachta indica*) is native to the arid regions of the Indian subcontinent (Schmutterer, 1990) and has been introduced to tropical and subtropical areas of Africa, America and Australia (National Research Council, 1992). The tree has been recognised as a multiple use forest crop with important roles in afforestation, the provision of fuelwood, pesticides, oils, tannins, timber and medicinal products (Pliske, 1983). In particular, neem leaves and seeds have been exploited traditionally for insect control and the major biologically active component, azadirachtin, has been identified from seed kernel extracts (Butterworth & Morgan, 1968). This highly oxidised triterpenoid is now well known for its insect antifeedant, growth regulatory and sterilant effects (Mordue(Luntz) & Blackwell, 1994). Neem formulations have been shown to be nontoxic to man and animals (Duke & duCellier, 1993) and consequently, their potential for controlling agricultural pests in an environmentally sound manner, has recently gained impetus. This is associated with commercial exploitation, with neem formulations being patented by Western companies (Pearce, 1993) and registered by the Environmental Protection Agency for both food and non-food crops. It is estimated that natural neem production will not meet increasing demands (Saxena, 1989) and although it is a fast growing tree, with fruit production occurring within 3-4 yrs, full productivity is not obtained until *c*. 10 years. In addition, azadirachtin itself is too complex to produce synthetically on a commercial basis and its synthesis *in vivo* is not fully understood.

Plant cell and tissue culture may provide a means of overcoming the constraints of traditional plant breeding and in the case of neem, offers an easier method for elucidating the complex synthetic pathway of azadirachtin. In addition it may also provide an alternative means for consistent and reliable production with the added potential of increasing yields in a particular cell line. Initial studies have shown that neem tissue cultures produce insect antifeedant compounds (Kearney *et al.*, 1994) and azadirachtin (Allan *et al.*, 1994). This paper will further develop the methodologies involved and will evaluate neem cell cultures, derived from trees from different geographical locations, for insect biological activity.

MATERIALS AND METHODS

Neem cell culture

Callus was derived from leaves of young trees (< 1 year old), which originated from Ghana and Niger, using the method of Kearney *et al.*, 1994. Suspension cultures were initiated by transferring friable callus (Ghana line, GH1) into 50 ml Maintenance Medium (Kearney *et al.*, 1994) in 250 ml flasks. These were incubated in the dark at 25 °C on an orbital shaker at 100 rev/min. In order to prevent the formation of large cell aggregates, subculture was achieved by discarding cells retained on a 2.83 mm Nubold mesh (Stanier & Co., UK) and subculturing the remaining cells, on a fresh weight basis, after filtering through Miracloth (Calbiochem Co., USA). Repeated subculture in this manner resulted in a more homogenous suspension. Ten flasks were thus subcultured using 4 g fresh weight material, four of these were used for monitoring growth by taking fresh and dry weight measurements (Allan, 1991) at regular intervals. Both the cells and growth medium were harvested from 3 flasks for extraction and subsequent insect bioassay at 7 and 10 d. Cells were filtered through Miracloth (Calbiochem Co., USA), immediately frozen, freeze dried and stored at - 20 °C (Electrolux, UK) until extraction. The resulting medium was also frozen prior to extraction. Growth kinetics were analysed using the method of Allan (1991) using the Minitab statistics package.

Extraction of cell cultures

Cell extraction methods

In order to compare extraction methods, ten week old freeze dried callus, line GH1, which had been subcultured 15 times was used. Triplicate 2 g samples were used to compare two different extraction methods viz. an ethanolic extraction (Kearney *et al.*, 1994, adapted from Govindachari *et al.*, 1990) and a water extraction (D.E. Morgan pers. comm.). Thus, 2 x 50 ml ethanol was used to extract the cells under reflux for 2 x 30 min. The resultant suspension was filtered through Whatman No. 1 paper prior to rotary evaporation and resuspended in 2 ml ethanol (Laboratory Grade). For the water extract, 2 g material was homogenised (MSE homogeniser) for 1 min in 50 ml distilled water. The resultant meal-water mixture was centrifuged at 5500 rev/min for 10 min. The meal was then homogenised and centrifuged a further 4 times using 50 ml fresh water each time. All the water extracts were then combined and an aliquot added to an equal volume of methanol. This extract was filtered through Floricil (Sigma, UK), if required, to remove any precipitate, and stored at -20 °C (Electrolux, UK). Prior to bioassay, extracts were rotary evaporated and taken-up in ethanol to 20 mg dry weight/ml. All other cell culture extractions (both callus and cells from suspension cultures) were undertaken by the water extraction method, using 0.5 g freeze dried material.

The medium was thawed prior to extraction and filtered (Whatman No. 1) to remove any precipitate and the resultant volume measured. One part medium was mixed with 0.25 parts dichloromethane in a separation flask and shaken for approximately 4 x 15 sec. After separation the lower (dichloromethane) phase was removed. This was repeated 3 times and all the dichloromethane fractions were combined. Any water was removed using approx. 10 g anhydrous magnesium sulphate which was discarded by filtration. Extracts were rotary evaporated at below 35 °C.

Bioassays

Antifeedancy assay

A choice feeding bioassay (Nasiruddin, 1993; Blaney *et al.*, 1990) was used to assay the antifeedant potency of cell culture extracts using 3-5 day old male and female fifth instar nymphs of the desert locust, *Schistocerca gregaria*. Prior to bioassay the insects were kept separately and their state of hunger was standardised. Individual insects were transferred into containers and given access to two glass fibre discs of 3.7 cm diameter (Whatman International Ltd., UK), a control and a test disc. Both discs were pre-treated with 350 µl 50 mM sucrose and allowed to oven dry at 37 °C overnight. A further 350 µl aliquot of appropriate extract or ethanol was added to the disc which was dried again. All discs were weighed before and after the experiment and results recorded as mg disc eaten. The percentage of antifeedance was calculated following the formula:

$$\frac{\text{Weight of Control disc eaten - Weight of Test disc eaten}}{\text{Weight of Control disc eaten + Weight of Test disc eaten}} \times 100$$

Results were analysed using oneway analysis of variance and Tukey's pairwise test (Minitab Statistic Package). All percentages were arcsine transformed and negative percentages indicating no antifeedant activity were given a value of 0.

Insect growth regulatory (IGR) assay

Nymphs of the large milkweed bug (*Oncopeltus fasciatus*) were obtained from a laboratory colony, maintained at 27 °C and 12:12 LD and reared on de-husked sunflower seeds. The IGR activity of azadirachtin and tissue culture extracts was assessed following topical application to newly moulted (within 24 h) fifth instar nymphs using the methods of

Isman *et al.*, (1990). Test solutions were applied to the abdominal dorsum in 1 μl acetone with a 10 μl Hamilton Syringe. Controls were treated with carrier alone. Bugs were kept in 10 cm diameter Petri dishes supplied with sunflower seeds and cotton wool soaked with water. The insects were monitored daily and assessed on day 10 by which time all the control insects had moulted (day 7). IGR activity was defined as any deviation from moulting to a morphologically normal adult, i.e. from deformity of the wings and/or legs, to death during moulting, and to a delay of the moult by 3 days or more. IGR results were corrected for control effects by Abbott's formula (Busvine, 1971).

RESULTS

Neem cell culture

The method of callus initiation and maintenance (Kearney *et al.*, 1994) previously shown to be successful for material of Ghanaian origin also proved successful for that derived from Niger with pale, friable, undifferentiated callus being obtained. All the callus lines used in this study had been in culture for at least 2 years. Growth of GH1 suspension culture (Fig. 1) showed an initial lag phase of 5 d followed by an exponential growth phase of 16 d at which time a maximum biomass of 25.1 g l^{-1} was obtained. On the basis of the dry weight, the specific growth rate of this line was 0.15 mg d $^{-1}$ with a doubling time of 110 h (4.6 d).

Extraction

Comparison of antifeedancy for callus extracted using two different methods showed that the water extraction resulted in a significantly higher recovery of antifeedant compounds than did the ethanol extraction procedure (Table 1).

TABLE 1. Antifeedancy of *Azadirachta indica* callus, line GH1 (10 mg callus dry weight/ml extract) using ethanol and water extraction methods against fifth instars of *Schistocerca gregaria*, in a choice feeding bioassay (n = 8-10/ replicate). Samples are significantly different (P < 0.05).

Type of Extract	Number of Replicates	Percentage Antifeedancy (Mean ± SE)
Ethanol	3	46.97 ± 5.12
Water	3	84.19 ± 2.80

Bioassays

A standard curve for % probit antifeedancy to azadirachtin was set up for the choice bioassay which gave an ED_{50} of 0.00035 ppm azadirachtin. Analysis of antifeedancy of 6

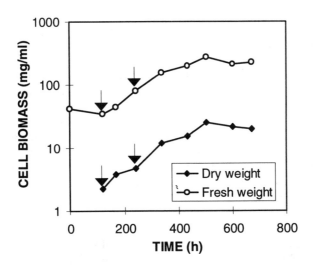

Fig. 1 Growth as measured by fresh and dry weight, of a Ghanaian *Azadirachta indica* cell suspension culture. Arrows indicate the 7 and 10 d times when cultures were harvested for insect bioassay.

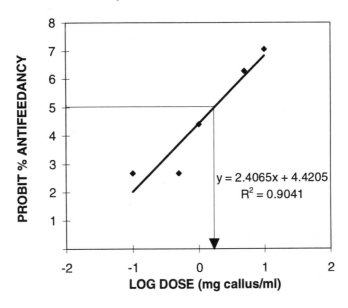

$$y = 2.4065x + 4.4205$$
$$R^2 = 0.9041$$

Fig. 2 Probit % antifeedancy of *S. gregaria* fifth instar nymphs presented with glass fibre discs with Niger callus extract and sucrose (50 mM) or sucrose alone in a choice bioassay, n=8-15. ED_{50}=1.74 mg callus/ml.

week old callus, line Niger 5A, showed a good dose response relationship with an ED_{50} of 1.74 mg dry weight callus/ml (Fig. 2).

Analysis of antifeedancy (AF) was also undertaken on the suspension culture during growth (Fig. 1). Suspension culture cells developed significant antifeedancy from day 7 to day 10 of culturing. Antifeedancy had increased significantly from day 7 to day 10 (26.7 ± 3.1 % and 55.1 ± 2.7 % AF, n = 19-21; P < 0.05) of culturing for 0.04 ppm dry weight of cell extract. A more detailed analysis of the cell suspension medium revealed that antifeedant products had been secreted into the medium and they too significantly increased from day 7 to day 10 of culturing (Table 2). Thus, whereas the cell biomass increased 38 %, on a dry weight basis, in that 3 day period of exponential growth, the antifeedant compounds released into the culture medium increased 20 fold.

TABLE 2. Antifeedancy of medium extracts of a cell suspension culture against fifth instars of *Schistocerca gregaria*, in a choice feeding bioassay (n = 8-10/ replicate). Medium was harvested at day 7 and day 10 of culture. (a-b, P < 0.05).

Sample	Dilution (ml medium/ml extract	Number of replicates	Percentage Antifeedancy (Mean ± SE)
Day 7	5.00	3	41.67 ± 3.73[a]
Day 10	5.00	3	100[b]
Day 10	0.50	3	91.92 ± 0.96[b]
Day 10	0.25	3	57.29 ± 6.84[a]

A standard curve for % probit IGR effects to azadirachtin was set-up for *O. fasciatus* nymphs. Abnormal and delayed moults occurred increasingly until there were no normal moults after a dose of 0.015 µg/insect. An ED_{50} of 0.0012 µg azadirachtin/insect was achieved.

DISCUSSION

Biomass production in the suspension culture used in this study, was much higher than that of Kearney *et al.*, (1994) with a max. biomass of 25.1 gl^{-1} compared to 14.7 gl.$^{-1}$ This presumably reflects the method used for routine subculture prior to the experiment which resulted in a more homogeneous culture. Both the max. biomass and the growth rate (as reflected by a doubling time, t_d, of 4.6 d), were also greater than has been obtained for other semi-tropical trees e.g. *Cinchona ledgeriana* max. biomass = 10 gl, $^{-1}$ t_d = 6.9 d (Scragg *et al.*, 1988) ; *Quassia amara* max. biomass = 6.2 gl^{-1} ; t_d = 5.1 d (Scragg *et al.*, 1990).

The extraction procedure of Kearney *et al.*, 1994 used an ethanolic reflux system which would expose azadirachtin to heat instability and therefore an alternative extraction procedure was considered. Water extraction of freeze dried callus produced twice as much product, in terms of antifeedancy, and hence has been adopted as the better protocol for routine use. Tissue cultures however contain less storage material compared with seeds where the water soluble components may interfere with efficient extraction of the azadirachtin products. In this instance, ethanol extraction may be preferable.

By using the water extraction methods and constructing a dose-response curve for callus cultures of Niger 5A, an accurate ED_{50} value for antifeedancy was obtained which could be equated to the azadirachtin standard curve. The yield of antifeedant compounds, using ED_{95} values, expressed as "azadirachtin equivalents" was 0.0001 % based on dry weight of callus and 0.0003% for suspension culture medium based on dry weight cells secreting into the medium. The ED_{95} values are similar to that of Allan et al., 1994 who found a yield of 0.0007 % using 100% AF and direct measurement of azadirachtin.

Antifeedancy tests of both cells and medium from cell suspension cultures revealed that antifeedant compounds are produced during growth and that substantial proportions are released into the medium. There is a marked difference in the growth kinetics found here compared to Kearney et al., 1994 who used a similar cell line. This difference may reflect the greatly reduced cell aggregation in the culture used in this experiment and more detailed analysis following production of antifeedant products throughout growth would be desirable. In terms of commercial production, the fact that antifeedant compounds are secreted is advantageous as it will allow alternative systems, such as immobilisation to be undertaken. Van der Esch et al., (1994) also showed that azadirachtin, as detected by HPLC, was present in both the cells and the medium of neem suspension cultures.

Our current work involves screening various neem cell lines for higher yields and the development of bioassays to measure both insect antifeedancy and growth disruption. These should yield results which will establish techniques to determine fundamental aspects concerning production of these compounds in planta in addition to evaluating the potential of plant tissue culture for commercial exploitation.

ACKNOWLEDGEMENTS

The authors would like to thank the following: Professor David E. Morgan for his advice on extraction methods; Faye Cameron and Brenda King for their help with the bioassays and Dr. E. Boa for the provision of seeds from Niger. We acknowledge the European Commission for funding for A. Zounos and the Agricultural Research Project of Sri Lanka for a studentship to I. R. Wickramananda.

REFERENCES

Allan, E.J. (1991) Plant cell culture. In: Plant Cell and Tissue Culture, A. Stafford and
    G. Warren (Eds), Milton Keynes: Open University Press, pp. 1-24.
Allan, E.J.; Eeswara, J.P.; Johnson, S.; Mordue(Luntz), A.J.; Morgan, E.D.; Stuchbury, T.
    (1994) The production of azadirachtin by in vitro tissue culture extracts of neem,

Azadirachta indica. Pesticide Science, **42**, 147-152.

Blaney, W.M.; Simmonds, M.S.J.; Ley, S.V.; Anderson, J.C.; Toogood, P.L. (1990) Antifeedant effects of azadirachtin and structurally related compounds on lepidopterous larvae. *Entomologia Experimentalis et Applicata*, **55**, 149-160.

Busvine, J.R. (1971) In: *A Critical Review of the Techniques for Testing Insecticides.* Slough, England : Commonwealth Agricultural Bureaux, pp. 345.

Butterworth, J.H.; Morgan, E.D. (1968) Isolation of a substance that supresses feeding in locusts, *Journal of the Chemical Society, Chemical Communication*, **1**, 23-24.

Duke, J.A.; duCellier, J.L. (1993) *Azadirachta indica* A. Juss. (Meliaceae) - Neem tree. In: *CRC Handbook of Alternative Cash Crops*, Boca Raton, Ann Arbor, London, Tokyo: CRC Press, pp. 62-66.

Govindachari TR.; Sandhya, G.; Ganeshraj, S.P. (1990) Simple method for the isolation of azadirachtin by preparative HPLC. *Journal of Chromatography*, **513**, 389-391.

Isman, M.B.; Koul, O.; Laczynski, A.; Kaminski, J. (1990) Insecticidal and antifeedant bioactivities of neem oils and their relationship to azadirachtin content. *Journal of Agricultural and Food Chemistry*, **38**, 1406-1411.

Kearney, M-L.; Allan, E.J.; Hooker, J. E.; Mordue(Luntz), A.J. (1994) Antifeedant effects of *in vitro* culture extracts of the neem tree, *Azadirachta indica* against the desert locust (*Schistocerca gregaria* Forskal). *Plant Cell Tissue and Organ Culture*, **37**, 67-71.

Mordue(Luntz), A.J.; Blackwell, A. (1993) Azadirachtin: An Update. *Journal of Insect Physiology*, **39**, 903-924.

Murashige, T.; Skoog, F. (1962) A revised medium for rapid growth and bioassays with tobacco cultures. *Physiologia Planta*, **15**, 473-479.

Nasiruddin, M. (1993) The effects of azadirachtin and analogues upon feeding and development in locusts. Ph.D. thesis. University of Aberdeen.

National Research Council (1992) *Neem: A tree for solving global problems.* Washington DC: National Academy Press.

Pearce, F. (1993) Pesticide patent angers Indian farmers. *New Scientist*, **1894**, p. 7.

Pliske, T.E. (1983) The establishment of neem plantations in the American tropics. *Proceedings of the 2nd International Neem Conference*, Rauischholzhausen pp. 521-526.

Saxena, R.C. (1989) Insecticides from neem. In: *Insecticides of Plant Origin*, J.T. Arnason, B.J.R. Philogene and P. Morand, P. (Eds), Washington : American Chemical Society, pp.110-135.

Schmutterer, H. (1990) Properties and potential of natural pesticides from the neem tree, *Azadirachta indica. Annual Review of Entomology*, **35**, 271-297.

Scragg, A.H.; Allan, E.J.; Morris, P. (1988) Investigation into the problems of initiation and maintenance of *Cinchona ledgeriana* suspension cultures. *Journal of Plant Physiology*, **132**, 184-189.

Scragg, A.H.; Ashton, S.; Steward, R.D; Allan, E.J. (1990) Growth and quassin accumulation by cultures of *Quassia amara. Plant Cell Tissue and Organ Culture*, **23**, 165-169.

Van der Esch. S.A.; Giagnacovo, G.; Lorenzetti, C.; Maccioni, O.; Vitali, F. (1994) Plant tisssue culture of *Azadirachta indica* (A. Juss) for the production of azadirachtin and related compounds. *Abstracts VIIIth International Congress of Plant Tissue and Cell Culture*, Firenze, June 12-17. p.240.

CEREAL-CLOVER BICROPPING, COULD IT AFFECT OUR FUNGICIDE DEPENDENCY?

M.J.SOLEIMANI, M.L. DEADMAN

Department of Agriculture, University of Reading, PO Box 236, Earley Gate, Reading, RG6 2AT

R.O.CLEMENTS

Institute of Grassland and Environmental Research, North Wyke Research Station, Okehampton, Devon, EX20 2SB

D.A.KENDALL

Department of Agricultural Sciences, University of Bristol, Institute of Arable Crops Research, Long Ashton Research Station, Bristol BS18 9AF

ABSTRACT

Field experiments were conducted during the 1993/94 and 1994/95 growing seasons to assess the impact of cereal-clover bicrops on the levels of wheat diseases in the growing crop. Results indicated that the severity of *Septoria* at growth stage 45 was reduced in bicrops relative to the disease level in monocrops, although this effect was not evident either earlier or later in the season. The inoculum level of both *Pseudocercosporella herpotrichoides* and pathogenic *Fusarium* was higher within the bicropping system where more fungal spores were produced on stubble under the clover canopy than on stubble within a direct drilled wheat monocrop. The number of colony forming units of *P. herpotrichoides* in soil was higher within bicropped plots relative to soil under direct drilled monocrops. The incidence of *Fusarium* was significantly lower in bicrops than in monocrops especially at conventional input levels. The incidence of eyespot showed no significant differences between bicrops and monocrops sown following ploughing although both treatments had a higher disease level than the direct drilled monocrops.

INTRODUCTION

Following implementation of major changes resulting from EC reforms of the Common Agricultural Policy, methods of crop production which use lower inputs require evaluation for their potential in reducing fertiliser and other agrochemical applications. A cereal bicropping system which utilises white clover (*Trifolium repens*) as a permanant understorey, has previously been shown to have considerable potential for growing winter cereals with greatly reduced inputs of agrochemicals (Jones & Clements, 1993).

The benefits of nitrogen released from white clover and made available to other crops either through a rotation system or through the grazing or cutting of crops have long been recognised (Cowling, 1982). However within a conventional arable system legume residues must first decompose and be mineralised before the nitrogen can become available (Ladd & Amato, 1986). Recent results however have shown that where a winter wheat crop is drilled directly into an established clover sward the response of successive cereal crops to nitrogen applications is diminished, implying a build-up of available soil nitrogen (Jones & Clements, 1993).

An additional benefit of the bicropping system is that clover survives through successive cereal crops and can be grazed before being redrilled with winter wheat (Jones & Clements, 1993). Furthermore, the bicropping system has been shown to reduce the level of attack by some important major cereal pests, such as aphids and slugs (Jones & Clements, 1991). However, little is known of the effects of cereal-clover bicropping on the incidence of the major cereal diseases, especially those caused by splash dispersed pathogens where an altered canopy structure might have a significant effect on spore dispersal and epidemic development through an altered microenvironment. This study aimed to evaluate the impact of cereal-clover bicropping, at conventional and reduced levels of agrochemical inputs, on the incidence and severity of *Septoria*, pathogenic *Fusarium* species and eyespot (*Pseudocercosporella herpotrichoides*).

MATERIALS AND METHODS

In an experiment at the Institute of Arable Crops Research, Long Ashton, a fine, firm seed-bed was prepared incorporating 75 kg/ha of P and K fertiliser respectively, but no N. White clover seed was sown at 10 kg seed/ha on 10 June 1993. The clover cv. Donna was established over most of the field, but the cv. Milkanova was established for areas destined for treatments 3 and 4. The developing swards of clover were cut for silage in August and early October. In late October of 1993 and 1994, plots (13 x 60 m) were marked out for each of the treatments (Table 1) in three replicate blocks. For treatments 7 and 8 plots were ploughed and sown with winter wheat cv. Hereward. Plots for other treatments were left unploughed and winter wheat cv. Hereward was direct drilled into the clover understorey (treatments 1-4) or directly into the soil, using a Hunter Rotaseeder, following chemical removal of the clover.

TABLE 1. Summary of treatments used to evaluate the effect of cereal-clover bicropping on disease severity on winter wheat cv. Hereward

Treatment	Clover cv.	Ploughed or not	Input level
1	Donna	No	Conventional
2	Donna	No	Reduced
3	Milkanova	No	Conventional
4	Milkanova	No	Reduced
5	None	No	Conventional
6	None	No	Reduced
7	None	Yes	Conventional
8	None	Yes	Reduced

During the growing season conventional input plots received a standard farm management regime of 140 kg N fertiliser, the growth regulators fluroxypyr, chlormequat and choline chloride, the herbicides chloridazon, ethofumesate and triclopyr, the fungicides tebuconazole, chlorothalonil, propiconazole and cyproconazole and the insecticide pirimicarb. The low input plots received no N fertiliser, no herbicides, no growth regulators, only one fungicide (propiconazole at 0.25 the recommended rate) and the insecticide primicarb (Clements *et al.*, 1994)

From November to the time of harvest plant samples were taken from one metre lengths of crop rows sited at random within each of the plots. From these samples mean tiller number and shoot dry weights were calculated. For disease assessments during the 1993/94 growing season 20 plants were taken at random from a diagonal transect of each plot. Cereal growth stage (Zadoks *et al.*, 1974) was noted and for each leaf the percentage area affected by *Septoria* was assessed using a disease leaf area key. Samples of infected leaf tissue were plated onto Czapek (Dox) agar to confirm pathogen identity. During the 1993/94 season stem base segments were surface sterilised in 10% sodium hypochlorite and placed onto PDA to allow *Fusarium* isolation and enumeration. During the 1994/95 season half of the stem base segments, collected, as during the 1993/94 season, were placed onto PDA, the other half placed onto a copper sulphate isolation medium (Sumino *et al.*, 1991) which allowed an accurate assessment of *P. herpotrichoides* incidence. Representative fungal isolates were sent to the International Mycological Institute for confirmation of identity. All wheat stem bases were also assessed visually for the severity of eyespot lesioning using the method of Scott & Hollins (1974).

During the 1994/95 season trash remaining on the surface of the three replicate reduced input direct drilled wheat monocrop plots (treatment 6) and the three replicate reduced input bicropped plots (clover cv. Donna, treatment 2) was collected and 10g of stubble was washed in 100 ml sterile distilled water containing a small amount of surfactant to aid spore removal. The numbers of *P. herpotrichoides* spores per g stubble (based on 3 replicated counts from each replicate plot) was calculated for each of the two treatments. Soil samples (10g) were collected from the same plots and were dilution plated on PDA. Following incubation at 15°C the number of *P. herpotrichoides* colony forming units was assessed.

RESULTS

Tiller counts and shoot dry weights of wheat per metre length of crop row on each sampling occasion during the 1993/94 growing season are shown in Table 2. There were no significant treatment effects on crop growth until May and June 1994, when the wheat in low input plots had fewer tillers and less dry weight per unit area than the crop in high input plots. There were only small, non-significant differences in tiller number and shoot dry weights between monocropped and bicropped wheat plants at the same input level.

TABLE 2. Tiller counts and shoot dry weight (g) of wheat (cv. Hereward), grown as monoculture or bicropped with white clover, per metre length of crop row on six sample dates (mean of three replicates)

	Sample date											
	Nov 93		Dec 93		Jan 94		Mar 94		May 94		Jun 94	
Treatment[1]	SN[2]	DW[3]	SN	DW	SN	DW	SN	DW	SN	DW	SN	DW
Conventional input												
1	66	1.2	78	1.2	146	3.4	149	7.6	127	208.0	129	385.0
3	60	1.1	74	1.2	137	3.3	150	6.7	109	187.0	123	343.0
5	74	1.4	88	1.5	174	4.0	206	10.7	139	266.0	110	386.0
7	66	1.0	85	1.3	167	3.7	157	9.3	130	218.0	116	439.0
Reduced input												
2	86	1.6	85	1.5	171	3.8	157	8.4	93	133.0	84	244.0
4	64	1.2	76	1.4	168	3.5	172	9.3	79	127.0	90	241.0
6	72	1.3	90	1.5	185	3.1	224	11.0	119	224.0	96	287.0
8	85	1.4	100	1.6	148	3.0	161	8.0	111	194.0	84	266.0

[1]: Treatments as in Table 1, [2]: Shoot number, [3]: Shoot dry weight (g).

The principal species of *Septoria* causing disease during the 1993/94 growing season was *S. tritici*. Mean *Septoria* severity on the upper leaves and heads for each treatment on each sampling date in the

1993/94 season is shown in Table 3. For all treatments on all sampling occasions the severity of disease was progressively lower on leaves higher up the plant. For the 9 June 1994 assessment, during stem elongation (growth stage 37, Zadoks *et al.*, 1974) the level of disease on corresponding leaves in different treatments did not differ significantly (Table 3). For the assessment on 1 July (growth stage 45) in the conventional input plots, bicrop treatments (1 and 3, Table 2) consistently had a lower severity of *Septoria*. This was significant ($P < 0.05$) on leaf 4 for both clover treatments compared with both monocrop treatments; on leaf 2 for clover cv. Milkanova (treatment 3) compared with the plough treatment, and on the flag leaf for both clover treatments compared with the ploughed plot. In the reduced input plots, bicrop treatments (2 and 4, Table 3) likewise showed consistently lower *Septoria* disease severities with significant differences on leaf 4, 3 and 2 for both clover treatments compared with the direct drilled plots and for the flag leaf for clover treatment cv. Milkanova (treatment 4) compared with the wheat crop grown on ploughed plots.

TABLE 3. Severity of *Septoria* on winter wheat cv. Hereward grown as monoculture or bicropped with white clover (mean of three replicates)

| Treatment[1] | Disease severity (%) on leaves and ears | | | | | | | | | |
| | June 9 | | | | July 1 | | | | July 18 | |
	L2[2]	L3	L4	L5	Flag	L2	L3	L4	Ear	Flag
Conventional input										
1	1.5	4.2	7.4	17.5	1.6	9.0	23.3	40.0	10.0	15.7
3	0.3	2.7	5.9	11.0	1.7	7.7	23.3	40.0	10.7	12.7
5	1.3	3.7	8.0	18.4	5.0	10.0	36.7	60.0	12.0	17.3
7	0.3	2.1	4.9	14.1	7.3	16.7	33.3	61.7	9.0	19.7
Reduced input										
2	0.5	3.1	6.9	18.6	4.3	9.7	20.0	35.0	10.7	16.3
4	1.1	4.0	7.3	14.1	3.0	11.0	21.7	38.3	7.3	11.7
6	0.4	2.5	5.4	13.0	7.3	20.0	38.3	61.7	13.7	25.0
8	1.1	3.6	7.1	16.6	6.0	15.0	31.7	48.3	5.7	17.7
LSD$_{0.05}$	1.3	2.0	3.3	9.5	3.4	8.8	14.1	17.0	6.4	10.2

[1]: Treatments as in Table 1. [2]: L2 - Leaf 2 (first leaf below the flag leaf); L3 - Leaf 3; L4 - Leaf 4; L5 - Leaf 5.

On 18 July, in the plots with conventional input levels, less disease was evident on the heads and flag leaves of the bicropped treatments (treatments 1 and 3, Table 3) although these differences were not significant. At reduced input levels the direct drilled wheat monocrop (treatment 6, Table 3) had higher *Septoria* levels ($P < 0.05$) on the ears and flag leaf than the bicrop with clover cv. Milkanova (treatment 4, Table 3).

The incidence of eyespot was assessed during the 1994/95 season on 16 June (Table 4). Direct drilling had a significant and reducing effect on disease incidence in monocrops (treatments 5 and 6) compared with conventional cultivations (treatments 7 and 8). The effect of bicropping was to increase eyespot incidence relative to the direct drilled treatment, although differences were not significant between bicropped plots and plots which had been conventionally cultivated. In monocrop treatments the level of input had no effect on the incidence of eyespot. An analysis of the severity of eyespot infection (assessed using the method of Scott & Hollins, 1974) showed that treatments 7 and 8 (conventional input and reduced input wheat monocrops sown following ploughing) had the greatest proportion of severe lesions, the treatments with the lowest numbers of severe lesions were 5 and 6 (conventional input and reduced input, direct drilled wheat monocrops).

TABLE 4. Incidence of eyespot on winter wheat cv. Hereward grown as monocrop and bicropped with white clover (mean of three replicates)

Treatment[1]	Eyespot incidence	Eyespot severity class[2] (bracketed figures are percent of total eyespot lesions)			
		0	1	2	3
Conventional input					
5	28.3	58	3(8)	20(48)	18(44)
7	46.7	60	2(4)	17(42)	22(54)
Reduced input					
2	41.7	72	7(23)	12(42)	10(36)
4	40.0	78	2(8)	12(54)	8(38)
6	21.7	47	8(16)	22(40)	23(44)
8	46.7	53	5(11)	18(39)	23(50)

[1]: Treatments as in Table 1; [2]: 0-no stem lesion, 1-slight lesions occupying less than half the stem, 2-moderate lesions occupying more than half the stem, 3-severe, stem girdled by lesion with tissue softened (Scott & Hollins, 1974).

An analysis of *P. herpotrichoides* spore production on trash showed that spore availability within the reduced input bicrop was 10 times greater than that within the reduced input direct drilled monocrop. Furthermore the *P. herpotrichoides* inoculum potential in soil from bicropped plots was significantly greater than that in monocropped plots at the two corresponding input levels (Table 5).

During the 1993/94 trial there was an increasing incidence of *Fusarium* in all plots during the course of the growing season. The majority of *Fusarium* isolates obtained from infected wheat stem tissue were of *F. avenaceum*. Expressed in terms of the area under the disease progress curve the highest disease levels were in treatments 5 and 7 (conventional input monocrops, both ploughed and directly drilled), disease levels in bicropped plots at the same input levels were significantly lower (Table 6). There were no significant differences in disease levels between monocropped and bicropped treatments at reduced input levels (Table 6).

DISCUSSION

The primary cause of *Septoria* infection on the wheat crop during the 1993/94 season was *S. tritici*. No significant differences were observed in *Septoria* severities between direct drilled and conventionally drilled wheat crops, between corresponding crops at contrasting input levels or between wheat crops above different clover varieties (Table 3). The most significant factor accounting for differences in disease levels between plots was the presence or absence of clover as an understorey. The presence of clover clearly reduced the severity of *Septoria* compared with the wheat monocrop treatment, although this effect was observed only at growth stage 45. The presence of the clover understorey did not significantly reduce the wheat shoot population (Table 2) and so differences in *Septoria* severities at growth stage 45 cannot be explained in terms of a reduced canopy density.

TABLE 5. Number of *P. herpotrichoides* colony forming units in soil samples taken from plots of winter wheat cv. Hereward grown as a direct drilled monocrop and bicropped with white clover at conventional and reduced input levels (mean of three replicates)

Treatment[1]	*P. herpotrichoides* colony forming units per g. soil
Conventional input	
1	52
5	3
Reduced input	
2	55
6	17

[1]: Treatments as in Table 1

Septoria is a typically splash-dispersed cereal pathogen where pycnidiospores produced from cirrhi by pycnidia on the leaf surface are dispersed by rain splash droplets to initiate new infections. Royle *et al.* (1986) suggested that the vertical movement of pycnidiospores from basal to upper leaves might be a limiting factor in the development of some *Septoria* epidemics. Royle *et al.* (1986) also suggested that because of the rapid emergence of new leaves during stem extension, in most cases initial infections on successively produced upper leaves are likely to have originated from inoculum on diseased leaves lower down on the plant. The severity of *Septoria* attack is dependant on the amount of inoculum on the lower leaves at the start of stem extension (growth stage 30-31), the suitability of weather for allowing infection, and the occurrence of rainfall heavy enough to move inoculum up through the crop (Royle *et al.*, 1986). The presence of a clover understorey at the base of the wheat crop might have acted to reduce the upward movement of spores and therefore disease in the present experiment. This effect could have been brought about by one or both of two mechanisms. Firstly, the clover may have prevented raindrops penetrating to those basal leaves from which most later infections originate, and secondly the clover canopy may have deflected or intercepted spore-carrying splash droplets on their upward path to the newly emerging foliage. In either case, the suggestion that fewer spores were being redistributed from *Septoria* cirrhi in the presence of the companion clover crop appears to be borne out by results from recent experiments conducted in controlled environment conditions (Cooke, B M & Bannon F, personal communication).

The results from the eyespot evaluations indicated that direct drilled monocrops suffered less disease than crops sown following ploughing. These results are broadly similar to those of Herrman & Wiese (1985) who found that a reduced tillage regime lessened eyespot by 50% compared with a conventional cultivation; no-tillage treatments were reduced by a further 50%. Herrman & Wiese (1985) suggest that the decrease in the level of disease under reduced and no-tillage treatments could be due to straw at the soil line separating the plant's lower stem from the pathogen located below in the soil. However results from the current study indicate that straw supports significant levels of spore production, this level being significantly increased in the presence of a clover canopy where the environmental conditions are likely to be more conducive to spore development. Although, as stated above, rainfall penetration to infected straw at the canopy base is likely to be reduced in the presence of clover, it would appear however that sufficient spores were being transferred to host tissue with the result that similar levels of disease occurred in bicropped plots as in ploughed, monocropped plots.

TABLE 6. Relative area under the disease progress curves for the incidence of *Fusarium* on winter wheat (cv. Hereward), grown as monoculture or bicropped with white clover (mean of three replicates)

Treatment[1]	Relative area under disease progress curve
Conventional input	
1	72[*]
3	73[*]
5	100
7	98
Reduced input	
2	62[*]
4	64[*]
6	50[*]
8	53[*]

[1]: Treatments as in Table 1, [*] Treatments differing significantly from treatment 5 at $P < 0.05$.

The altered canopy structure may also have had an effect on the levels of *Fusarium* observed during the 1993/94 growing season. However the disease measures as indicated by the areas under the disease progress curves also appear to indicate that the level of nitrogen has an important effect as in both monocrops and bicrops those plots receiving higher levels of nitrogen had greater levels of disease. It is also worth noting that the *Fusarium avenaceum*, the species most frequently isolated from the wheat stems is also pathogenic on clover and was frequently isolated from clover stem tissue collected from the bicropped plots. The degree of cross infection between clover and wheat is uncertain and so the extent to which clover could act as a reservoir for *Fusarium* infections requires further investigation.

ACKNOWLEDGEMENTS

The work reported was conducted in collaboration with a project commisioned by the Ministry of Agriculture, Fisheries and Food, and was carried out whilst M.J. Soleimani was in receipt of an Iranian Government, Ministry of Higher Education and Culture Scholarship.

REFERENCES

Clements R O, Kendall D A, Asteraki E J, George S. (1994) Progress in development of a clover-cereal bi-cropping system. *Proceedings of the British Grassland Society Research Meeting, Reading 1994*, pp. 77-78.

Cowling D W. (1982) Biological nitrogen fixation and grassland production in the UK. *Philosophical Transactions of the Royal Society, London* **B 296**:397-404.

Deadman M L, Soleimani M J. (1994) Cereal-clover bicropping: implications for crop disease incidence. *Proceedings of the British Grassland Society Research Meeting, Reading 1994*, pp. 79-80.

Herrman T, Wiese M V. (1985) Influence of cultural practices on incidence of foot rot in winter wheat. *Plant Disease* **69**:948-950.

Jones L, Clements R O. (1991) Cereals in clover. *Crops, 24 August 1991*, pp. 16-17.

Jones L, Clements R O. (1993) Development of a low input system for growing wheat (*Triticum vulgare*) in a permanent understorey of white clover (*Trifolium repens*). *Annals of Applied Biology* **123**:109-119.

Ladd J N, Amato M. (1986) The fate of nitrogen from legume and fertiliser sources in soils successively cropped with wheat under field conditions. *Soil Biology and Biochemistry* **18**:417-425.

Royle D J, Shaw M W, Cook R J. (1986) Patterns of development of *Septoria nodorum* and *S. tritici* in some winter wheat crops in Western Europe, 1981-83. *Plant Pathology* **35**:466-476.

Scott P R, Hollins T W. (1974) Effects of eyespot on the yield of winter wheat. *Annals of Applied Biology* **78**: 269-279.

Sumino A, Kondo N, Kodama F. (1991) A selective medium for isolation of *Pseudocercosporella herpotrichoides* from soil. *Annals of the Phytopathological Society of Japan* **57**: 485-491.

Zadoks J C, Chang T T, Konzak C F. (1974) A decimal code for the growth stages of cereals. *Weed Research* **14**:415-421.

AN EVALUATION OF A CHITIN BASED FERTILISER AGAINST POTATO CYST NEMATODE, CABBAGE ROOT FLY AND RHIZOCTONIA SOLANI

S. A. ELLIS, M. L. HALLAM AND C. J. OTTWAY

ADAS, Lawnswood, Otley Road, Leeds, LS16 5PY, UK

D. WINTERS

Ocean Organics, Factory Road, Blaydon Hough Industrial Estate, Tyne and Wear, NE21 5SA, UK

ABSTRACT

Three pot trials were established to investigate the control of wirestem on cauliflowers, caused by the fungus *Rhizoctonia solani*, cabbage root fly (*Delia brassicae*) on cauliflowers, and the white potato cyst nematode (*Globodera rostochiensis*) on potatoes. The relative efficacy of a chitin based fertiliser, at the recommended and three times recommended rate (270 and 810 g/m^2), incorporated up to 12 weeks before and/or applied on the day of inoculation with the pest or pathogen, was compared with tolclofos-methyl, aldicarb 10G and chlorpyrifos. The effectiveness of the test products against wirestem was assessed in terms of seedling emergence and presence of disease symptoms. Control of potato cyst nematode was measured in numbers of cysts per g of soil and control of cabbage root fly was assessed in terms of larval/pupal numbers and root damage.

All synthetic pesticides gave significantly better control of the test organisms than the chitin fertiliser. However, there was evidence that the fertiliser gave some control of potato cyst nematode, wirestem and cabbage root fly.

The potential for a chitin based fertiliser as a component of an integrated control strategy is discussed and suggestions are made for future research.

INTRODUCTION

Actinomycetes growing on chitin in oyster shell powder have been shown to control *Sclerotinia sclerotiorum* in vegetables (Lin *et al.*, 1990) and adding crustacean chitin to soil reduced cyst numbers of the soya bean cyst nematode (*Heterodera glycine*) (Rodriguez-Kabana *et al.*, 1984). As crab shell chitin is a major constituent of the fertiliser Ocean Supermix, it is possible that it may provide some control of soil pests and diseases, by stimulating the development of antagonistic micro-organisms. The present work was designed to evaluate the efficacy of this product against wirestem, caused by the common soil fungus *Rhizoctonia solani* (Telemorph = *Thanatephorus cucurmeris*), cabbage root fly (*Delia radicum*) and the white potato cyst nematode (*Globodera pallida*).

MATERIALS AND METHODS

In all experiments, the rates of fertiliser studied were 270 and 810 g/m², this being the recommended and three times the recommended dose rate. All test crops were grown in a sandy loam soil (Wigton Moor soil association, Quorndon Series). Fertiliser application was made at range of timings before and on the day of planting or sowing of the test crop and inoculation with the pest or disease. Incorporation of fertiliser before pest/disease inoculation was included to determine whether chitinolytic microbes could be stimulated prior to introduction of the test organism and so improve control.

Experiment 1: *Rhizoctonia solani*

Hand incorporation of the fertiliser, at 270 or 810 g/m², was undertaken 45, 31 or 17 days prior to sowing of ten viable non-fungicide treated cauliflower seeds (cv. Dok Elgon) per pot. Tolclofos-methyl as 'Basilex' was used as the standard fungicide in comparison with the fertiliser. An untreated control treatment with no fungicide or fertiliser was also included. Each treatment was replicated six times to provide 48 pots (9 cm diameter) in total. Soil known to be infected with *R. solani* was baited with cauliflower seedlings to provide a pure culture of the fungus. Isolates were prepared from the infested seedlings and cultured in a mixture of vermiculite and maize meal for seven days at ambient room temperature. Equal sized pot doses were prepared by subdividing this medium and these were inoculated 17 days before sowing the cauliflower seed. The fungicide was prepared as a 2 g/l suspension and 6.4 ml applied to the surface of the soil in each pot with a syringe. Pots were maintained in a glasshouse in a randomised block design at 20°C and watered as necessary. Fourteen days after sowing the numbers of emerged seedlings were counted.

Experiment 2: Potato cyst nematode

The fertiliser was incorporated by hand at 270 or 810 g/m² either 28, 14, or 0 days before inoculation with cysts of potato cyst nematode. Aldicarb, as 'Temik' was used as the standard nematicide in comparison with the fertiliser. Individual pot doses of this (0.002 g AI/pot equivalent to 3.36 g AI/m², the standard rate for maincrop potatoes) were prepared and mixed thoroughly with pot soil. An untreated control was also included and each of the eight treatments replicated six times to give a total of 48 pots (9 cm diameter). Soil known to be infested with the white potato cyst nematode was used to provide cysts of the pest and twenty of these were inoculated per pot. Each pot was planted with a single potato sprout (cv. Desiree) and then buried to just below its rim in a larger 18 cm-diameter pot containing general purpose potting compost. This gave an additional area into which potato roots could grow and obtain moisture (Cotten, 1967). Pots were then arranged in a randomised block design in an outdoor insectary. Twelve weeks later the 9 cm diameter pots were assessed for numbers of cysts of potato cyst nematode using a Fenwick can technique (Fenwick, 1948).

Experiment 3: Cabbage root fly

Fertiliser at 270 and 810 g/m² was incorporated by hand either once, twelve weeks before planting or twice, 12 weeks before and at planting. The double application was investigated to determine whether it would enhance pest control. This was compared with the insecticide chlorpyrifos, 'Dursban 4', made up as a solution of one ml in one litre of water, the rate approved as a drench treatment in the field. An untreated control was also included. There were six replicates of each treatment giving 36 pots (9 cm diameter) in total. A cauliflower plant (at the four leaf stage) was planted in each pot. Cabbage root fly eggs were extracted from soil samples taken from around the base of field brassica plants using a Fenwick can technique (Fenwick, 1948) and 10 of these were inoculated around the base of each cauliflower plant. A syringe was then used to apply the chlorpyrifos treatment with 70 ml of the insecticide solution used per plant. Pots were arranged in a randomised block design in a glasshouse at ambient temperature and watered as necessary. After six weeks the numbers of cabbage root fly larvae or pupae in each pot were recorded following flotation of the compost in concentrated magnesium sulphate solution.

RESULTS

Experiment 1

Seedling emergence was significantly greater in pots treated with tolclofos-methyl than in those receiving the fertiliser or in the control (P<0.005, Table 1). There was no significant difference between fertiliser treatments but all had significantly more seedlings than the control (P<0.05).

TABLE 1.   Emergence of cauliflower seedlings after treatment with a chitin fertiliser or tolclofos-methyl

Treatment	Application rate (g/m²)	Days before sowing treatment incorporated	Seedling emergence (mean numbers)
1. Untreated control	-	0	1.8 a
2. Tolclofos-methyl	6.4 mg AI/pot	0	9.3 c
3. Fertiliser	270	17	4.8 b
4. Fertiliser	270	31	4.2 b
5. Fertiliser	270	45	4.8 b
6. Fertiliser	810	17	4.3 b
7. Fertiliser	810	31	5.3 b
8. Fertiliser	810	45	5.0 b

SED (35 df) = 0.88

a, b and c are Duncan's Multiple Range Test indices, values followed by the same letter are not significantly different, P<0.05.

Experiment 2

Aldicarb gave good control of potato cyst nematode and significantly fewer cysts were recovered from pots which received this product than from all other treatments (P<0.05, Table 2). In general, pots treated with fertiliser had fewer cysts than the untreated control but this difference was only significant (P<0.05) where 810 g/m² was incorporated 14 days before pest inoculation.

Table 2.   Mean numbers of cysts of potato cyst nematode after treatment with a chitin fertiliser and aldicarb

Treatment	Application rate (g/m²)	Days before planting treatment incorporated	Mean cyst numbers/g soil
1. Untreated control	-	0	2.54 c
2. Aldicarb	3.36g AI/pot	0	0.15 a
3. Fertiliser	270	0	1.65 bc
4. Fertiliser	270	14	1.87 bc
5. Fertiliser	270	28	2.25 bc
6. Fertiliser	810	0	2.23 bc
7. Fertiliser	810	14	1.42 b
8. Fertiliser	810	28	2.31 bc

SED (33 df) = 0.403

a, b and c are Duncan's Multiple Range Test indices, values followed by the same letter are not significantly different, P<0.05.

Experiment 3

Significantly fewer cabbage root fly larvae were found in pots treated with chlorpyrifos than in the untreated control or where the fertiliser was incorporated at 270 g/m², 12 weeks before pest inoculation (P<0.05, Table 3). Fertiliser incorporation at 810 g/m 12 weeks before and on the day of pest inoculation also resulted in significantly smaller numbers of larvae in comparison with 270 g/m² incorporated 12 weeks pre pest inoculation (P<0.05).

TABLE 3.   Mean number of cabbage root fly larvae ($\sqrt{(x + 0.5}$ values) following treatment with a chitin fertiliser and chlorpyrifos. Values in brackets are back-transformed data.

Treatment	Application rate g/m²	Days before planting treatment incorporated	Numbers of larvae	
1. Untreated control	-	-	1.53 bc	(1.84)
2. Chlorpyrifos	0.34g AI/plant	0	0.71 a	(0)
3. Fertiliser	270	84	1.75 c	(2.56)
4. Fertiliser	270	84 and 0	1.41 abc	(1.49)
5. Fertiliser	810	84	1.37 abc	(1.38)
6. Fertiliser	810	84 and 0	0.90 a	(0.31)

SED (25 df) = 0.314

a, b and c are Duncan's Multiple Range Test indices, values followed by the same letter are not signficantly different, P<0.05.

DISCUSSION

In all experiments the standard pesticide treatment was more effective at controlling *Rhizoctonia solani*, cabbage root fly and potato cyst nematode than the chitin based fertiliser. However, there was evidence that the fertiliser gave some suppression of all test organisms.

Other studies with *R. solani* (Ellis *et al.*, 1993a) have shown that the fertiliser is less effective when incorporated with the fungus at sowing. Therefore it is possible that inoculation of the fertiliser with the fungus, as in the present experiment, allows antagonistic micro-organsims to develop and start to control the pathogen so that its virulence is reduced.

Although the fertiliser appeared to give some control of potato cyst nematode, results were inconclusive. Development of the antagonistic microflora may not be rapid enough to influence cyst production. However, as chitin is a component of nematode egg shells it is possible that the new generation of cysts would be less viable and contain low numbers of eggs. Ellis *et al* (1993b) showed that egg numbers per cyst were lower where a chitin fertiliser was applied in comparison with aldicarb although the differences were not statistically significant.

In general, fewer cabbage root fly larvae were recorded where 810 g/m² of fertiliser was used as opposed to 270 g/m² and also where two applications were made compared with one. A high level of substrate is likely to result in a high population of chitinolytic microbes. Also, where repeat applications are made it is possible that antagonistic organisms develop more rapidly. Such a situation has evolved through continued use of carbamate insecticides to control insect pests in field vegetables. Some soils now show enhanced degradation (Suett, 1990) such that carbamate pesticides are rapidly broken down by the soul microflora. Continued use of a chitin based fertiliser may also enhance the development of populations of chitinolytic micro-organisms and pest/disease control.

In summary, results suggest that a chitin based fertiliser may be effective as part of an integrated strategy incorporating a range of cultural or biological methods of control of plant pathogens. In this situation the pest/disease pressure is likely to be considerably less than experienced in the current studies.

ACKNOWLEDGEMENTS

We thank Royal Sluis Ltd, for providing the cauliflower seed.

REFERENCES

Cotten, J. (1967) Cereal root eelworm pathotypes in England and Wales. *Plant Pathology*, **16**, 54-59.

Ellis, S. A.; Ottway, C. J.; Winters, D. (1993a) Evaluation of Ocean Supermix against *Rhizoctonia solani* in cauliflowers. *Tests of Agrochemicals and Cultivars* **No. 14** (*Annals of Applied Biology* **122** *Supplement*), 30-31.

Ellis, S. A.; Hallam, M. H.; Winters, D. (1993b) Evaluation of Ocean Supermix against the white potato cyst nematode (Globodera pallida). *Tests of Agrochemicals and Cultivars* **No. 14** (*Annals of Applied Biology 122 Supplement*), 50-51.

Fenwick, D. W. (1940) Methods for the recovery and counting of cysts of *Heterodera schactii* from soil. *Journal of Helminthology* **18**, 155-172.

Lin, X. S; Sun, S. K; Hsu, S. T; Hsieh, W. H. (1990) Mechanisms involved in the control of soil-borne plant pathogens by S-H mixture In: *Biological control of soil-borne plant pathogens* D. Hornby (Ed.), Wallingford, CAB International, 249-259.

Rodriguez-Kabana, R; Morgan-Jones, G; Gintis Ownley, B. (1984) Effects of chitin amendments to soil on Heterodera glycines, microbial populations and colonisation of cysts by Fungi *Nematotropica*, **14**, 10-25.

Suett, D. L. (1990) The threat of accelerated degradation of pesticides - myth or reality? *Brighton Crop Protection Conference - Pests and Diseases 1990*, **3**, 897-906.

INTEGRATED CROP PROTECTION IN COTTON: THE EXAMPLE OF ZIMBABWE

R.J. HILLOCKS

Natural Resources Institute, Chatham Maritime, Chatham, Kent, ME4 4TB

M.J. DEADMAN

University of Reading, Department of Agriculture, Earley Gate, Reading, Berks, RG6 2AT.

ABSTRACT

Cotton was one of the first crops for which IPM programmes were developed in
response to increasing occurrence of resistance to conventional insecticides
in pest populations. It is therefore appropriate that cotton should provide
a model system for the expansion of the IPM concept to one of ICP (Integrated
Crop Protection), to include control of diseases and weeds in addition to
the insect pests. The poster describes some of the management practices which
constitute an integrated crop protection system developed at the Cotton
Research Institute in Zimbabwe.

INTRODUCTION

Cotton is grown in 32 countries in Africa with Egypt and Sudan the two largest producers and Tanzania
and Zimbabwe the largest producers south of the Sahara. The crop is grown on a large scale in plantations,
or aggregated small holdings in Egypt and Sudan but in sub-Saharan Africa, most of the crop is produced by
smallholders often with minimal inputs. In the north African producer countries, a high proportion of the
crop is derived from the long and extra long staple varieties of *Gossypium barbadense*, while in the rest of
Africa, the vast majority of the crop is medium staple, derived from African Upland varieties (*G. hirsutum*).

In Zimbabwe, cotton producers have traditionally been large scale farmers with average yields of 1200
kg/ha of seed cotton for the rain grown crop and 2500 kg/ha for the irrigated crop. Since independence, the
smallholder sector has made an increasing contribution to national production and now accounts for more
than 50% of the annual crop of around 400,000 bales of lint (200kg bale), with average seed cotton yields of
around 800 kg/ha. Intensive use of insecticides, beginning with DDT and later carbaryl for bollworm control
in the large scale production sector, has led to an insect pest complex which is partly pesticide induced
(Brettell, 1986). This is reflected by the relative importance of red spider mite and sucking insects in the pest
complex, compared to their relative unimportance in countries such as Tanzania and Malawi, where insect
control has been less rigorous. However, these indicators provided an early warning of the potential for
unregulated spraying to create crop protection problems and measures were taken to limit the risk of
pesticide resistance by regulating the use, first of acaricides and then the synthetic pyrethroids (SPs).
Restrictions on the use of SPs right from their inception in Zimbabwe has allowed their continued use
without the widespread appearance of resistance (Brettell, 1986).

To maximise the yield potential of current varieties, recommendations from the Cotton Research
Institute (CRI) at Kadoma in central Zimbabwe, should be followed as an integrated crop management
package in which crop protection is viewed within the context of crop agronomy as a whole. This is an
approach rather than the application of particular technologies, extending the concept of IPM from a narrow
one of integrating the technology of insecticide application and biological control, in order to prolong the
effective life of conventional insecticides.

AGRONOMIC PRACTICE

It may be obvious to the large scale farmer that money spent on crop protection is better spent the more crop per unit area you have to protect. It is not perhaps so clear to the small producer who's yields may already be so limited by agronomic factors such as late planting, poor weed control and factors beyond his control such as drought, that pest control becomes a costly exercise in revenge. The implications of good agronomic practice to crop protection can be seen in the case of failure to remove weeds which are alternative hosts for pests and diseases or, the upsurge in whitefly infestations that often occurs when irrigation becomes available and nitrogen levels increased. Another specific example from Zimbabwe is the case of predisposition to *Alternaria* leaf spot by potassium deficiency. Increased yield in recent years has led, on certain soil types, to a depletion of available potassium because recommendations based on different soil types and lower yields were inadequate to replace potash taken out by the crop. The result was premature senescence and predisposition to the leaf spot disease which caused rapid and dramatic defoliation during periods of wet weather (Hillocks & Chinodya, 1989).

WEED CONTROL

Cotton is slow to establish and competes poorly with weeds during the early stages of crop development. Recommendations for early weed control in Zimbabwe include the application of an incorporated herbicide such as trifluralin, targeted mainly at the grass weeds. This is followed by a pre-emergence application of a broad spectrum herbicide such as cyanazine. Later in the crop cycle, weeds can be removed by inter-row cultivation or by the application of a lay-by treatment of cyanazine for instance, and once the crop is well established and the stem has become woody, paraquat may be used. These operations are easily carried out where land preparation, planting and cultivation are mechanised and where irrigation is available. However, where there is a shortage of labour, smallholders can also benefit from the use of herbicide applied with a knapsack sprayer for early season weed control. The design of an integrated crop protection plan requires some knowledge of the role of weed species both as alternative hosts for pests and as potential reservoirs of beneficial insects. The vegetation around the field margin is a source of populations of insects which have not been exposed to pesticide and therefore do not contain pesticide resistant individuals. From the perspective of the African smallholder the definition of a weed is unclear, for many of the most common species have culinary, medicinal or other household uses. The farmer is therefore reluctant to weed out the more useful weed species and they may even be cultivated within or near to the main crop species. In non-intensive farming in Africa, areas of uncultivated land provide a reservoir of beneficial insects but this is less so under intensive farming, where large areas are sown to cotton and are subject to regular spraying. In the latter situation, there may be a case for the inclusion of uncultivated areas if predators and parasites of the common insect pests are to be encouraged as part of an IPM programme.

DISEASES

The main diseases of cotton in Zimbabwe are seedling disease (mainly *Rhizoctonia solani*), bacterial blight (*Xanthomonas campestris* pv. *malvacearum*), wilt (*Verticillium dahliae*) and leaf spot (*Alternaria macrospora*). Seedling disease is of sporadic occurrence depending much on weather conditions between sowing and the first two weeks after emergence. Conditions which slow down emergence will allow the hypocotyl to be attacked by *R. solani* and other fungi such as *Fusarium* spp. and *Colletotrichum* spp. The application of fungicides for seedling disease control, either by in-furrow treatment, or, as a seed dressing, is a cost effective means of protecting the crop from early stand loss. A nozzle for fungicide application can be mounted on the planter so that the chemical is applied to the seed and covering soil as the seed goes into the ground. If the seed is hand planted, seed and surrounding soil can be treated using a conventional knapsack sprayer. Seed dressing requires only around 500g of compound per 100kg of seed and can be applied to the seed, simply by shaking the seed with the chemical in a drum. Two of the recommended seed treatment fungicides in Zimbabwe are carboxin and tolclofos-methyl.

Bacterial blight is controlled through the use of resistant varieties. All material under selection within the cotton breeding programme at CRI is screened for blight resistance at each stage (Hillocks & Chinodya, 1988). The upland cottons grown in most east and southern African countries have traditionally been derived from Albar 51, a blight resistant selection produced in Uganda in 1951. The modern commercial cultivars grown in Zimbabwe still contain selections from Albar 51 in their background. The approach to bacterial blight control in Zimbabwe has always been an integrated one, combining the use of resistant varieties with, rotation and phytosanitary controls. A break of one season without cotton is usually sufficient to greatly decrease the bacterial inoculum which requires crop residues in the soil for its survival in the absence of cotton plants. Also in place, are regulations for uprooting and stalk destruction to allow a break between one seasons crop and another. These measures were introduced to decrease the carry-over of pink bollworm but also help to control bacterial blight by eliminating the possibility of infection directly to the new crop from the previous seasons crop or undestroyed residues. The bacterium can be seed borne and is transmitted to the seed by insect pests which feed on the developing boll. The main insect vector is the stainer (*Dysdercus* spp.). The stainer feeds directly on the seed and high levels of yellow stained lint and seed infected with the bacteria are associated with failure to control late season stainer populations.

Verticillium wilt is controlled by varietal resistance and as with blight, all material in the pedigree line breeding programme is screened for wilt resistance in a sick plot at CRI and at other sites (Hillocks, 1991a). The pathogen is capable of survival in the soil in the form of microsclerotia which may remain viable for many years in the absence of cotton. However, the fungus can infect the roots of a number of common weeds, particularly billy goat weed (*Ageratum conyzoides*) (Hillocks, R.J., unpublished). The ability of *V. dahliae* to maintain inoculum levels in the absence of cotton may be enhanced by the presence of secondary hosts among the weed flora.

Alternaria leaf spot disease can be prevented in upland cotton by ensuring that potassium replacement after each crop is sufficient to prevent the development of potassium deficiency. The initial focus of infection is crop residues but this source of infection can be greatly minimised by a one year break between cotton crops.

INSECT PESTS

When cotton was first cultivated in eastern and southern Africa, the most destructive pest was the jassid (*Jacobiasca* spp.). Once it was shown that leaf hairiness conferred resistance to jassids, the pest could be readily controlled by ensuring that selection for the hirsute leaf character became a routine component of the breeding programme (Gledhill, 1979). Bollworms, particularly American bollworm (*Helicoverpa armigera*)(ABW) are the main insect pests in Zimbabwe (Brettell, 1986) as they are in many other countries in Africa. Populations of ABW may be increased over the long term where spraying has been carried out consistently since the time when DDT was the main chemical for its control. This is due to the damaging effect broad spectrum insecticides have on populations of natural enemies. In countries such as the USA and Australia where the use of synthetic pyrethroids was insufficiently regulated, large numbers of sprays were used and populations of the target insect pests increased due to their resistance to SPs. These compounds are still effective in Zimbabwe due to the policy of resistance management which was initiated when SPs were first introduced. The use of pyrethroids is restricted to a nine week period around flowering and boll set and prohibited during the early and later part of the season. Other compounds such as thiodicarb and endosulphan may be used during the closed season for SPs. As part of the IPM programme, it is recommended that the first spray with any chemical be delayed as long as possible and where pests other than bollworms reach damage threshold levels, selective insecticides should be used such as pirimicarb, in the case of early season aphid attack. Delayed use of insecticides until damage threshold are reached is fostered in order to encourage the build up of natural enemy populations early in the season. Much emphasis is placed on the selection of insecticides which have least effect on populations of the main beneficials, the green lacewings (*Chrysopa* spp.) (Brettell, 1984). The SPs recommended by CRI are those which are most effective against the bollworm complex but are also those which do not exacerbate the red spider mite problem. Some SPs have been found to have non lethal effects on spider mites but have a repellant action, dispersing them to reinfest at a later stage (Brettell, 1986).

Red spider mites (RSM) (*Tetranychus* spp.) have now become a serious pest in Zimbabwe, appearing initially in patches in the field, during periods of hot, dry weather. A policy of resistance management was introduced in 1973 which has helped to maintain the efficacy of recommended acaricides. The acaricide rotation scheme, as it is known, is based on the division of the cotton growing area into three, each of which may use an acaricide from the same chemical group for only two consecutive seasons (Dunscombe, 1973).

Pesticide application based on damage threshold levels requires that scouting be carried out on a regular basis. This is an issue which has caused some controversy because while it is a practise accessible to relatively well educated, larger scale producers, it is more difficult for the sometimes illiterate smallholder. However, in Zimbabwe, considerable success has been achieved in introducing scouting to the smallholders through the adoption of the pegboard system of recording insect numbers and by offering instruction courses to targeted farmers in each area (Burgess, 1983). The alternative to scouting and spraying based on pest thresholds is to spray on a fixed calendar basis, usually once a month for the six critical months of the growing season. The disadvantage of this is that sprays may be applied when they are not needed with unnecessary damage inflicted on the environment and on the natural enemy population. Among resource-poor farmers the high cost of spraying their crop five or six times, dissuades them from spraying at all, or, they carry out one or two sprays only when they see substantial damage or pest infestation.

TABLE 1. Components of the integrated crop protection system in Zimbabwe

Management practice	Rationale	Target organism
Control of seed production and	Provision of high quality seed free of pathogens	Seed-borne fungi and *Xanthomonas*
Crop rotation	Minimize carry-over of pathogens	*Xanthomonas* and *Alternaria*
Herbicide application	Reduce competition from weeds during early stage of crop development	All weed species
Seed dressing or in-furrow treatment with fungicide	Reduce losses to seedling disease	*Rhizoctonia solani*
Plant resistant varieties	Control of main diseases and some pests	*Jacobiasca* spp. *Xanthomonas campestris* *Verticillium dahliae*
Scouting	Minimal pesticide use by spraying only when damage thresholds are reached	All insect pests
Selective insecticides	Minimize non-target effects to protect beneficials	e.g. *Aphis gossypii* with pirimicarb
Restricted use of SPs	Resistance management	e.g. *Helicoverpa*
Acaricide rotation	Resistance management	*Tetranychus* spp.
Control late season sucking pests	Reduce boll disease and lint stain	*Dysdercus* spp. *Bemisia tabaci*
Stalk destruction	Closed season between successive cotton crops	*Pectinophora gossypiella*

CONCLUSIONS AND FUTURE DEVELOPMENTS

The adoption of resistance management strategies in Zimbabwe has demonstrated that where the use of conventional pesticides is closely regulated, their long-term efficacy can be maintained. However, concern about the implications for the environment, of continued use of highly toxic chemicals and for the health of those using them, together with their increasing cost, continues to provide an impetus towards decreased reliance on these compounds. In order to decrease the use of conventional pesticides without sacrificing yield, a large number of alternative practices may need to be deployed which are effective when properly integrated but give poor control when used on their own. A fully integrated crop protection system also requires that control of insect pests, disease and weeds be viewed together (see Table 1.), within the context of crop husbandry as a whole and that the crop itself be considered as part of the local vegetation complex.

Most of the measures which can be adopted to reduce dependency on conventional insecticides (Table 2.) are directed at providing some control of the target pests with minimal effects on populations of their natural enemies (Rosier, 1990; Menn, 1991). This requires that the main beneficials be identified and that candidate insecticides are screened for their effect on beneficials such as lacewings. Some of the recently developed soil-applied systemic insecticides are compatible with IPM systems because only insects feeding on the crop are targeted. Some of the traits which confer a degree of insect resistance such as the okra leaf and nectariless characters may be increasingly exploited. The toxin from *Bacillus thuringiensis* or plants engineered to produce the toxin are also being increasingly used to supplement conventional insecticides. Pheromones can be deployed in lure and kill strategies or to disrupt mating activity. Phagostimulants or trap crops can be used to attract insects away from the main crop. There is much still to be done to evaluate potential component technologies for IPM systems. Although the market for organic cotton remains relatively small, the requirement for chemical-free crop production provides an excellent opportunity to test alternative pest control methods.

TABLE 2. Some alternatives to conventional insecticides

Measure	Target pest or rationale
Host plant resistance	
Pubescence	jassid
Nectarilessness	bollworms
Frego bract	increased spray penetration
High gossypol	bollworms
Disease resistance	bacterial blight verticillium wilt fusarium wilt
Biopesticides	
Nuclear polyhedrosis virus	bollworms
Bacillus thuringiensis	American bollworm (ABW) and other pests
Pheromones	pink bollworm
Natural enemies	
Green lacewings (*Chrysoperla* spp.)	red spider mite ABW eggs
Ladybirds (*Cheilomenes* spp.)	aphids, bollworm eggs
Hoverfly larvae (*Melanostoma* spp.)	aphids
Assassin bugs (*Phonoctonus* spp.)	stainer

REFERENCES

Brettell, J.H. (1983) Strategies for cotton bollworm control in Zimbabwe. *Zimbabwe Agricultural Journal*, **80**, 105-108.

Brettell, J.H. (1984) Green lacewings (Neuroptera: Chrysopidae) of cotton fields in Zimbabwe. 3. Toxicity of certain acaricides, aphicides and pyrethroids to larvae of *Chrysopa boninensis* Okamoto, *C. congrua* Walker and *C. pudica* Navas. *Zimbabwe Journal of Agricultural Research*, **22**, 133-139.

Brettell, J.H. (1986) Some aspects of cotton pest management in Zimbabwe. *Zimbabwe Agricultural Journal*, **83**, 41-46.

Burgess, M.W. (1983) Development of cotton pest management in Zimbabwe. *Crop Protection*, **2**, 247-250.

Duncome, W.G. (1973) The acaricide spray rotation for cotton. *Rhodesian Agricultural Journal*, **70**, 115-118.

Gledhill, J.A. (1979) The Cotton Research Institute, Gatooma. *Rhodesia Agricultural Journal*, **76**, 103-118.

Hillocks, R.J. (1991a) Screening for resistance to verticillium wilt in Zimbabwe. *Tropical Agriculture*, **68**, 144-148.

Hillocks, R.J. (1991b) Alternaria leaf spot of cotton with special reference to Zimbabwe. *Tropical Pest Management*, **37**, 124-128.

Hillocks, R.J.; Chinodya, R. (1988) Current status of breeding for resistance to bacterial blight in Zimbabwe. *Tropical Pest Management*, **34**, 303-308.

Hillocks, R.J.; Chinodya, R. (1989) The relationship between Alternaria leaf spot and potassium deficiency causing premature defoliation of cotton. *Plant Pathology*, **38**, 502-508.

Menn, J.J. (1991) Prospects and status for development of novel chemicals for IPM in cotton. *Crop Protection*, **10**, 347-353.

Rosier, M.J. (1990) Cotton. *Pesticide Outlook*, **1**, 19-23.

APPLICATION AND IMPORTANCE OF INSECTPATHOGENIC VIRUSES IN RESPECT TO INTEGRATED PEST MANAGEMENT (IPM)

K. GEISSLER

Martin-Luther-University Halle-Wittenberg, Institute for Plant Breeding and Plant Protection, Working Group Aschersleben, Theodor-Roemer-Weg 4, D-06449 Aschersleben, Germany

ABSTRACT

This work summarises a survey about the possibilities of using insectpathogenic viruses as biological pesticides within Integrated Pest Management Systems (IPM). Results of controlled experiments are reported. The merits and disadvantages of insectpathogenic viruses are discussed.

INTRODUCTION

Regulating the population densities of animal pests and microbial diseases on cultivated plants today will never result in their total elimination. Modern pest management maintain plants in good health and control measures will primarily begin when the population density of a pest will reach the specific damage threshold.

The FAO defines 'Integrated Pest Management (IPM)' as "... a system of regulating pests and diseases. In correspondence to the whole ecosystem and the population dynamics of the pests all usefull methods and procedures are combined to keep their population density below the economic threshold of danger. ..."

Biological control strategies and especially using insectpathogenic viruses will therefore play as major role as other essential control methods. They should be used in such areas of pest control systems in which they will bring clear advantages in respect to other methods.

POSITION OF INSECTPATHOGENIC VIRUSES IN THE SYSTEM OF INTEGRATED PEST MANAGEMENT

Cook & Baker (1983) define 'Biological Pest Control' as "... a restrained use of organisms (and viruses) and their effects for protecting plants and animals (including humans) against biotic and nonbiotic damaging factors. ..."

This was the first time that viruses were mentioned as an alternative to other organisms for control pests and diseases in ecosystems. The prospect of using insectpathogenic viruses in comparison to other biological pesticides is that insectpathogenic viruses will show a high selectivity and specifity of infestation and as a rule they will infect only a single host species. Further these pathogens will not cause any danger to plants, other animals including humans, or groundwater.

The most important group of insectpathogenic viruses in respect to their qualification as biological pesticides are the Baculoviridae with the nucleopolyhedrosis viruses (NPV) and the granulosis viruses (GV). Viruses with insect hosts also can be found in other virus families. However only the Baculoviridae contain no plant pathogenic viruses or vertebrate pathogenic viruses including human pathogens.

As a result of their stability in correct storage, and their harmless nature in case of correct application insect-pathogenic viruses will be applied favourably in biological pest control. But it is essential to have an accurate and full knowledge with respect to the mannifold interactions in specific ecosystems for eliminating unforeseen reactions, and also of the biology of the pest being controlled to determine the optimal term of application.

Merits of insectpathogenic viruses can be stated as follows:
- High selective effects against the specific host.
- Virtually no effects on other members of the ecosystem

At the same time in contrast we must also take account of their disadvantages:
- Only a small size of application and therefore a small volume of production is possible.
- Requirement of low temperature about -18° for lengthy storage of virus preparations.
- High amount of materials and working time for mass rearing of caterpillars as the sole economic possibility of virus production in vivo (see also Table 1).

According to Huber (1987) about 30% of worldwide known pests of cultivated plants of agricultural importance could be combated principally by biological means, and especially by insect pathogenic viruses.

POSSIBILITIES OF APPLICATION OF INSECTPATHOGENIC VIRUSES

In our opinion the application of virus preparations in the framework of IPM efforts will be concentrated on the following areas:
- Cultivated plants with large areas of cultivation with occurence of animal pest populations resistant to chemical insecticides (e.g. maize and cotton = Heliothis armigera).
- Perennial ecosystems, such as fruit plantations or forests with a relatively high stability.
- Plants in greenhouses.

A centre of activity for using biological agents including viruses to minimize the occurence of pests in future will surely be for minor uses on minor crops, such as vegetables, ornamental plants, hop, fruits, tobacco, vine, officinal herbs, in crops for producing foods for special dietary purposes, and - last but not least - forests.

TABLE 1. Adventages and disadventages of insect pathogenic virus preparations for controlling pests in frame of IPM.

Attributes of virus preparations in plant control	
positive	negative

- Selectivity
  sparing usefull origins
  few problems with secondary
  pests
	- Selectivity
	  different viruses are needed
	  for different pests

- Good integration in IPM	- Application only on small
	  scale possible, therefore
	  high costs of production

- No negative effects on
  environment
	- Relatively slow increase of
	  older stages of pest species
	  only with little sensitivity

- Increase and spread in pest
  populations, longtern
  effect
	- Little persistence of the
	  viruses, sensitive for UV-
	  rays of sun light

- No selection of resistant
  individuals
	- Inefficient in case of
	  incorrect application

- No residues	
- Also effective with little expense	
- Known technique for appli- cation possible	

An important field for application of insect pathogenic viruses is seen in greenhouse cultures; biological control of insects and mites has been demonstrated. For example: Controlling of insecticide resistant Spodoptera exigua with the specific virus in chrysanthemum cultures in the Netherlands (Vlak et al., 1982).

In our experiments using Mamestra brassicae in greenhouse cultures of roses and paprika, mortality rates of 80 to 100 % after application of the specific nucleopolyhedrosis virus (MbKPV) were obtained. Using chemical insecticides was not possible, while the whole pest management system (e.g. control of aphids and spider mites) was done by biological agents (predators) (Geißler et al., 1989). In field experiments in cabbage, 69 to 100 % of the caterpillars of M. brassicae died after application of MbKPV (Geißler et al. 1991).

An application of the granulosis virus of Agrotis segetum (AsGV) in asters killed 91 to 94 % of the larvae (Geißler and Schliephake, 1991).

In a four-year experiment under natural infestation conditions we showed, that it was possible, by two treatments of peas (<u>Pisum</u> <u>sativum</u>) with the granulosis virus preparation 'Granupom' of <u>Cydia</u> <u>pomonella</u> in the middle and at the end of the flowering period to combat <u>Cydia</u> <u>nigricana</u>. We found mortality rates up to 98 % (Geißler, 1995).

In all these experiments the viruses were as effective as standard, chemical insecticides.

In the past few years the breakdown of populations of <u>Diprion</u> <u>similis</u> (Hymenoptera - Tenthredinidae) and <u>Lymantria</u> <u>dispar</u> (Lepidoptera - Lymantriidae) in different territories of Germany was caused by participation of specific nucleopolyhedrosis viruses.

Specific nucleopolyhedrosis viruses in different concentrations and frequencies, were found in larvae and pupae of successions of forest pests like <u>Lymantria</u> <u>monacha</u>, <u>Dendrolimus</u> <u>pini</u>, <u>Dasychira</u> <u>pudibunda</u>, <u>Bupalus</u> <u>piniarius</u>, <u>Thaumetopoea</u> <u>processionea</u>, <u>Pristiphora</u> <u>laricis</u>, and <u>Gilpinia</u> <u>frutetorum</u> from different host plants and forest boards of Germany since 1991 (Geißler, In Press) (see also Tables 2 and 3). Therefore it should be possible to utilize them for preventing or delaying mass increases of forest pests.

TABLE 2. Infection of different Hymenopteran forest pests by their specific nucleopolyhedrosis viruses.

Species	Stage	Differences in infestation rate (%)
<u>Diprion similis</u>	larvae (living)	9...78
	larvae (dead)	29...65
	pupae	75...96
<u>Pristiphora laricis</u>	larvae (living)	57
<u>Gilpinia frutetorum</u>	larvae (dead)	57...76

TABLE 3. Infection of different Lepidopteran forest pests by their specific nucleopolyhedrosis viruses.

Species	Stage	Differences in infestation rate (%)
Lymantria dispar	caterpillars (living)	32...100
	caterpillars (dead)	26... 60
	pupae (living)	71
	pupae (dead)	100
Lymantria monacha	caterpillars (living)	59... 70
Bupalus piniarius	caterpillars (living)	15... 47
Thaumetopoea processionea	caterpillars (living)	37... 75
Dasychira pudibunda	caterpillars (living)	36
Dendrolimus pini	caterpillars (living)	92

REFERENCES

Cook, R.J.; Baker, K.F. (1983) The Nature and Praxis of Biological Control of Plant Pathogens. American Phytopathological Society, St. Paul, Minnesota, 539 pp.

Geißler, K. (1995) Anwendung und Bedeutung insektenpathogener Viren im integrierten Pflanzenschutz. Archiv für Phytopathologie und Pflanzenschutz, 31, (In Press).

Geißler, K. Untersuchungen zum Vorkommen und zur epidemiologischen Bedeutung insektenpathogener Viren bei aktuellen Forstschädlingen in Sachsen-Anhalt, Brandenburg und Mecklenburg-Vorpommern. Archiv für Phytopathologie und Pflanzenschutz (In Press).

Geißler, K.; Schliephake, E.; Lehmann, W.; Erfurth, P. (1989) Einsatz des Kernpolyeder-Virus (MbKPV-D) gegen die Kohleule (Mamestra brassicae L.) als Schädling an Rosen im Gewächshaus (Kurze Mitteilung). Archiv für Phytopathologie und Pflanzenschutz, 25, 615 - 617.

Geißler, K.; Schliephake, E. (1991) Erste Ergebnisse von Freilandprüfungen des Granulose-Virus der Wintersaateule (Agrotis segetum Schiff.) zur Bekämpfung des Schädlings in Zierpflanzen (Kurze Mitteilung). Archiv für Phytopathologie und Pflanzenschutz, 27, 79 - 80.

Geißler, K.; Schliephake, E.; Rutskaja, Valentina (1991) Untersuchungen zur insektiziden Wirksamkeit des Kernpolyeder-Virus der Kohleule (Mamestra brassicae L.). Archiv für Phytopathologie und Pflanzenschutz, 27, 157 - 161.

Huber, J. (1987) Hochselektive Pflanzenschutzmittel am Bei-
    spiel der Insektenviren. In: Bundesministerium für
    Ernährung, Landwirtschaft und Forsten (Ed.): <u>Biologi-</u>
    <u>scher Pflanzenschutz</u>, Landwirtschaftsverlag, Münster-
    Hilstrup. (Schriftenreihe des Bundesministeriums für
    Ernährung, Landwirtschaft und Forsten, Reihe A: Ange-
    wandte Wissenschaft, **344**), 73 - 80.
Vlak, J.M.; denBelder,E.; Peters, D.; van de Vrie, M. (1982)
    Bekämpfung eines eingeschleppten Schädlings, <u>Spodoptera</u>
    <u>exigua</u>, in Gewächshäusern mit dem autochthonen Virus.
    <u>Mededelingen van de Faculteit Landbouwwetenschappen</u>
    <u>Rijksuniversiteit</u>, Gent, **47**, 1005 - 1016.

Session 4

Molecular Biology and Genetics

Chairman   Dr W SPOOR

Session Organiser   Dr R HARLING

BIOTECHNOLOGICAL METHODS TO PROVIDE MORE SUSTAINABLE PEST, DISEASE AND HERBICIDE RESISTANCE

T M A WILSON

Scottish Crop Research Institute, Invergowrie, Dundee DD2 5DA

No script was provided by this speaker.

DETECTION TECHNOLOGY FOR PLANT PATHOGENS

R. T. V. FOX

School of Plant Sciences, University of Reading, 2 Earley Gate, Reading, RG6 6AU

ABSTRACT

The requirements expected of a test to detect plant pathogens are reviewed. Although a number of technologies are now used, effective diagnostic tests must be simple, accurate, rapid and safe to perform, yet be sensitive enough to avoid "false positives". Often the presence of a disease is hard to identify, quantify or even detect visually, particularly by the inexperienced. Some highly sensitive methods of diagnosis are substantially slower and more laborious than the traditional visual inspection of crop plants for disease symptoms. However these newer techniques require little training to give routine dependable results relatively quickly. A comparison between the different methods of diagnosing plant disease, shows that many methods are complementary rather than alternative options. There is no exclusive or reliably simple method of identifying pathogens or the diseases that they cause, so it is likely that most diagnostic methods will continue to be used or co-exist in some form in the future. The major question is how much the traditional methods, such as identification by visual inspection of pathogens *in situ* or *in vitro* in pure cultures by microscopic examination, will become less widely used if the methods based on pathogen biochemistry, microscopy, immunology and DNA hybridization become more widespread.

REQUIREMENTS OF A DIAGNOSTIC TEST

An effective diagnostic test must be simple, accurate, rapid and safe to perform, yet be sensitive enough to avoid "false positives". Highly sensitive modern methods of diagnosis have usually been adapted from other branches of biology. Most are substantially slower and more laborious than the traditional visual inspection of crop plants for disease symptoms (Fox, 1990a. & b.). Yet these newer techniques require little training to give dependable results on a routine basis.

Usually a quick diagnosis is essential, so the choice of the most appropriate diagnostic technique is often vitally important. Few diagnostic tests can be as quick as an expert examination of specimens visually for symptoms. In practice it is also important to recognise that the value of even the most rapid diagnostic procedure can be wasted if sampling is slow or the result is not immediately available. For example the time that it takes for samples to be mailed or for an expert to travel to a site should be taken into account when comparing different methods for detecting disease outbreaks in the field. There are some diseases, usually foliar, which are usually fairly easily recognised by farmers and growers, but only after the disease has caused the damage that results in the symptom. Also the identification of other types of disease is rarely so simple, even when well advanced (Fox & Hahne, 1988). Diseased roots take longer to examine than foliage because soil obscures the symptoms and wilted plants are often very difficult to diagnose as the pathogen is so deep seated (Fox, et al.,1994). Often fertile agricultural soils contain several different species of pathogens causing similar symptoms. Consequently many soil-borne fungi causing root diseases in plants are rarely quantified easily, even when recognized (Fox & Dusunceli, 1992).

EVALUATION OF CURRENT METHODS TO DIAGNOSE PLANT DISEASE

When a comparison is made between different methods of diagnosing plant disease, it is clear that many methods are complementary rather than alternative options (adapted from Fox, 1993a.).

Visual inspection (including remote sensing)

Advantages
1.   Quick when symptoms are distinct and clearly exposed.
Disadvantages
1.   Symptoms must clearly conform to one of the known syndromes.
2.   Soil obscures symptoms.
3.   Requires much prior knowledge and expertise on the part of the inspector.

Identification of pure cultures of pathogens

Advantages
1.   Morphological taxonomic characters are generally well documented.
2.   Anastomosis and interfertility testing is not difficult and permits separation from otherwise practically indistinguishable related strains or species by plating out mycelium of the test isolate alongside pure cultures of fungi known to be closely related.
Disadvantages
1.   Although occasionally the pathogen may be coaxed into producing sexual or asexual reproductive structures in situ, the production of pure cultures in vitro is required, which is neither rapid nor completely reliable (especially if the person sampling is untrained).
2.   Identification is not always straightforward if literature is unavailable
3.   Specific growth media may not be available.
4.   Anastomosis and interfertility testing requires suitable facilities.

Biochemical methods

Advantages
1.   Substrate utilisation has been well developed for bacteria of medical importance and hence biochemical methods have much potential to diagnose bacterial pathogens in plant pathology.

2. Chromatographic methods are now mature technology, including Polyacrylamide-Gel Electrophoresis (PAGE) which is well established for comparison of protein differences between species previously classified on the basis of their morphological characters.
3. Some distinct protein bands between proteins from related species of pathogens demonstrated by sodium dodecyl sulphate-PAGE may be used as immunogens.

Disadvantages
1. Substrate utilisation has not yet been widely used for fungi.
2. Sufficient volume of an unknown isolate must be produced in pure culture for some chromatographic techniques including SDS-PAGE.
3. General protein or isozyme profiles can only be compared with those of the limited range of pathogens already described and even then differences are frequently slight with quantitative variations in bands.
4. These methods are neither very rapid nor designed to be readily used in the field.

Microscopic examination

Advantages
1. Depends on the recognition of well documented morphological taxonomic characters.
2. Viruses and bacteria can be examined by electron microscopy.

Disadvantages
1. Requires careful expert inspection and equipment.
2. Although fruit bodies and spores may be absent hyphae abound in the host tissue and microscopic differences between them may often aid preliminary identification. However these are rarely acceptable as the sole method of separating a pathogen from a range of similar saprophytes under field conditions.
3. There is a lack of diagnostic stains for fungi.
4. Electron microscopy requires expertise.
5. Microscopy is expensive.
6. Pathogens difficult to locate in a section or on a coated grid if no immunological or specific stain has been used.

Immunological Methods

Advantages
1. Most are simple techniques that require little expertise.
2. Most methods are quick.
3. The results are clear.
4. An accurate result may be obtained.
5. Pathogens which cause diseases with variable or latent symptoms on the host plant can be separated.
6. Pathogens with an indistinct structure or an undistinguished morphology such as in many groups of viruses and bacteria may be distinguished.
7. A number of commercial kits are available.

8. There is an almost unlimited potential for more kits to be produced.
9. Specificity to a particular strain, species, genus or any other taxon may be chosen.
10. Since hybridomas are potentially immortal, an ample source of highly specific monoclonal antibodies may be assured.
11. Selection of hybridomas by monoclonal antibody enzyme-linked immunosorbent asssay is rapid and staightforward, so preliminary purification is not essential for immunisation prior to producing monoclonal antibodies.
12. Adaptable for use in the field in simple monoclonal antibody ELISA kits based on use of filters or magnetic beads coated with antigen and chromogen conjugated to specific antibodies.

Disadvantages
1. Animal handling is still necessary requiring expertise (and a Home Office Licence in UK) despite development of in vitro systems.
2. Specific methods have not yet been developed for most diseases.
3. Too many antigens occur in common between fungi, bacteria and plants to permit polyclonal antibodies to effectively diagnose micro-organisms in host tissues even when unwanted cross reactions have been reduced by using pure antigen.
4. Not effective for viruses that lack a protein coat.
5. Mice have to be immunized with the immunogen preparation 6 months before fusion.
6. Only a few percent of hybridomas can be expected to produce a valuable monoclonal antibody.

Nucleic Acid Techniques

Advantages
1. The maximum sensitivity of detection for most standard versions of hybridization tests is comparable with ELISA.
2. Nucleic acid probes have already been prepared to a range of viral plant pathogens.
3. Nucleic acid probes can detect any part of the genome whereas serological tests are specific to proteins and polysaccharides which are not always accessible.
4. Hybridization tests are useful in quarantine for detecting unknown pathogens (including viroids).
5. A single suitable nucleic acid probe can detect a range of strains.

Disadvantages
1. Immunological tests are more widely used and unlikely to be supplanted.
2. Hybridization tests are not yet widely used against many fungi and bacteria.
3. Hybridization may initially require the prior extraction of nucleic acid from the test sample.
4. Most hybridization has been done with radioactive probes and filter-bound nucleic acids regarded as time-consuming, unsafe and

troublesome to perform even by experts although the sandwich assay has now largely displaced the original awkward laboratory tests and non-radioactive labels are being developed.

5.   DNA hybridization "dot blot" tests are likely to continue to be carried out in a laboratory.

FUTURE TRENDS IN PLANT DISEASE DIAGNOSIS

It is likely that most diagnostic methods will continue to be used or co-exist in some form in the future since there is no exclusive or reliably simple method of identifying pathogens or the diseases that they cause. The major question is how widespread will the methods based on pathogen biochemistry, microscopy, immunology and DNA hybridization become compared to traditional methods such as identification by visual inspection of pathogens in situ or in vitro in pure cultures by microscopic examination (Fox & Cook, 1992; Fox & Hart, 1993).

The inspection of a specimen visually for symptoms by an expert is far quicker than most other diagnostic tests and was until recently, freely available in Britain as well as some other countries. Now most farmers and growers have lost the free advisory support from trained experts, leading to an expansion by consultants who charge for their services.

The identification of many types of diseases is not simple. The diagnosis of wilted plants requires the destruction of the plant as the pathogen is usually deep seated. Diseased roots take longer to inspect than foliage because the plants have to be dug out or pulled up first, and even then soil frequently conceals the symptoms.

Identification of a disease of one of the economically more important crops generally is quite straightforward, as these plants usually have readily accessible and more complete disease descriptions that also describe the pathogens in some detail. Cummins (1969) outlined a diagnostic procedure normally used to identify diseases in which a relatively crude identification of the causal pathogen is usually regarded as authenticated if the symptoms of the disease also correspond to the description in the host index, or simply the index present in the disease literature on the crop. Unfortunately details of many exotic organisms are frequently difficult to find in the literature. This omission is serious as the European Single Market now allows in a greater variety of produce and with it pathogens, including some of those resistant to fungicides (Fox, 1993b.).

At present many pathologists, and even more farmers and students, complain that the once familiar names of the pathogens of common diseases become changed apparently endlessly (and needlessly!). Taxonomists (Hawksworth & Kirsop, 1988) claim that this is largely the result of the inadequate level of knowledge of many genera and species, mainly arising from a shortage of mycologists with the skills needed to develop more satisfactory taxonomic systems. The introduction of improved rules of nomenclature should favour longterm stability, though at the expense of short-term instability. In 1986, the International Commission of the Taxonomy of Fungi

(ICTF) of the International Union of Microbiological Societies (IUMS) started to publish current changes in the names of fungi of importance in the IUMS journal, Microbiological Sciences (Cannon, 1986). These publications also provide the reasons for changes and guidance on their adoption. The ICTF has also prepared a Code of Practice for mycological taxonomists to minimise the changes due to bad practice (Sigler & Hawksworth, 1987) and promote stability. At the same time, well-used names for fungi may be saved under a procedure known as 'conservation', designed to ensure the maintenance of well-known generic names which would otherwise have to be changed by a strict application of the ICBN by review and vote by the Special Committee.

Laboratory based tests such as the methods based on the identification of pure cultures of pathogens, biochemistry, microscopy, immunology or DNA hybridization, do not allow a direct opportunity for Koch's Postulates to be satisfied to provide proof of the pathogenicity of the suspected organism. With classical techniques, pathogenicity must be established before the cause of the disease can be authenticated. In general, this extra stage should be introduced, unless the microorganism is already familiar or its pathogenity is otherwise clearly evident. However conventional pathogenicity tests have the disadvantage of consuming time, space and materials, as well as being subject to environmental conditions that affect symptom expression or even the characteristics of the pathogen.

Once the pathogenic nature of the disease has been corroborated, the keys and descriptions that are used to classify fungi are useful for diagnosis. However these pose some problems for methods based on identification by visual inspection of pathogens in situ or in vitro in pure cultures by microscopic examination. Isolating pure cultures of pathogens to coax the pathogen into producing sexual or asexual reproductive structures, is neither rapid nor completely reliable especially when done by an inexperienced mycologist to whom identification is also not always straightforward. Often the isolation the causal agent of a disease, fungal, bacterial or viral, from the host is not without problems. If the exact region of infection is not clearly defined, then the whole plant must be thoroughly examined for the pathogen, including roots plus attached soil, as well as the aerial parts of the plant. It may be possible to use a non-specific scanning system to locate the presence of the pathogen similar to the infra-red detectors used to detect breast cancer in humans. In future isolation should also be made easier and more conventional by using a wider range of standard media.

The size of the task of searching the literature and the need for current awareness, it is probable when investigating an unfamiliar disease in future, that information from reference books and experienced plant pathologists on common diseases of the crop will be supplemented by data that is electronically stored and retrieved. The most appropriate way forward here seems to be the publication of more information using the new technology offered by the video disc. This system can provide a library of specialist information on a single disc. Since it is possible for even modest personal computers linked to a CD-ROM (compact disk-read only memory) to tap an extensive library of information on the literature including illustrations, it could become possible to connect this well ordered memory with an intelligent scanning system such as used by the police to scan fingerprints to produce a semi-automated system. In medicine this technology has already been coupled to an intelligent computer program to produce an interactive diagnosis "key" for general practitioners.

Electronic systems could be quick but the absence of fruiting bodies in a sample would still require the isolation and growth of the microorganism on specialized media under controlled conditions. Although this may eventually encourage the formation of reproductive structures to be induced, not all fungal pathogens produce fruiting structures. At present if no reproductive structures can be found, if the mycelium is nonseptate, records of Oomycetes or Zygomycetes should be examined; but if septate it is often possible to separate Ascomycetes from Basidiomycetes by transmission electron microscopy, although this is rather slow. This sort of microscopic examination is also largely restricted to pathogens whose morphology is sufficiently well-defined to detect distinct taxonomic traits. Although electron microscopy can be used to identify viruses and bacteria in this way, it is expensive and pathogens can prove difficult to locate in a section or on a coated grid if no immunological or specific stain is available. In future it is likely that a range of labelled antibodies will become available. Nevertheless, some consideration should be given to discover techniques to improve on the use we make of hyphae to aid preliminary identification, since at present they are seldom acceptable as the sole basis for diagnosis. One area for improvement would be the development of a range of simple diagnostic stains for the light microscopy of fungi similar to those used in bacteriology, because at present, apart from those based on immunology, such specific fungal stains are generally lacking.

Traditional methods generally still need more experience on the part of an investigator than do tests based on differences in nucleic acids and immunology, where knowledge is increasingly becoming replaced by expensive equipment and reagents. Visual inspection can be instantaneous when symptoms clearly conform to a well known syndrome but is difficult when they are not. Historically the primary route to identification in the laboratory was cultural isolation followed occasionally by biochemical and/or immunological tests. Immunoassays have been revolutionized by the introduction of monoclonal antibodies and Enzyme-Linked Immunosorbent Assays (ELISA) that have made them routine, thus allowing completion in several hours instead of the days or even weeks taken by culturing. The main drawback of ELISA is the level of nonspecific binding. This problem led to efforts to find a method of rapid diagnosis of *Armillaria* by monoclonal antibody ELISA (Fox & Hahne 1988) being replaced by an investigation using PCR.

Apart from techniques based on immunology, most laboratory diagnoses have proved ill-suited for field use as they are neither sufficiently flexible nor portable.

SYSTEMS OF DIAGNOSIS IN FUTURE

While traditional methods are sure to be used into the future probably with increasing help from electronic aids, it is clear we are already entering a period of great change. The range of choice of relatively cheap, easy to use diagnostic kits now being developed should allow farmers and growers to monitor low levels of disease on the spot under field conditions (Klausner, 1987; Miller and Martin 1988; Miller et al., 1988, 1990; MacAskill, 1989). Both immunological and nucleic acid hybridization techniques are increasingly becoming developed for the rapid detection of many of those pathogens of plants which cannot be easily identified by other routine ways. For example these methods can quickly and accurately identify pathogens which cause diseases with variable or latent symptoms on the host plant. Equally, pathogens with an indistinct structure or an undistinguished morphology, such as in many groups of viruses, bacteria and fungi, particularly imperfect fungi (especially those spreading as a sterile mycelium) can now be reliably detected and identified in host tissue at an early stage and hence more effectively eradicated. They should also allow changes in the strains and races of a pathogen to be monitored quickly.

Mycotoxins often need to be monitored but since many are simple non-antigenic chemicals, a branch of diagnostics has to be used based on hapten technology, in which the mycotoxin is bound to a known antigen (Klausner, 1987; Candlish et al., 1989). Antibodies produced by such techniques are likely to become increasingly important in crop protection (Klausner, 1987). Minute levels of pesticide residues may be detected by similar kits without the need for expensive laboratory equipment (Niewola et al., 1983, Van Emmon et al., 1987; Coxon et al., 1988; Tomita et al., 1988). As such detection methods are simple to use, groups of consumers who are worried about pesticide residues in their food and environment, could have direct access to reliable assay facilities for the first time. At the same time methods based on ELISA are being used to develop the rapid detection of pathogenic fungi resistant to fungicides based on carbendazim (Groves et al., 1988; Martin et al., 1992a. & b.).

Immunology has already provided cheap kits for the pesticide industry that are so sensitive, farmers could detect and hence treat lower levels of pathogens than previously.

Although the majority of nucleic acid hybridization "dot blot" tests are still likely to continue to be confined to the laboratory, new developments such as immunocapture could allow increased portability in the future. The handicap of non-mobility must be overcome as it can only reduce the usefulness of this valuable technique to practical plant pathologists, who often need more accessible methods of diagnosis. Market forces should ensure that other analytical methods based on molecular hybridization will continue to be developed that are no more restricted than immunological tests. Unless this happens nucleic acid hybridization seems destined to remain somewhat longer only in the hands of the advisory or consultancy services, rather than those of the field worker or farmer.

Both immunological and nucleic acid hybridization techniques are increasingly becoming developed for the rapid detection of many of those pathogens of plants which cannot be easily identified by other routine ways. The prospect of cheap tests may not benefit the advisory and consultancy services, but when tests become even cheaper they should be affordable for countries in the developing world.

REFERENCES

Candlish, A.A.G.; Stimson, W.H.; Smith, J.E. (1992) Assay methods for mycotoxins using monoclonal antibodies. In: *Techniques for the rapid diagnosis of plant pathogens*. Ed. for BSPP by Duncan, J.M. & Torrance, L. Blackwell, Oxford., 63

Candlish, A.A.G.; Stimson, W.H.; Smith, J.E. (1989) Assay methods for mycotoxins using monoclonal antibodies. *Abstracts of Papers, The British Society for Plant Pathology/British Crop Protection Council Conference on Techniques for the Rapid Diagnosis of Plant Disease*. University of East Anglia, Norwich, 11.

Cannon, P. F. (1986). Name changes in fungi of microbiological, industrial and medical importance, I-II. *Microbiological Sciences* **3**, 168-171, 285-287.

Coxon, R.E.; Rae, C.; Gallacher, G.; Landon, J. (1988) Development of a simple fluoroimmunoassay for paraquat. *Clinica Chimica Acta*, **175**, 297-305.

Cummins G.B. (1969) Identification of fungal pathogens In: Tuite, J. *Plant pathological methods, fungi and bacteria*. Minneapolis: Burgess Publishing Company.

Dusunceli F.; Fox R.T.V. (1992) The accuracy of methods for estimating the size of Thanatephorus cucumeris populations in soil. *Soil Use & Management* **8**, 21-26.

Fox, R.T.V. (1990a) Rapid Methods for Diagnosis of Soil Borne Plant Pathogens. In: *Soil-borne Diseases*. Ed. Hornby, D. Special Issue. Soil Use & Management **6**, 179-184.

Fox, R.T.V. (1990b) Diagnosis and control of Armillaria honey fungus root rot of trees. *Professional Horticulture* **4**, 121-127.

Fox, R.T.V. (1993a) *Principles of Diagnostic Techniques in Plant Pathology*. CAB International: Wallingford.

Fox, R.T.V. (1993b) Prospects for the diagnosis and control of soil-borne pathogens. *International Symposium on Plant Health and the European Single Market*, 409-412.

Fox, R.T.V.; Manley, H.M.; Culham, A.; Hahne, K.; Tiffin, A.I. (1994) Methods for detecting Armillaria mellea. In: *Ecology of Plant Pathogens*. Blakeman, J.P. (ed.). Blackwells/BSPP: Oxford.

Fox, R.T.V.; Cook, R.T.A. (1992) Powdery mildew on faba beans and other legumes in Britain. *Plant Pathology* **41**, 506-512.

Fox, R.T.V.; Hahne, K. (1988) Prospects for the rapid diagnosis of Armillaria by monoclonal antibody ELISA. *Proceedings of the 7th International Conference IUFRO Working Party S2.06.01 on Root and Butt Rots of Forest Trees*, Vernon & Victoria, British Columbia, Canada. August 9-16 1988, 458-468.

Fox, R.T.V.; Hart, C.A. (1993) Microscopy in the study of plant disease. *Microscopy* **39**, 64-72.

Groves, J.G.; Fox, R.T.V.; Baldwin, B.C. (1988) Tubulin from Botrytis cinerea and the potential for development of an immunodiagnostic for benzimidazole resistance. *Proceedings 1988 Brighton Crop Protection Conference - Pests and Diseases* **1**, 415-420.

Hawksworth, D.L.; Kirsop, B.E. (1988) (Ed.) *Filamentous fungi. Living resources for biotechnology.* Cambridge: Cambridge University Press.

Klausner, A. (1987) Immunoassays flourish in new markets. *BioTechnology* **5**, 551-556.

MacAskill, J. (1989) Diagnosic kits for cereal diseases. *Abstracts of Papers, The British Society for Plant Pathology/British Crop Protection Council Conference on Techniques for the Rapid Diagnosis of Plant Disease.* University of East Anglia, Norwich, 10.

Martin, L-A,; Fox, R.T.V.; Baldwin, B.C.; Connerton, I.F. (1992a.) Rapid methods for the detection of MBC resistance in fungi: II. Use of the polymerase chain reaction as a diagnostic tool. *Proceedings 1992 Brighton Crop Protection Conference - Pests and Diseases* **1**, 207-214.

Martin, L-A,; Fox, R.T.V.; Baldwin, B.C.; Connerton, I.F. (1992b.) Rapid methods for the detection of MBC resistance in fungi: II. Use of the polymerase chain reaction as a diagnostic tool. *Proceedings 1992 Brighton Crop Protection Conference - Pests and Diseases* **1**, 207-214.

Miller, S.A.; Martin, R.R. (1988) Molecular diagnosis of plant disease. *Annual Review of Phytopathology* **26**, 409-432.

Miller, S.A.; Rittenburg, J.H.; Petersen, F.P.; Grothaus, G.D. (1988) Application of rapid, field useable immunoassays for the diagnosis and monitoring of fungal pathogens in plants. *Brighton Crop Protection Conference - Pests and Diseases,* 795-803. British Crop Protection Council; Thornton Heath.

Miller, S.A.; Rittenburg, J.H.; Petersen, F.P.; Grothaus, G.D. (1990) Development of modern diagnostic tests and benefit to the farmer. In: *Monoclonal Antibodies in Agriculture* pp 15-21. Ed. A. Schots. Pudoc; Wageningen, Netherlands.

Niewola, Z.; Walsh, S.T.; Davies, G.E. (1983) Enzyme-linked immunosorbent assay (ELISA) for paraquat. *International Journal of Immunopharmacology* **5**, 211-218.

Sigler, L.; Hawksworth, D.L. (1987). Code of practice for systematic mycologists. Microbiological Sciences **4**, 83-86, Mycopathologia **99**, 3-7, Mycologist **21**, 101-5.

Tomita, M.; Suzuki, K.; Shimosato, K.; Kohama, A.; Ijiri, I. (1988) Enzyme-linked immunosorbent assay (ELISA) for plasma paraquat levels of poisoned patients. *Forensic Science International* **37**, 11-18.

Van Emmon J.; Seiber, J.N.; Hammock, B. (1987) Application of an enzyme-linked immunosorbent assay (ELISA) to determine paraquat residues in milk, beef and potatoes. Bulletin of Environmental Contamination and Toxicology **39**, 490-497.

MOLECULAR PERSPECTIVES IN CROP PROTECTION:INTEGRATION WITH ARABLE CROP PRODUCTION

G MARSHALL

Department of Plant Science, Scottish Agricultural College, Auchincruive, Ayr, KA6 5HW

ABSTRACT

This paper reviews selected areas where molecular techniques are presently making an impact upon crop protection including biorational pesticide design and transgenic crops for improved resistance to diseases, pests and herbicides. The scientific, environmental, and commercial aspects surrounding the adoption of herbicide-resistant crops are described using examples from current Canadian cropping practices. These examples illustrate the need to bridge the field trial experience between regulatory/research experimentation and farmer-orientated trials which incorporate an appreciation of crop rotational systems, the associated weed flora and herbicide use.

INTRODUCTION

Molecular biology has its origins in applied biology and chemistry progressing via the structure and function of DNA to our present knowledge of molecular genetics. It is from this evolving scientific base that molecular biology has been harnessed to hasten progress in a range of disciplines including crop protection. The impact of molecular biology on crop protection has been subject to several reviews including Marshall and Atkinson (1991), Gatehouse et al. (1992) and Marshall & Walters (1994). The object of this review is to consider the current status of the principle applications of molecular biology in crop protection: design of chemical crop protection agents, the development of crop cultivars resistant to herbicides, diseases and pests and assessing the environmental impact of genetically modified organisms. The practical integration of herbicide resistant crops into low input arable crop production systems will be examined by the use of case studies.

BIORATIONAL DESIGN OF NEW PESTICIDES

Crop protectionists continue to be required to apply all their ingenuity and skill to improve the food supply for arable agriculture given the ever increasing demands from world population growth. This requirement is now framed in a background where it is increasingly difficult and expensive to screen, identify and market a useful new pesticide. Furthermore, each year the loss of approved pesticides via unfavourable toxicological properties is rarely balanced by the gain of new products. New pesticides must be effective at low rates, provide crop safety, possess minimal environmental impact and favourable toxicological properties. While traditional methods of pesticide synthesis and subsequent screening are likely to remain the primary source of crop protection products, advances in our knowledge related to the mode of action of existing pesticides can be used to probe novel biochemical sites of action. Accordingly, new research based upon our biochemical knowledge can be described as biorational (reviewed by Pillmoor & Foster, 1994). Essentially, biorational design of pesticides can be viewed as a logical adjunct to traditional approaches for identifying and developing new pesticides.

In biorational design the starting point may be the identification of a new biochemical target. It may be possible to determine the effect of inhibition of that target site by reference to the use of a traditional mutant organism which has previously been characterised e.g. *Arabidopsis*. A molecular approach might also be applied where the gene has been isolated which is responsible for a specific enzyme in a plant or fungus. In plants, the gene may be nullified in its effect by using anti sense RNA technology, best known for the delayed ripening tomato. The effect of the enzyme system on physiological and metabolic plant processes can then be studied. In selected fungi the analogous process is known as gene disruption and it has application in the study of mutation vs pathogenicity (Stahl & Schafer, 1992).

Thus the design of new chemical inhibitors relies upon our understanding of a particular enzyme. Knowledge about known inhibitors is often the starting point for further investigations e.g. metabolism-based herbicide selectivity (Brown *et al.*, 1991). In practice the most fruitful approach to inhibitor design has been through a consideration of the chemical mechanism employed by the enzyme (Pillmoor *et* al., 1991). Still, biorational design of new pesticides is still developing and evolving since to date there are no commercial examples where this approach to new pesticide discovery has succeeded. However, in the pharmaceutical area this approach notably for antibacterial and anticancer treatments has become productive (Kuyper, 1990).

TRANSGENIC CROPS

It is a prerequisite of sustainable systems of arable agriculture that continuous improvements are made in the provision of new varieties. Traditional technologies employed by plant breeders have over the past two decades been supplemented by new biotechniques including the adoption of genetic engineering. While the general breeding objectives in crop cultivars have seen trends towards a greater emphasis upon crop quality and resistance to pests and diseases it is in these target areas that cell and molecular biology techniques can be exploited. It is now technically possible to identify, isolate, clone and transform single genes into a range of crop plants.

Transgenic crops are already a practical reality and have been released for controlled field experiments in a wide range of countries around the world. The dominant themes are resistance to virus diseases, insect pests and herbicides. Examining the published release permits for trials around the world (Table 1) provides a clear indication of the future opportunities (Beck & Ulrich, 1993).

Table 1.   Number of approvals granted by crop and trait to 1993*

	Traits							
	Resistance to:							
Crop	Herbicides	Viruses	Fungal diseases	Insects	Crop quality	Crop fertility	Stress resistance	% Total
Rapeseed	242	2	2	1	21	27	2	37.2
Potato	14	46	19	29	17		2	16.0
Maize	42	8	1	13	3	5		9.1
Tomato	18	14	5	13	19			8.7
Flax	45							5.7
Cotton	26			15				5.2
Soybean	35		1		3			4.9
Sugar beet	21	8						3.7
Alfalfa	11	6				1	4	2.8
Others	8	34		5	4	4		7.0

* After Beck and Ulrich (1993)

Resistance to diseases and pests

It is apparent that our understanding of host-pathogen interactions is far from complete. Therefore only after very detailed studies with plant-virus interactions has it been possible to engineer pathogen-derived resistance in plants (reviewed by Ward et al., 1994). In this resistance strategy the functions of the viral genomes are transferred to the host plant in order to interfere with the normal life cycle of the virus. Thus host expression of pathogen-derived genes is responsible for the protection against the specific virus disease.

Although the plant-virus interaction system has been well characterised it is clear that understanding resistance to fungal pathogens involves several extra levels of complexity. Therefore defining the genes of critical importance in the host response will not be a simple matter but rather will require an increased understanding of the biochemistry and molecular genetics of the host-pathogen response (Ward et al., 1994). For the forseeable future it is likely that crop cultivars will rely for fungal disease resistance upon traditional breeding technologies.

Protection against insect pests which cause crop damage and may transmit viruses has conventionally been achieved by applying pesticides. Now however the use of pesticides can be reduced where insect-resistant crops are adopted. In general terms there are three strategies currently available (Gatehouse & Hilder 1994). First, the use of plant derived insecticidal genes such as proteolytic enzymes; second, insect-resistant transgenic plants expressing plant derived genes e.g. cowpea trypsin inhibitors (CpTI)- see Hilder et al. (1993); third, insect-resistant transgenic plants expressing the insecticidal toxin normally produced by *Bacillus thuringiensis* (*B.t.*), reviewed by Peferoen (1992) and Barton & Miller (1993).

One of the emerging problems in relation to field studies with this strategy for insect resistance is the development of resistance by insects to the *B.t.* crystal proteins (Tabashnik *et al.*, 1990). Gatehouse & Hilder (1994) concluded that insect resistant transgenic plants are a viable means of producing crops with significantly enhanced levels of resistance and the adoption of the technology was not limited by suitable genes but rather regulatory barriers and consumer acceptability.

Herbicide resistant crops: principles and current practices

The use of herbicides in arable crop production has revolutionised our ability to manipulate the availability of water, minerals, light and space in favour of crops while weeds are controlled. In addition, selective herbicides have evolved by the efforts of industry to possess low toxicity to non-target organisms and dissipate rapidly in the environment. Our knowledge of the biochemical, physiological and genetic basis of herbicide mode of action is now advanced to the extent that many of the world's major crops can now be transformed to confer herbicide resistance. Ambitions to produce herbicide-resistant crops are driven via two principal mechanisms. First, herbicides are frequently well characterised in terms of the biochemical basis of gene function. Therefore gene isolation, cloning and transformation of plants together with a readily-selectable morphological marker provide a challenging academic system to investigate. Clearly, where single genes can be manipulated in this fashion an excellent model system is established to provide a guide for other gene acquisitions in plant improvement. Second, the vast majority of this research has been funded by the agrochemical industry and its plant breeding or biotechnology-related partners. The private sector have essentially used modern environmentally benign herbicides which are not off patent to produce an opportunity to maximise the return on their research investment in both the herbicide and novel plant varieties. The practical consequences of this approach are outlined in a later section.

The scientific background, techniques and current state of the art in herbicide resistant crops are reviewed by Gressel (1993) and Cole (1994). A summary of the anticipated launch years for a selection of world crops is presented in Table 2. Clearly the current emphasis is on the development of crop resistance to two non-selective herbicides, glyphosate with its renowned translocation ability and glufosinate (phosphinotricin) for its contact and limited translocation properties. Within the next few years farmers in the UK are likely to have the opportunity to grow herbicide resistant rapeseed and perhaps sugar beet. As Canadian farmers are planting herbicide resistant rapeseed this year it is a useful case study to consider with a view to examining the integration of this development in arable agriculture.

Herbicide-resistant crops in low-input production systems

Rapeseed is grown on some 3.0 million ha in Canada. The production system uses spring-sown cultivars only and by comparison with rapeseed production in the UK can be considered low-input. The only significant crop protection chemical applied is

Table 2. Anticipated commercial availability of herbicide-resistant crops

Crop	Country	Phosphinothricin	Glyphosate	Imazethapyr	Chlorimuron
Rapeseed	Canada	1995	1995	1995	-
	Europe	1997-98	1999-2000	-	-
Soybeans	USA	1997	1996	1995	1993
Maize	USA	1997-98	2000	1991	-
Cotton	USA	1998	1998	-	-
Sugar Beet	USA	2000	2000	-	-
	Europe	2001	1998	-	-
Wheat	USA/Europe	>2000	>2000	-	-

herbicide. Although rapeseed is a competitive crop, uncontrolled cruciferous weeds, wild oats, *Setaria* species and cereal volunteers can reduce the crop yield and quality significantly. Traditional weed control programmes relied upon trifluralin for broad/grass weeds with a follow-up post-emergence graminicide application. In 1990 a new selective sulfonylurea herbicide (ethametsulfuron) was approved for use specifically to control the ubiquitous *Sinapis arvense* (wild mustard). Recently Canadian farmers have become aware of widespread resistance of grass weeds to the popular graminicides (acetolactate synthase or ALS inhibitors) and the introduction of herbicide-resistance rapeseed will provide a new management option.

For 1995 the Canadian farmer has three options with herbicide resistant rapeseed. This concept is not revolutionary since triazine-tolerant rapeseed varieties were used in Canada during 1985-90 (Marshall, 1987). The first option open to selected farmers is the Roundup Ready® Canola (rapeseed) to be 'trial-grown' on 800 ha. Monsanto will oversee the crop production, harvest and seed crushing. This introductory field production will serve to create awareness of the product and will undoubtedly generate subsequent demand for seed from farmers, assuming the variable costs are in line with maintaining the gross margin for the crop.

Glyphosate will be recommended for application at the 0-6 leaf stage of crop growth with use rate of *ca.* 356 g a.i./ha. Repeat applications may be required to control late weed growth especially since glyphosate's spectrum of activity favours grass weeds rather than broad leaved species at these low rates of application. The level of resistance to glyphosate is moderate only therefore transient crop yellowing may be noted. In addition only one Roundup Ready® canola cultivar is presently available based upon the previously popular cultivar Westar. Thus the agronomic performance of this transgenic cultivar (ignoring herbicide-resistance) will be generally inferior to currently available non-transgenics. To counter this initial lack of choice for farmers, Monsanto hope by 1997 to have 8 other cultivars available all with glyphosate resistance.

The second option is the use of the AgrEvo herbicide/cultivar package which is based upon the canola cultivar Innovator and Liberty Link® (glufosinate resistance). Again, the agronomic performance of the cultivar is not on a level with existing non-transformed genotypes but 4-5 new cultivars are awaited for 1997. Seed is available to treat some 16,300 ha in 1995. The resultant canola must be segregated from other rapeseed and sold only into the North American market. Crop safety following the use of glufosinate (at any stage of crop growth) is excellent although at the rates of use proposed (300 g a.i. ha) repeat applications will probably be required in one growing season especially since volunteer cereals and small perennial weeds may prove difficult to control.

The third option is a non-transgenically produced herbicide resistant canola, cv. Pursuit Smart® released via collaboration by American Cyanamid and Pioneer Hi-Bred companies. The cultivar is resistant to the imidazolinone herbicide inazethapyr (post emergence, selective only in legume crops, residual and translocated activity). Over 57,000 ha of Pursuit Smart canola could be planted in 1995. This herbicide has the advantages of requiring only one application per season and crop tolerance is good. Imazethapyr is however relatively weak on volunteer cereals and will not control ALS resistant weed biotypes which are already part of Canadian prairie agriculture. The agronomic performance of this cultivar appears to significantly better than Westar upon which most of the transgenic canolas are based.

The integration and adoption of these herbicide-resistant cultivars will depend on both economic and agronomic factors. At the moment a traditional herbicide programme (trifluralin or ethylfluralin or ethametsulfuron or clopyralid followed by a graminicide) would cost about $45-75/ha plus seed costs $11-50 /ha (mean cost $62-80 /ha). By contrast estimates for Pursuit Smart® are $45/ha for seed and $45/ha for herbicide, for Innovator/Liberty Link® $42/ha seed and $45-90/ha for herbicide and finally Roundup Ready® canola $87 for seed and $5-10/ha for herbicide. Overall, the extra cost of adopting the new herbicide resistant canola will be some $25-50/ha plus the disadvantage that some of the present herbicide-resistant cultivars may not show the same yield, quality and disease resistance as recent non-herbicide resistant cultivars.

Therefore it is evident that rather than becoming an overnight success and relegating traditional production systems to a more minor role, herbicide resistant rapeseed cultivars will occupy a specialist niche in Canadian agriculture. It may indeed appear ironic to those who considered new herbicide resistant crops would increase the risk of spreading resistance genes in the environment to discover their utility in weed control programmes designed to reduce the impact of existing herbicide resistant weeds. Certainly these remarks apply for the non-selective glyphosate and glufosinate-resistant rapeseed varieties. However, the use of imidazolinone-resistant rapeseed in areas where ALS-resistant weeds were present could not be recommended. The opportunity to shift the emphasis of soil-applied herbicides such as trifluralin towards post-emergence herbicides made possible by the herbicide-resistant rapeseed will be welcome as a means of minimising unnecessary tillage thus preventing soil erosion and enhancing moisture conservation. Similarly, herbicide carryover from one season to the next will not be a problem for either glyphosate or glufosinate.

A second Canadian example of the integration of a transgenic crop into a traditional cropping programme has recently been described by McHughen and Holm (1995). In this field study the concerns raised about the commercialisation of transgenic herbicide-resistant crops (increased useage of herbicides, non-sustainable practices, lack of gene expression in the field or agronomic penalties) were addressed in a three year field trial using sulfonylurea-resistant linseed cultivars. The results showed that at least one transgenic line was fully resistant to the field rates of herbicide, no agronomic penalties were shown in the presence or absence of herbicide and the adoption would lead to reduced chemical usage and more sustainable agronomic practices in commercial production.

Herbicide resistant crops: future issues

The above represents an interpretation of the immediate impact following release of these herbicide-resistant rapeseed cultivars. There remains however some longer-term issues which are not so easy to resolve or predict with certainty. Pricing policies of the vendors of the herbicide-resistant seeds and the associated herbicides will undoubtedly have a major influence on the adoption of these crops by farmers. Similarly unless the agronomic performance of these cultivars sold at 'premium' prices can more closely compete with the best of traditional cultivars they will remain as minor use or relegated to obscurity. It will also be interesting to see if the market and consumer loyalty for the high quality image of Canadian rapeseed (canola) will remain unmoved by the introduction of the transgenic herbicide resistant cultivars.

The remaining environmental issue which is presently incompletely resolved with universal satisfaction is that of the possible introgression of herbicide resistance genes from rapeseed into weedy relatives such *Sinapis arvensis*. Controlled and natural interspecific crosses were performed by Downey *et al.* (1991) among four Brassica species and *S. arvensis*. These authors concluded that gene transfer from the three major oilseed species to *S. arvensis* was not achieved under the most favourable conditions, and no hybrids were identified from natural crossing of these species when they were co-cultivated in field plots over a three year period. Still these authors acknowledged that although gene transfer among the oilseed-brassicas under natural conditions can and probably does occur, the natural barriers for such gene flow in the weedy species is formidable and would not occur. Similarly Darmency (1994) concluded that hybrids between rapeseed and *S. arvensis* set no seeds, however those between the crop and wild radish (*Raphanus raphanistrum*) set 0.3 viable seeds per hybrid in the first backcross generations. These results show that gene introgression in wild Brassica populations can occur at different rates in different species.

Clearly the opportunities for introgression of herbicide-resistance genes are going to depend upon the local associated vegetation, the flowering dates of the species and reproductive behaviour of the various plants. To date risk assessment field studies have been criticised for their lack of attention to the dynamics of pollen flow (Mellon and

Rissler, 1995) although the invasiveness of transgenic rapeseed in 12 different habitats in the UK has been reported by Crawley *et al.* (1993). While the transgenic rapeseeds included in this research proved no more invasive than non-transformed rapeseed the authors cautioned that risks for other transgenics must be assessed on a case by case basis.

Volunteer crops represent some of the potentially most serious weed control problems and herbicide resistant crops might potentially reduce the herbicide choice which farmers have in their control. With a glyphosate, glufosinate or imidazolinone-resistant rapeseed, volunteer control should be possible by the application of a phenoxyalkonoic herbicide similar to non-transgenic rapeseed. If however, glyphosate-resistant potato cultivars were introduced, volunteers would present a serious weed problem since glyphosate is presently a preferred method of volunteer potato control. It is obvious that the introduction of a herbicide-resistant crop cultivar must be carefully considered with respect to the existing cropping regimes and herbicide availabilities for a region or country.

CONCLUSIONS

Within the rapid evolution of techniques in molecular biology there can be no doubt that many aspects within the food and fibre production chain can benefit scientifically from their application. The residing uncertainties in terms of the benefits which will be accrued in practical crop production are principally concerned with the unchartered territory between laboratory or researcher trials and commercial production. Present world-wide trialling of transgenic crops has its focus on herbicide resistance conferred by single genes, but in years to come should the transformation of polygenes become a reality this present development will become eclipsed. As we adopt such high-technology crops into our traditional systems with all their heritage of regulated trialling and release, Dyer (1994) asks the prudent question will anyone monitor the use of herbicide resistant crops ? In the UK if we are to integrate the benefits which molecular technologies can bring to sustainable systems of crop production we need to consider which transgenic crops are most suitable, which transgenes should be used and which should be rejected as unsuitable for our cropping systems. Perhaps the real test for the products of molecular biology is just about to begin in earnest.

REFERENCES

Barton, K A; Miller M J (1993). Production of *Bacillus thuringiensis* insecticidal proteins in plants. In: *Transgenic Plants*, Volume 1, S Kung & R Wu (eds). Academic Press, London. pp. 297-315.

Beck C I; Ulrich T H (1993). Environmental release permits, valuable tools for predicting food crop developments. *Bio/Technology* **11**, 1524-1528.

Brown H M; Deitrich R F; Kenyon W H; Lichter F T (1991). Prospects for the biorational design of crop selective herbicides. In: *The Brighton Crop Protection Conference - Weeds* - 1991 **2**, 847-856.

Cole D (1994). Molecular mechanisms to confer herbicide resistance. In: *Molecular Biology in Crop Protection*, G Marshall & D R Walters (eds). Chapman and Hall, London. pp. 146-176.

Crawley M J; Hails R S; Rees M; Kohn D; Baxton J (1993). Ecology of transgenic oilseed rape in natural habitats. *Nature* **363,** 620-623.

Darmency H (1994). The impact of hybrids between genetically modified crop plants and their related species: introgression and weediness. *Molecular Ecology* **3,** 37-40.

Downey R K; Bing D J; Rakow G F W (1991). Potential of gene transfer among oilseed *Brassica* and their weedy relatives. In: *Proceedings of the 8th International Rapeseed Congress*, Saskatoon, Canada. pp.1022-1027.

Dyer W E (1994). Herbicide-resistant crops: a weed scientist perspective. *Phytoprotection* **75** (suppl), 71-77.

Gatehouse A M R; Hilder V A (1994). Genetic manipulation of crops for insect resistance. In: *Molecular Biology in Crop Protection,* G Marshall & D R Walters (eds). Chapman and Hall. pp. 177-201.

Gatehouse A M R; Hilder V A; Boulter D (eds) 1992. Plant Genetic Manipulation for Crop Protection. CAB International. pp. 1-260.

Gressel J (1993). Advances in achieving the needs for biotechnologically derived herbicide resistant crops. *Plant Breeding Reviews* **11,** 155-198.

Hilder V A; Gatehouse A M R; Boulter D (1993). Transgenic plants for conferring insect tolerance: protease inhibitor approach. In: *Transgenic Plants* Volume 1, S Kung & R Wu (eds). Academic Press, London. pp. 317-338.

Kuyper L F (1990). Receptor-based design of dihydrofolate reductase inhibitors. In: *Protein design and the development of new therapeutics and vaccines*, J B Hook and G Poste (eds). Plenum, New York. pp. 297-327.

Marshall, G (1987) Implications of herbicide-tolerant cultivars and herbicide-resistant weeds for weed control management. In: *Proceedings of the 1987 British Crop Protection Conference - Weeds.* pp. 489-498.

Marshall G; Walters D R (eds) (1994). Molecular Biology in Crop Protection. Chapman and Hall, London. pp. 1-283.

Marshall G; Atkinson D (1991). Molecular biology: its practice and role in crop protection. BCPC Monograph No 48, pp 1-93.

McHughen, A; Holm F A (1995). Transgenic flax with environmentally and agronomically sustainable attributes. *Transgenic Research* **4,** 3-11.

Mellon M; Rissler J (1995). Transgenic crops: USDA data on small-scale tests contribute to commercial tisk assessment. *Bio/Technology* **13,** 96.

Peferoen M (1992). Engineering of insect-resistant plants with *Bacillus thuringiensis* crystal protein genes. In: *Plant Genetic Manipulation for Crop Protection - Biotechnology in Agriculture No 7.* A M R Gatehouse; V A Hilder & D Boulter (eds), CAB International, Wallingford. pp. 135-153.

Pillmoor J B; Foster S G (1994). Molecular approaches to the design of chemical crop protection agents. In: *Molecular Biology in Crop Protection,* G Marshall & D R Walters (eds). Chapman and Hall, London. pp. 41-67.

Pillmoor J B; Wright K; Lindell S D (1991). Herbicide discovery through rational design: some experiences. In: *The Brighton Crop Protection Conference - Weeds - 1991.* **2,** 857-866.

Stahl D J; Schafer W (1992). Cutinase is not required for fungal pathogenicity on pea. *The Plant Cell.* **4,** 621-629.

Tabashnik B E; Cushing N L; Finson N; Johnson M W (1990). Field development of resistance to *Bacillus thuringiensis* in diamond back moth (Lepidoptera: Plutellidae). *Journal of Economic Entomology,* **83**, 1671-1676.

Ward E; Uknes S; Ryals J (1994). Molecular biology and genetic engineering to improve plant disease resistance. In: *Molecular Biology in Crop Protection* G Marshall & D R Walters (eds). Chapman and Hall. pp. 121-145.

Session 5
Landscape Management

Chairman and
Session Organiser   Dr L FIRBANK

LANDSCAPE ECOLOGICAL PRINCIPLES AND SUSTAINABLE AGRICULTURE

G.L.A. FRY

Norwegian Institute of Nature Research, PO Box 5064, The Agricultural University of Norway, N-1432 Aas, Norway.

ABSTRACT

The management of agricultural land to meet the goals of sustainability cannot be based on field or farm level decisions alone. A landscape approach offers new perspectives on pest control, soil erosion and hydrological management. Reducing external inputs of energy and chemicals or replacing them with ecological services will also demand a landscape-scale planning perspective. In this paper I explore the potential consequences of introducing landscape ecological principles to farm management with special emphasis on reducing environmental problems, increasing nature conservation interests and enhancing integrated pest management.

INTRODUCTION

A landscape ecology perspective on integrated crop management and sustainability can be stated quite simply as 'space is the final frontier'. This famous phrase was also used by Liebhold, et al., (1993) in a paper which described the rapidly developing fields of spatial statistics and geographic information systems (GIS) in insect ecology. In this paper I argue the case for a greater appreciation of the role spatial process cam play in solving current and future crop management problems. The field of ecology that specifically deals with spatial processes is landscape ecology. The name is a little misleading as the content matter deals with many scales not just that of human landscapes. The landscape of cabbage root fly larva, for example, is rather small, whereas that of a sheep grazing on moorland is clearly much larger.

Problems associated with the enhancement of beneficial insects at the farm or regional levels cannot be solved at the site level alone. Studies on the movement of arthropods in agricultural landscapes have shown that larger-scale dispersion can seriously limit the interpretation of small-scale plot experiments, e.g. to evaluate the effects of biocides or habitat manipulation on the density of beneficials. Since both pests and their potential control agents posses spatial dynamics operating at larger scales than the field or farm, we clearly need to improve our understanding of these processes. There are, of course, many constraints to doing this. First there is the cost and commitment to undertaking large-scale and long-term studies. Next there are very real practical problems with establishing landscape -scale research. These are not just related to the levels of resources needed to work at larger scales, but to the difficulties in setting up experimental protocols to satisfy statisticians and provide the necessary controls and replication. These problems are both real and in some ways insurmountable. Nevertheless, through methods such as case studies used for hypothesis generation advances are being made, and agroecologists are slowly moving away from small-plot scale experiments to study larger-scale processes. This is clearly evidenced by the research interest in full scale integrated arable farming systems (Holland, *et al.*, 1995).

LANDSCAPE ECOLOGY - FARMERS ALREADY DO IT!

Landscape ecology has become a fashionable aspect of conservation biology, and landscape approaches are also more and more prominent in agricultural research papers. In reality agricultural research was there first. Many of the very basic ideas now incorporated into modern landscape ecology resulted from agricultural research. This should not be a great surprise since farming systems are concerned with landscape management. Farming can still learn from landscape ecology, but we should not forget that there is a wealth of landscape expertise already working in agro-ecosystem studies. Landscape ecologists could also benefit from the battery of methods developed by agricultural research especially field trials to test the results of more systematic small-scale experiments.

LANDSCAPE ECOLOGY - THE BASICS

Landscape ecology is essentially all about the patches of crop and non-crop habitats in a landscape and how their spatial arrangement affects the survival or well-being of species living in a landscape. Landscape ecology also quite validly incorporates man and socio-economic interactions into its framework. A simple diagram Fig 1. shows the basic components of landscape ecological theory; structure, flows and dynamics (Forman, 1995; Forman & Godron, 1986).

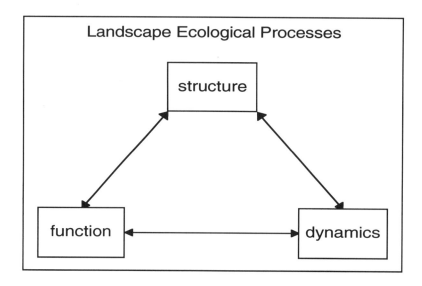

Figure 1. The basic landscape ecological processes.

These major components of landscape ecology will now be examined in relation to crop production systems.

Landscape structure

Structure is a way of describing spatial patterns formed by fields, remnant patches of wildlife habitat, woodlands, buildings, drainage networks, field boundary systems, and roads etc. These landscape elements will have certain physical attributes which can be measured such as their size, shape, distance apart, diversity of type and size classes, the amount of edge habitat, and how isolated or connected patches are to other similar or dissimilar patches. Farmers are continually re-arranging the structure of farms, e.g. fields of carrots may be widely spaced to avoid pest problems or field boundaries placed along contours to intercept run-off of nutrients and particulates.

Landscape function

Function is related to structure but more concerned with the biological, chemical and physical processes affected by landscape structure. Species diversity and distribution, population processes such as immigration and emigration, birth and death rates, population genetics, plant-animal interactions including both pests and pollinators, the spread of domestic and wild species in landscapes and the amenity value to man are all influenced by the structure of landscapes. Flows of energy, species and materials are also important landscape functions. The way farm landscapes are arranged, e.g. the size and shape of fields, the proportions of crop and non-crop areas all influence the flows of species, energy and nutrients in the landscape.

Landscape dynamics

All landscapes are undergoing a process of change. Land use may change from year to year with crop rotations and more permanent vegetation may undergo ecological succession. Most European farmed landscapes have a long history, sometimes involving cycles of intensive use followed by abandonment and then return to production, or more regular cycles between cropping and grazing. Some of today's environmental problems such as nitrogen in ground water, or soil compaction have developed for a long time. Farming systems themselves are very dynamic and the romantic notion of a stable mixed farming landscape is often not supported by agricultural statistics. Farmers have always needed to follow trends in production, markets and policy. In the second half of this century, several European countries have experienced increasing rates of change in policy, subsidies and constraints. Farming has to follow these to remain viable and hence land use change can be very rapid. Policy changes bring about physical changes in farm structure and hence the ways landscapes function.

An example of the way history leaves its imprint on farmland can be found in Petit (1994). This study provides clear evidence that species on farmland are responding to dynamics over quite long time scales. In her study of carabid beetle distribution, Petit (1994) found that their current distribution was best explained by historical patterns of hedgerow networks rather than current hedge distribution.

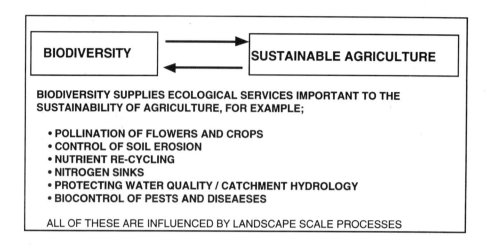

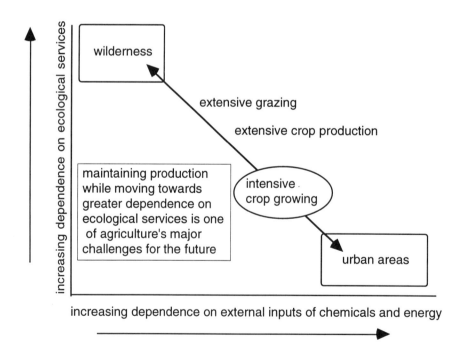

Figure 2. Natural, urban and farming landscapes seen in relation to the degree of dependence on external inputs of chemicals and energy and dependence on ecological services.

Having established that landscape ecology is nothing new to agriculture, I will argue that landscape perspectives were not given a high priority in the 1950-1990 period since the emphasis was more on ecological engineering (how to make the farm fit the crop) rather than working with nature. The sustainability debate has altered this focus see Fig. 2. and given a new impetus to landscape ecological approaches. Plant protection has been a field of research where the spatial relationships between plants and their pests has always been recognised as an important component of pest outbreak dynamics. Likewise the ability of biocontrol agents to find their prey and for newly introduced predators or parasitoids to be able to maintain viable populations are landscape ecological problems.

SPATIAL SCALE

Agricultural problems are often dealt with at a single scale whereas a more optimal solution might be found by operating at multiple scales where each level in the hierarchy from field to geographic region is dependent on both larger and smaller scale processes. Problems may be impossible to solve by solutions concentrating on one scale, whereas a multi-scale solution might achieve the desired goals. This holds true for achieving ground water nitrogen targets, biodiversity goals or production targets on farmland.

In Fig. 3 a hierarchy of scales from individual plants to regions is shown. At each level different ecological social and economic processes operate. If we take as an examples, the selection of crop, cultivar, spacing, inter-cropping and spot herbicide treatment, these are at a scale from individual plant to field. At the field level we can consider techniques of sowing and harvesting, pesticide regimes. The impact of the size and shape of field and its boundaries on the run-off of sediment and nutrient salts, and the density and dispersal of beneficial organisms are important considerations. As we go up in scale to the landscape level we focus on the between-field movement of pests and their natural enemies, the dynamics of co-ordinated pesticide application versus a field by field approaches, the benefits of crop rotation systems and the area-wide introduction of biocontrol agents. At even larger scales we may need to consider the long-range dispersal of pests and diseases, re-invasion of treated crops from surrounding regions and the effects of crop specialisation and agricultural policy. A well researched case study using a multiple-scale appoach and focussing on crop damage by starlings is offered by Clergeau (1995). In this study it is clear that a multi-scale solution is the only one likely to work and both the need for and the difficulties of working across scales are made explicit.

Scale also interacts with the dynamics of the social, economic and policy setting of farming and land-use planning. Even if we accept that certain problems occur at multiple scales, at some point between the field (parcel) level and the landscape level, decisions move from the farmer to higher administrative levels limiting landscape planning.

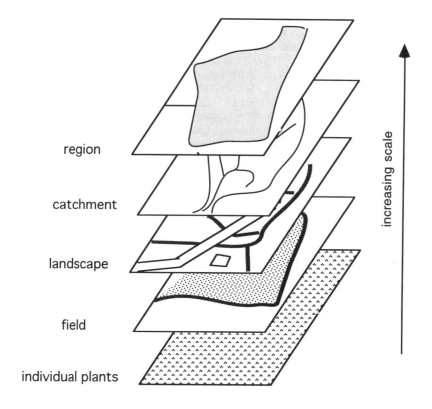

region

catchment

landscape

field

individual plants

increasing scale

Figure 3. A series of scales that may be necessary to consider in integrated crop protection strategies. (See text for further explanation.)

DISPERSAL STUDIES AND LANDSCAPE ECOLOGY.

The key to understanding many aspects of both conservation biology and crop protection in agroecosystems is the movement of organisms in real landscapes. Dispersal is one of the major life-history characteristics of both pests and beneficial organisms. It is clear that no matter how potentially effective a predator or parasitoid may be, if it does not reach the prey pest then it will be of little significance as a biocontrol agent. Our ability to design landscapes in such a way that beneficials are enhanced and dependency on pesticides minimised will require additional knowledge on how the spatial arrangement of farmed landscapes affects the movement of organisms. Dispersal also can be viewed at various scales (Fry, 1995) from trivial movements within a single plant to life-time displacement. Since most studies of beneficial organisms have used one or other form of mark-release-recapture method we have information on where these individuals arrive but very little about how they get there. This is an important lack of knowledge, as it is the key to developing design criteria on where to locate reservoirs of natural enemies to achieve biocontrol targets for specific farms.

It is quite beyond the realms of research funding to consider projects to study the dispersal of all interesting species. Instead we may be able to concentrate efforts on identifying a few major life-history traits and ecological characteristics that suffice for most modelling purposes. Two major traits found useful in this context are (1) dispersal ability, and (2) niche breadth. The approaches of Hodgson (1993) and Jepson (1989) are good indicators of the value of this approach showing how the analysis of a few ecological traits, including movement characteristics, have provided indicators of the vulnerability of selected arthropod groups to landscape or farm management.

Effects of different farm landscapes on beneficials

Most movement data has been acquired through the study of beneficials (mostly carabid beetles) within crops (cereals). Little research has been done concerning their movement within non-crop habitats such as field margins (Frampton, et al., 1994) nor on the role of farming systems on dispersal. We also have little information on the effects of movement barriers on dispersal and survival. Since the dispersal of individuals between sub-populations is a key factor in the long term survival of spatially structured populations (meta-populations) the impact of landscape structure on dispersal is essential for understanding large-scale processes and advising on pest control strategies (Sherratt & Jepson, 1993). Immigration and emigration rates and especially the probability of individuals reaching from one (sub) population to another are essential characteristics of population models. Yet for most beneficials we have very scant or no information on these rates of inter-population movement in real landscapes.

Some priorities for landscape studies

As most existing data on landscape effects is for a narrow range of species and taxonomic groups, we would gain valuable insights by including other important beneficials, e.g. parasitoids, spiders and hoverflies. Research on the landscape ecology of beneficials in Europe has been carried out predominantly in cereal crops and should be expand to other important crops especially vegetables. We need to determine what are barriers to the dispersal of beneficials within and between crop fields. As a product of this work the aim should be to produce guidelines on the proportion and spatial pattern of crop and non-crop habitat required to maximise the efficacy of beneficials at the farm and landscape levels.

CONCLUSIONS

My aim has been to draw attention to the spatial processes operating at various scales on farmland and use integrated crop protection as a case study. The results of recent reviews of research on integrated farming systems place great emphasis on landscape processes (Glen, et al., 1995). The shape and size of current fields may well be sub-optimal for current farming and environmental goals. New farm landscapes are certain to be treated with suspicion, but they are the logical outcome of recent studies in a wide range of agricultural sciences. Long, narrow fields with margins following contours seem to fit both machinery needs, soil conservation requirements and enhance the dispersal of beneficials into crop fields (Fry, 1994). Careful study of landscape ecological processes, especially of both pests and their natural enemies, will undoubtedly lead to new techniques which will become important components of future integrated farming systems.

REFERENCES

Clergeau, P. (1995) Importance of multiple scale analysis for understanding distribution and for management of an agricultural bird pest. *Landscape and urban Planning*, **31**, 281-289.

Forman, R. F. (1995) Some general principles of landscape and regional ecology. *Landscape Ecology*, **10**, 133-142.

Forman, R. T. T.; Godron, M. (1986) *Landscape Ecology*. New York: John Wiley & Sons.

Frampton, G.; Cilgi, T.; Fry, G. L. A.; Wratten, S. D. (1994) Effects of grassy banks on the dispersal of some carabid species (Coleoptera: carabidae) on farmland. *Biological Conservation*, **71**, 347-355.

Fry, G. (1995) Landscape Ecology of Insect Movement in Arable Ecosystems. In Glen, D. M., Greaves, M. P., & Anderson, H. M. (Ed.), *Ecology and Integrated Farming Systems* (pp. 177-202). Chichester: Wiley.

Fry, G. L. A. (1994) The role of field margins in the landscape. In Boatman, N. (Ed.), *Field Margins: Integrating Agriculture and Conservation*, Monograph no. 58 (pp. 31-40). Warwick, England: British Crop Protection Council.

Glen, D. M.; Greaves, M. P.; Anderson, H. M. (Ed.), (1995) *Ecology and Integrated Farming Systems*. Chichester: Wiley.

Hodgson, J. G. (1993) Commonness and rarity in British butterflies. , **30**, 407-427.

Holland, J. M.; Cilgi, T.; Frampton, G. K.; Wratten, S. D. (1995) Arable acronyms analysed - a review of integrated arable farming systems research in Europe. *Annals of Applied Biology*, **125**, (in press).

Jepson, P. C. (1989) The Temporal and Spatial Dynamics of Pesticide Side Effects on Non-Target Invertebrates. In Jepson, P. C. (Ed.), *Pesticides and Non-Target Invertebrates* (pp. 95-127). Wimborne: Intercept.

Liebhold, A. M.; Rossi, R. E.; Kemp, W. P. (1993) Geostatistics and geographic information systems in applied insect ecology. *Annual Review of Entomology*, **38**, 303-327.

Petit, S. (1994) Metapopulations dans les Reseaux Bocagers: Analyse Spatiale et Diffusion. PhD, Rennes.

Sherratt, T. N.; Jepson, P. C. (1993) A metapopulation approach to modelling the long-term impact of pesticides on invertebrates. , **30**, 696-705.

REGIONAL CROP ROTATIONS FOR ECOLOGICAL PEST MANAGEMENT (EPM)
AT LANDSCAPE LEVEL

J. HELENIUS

MTT Agricultural Research Centre, Farmland Ecology Unit, FIN-31600 Jokioinen, Finland

ABSTRACT

Spatiotemporal scales and landscape heteregeneity are known to play crucial role in
(meta)population dynamics of many kinds of organisms. The theory of conservation
of species can be reversed for purposes of EPM, Ecological Pest Management. Many
pest species of annual arable crops are essentially species of early successional
habitats. In addition, many have low to moderate dispersal ability. Such pest species
can be expected to form true metapopulations of the Levins's 1969 classic model, in
the sense that the persistence is dependent on between-population rather than
within-population processes. Adjustment of the spatial and temporal scale of rotation
to help drive pests into local extinction is a management alternative in annual arable
cropping. In such ephemeral habitats, true regulation of populations by specialist
natural enemies is not achievable. By planning at a landscape level, benefits of
regional pest management as an essential part of EPM can be combined to bring
benefits of improved logistics and economy of scale, which may provide sufficient
incentive for the necessary local cooperation between farmers.

INTRODUCTION

The term metapopulation was first introduced by Levins (1970) to mean a population of
populations which go extinct locally and recolonize. This conceptual innovation was published
a year earlier, when Levins (1969) presented his metapopulation model. For twenty years, the
concept was largely ignored, but then the interest was renewed (for the history, see Hanski
& Gilpin, 1991). The main area of motivation for theory development and the main area of
applications has been for several years in conservation biology. However, Levins's (1969)
concern was in pest control: he noticed that local populations would fluctuate in asynchrony,
and the model predicted that control should be concerted throughout a region in order to
decrease the size of metapopulation and hence, in order to reduce future control investment.
Levins emphasized this area of application again later (1970), by stating that extinction is
fundamental to any theory to pest control, and that it is in the field of economic entomology
that long term studies over wide areas of real populations can be done.

It is obvious that much of the theory development done in the field of conservation
biology is directly applicable to pest control problems, just by reversing the goal of
conservation into the goal of extinction. Moreover, there seems to be more scope and more
urgent need for such applications in Ecological Pest Management (EPM), in which chemical
'quick fixes' are at present not an option, than in Integrated Pest Management (see also
Helenius, 1995a).

ECOLOGICAL PEST MANAGEMENT, EPM

EPM is defined here as a pest management approach that sets ecological sustainability as a starting point and ecology as a scientific base. It is fundamentally different to Integrated Pest Management, IPM, in the sense that it does not compromise over the primary criteria. Towards use of pesticides, IPM takes a liberal view: 'gains must be greater than losses'. It represents understandable consensus among intrests of pesticide industry, conventional agriculture and applied science. In EPM, the cumulated knowledge of ecological side-effects of pesticide use is enough for taking a more strict view. Taking sustainability seriously, the short-term economic gains must not dictate the strategy.

EPM is a scientific approach suited to all production systems: of course, to organic (ecological) farming, but also to integrated farming, pesticide free farming and even to conventional farming. Chemical options are not excluded, but a safe pesticide is awaiting. The landscape level pest control strategy described in this paper serves as an example of EPM-oriented approach.

REGIONAL STOCHASTICITY AND METAPOPULATIONS OF PESTS

Regional stochasticity (Hanski, 1991) is a metapopulation concept that refers to chance effects that are regionally correlated between local populations. In general, metapopulation persistence is decreased by increasing regional stochasticity (Hanski, 1991). Conventionally, crop fields are managed within units of single farms: at any level that exceeds single farm boundaries, there is little or no regional stochasticity in habitat availability to pest organisms. Crop fields that are suitable to a pest blink on and off from season to season in a spatially uncorrelated manner.

In Levins' (1969) model, dynamics of $p(t)$, the fraction of habitat patches occupied by the species at time t is described as

$$\mathrm{d}p/\mathrm{d}t = mp\,(1\text{-}p) - ep,$$

where e and m are the rates of local extinction and colonization of empty patches, respectively (notation from Hanski & Gilpin, 1991). In this model, regional stochasticity is affected by manipulation of e. Best control would be achieved by maximizing temporal variance of a regionally uniform extinction rate, in other words, by increasing regional stochasticity. From this result, Levins (1969) outlined a strategy of applying chemical control measures over large regions simultaneously. This would result in decrease in average size of the metapopulation of the pest.

The process outlined by Levins (1969) is not, however, quite analogical to such a regional stochasticity that is achieved by regionally correlated crop rotation. In regional rotations, local extinctions are achieved by crop (habitat!) patch removals and rearrangements, not by control of local populations within patches. Of course, patch removals produce local extinctions, and if these are synchronized temporally over the region, then at an extreme e approaches 1, control is complete in one season, and no attention needs to be paid into manipulation of m.

EFFECTS OF PATCH SIZE, PATCH NUMBER AND PATCH ISOLATION

It is quite obvious that increased distance (isolation) between local pest populations, i.e. increased distance between occupied habitat patches, decreases naturally occurring regional stochasticity; for example, catastrophes due to extreme weather seldom cover wide regions. But let us focus on colonization for a while.

In practice, the colonization parameter m is also subject to manipulation by regional rotations. The distance between crop patches is directly affected. Any effects on local population densities is likely to affect m as well, through the number of potential colonists. Levins (1969) already noted, that if the strategy relies in reducing migration rate, the recommendations for control are opposite of those for control by increasing local extinction rate.

Using model (1) as a starting point, Hanski (1991) studied the effects of patch size, number and isolation in relation to metapopulation extinction. The most fundamental cause for extinction would be lack of positive equilibrium point in model (1): when p is small, m is lower than e. In this scenario, decreasing the size of habitat patches increases extinction rate, and increasing the isolation of habitat patches decreases colonization rate (number of colonists produced is assumed to be directly proportional to patch size).

As a metapopulation parallel to demographic stochasticity (chance events of death and birth that are uncorrelated between individuals in a local population), Hanski (1991) coined the term immigration-extinction stochasticity to mean the chance extinctions of local populations that are uncorrelated between the populations. As a cause of metapopulation extinction, immigration-extinction stochasticity is dependent on number of habitat patches. If the number of habitat patches is small and extinction probability is not negligible, a metapopulation may go extinct simply because all local populations happen to go extinct at the same time.

For more details of the theoretical background, see Hanski (1991) and references therein.

CROP ROTATION AT A LANDSCAPE LEVEL FOR PEST CONTROL

Perhaps the most obvious alternative for regional rotation is to reduce number of host crop fields by aggregating the desired area of production into a small number of more isolated but, inevitably, large crop field patches. The idea would be to generate aggregates of field parcels within which, the pest would form single local populations (see Helenius, 1995a), and to rotate these aggregates regionally to new sites each season.

Replacement of the large number of within-farm rotated crop patches with small number of regionally rotated patches of same total area of production increases isolation. In theory, isolation increases metapopulation persistence by reducing regional stochasticity and by increasing local population size, and at the same time, it decreases persistence by increased migration distances and migration losses. However, regional stochasticity from natural causes is likely to be of minor importance: the 'managed' regional stochasticity at its extreme operates through concerted removal and spatially unpredictable re-establishment of the habitat itself, in each season.

The latter two effects are relevant to the 'no positive equilibrium'-scenario of metapopulation extinction. In the theory, 'local population size' refers to number, not to density of specimens. Even if in nature, large patches support large populations least prone to extinction, does this undermine the 'aggregate-rotation' strategy? The theory refers to stable habitat patches or patch mosaics with low turnover rate (in relation to generation turnover of the species). However, the rotated habitat 'patches', i.e. the crop field aggregates of the regional rotation are highly ephemeral. Obviously, the point of larger populations being supported by large patches is of no or minor relevance to the application discussed here. On the contrary, positive implications of increased size of crop blocks may include 1) dilution effect, as the remaining crop colonizers are overwhelmed by the abundance of host plants, and 2) reduced edge effect (tendency of many pest problems concentrating to crop margins) due to decreased area to perimeter ratio.

Spatial and temporal scales and patterns of landscape level rotations for pest control combine to a large potential set of designs. From general ecological and agricultural knowledge, only broad rules can be derived for what would be a good strategy for each individual case. Much depends on properties of the crop and pest in question.

FOR WHAT KINDS OF PESTS AND CROPS?

First, it is important that the pest does not have large source populations in natural habitats scattered among the crop habitats. The second important criteria is the dispersal range and rate of the pest. In most circumstances, regional management is more likely to succeed with species with low or moderate dispersal range, than with species cabable of dispersing across the entire landscape in a short period of time. The third criteria is the frequency and regularity of outbreaks: most effort should be spent into pests that are common and cause regular damage. However, occasional 'rare' pest species may well turn out to be best targets for region-wide 'eradication' schemes.

Fourth and equally important; the spatial range and distribution and temporal range and regularity of the crop host must be considered. Widely grown, uniformly distributed major crops cannot be rotated regionally: good examples are cereal crops in most areas worldwide. High-valued crops that are common but occupy a relatively small proportion of the land may in most cases best suit to regional rotation strategy. Many vegetable crops may be well suited to such an approach.

COMPATIBILITY WITH BIOLOGICAL CONTROL

The role of natural enemies depends on the spatial and temporal patterns of the pest and the crop. May (1994) developed a simple predator-prey metapopulation model that supports the intuition that habitat removal decreases persistence of specialized natural enemies more than persistence of the pest. May's (1994) model predicts that decreasing the number of habitat patches in a landscape results in steady decrease in proportion of remaining patches being occupied by the specialist enemy, until at some stage, defined by the model parameters, the natural enemy is extinguished. The proportion of remaining patches occupied by pest only first increases, but after the enemy goes extinct, due to extinction rate of local host populations, the proportion of patches occupied by the host declines until the host also goes regionally extinct.

In the regional rotation based on aggregating the crop fields into few large patches, the total area of habitat does not decrease, as it does in May's (1994) model. The model does not account for a situation in which the number of patches decreases, but at the same time, the size of the patches increases. However, the modelling exercise warns about problems that various regional rotation schemes may cause to biological control. How real is this threat?

Most pest species of annual arable crops are exploiters of early successional habitats. In cases when the dispersal ability of the species is relatively low in the regional scale, such species can be expected to form 'true' metapopulations (*sensu* Harrison, 1991) of the Levins's (1969, 1970) classic model. In other words, the persistence of the species is dependent on between-population rather than within-population processes (Harrison, 1994). This conclusion has three important implications in favour of management by regional rotatation.

The first is that only metapopulations that behave according to the Levins model lend themselves to regional management, be it conservation (Harrison, 1994) or pest control. Secondly, according to the synoptic model of how predation may interact with habitat stability in the population dynamics of a pest (Southwood & Comins, 1976), for pest species of ephemeral arable crops that are from the r-end of the r-K continuum, the reproductive numerical response of natural enemies to 'booming' pest density is limited. Thus, within the 'natural enemy ravine', i.e. the range of population growth rate of the pest within which biocontrol is effective, the significant predators are often polyphagous and large relative to their prey, and occur at relatively low densities (Southwood and Comins, 1976). This implies that the key predator species would be large voracious generalists. Generalists are not coupled to a one pest as prey and, as a consequence, would not be suppressed by regional rotation.

The third implication is related to the previous ones: because individual fields of annual arable crops are maintained in nonequilibrium, there are no grounds for exploring equilibrium solutions for pest-enemy interactions anyway (Murdoch, 1975). This view was also supported, although perhaps implicitly, by Beddington *et al.* (1978) in their characterization of successful natural enemies (for further discussion, see Helenius, 1995b). Good empirical evidence comes from records of success in classical biological control (Hokkanen, 1985).

Thus, the pest species that are likely candidates to management by regional crop rotation are, fortunately, the ones not regulated by specialist enemies. In fact, the prerequisites for successful regional management contrast the prerequisites for natural control by specialist enemies. The host crop-pest systems most promising for the strategy are the least promising for classical biological control. Of course, 'buffering' the crops against colonizing pests by enhancement of generalist predator complex is compatible with regional rotation. For conclusions concerning traits of an effective predator in control of model metapopulations of pests, see Levins (1969)!

CONCLUSIONS

Regional crop rotation for pest control is an example from a wide array of strategies based on landscape level management. If appropriately planned, it can increase energy efficiency and improve logistics of crop production as a whole.

Questioning the role of single farms as optimal management units has several socioeconomic implications. However, the effort should be made as, traditionally, the optimal use of natural resources has not been the criteria for land allocation. Modern techniques of landscape ecology facilitate the design of idealized landscapes on several simultaneous criteria (Lenz & Stary, 1995). Of course, the implementation of landscape level strategies requires political will and new forms of cooperation between farm enterprises.

More research is required on metapopulation dynamics of pest species and their natural enemies than is currently in progress. Modelling approaches will be necessary, because experimentation with farmers' fields at a regional level is seldom possible. Good evidence of effectiveness and prediction of how many seasons is required before the pest decline reaches the desired level are required before introducing such schemes for widespread practice. However, as no great risks are involved, farmers' own experimentation can be encouraged.

Recent development in population biology and its application to conservation of species certainly points towards a novel pest management strategy. The strategy should prove especially useful for ecological pest management in organic (ecological) farming where pesticides are abandoned and sustainability is emphasized.

ACKNOWLEDGEMENT

I thank I. Hanski for discussion and and L. Firbank for comments on the draft.

REFERENCES

Beddington, J.R.; Free, C.A.; Lawton, J.H. (1978) Characteristics of successful natural enemies in models of biological control of insect pests. *Nature*, **297**, 513-519.

Hanski, I. (1991) Single-species metapopulation dynamics: concepts, models and observations. *Biological Journal of the Linnean Society*, **42**, 17-38.

Hanski, I.; Gilpin, M. (1991) Metapopulation dynamics: brief history and conceptual domain. *Biological Journal of the Linnean Society*, **42**, 3-16.

Harrison, S. (1991) Local extinction in a metapopulation context: an empirical evaluation. *Biological Journal of the Linnean Society*, **42**, 73-88.

Harrison, S. (1994) Metapopulations and conservation. In: *Large Scale Ecology and Conservation Biology*, P.J. Edwards, R.M. May and N.R. Webb (Eds), Oxford: Blackwell, pp. 111-128.

Helenius, J. (1995a) Spatial scales in ecological pest management: importance of regional crop rotations. *Biological Agriculture and Horticulture*, submitted.

Helenius, J. (1995b) Enhancement of predation through within-field diversification. In: *Enhancing Natural Control of Arthropod Pests Through Habitat Management*, C.H. Pickett and L.R. Bugg (Eds), Davis: AgAccess, in press.

Hokkanen, H.M.T. (1985) Success in classical biological control. *CRC Critical Reviews in Plant Sciences*, **3**, 35-72.

Levins, R. (1969) Some demographic and genetic consequences of environmental heterogeneity for biological control. *Bulletin of the Entomological Society of America*, **15**, 237-240.

Levins, R. (1970) Extinction. *Lectures on Mathematics in the Life Sciences*, **2**, 77-107.

Lenz, R.J.M.; Stary, R. (1995) Landscape diversity and land use planning: a case study in Bavaria. *Landscape and Urban Planning*, **31**, 387-398.

May, R.M. (1994) The effects of spatial scale on ecological questions and answers. In: *Large Scale Ecology and Conservation Biology*, P.J. Edwards, R.M. May and N.R. Webb (Eds), Oxford: Blackwell, pp. 1-17.

Murdoch, W.W. (1975) Diversity, complexity, stability, and pest control. *Journal of Applied Ecology*, **12**, 795-807.

Southwood, T.R.E.; Comins, H.N. (1976) A synoptic population model. *Journal of Animal Ecology*, **45**, 949-965.

THE EFFECTS OF ORGANIC AND CONVENTIONAL SYSTEMS ON THE ABUNDANCE OF PEST AND NON-PEST BUTTERFLIES

R.E. FEBER; P.J. JOHNSON; D.W. MACDONALD

Wildlife Conservation Research Unit, Department of Zoology, University of Oxford, South Parks Road, Oxford OX1 3PS.

L.G. FIRBANK

Institute of Terrestrial Ecology, Monks Wood, Abbots Ripton, Huntingdon, Cambridgeshire, PE17 2LS.

ABSTRACT

Butterfly transects were conducted on eight pairs of organic and conventional farms in England in 1994. The abundance of each species of butterfly was recorded. Organic systems increased overall butterfly abundance, and more non-pest butterflies were recorded on the margin than the crop. By contrast, there was no significant difference in the abundance of two pest species, *Pieris brassicae* (the large white) and *Pieris rapae* (the small white) between the two systems on either the crop or the margin habitat. Patterns of crop use by pest and non-pest butterflies differed significantly.

INTRODUCTION

Recent years have seen shifts in agricultural policy in Europe, resulting primarily from overproduction, together with consumer concerns about the environment and the way in which food is produced. Organic farming is the extreme expression of low-input agriculture and has become a small, but established, part of the agricultural scene. While organic agriculture continues to attract technical research (summarised in Lampkin, 1990), studies of the effects of organic farming systems on wildlife populations in the UK have so far largely been restricted to birds and their insect food resources (e.g. Wilson & Browne, 1993). In this paper we report on a study of butterfly abundance on organic and conventional farms in 1994.

Of the butterfly species resident in Britain, only the large white (*Pieris brassicae*) and the small white (*Pieris rapae*) are significant agricultural and horticultural pests. The damage inflicted by the larvae of both of these species on their cruciferous hosts may run to millions of pounds sterling annually (Feltwell, 1980). The remainder of the British butterfly species are agriculturally benign and, by contrast to the two pest species, are frequently targeted for conservation measures within habitats such as woodland and grassland reserves (Thomas 1984). Increasingly, butterflies are also the subject of conservation interest in the wider countryside (Dover *et al.*, 1990; Feber & Smith, 1995; Feber *et al.*, 1994) where agricultural intensification has led to large-scale losses of semi-natural habitats (Anon., 1984). In this paper we present results which show the effects of organic and conventional farming management on the abundances of non-pest and pest butterflies within agricultural systems.

METHODS

Butterfly abundance was recorded on eight pairs of organic and conventional farms between June and September in 1994. Farm pairs were located across England in an area roughly bordered by Dorset, Shropshire, Lincolnshire and Essex. Butterflies were recorded by experienced volunteer recorders at approximately fortnightly intervals during the summer, following methods modified from those described by Pollard (1977) and Pollard *et. al.* (1975), which are used in the National Butterfly Monitoring Scheme. Volunteers walked a fixed transect route which was divided into sections corresponding to crop and/or boundary type. For each section, all butterfly species seen, and the abundance of each species, was recorded. Butterflies seen over the crop edge were recorded separately from those seen over the uncropped field boundary. Details of management type, crop and boundary were recorded for each section of the transect route.

Analysis

The dependent variable was defined as the total butterfly count for the season for each management type (organic or conventional) in each farm pair, on both the crop edge and the uncropped field boundary, standardised to a count per unit length of transect walked. Data were log(x+1) transformed for analysis.

Two separate analyses were carried out; the first on the total number of butterflies regardless of their pest status, and the second on the data partitioned into pest (*P. brassicae* and *P. rapae*) and non-pest (all other) individuals.

For the first analysis, a two-way analysis of variance was carried out, including management type (subsequently referred to as "management") and crop or margin ("habitat") as effects, with repeated measures for each farm pair ("site") on both effects (SAS PROC GLM; SAS Institute 1988). In the second analysis the pest status of the butterflies ("status") was included as a third effect. When significant interactions were detected, the data were stratified within the levels of the appropriate effect. Further analyses were then applied to clarify these interactions.

RESULTS

Effects of management on overall butterfly abundance

Total butterfly abundance was significantly higher on organic farms than on conventional farms ($F_{(1,7)}$=8.32, P=0.024; Figure 1). Significantly more butterflies were recorded on the uncropped field boundary than on the crop edge ($F_{(1,7)}$=12.91, P=0.009). There was a significant interaction between the two factors ($F_{(1,7)}$=6.28, P=0.041) with the difference in butterfly abundance between crop and margin being greater in conventional than in organic systems.

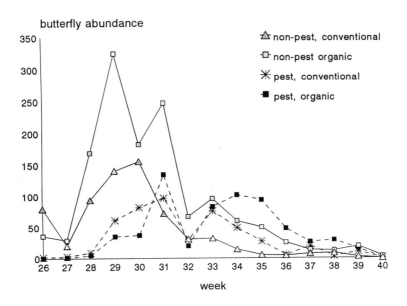

FIGURE 1. Mean pest and non-pest butterfly abundance per kilometre per site per week, on organic and conventional farmland in 1994.

The relationship between pest status and management, and butterfly abundance

The two-way interactions between pest status and management, and between status and habitat, were significant ($F_{(1,7)}=8.42$, $P=0.023$ and $F_{(1,7)}=35.23$, $P<0.001$ respectively). One-way analyses were therefore carried out separately for both levels of pest and non-pest status butterflies.

Effects of management on non-pest butterfly abundance

The abundance of butterflies, not including *P. brassicae* and *P. rapae*, was significantly higher in organic than conventional systems ($F_{(1,7)}=28.43$, $P<0.001$; Figure 1). There was a highly significant effect of habitat on the abundance of non-pest butterflies ($F_{(1,7)}=73.31$, $P<0.001$), with more butterflies recorded on the boundary than the crop. The interaction between these factors approached significance ($F_{(1,7)}=5.59$, $P=0.050$). The management of the uncropped boundary had a significant effect on non-pest abundance, with organic boundaries attracting higher numbers of butterflies than conventional boundaries. Similarly, organic management increased the abundance of non-pest butterflies within the surveyed cropped habitats ($F_{(1,7)}=28.25$, $P<0.001$)

Effects of management on pest butterfly abundance

By contrast with the non-pest species, there was no significant difference in the abundance of *P. brassicae* and *P. rapae* between the two management systems ($F_{(1,7)}=0.19$, $P=0.672$; Figure 1). The abundance of these species did not differ significantly between crop

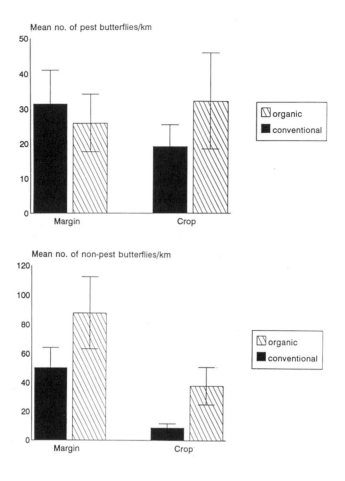

FIGURE 2. Abundance of (a) pest and (b) non-pest butterflies on crop and boundary habitats, on organic and conventional farmland.

and boundary habitat ($F_{(1,7)}$=2.18, P=0.183) and there was no significant interaction between the two factors ($F_{(1,7)}$=1.88, P=0.213). There was no significant effect of management on pest butterfly abundance within either the surveyed boundary ($F_{(1,7)}$=0.17, P=0.692) or crop ($F_{(1,7)}$=0.76, P=0.412) habitats.

Effects of crop type on pest and non-pest butterfly abundance

The cropping patterns on the surveyed sites differed considerably between the organic and conventional components. For example, approximately six times as much grass ley was surveyed on the organic areas than the conventional areas, while oilseed rape was not encountered on any organic area (Table 1).

TABLE 1. The lengths of each crop type surveyed, and the mean abundances of pest and non-pest butterflies per kilometre, per transect, on each crop type. Crop types not present on both organic and conventional not included in the analysis.

Crop	System	Length of crop type surveyed (km)	Mean pest abundance	Mean non-pest abundance
winter wheat	organic	3.16	2.95	6.59
	conventional	3.42	2.16	1.61
barley	organic	-		
	conventional	0.18	6.04	0.00
oats	organic	0.66	7.71	20.43
	conventional	-		
linseed	organic	-		
	conventional	0.68	13.42	0.15
oilseed rape	organic	-		
	conventional	0.62	7.16	4.48
beans	organic	0.80	8.50	1.61
	conventional	2.40	17.79	4.93
grass ley	organic	7.14	2.12	13.70
	conventional	1.47	1.13	4.48
set-aside	organic	0.69	12.63	24.29
	conventional	0.12	6.97	15.10

The patterns of crop use by pests and non-pests were significantly different (interaction between status and croptype: $F_{(3,4)}=20.98$, $P=0.002$). Pests were more abundant on beans than non-pests, while non-pest individuals were more abundant on grass leys than pests (Table 1). Set-aside and cereal crops attracted similar numbers of pest and non-pest individuals. Oil-seed crops (linseed and rape) attracted higher numbers of pest than non-pest butterflies.

There was no significant main effect of management on the abundance of either pest or non-pest individuals for any crop type ($F_{(1,4)}=0.07$, $P=0.806$).

DISCUSSION

The majority of non-pest butterflies recorded on farmland are common species, usually associated with hedgerow or grassland habitats. A number of these species are relatively sedentary, and may have specific larval food requirements. The pest butterflies, *P. brassicae* and *P. rapae*, however, are highly mobile, and their larvae will feed on a wide range of cruciferous and related foodplants.

In our study, most of the non-pest butterflies were associated with the uncropped field boundary habitat. The nature of the boundary varied considerably between fields and farms but, overall, non-pest butterfly abundance was significantly higher on organic than on conventional boundaries. Absence of herbicide application to the hedge base, and better hedgerow management, are likely to account for this result. This suggests that targeting uncropped field boundaries is perhaps the most productive conservation measure for butterflies on farmland, and this can be achieved under a range of farming systems.

By contrast with the boundary habitat, non-pest butterflies abundance did not differ significantly for any given crop type between systems, although our small sample size at this level of analysis warrants caution in interpretation. Effects of different crop management on butterflies may be complex and subtle. Weedier crops may provide more nectar resources for adult butterflies (e.g. Conservation Headlands: Dover *et al.* 1990). They may also contain larval foodplants on which butterflies oviposit but, unless the larvae can develop and hatch during the summer (only a few species complete their life cycle in such a short period), there may be considerable larval mortality at harvest. The polyphagous predator web is also affected by management. Studies have shown significantly increased predation on *P. rapae* by the presence of weeds within the crop (Dempster, 1969).

Organic systems, however, did increase the abundance of non-pest species in the cropped habitats overall. The most likely reason for this is the different proportions of crop types between the two systems. Organic farms have a higher proportion of grass leys than conventional farms. Species such as the meadow brown (*Maniola jurtina*) and the gatekeeper (*Pyronia tithonus*) feed on a range of grasses and may breed successfully on such leys if they are in place for more than one year. Thus, at a landscape level, organic farming systems may have positive implications for the conservation of butterflies on farmland.

In terms of absolute numbers, there was no significant difference between the two systems for pest butterflies. There was no significant difference between crop or boundary habitat use by the two pest species, and no effect of farming system on their abundance. Some contrasting features of the two systems, though, may account for a difference in the proportion of butterflies which were pests between the two systems. Oilseed rape, for example, is rarely found on organic farms because of its low premium and high nutrient demand. This cruciferous crop is very attractive to white butterflies and is common in conventional systems. Weedy crucifers may be out-competed in grass clover leys which form a large proportion of most organic farms. The high mobility of the pest white butterflies may allow them to use the cropped habitat to a greater extent than non-pest species, and increase the species' resilience to farming practices common in conventional systems, such as widespread herbicide use. The pest white butterflies showed no significant association with the uncropped field boundary, and so any conservation measures applied to this habitat are unlikely to increase the abundance of pest whites.

In summary, our results have shown that organic systems increased non-pest butterfly abundance overall without increasing the abundance of pest butterflies. This is likely to be due to a combination of factors, in particular, differences in cropping patterns and boundary management, and the ecological characteristics of the pest and non-pest butterflies. Whilst the rotations are intrinsic to organic agriculture, we suggest that changes in approaches to boundary management may have conservation benefits for non-pest butterflies, and be acceptable within the constraints of conventional farming systems.

ACKNOWLEDGEMENTS

This work was funded by NERC as part of the BBSRC/ESRC/NERC Organic Farming Study, with additional financial support from the Worldwide Fund for Nature (UK) and the SAFE Alliance. We are grateful to the British Trust for Ornithology and Butterfly Conservation for advice and assistance. Our thanks to the volunteer recorders who contributed data to the Farmland Butterfly Survey, and to the farmers who allowed us access to their land.

REFERENCES

Anon. (1984) *Nature Conservation in Britain*. Nature Conservancy Council, Peterborough.

Dempster, J.P. (1969) Some effects of weed control on the numbers of the small cabbage white (*Pieris rapae*) on brussels sprouts. *Journal of Applied Ecology*, **6**, 339-345.

Dover, J.W.; Sotherton, N.; Gobbett, K. (1990) Reduced pesticide inputs on cereal field margins: the effects on butterfly abundance. *Ecological Entomology*, **15**, 17-24.

Feber, R.E.; Smith, H. (1995) Butterfly conservation on arable farmland. *Ecology and conservation of butterflies* (Ed. A.S. Pullen). Chapman & Hall, London.

Feber, R.E.; Smith, H.; Macdonald, D.W. (1994) The effects of arable field margin restoration on the meadow brown butterfly (*Maniola jurtina*). *Field Margins - Integrating Agriculture and Conservation. BCPC Monograph No. 58*. Farnham: BCPC Publications, pp. 295-300.

Feltwell, J. (1982) *The large white butterfly*. W. Junk, The Hague.

Lampkin, N. (1990) *Organic Farming*. Farming Press Books, Ipswich.

Pollard, E. (1977) A method for assessing changes in the abundance of butterflies. *Biological Conservation*, **12**, 115-134.

Pollard, E.; Elias, D.O.; Skelton, M.J.; Thomas, J.A. (1975) A method of assessing the abundance of butterflies in Monks Wood National Nature Reserve in 1973. *Entomologist's Gazette*, **26**, 79-88.

SAS Institute Inc. (1988) SAS/STAT User's Guide. Release 6.03 Edition. SAS Institute Inc., Cary, NC, USA.

Thomas, J.A. (1984) The conservation of butterflies in temperate countries: past effort lessons for the future. Symposium of the Royal Entomological Society of London No. 11: *The Biology of Butterflies* (Eds R.I. Vane-Wright & P.R. Ackery), pp. 333-353. Royal Entomological Society, London.

Wilson, J.D.; Browne, S.J. (1993) Habitat selection and breeding success of skylarks *Alauda arvensis* on organic and conventional farmland. *BTO Research Report 129*. British Trust for Ornithology, Thetford, Norfolk.

FINANCIAL ANALYSIS OF A POSITIVE FIELD MARGIN STRATEGY ON ORGANIC AND CONVENTIONALLY MANAGED LAND.

A.B.Davies, R.Joice, and L.Noble

Rhône-Poulenc Agriculture Ltd, Fyfield Road, Ongar, Essex, CM5 OHW

ABSTRACT

Boarded Barns Farm field margins have been managed by leaving at least a 1m wide grass strip at the base of hedges and where necessary an additional 3m wide mown access path. Given the fact that weed species increase with proximity to hedges particularly species such as *Galium aparine* (Cleavers), cropping these areas carries a yield penalty.

By the careful use of conventional herbicides and grassed margins, farmers can crop the more responsive field areas whilst maintaining valuable habitats without incurring a financial penalty.

INTRODUCTION

Boarded Barns Farm has been managed and farmed over the last twenty years on behalf of the owners by Rhône-Poulenc, although there were links between the farm and the Company prior to this period. The farm adjoins the Rhône-Poulenc Agriculture, Ongar, Essex site and slopes gently down to the River Roding on the Eastern boundary.

Covering a total area of 57 hectares the land is cropped with an arable rotation on 43.6 hectares, the remainder of the land being down to permanent pasture, woodland, ponds, access paths and buildings.

The soil is a medium loam over chalky boulder clay, which can produce very good yields provided care is taken to work the soil in optimum conditions to minimise soil damage by compaction. A comprehensive drainage system allows limited working in a wet season. The land is farmed with a comprehensive range of modern machinery utilising proven crop husbandry and pest control techniques on the conventionally cropped section of the farm. In the past twenty years the crop rotation had involved wheat, barley, oilseed rape, beans and peas. The emphasis on recent cropping has been to produce quality bread milling wheat with break crops of beans and oilseed rape.

From 1989, following discussions and detailed planning, part of the farm was managed with a system complying with recognised organic farming standards and is registered with United Kingdom Register of Organic Food Standards (UKROFS) as an organic arable production unit. Five fields, Shelley, Brook, Further, Well East and Well West were devoted to the organic study. Another five fields with similar soils, Folyats, Stocklands South, Stocklands North, Barn North and Barn South continued to be managed using conventional methods.

Following a mandatory two year conversion period our first organic crop of Mercia Winter wheat was grown on Shelley field in 1991/92. Brook and Further fields were sown with organic crops for 1992/93, and Well field is fully converted to organic food production for the 1993/94 season, giving a total of 15.2 hectares of crops in the organic rotation.

This paper outlines the field margin management at the farm, the opportunity offered by the organic system to monitor weed levels and describes the financial balance between cropping every last square metre of field and establishing grass field margins.

FIELD MARGINS MANAGEMENT AT BOARDED BARNS FARM

Field margins are managed to provide maximum advantage for the agricultural operations on the farm in conjunction with enhancement and return from the environment.

Hedge and ditch boundaries around all fields have a one metre rough vegetation strip left between the crop and the field edge. By not ploughing and cultivating at the immediate edge of the field erosion and even collapse of ditch banks is avoided. Experience has shown that these rough strips must be trimmed once per year or encroachment of vegetation from the ditch bank or hedge will occur.

To ensure that the one metre strip is not sprayed or fertilised the operator at Boarded Barns Farm is well trained and uses modern well maintained equipment with a good "cut-off" at the end of the sprayer or spreader pattern.

Farm operator involvement and a full understanding of the importance of all conservation matters on the farm is imperative if full advantage is to be gained from any conservation project.

The margin strip is a haven for wildlife including small mammals and birds which can benefit from the cover provided. Beneficial insects such as rove beetles, ground beetles, parasitic wasps, hover flies and also hunting and money spiders can over winter in the diverse cover. (Brown. R, 1995).

The grass access paths are quite resistant to mechanical wear and are mown 2 to 3 times a year to keep the grass to a reasonable height, this allows birds and small mammals an open area for drying, preening and feeding.

Access paths at Boarded Barns Farm are used for both machinery, equestrian and pedestrian purposes and are sown with a hard wearing grass mix, the horse gallop mix is ideal on this particular soil type.

IMPACT OF MARGIN MANAGEMENT ON WEED POPULATIONS

As part of the environmental impact monitoring on the study the weed populations were measured. Paired sets of permanent quadrats were established every 10 metres around the edge of each winter wheat field, one and six metres into the cropped area.

The organic system offered the opportunity of monitoring weed numbers and species. Under this system only mechanical cultivations are permitted and so species diversity and densities were well developed (Table 1). Evaluating weed distribution and dynamics became a possibility.

Table 1 -   Assessment of "Margin" flora or an organically farmed field
             Shelley Field : 1991 - 1992
             Order of frequency (occurrences) * at 1m from field boundary

Date of Assessment	28/11/91	05/03/92
Species	Before harrowing	After harrowing
Stellaria media	10	8
Veronica persica	7	5
Poa annua	6	8
Trifolium repens	5	4
Sinapis arvensis	4	0
Matricaria spp.	3	1
Rumex spp. **	3	6
Cichorium intybus	2	7
Galium aparine	2	8
Elymus repens	0	3
Lamium purpureum	0	2
Lolium sp.	0	2
Myosotis arvensis	0	2
Agrostis stolonifera	0	1
Anthricus cerefolium	0	1
Capsella bursa-pastoris	0	1
Poterium sanguisorba	0	1
Veronica hederifolia	0	1
Total No. of species present	9	17

* Total of 10 quadrats assessed (does not include the start and finish transects on each diagonal transect

No uncultivated controls left on field margins

** Rumex spp. - mainly *Rumex obtusifolius*, however, a few *Rumex crispus* also found, some plants also appeared to be hybrids of the two.

On the organically managed fields the only weed control practice carried out was the use of a harrowcomb weeder. The original weed species present and the effect of a passage of the harrowcomb at 1m into the crop.

Agronomic assessments were employed to decide when the weeder should be used. On average 2/3 passages are used per season which is similar to that reported by Rasmussen (1991). *Stellaria media* is the most frequent occurring species on both occasions. By March the second most frequently occurring species changed from *Veronica* to *Poa annua*. The number of species increased from 9 to 17 between November and March.

The species present 6m into the crop (Table 2) show a similar trend of increasing species present with time despite the use of a harrowcomb weeder.

Table 2 -   Assessment of "Margin" flora of an organically farmed field
             Shelley Field : 1991 - 1992
             Order of frequency (occurrences) * at 6m from field boundary

Date of assessment	28/11/91	05/03/92
Species	Before harrowing	After harrowing
Stellaria media	10	8
Trifolium repens	10	9
Veronica persica	9	4
Matricaria spp.	5	3
Poa annua	5	7
Geranium molle	1	0
Polygonum persicaria	1	0
Veronica hederifolia	1	1
Galium aparine	0	3
Lolium sp.	0	4
Myosotis arvensis	0	2
Papaver rhoeas	0	1
Rumex spp. **	0	1
Total No. of species present	8	11

* Total of 10 quadrats assessed (does not include the start and finish quadrats on each diagonal transect).

No uncultivated controls left on field margins.

** *Rumex spp.* in this case *Rumex obtusifolius.*

There was no clear pattern of annuals or perennial species with distance from the full edge. The only exception was *Galium aparine*. During Autumn only two quadrats contained *Galium* and these were in the field edge. This spring germination of *Galium* conforms to expected pattern. Use of the Harrowcomb in the autumn has not been repeated on the study because it was thought to have stimulated a new flush of weed seedlings.

By March the population had expanded both 1m and 6m into the field. The 1m sample contained a higher frequency of plants. This confirms that established picture for *Galium* where it colonises the field edges and then progressively moves into the cropping area.

As you move away from the field margin the number of weed species declined. The more pernicious weeds like *Galium* follow this pattern which has implications for margin and crop management.

FINANCIAL ASPECTS OF FIELD MARGIN MANAGEMENT

Farmers always have a choice between using every last square metre of available field space or allowing space next to hedges.

By planting the last 1m under a hedge or around the field margin the following costs are incurred per kilometre:-

Crop inputs (winter wheat) 1994 harvest per ha (Noble. L, 1994).

Seed	£6.00
Fertiliser	£9.00
Chemicals	£15.00
Cultivations	£25.00
	£55.00

For this investment in seeds, fertiliser, chemicals and cultivations the return across the field as a whole in terms of gross margin was £800. However as a result of increased competition for water, light, fertiliser, compaction and the predation of birds and mammals the crop yields at the margin is substantially lower than the average for the field. Experience of the RP Farm Manager leads to the estimate that yields are about 50% of the overall average. On this basis the return per hectare of the 1m wide field margin is £400 or £40 per kilometre. The overall financial result is a loss of £150/ha or £15.00 per kilometre of field margin for those farmers who crop right up to the hedge bases.

At Rhône-Poulenc's Farm Study, in addition to the 1m wide uncut grass area at the hedge base, a 3m wide mown grass access strip is also used.

The cost of establishing these access paths is:-

Grass seed per hectare	£50.00
Cultivations per hectare	£60.00
	£110.00

It is quite conceivable that yield reductions even during the establishment year will off-set these costs.

By adopting a strategy of grassed hedge bases at twice annually trimmed grass access paths/roadways can make financial sense.

SUMMARY

Margins can be managed to give environmental benefits and improve both operational and public access. The costs involved in establishing and maintaining margins can be more than off-set by not losing yields. Weeds within the cropped areas can be managed as shown by the conventional study area (Turner, 1995.) Careful use of herbicides can enable farmers to grow high quality crops whilst maintaining the stability of field margins.

REFERENCES

Brown, R. (1995)   Small Mammals Study in Rhône-Poulenc Farm Management Study 5th Annual Report, 1995.

Noble, L. (1994)   Farm Management Report in Boarded Barns 4th Annual Report, 1994.

Turner, M. (1995)   Field and Margin Flora study in Rhône-Poulenc Farm Management Study 5th Annual Report 1995.

Rasmussen, J. (1991)   Optimising the intensity of harrowing for mechanical weed control in winter wheat. *Brighton Crop Protection Conference. p.177*

LANDSCAPE MANAGEMENT - CENTRAL TO THE WHOLE FARM POLICY

C.J. DRUMMOND

LEAF (Linking Environment And Farming), National Agricultural Centre, Stoneleigh, Warwickshire, CV8 2LZ

ABSTRACT

Integrated Crop Management, (ICM), offers a realistic solution for positive landscape management. Practical examples, using the LEAF Environmental Audit as a management tool, show just how farmers are taking action, by ensuring the decisions for landscape management are fully integrated into the whole farm philosophy and the flow of land is paramount to the vision of the farm and the fabric of the countryside.

INTRODUCTION

The countryside is at the heart of the sustainability debate, and our long term aims are for an environmentally healthy countryside, a beautiful countryside, a diverse countryside, an accessible countryside and a thriving countryside.

Sustained development is impossible without sustainable agricultural land use. Indeed land is one of our most precious natural resources. Our landscape needs to be managed as part of the whole farm system as inappropriate management can represent a permanent and irreversible loss of an irreplaceable asset, resulting in a loss of valued countryside and reducing the long term capacity of the UK agriculture sector to produce both agricultural and environmental goods.

Integrated Crop Management (ICM) is a whole farm philosophy that is site specific, bringing together the farming system and the positive management of landscape and wildlife and habitat features. In the management of the landscape one of the key factors to address is the flow of the land and its features. The starting point is appropriate planning of the area and this is where the good housekeeping and the discipline of an ICM system can help, through identifying the already existing features and mapping them and their value. From here there is a value in aiming to maintain or enhance the existing features. It is essential that if the landscape is altered assessing the impact that any changes may make is vital. This means the consideration of cropping, tree planting, creation of hedgerows or beetle banks as well as the consideration of more dramatic land changes, such as merging existing and new buildings into the environment, excavations, gravel digging etc. Although these irreversible losses of land and landscape are to be avoided.

WORLD DEMAND

From the Earth Summit 1992 (Quarrie, 1992) in Rio the priority for agriculture was set to maintain and improve the capacity of higher potential agriculture lands to support an expanding population, whilst conserving and rehabilitating the natural resources on lower potential lands in order to maintain sustainable man/land ratios.

This included the importance of management related activities such as the need:
- To formulate and implement integrated agricultural projects that include other natural resource activities;
- To ensure people's participation and promote human resource development for sustainable agriculture.

This component bridges policy and integrated resource management. The greater the degree of community control over the resources on which it relies, the greater will be the incentive for economic and human resources development. It is on these policies that Integrated Crop Management (ICM) offers a positive approach to farming, environment and landscape management.

THE LANDSCAPE

Land means different things to different people. This makes it difficult to find a single emotive and meaningful indicator. To many people the primary purpose for landscape is as a resource for the production of food and other products and as an ecosystem for vital organisms. However, overall, the need to protect the countryside for its wildlife value, landscape, natural resources, recreational and agriculture and to balance the competing demands for urban land is of prime importance (Baldock 1992).

It is possible to identify two broad but distinct types of concern; a question of the quality of the resource (soil quality), and a question of quantity (land use). It is clear that valued features of the countryside have been lost at an alarming rate. This includes the removal of hedges, ponds and ancient woodland. But despite the facts and the concerns of many there is much action taking place and in some incidences this has been going on for years, there is in effect a *quiet revolution*. This is where ICM comes into its own.

In order to achieve a healthy soil and environment a fully integrated system for farming needs to be adopted taking account of the environment in every agronomic decision whilst ensuring an economically viable return. Agriculture has not been in a more financially healthy situation for a long time and although there is uncertainty as to the long term nature of policy decisions all fingers point in the direction of future support being more in line with the need to supply environmental criteria. Indeed we need only look to the new countries that have joined us in the EU to see where they are coming from. Sweden and Austria, both have a very strong commitment to the environment. Furthermore Denmark insists that new farmers and tenants take up training for environmental and agricultural skills.

Whether we like it or not our countryside has been fashioned by the changes over centuries, and more often than not we see history repeat itself. Today, however we have that historic knowledge and we have technology to help us overcome the mistakes of the past. It is important that we are realistic about the demands of the future. We need a thriving countryside, and the landscape that we fashion is a function of the health of the countryside.

AN INTEGRATED SYSTEM

Integrated production systems accept moderate and environmentally appropriate use of external inputs - mainly fertilisers and pesticides - in combination with measures to make maximum use of internal nutrient recycling, biological pest control and symbiosis between different crops and livestock. Integrated pest management, integrated plant nutrition, soil conservation, and efficient and non-exhaustive water use are mutually reinforcing measures. They should be applied together within the farming system at levels appropriate to the physical and socioeconomic conditions; the optimum strategies are highly specific to individual farms. However given the diversity of ecosystems, land resource potentials, preferences of farmers and consumer demands there is opportunity for the co-existence of different systems - organic and integrated - in a manner that respects the environment.

The overall aim is to develop and adopt diverse and improved plant production systems in which the optimal utilisation of inputs maximises economic returns to the farms and protects the environment.

THE AGENDA FOR ACTION TO ACHIEVE AN INTEGRATED SYSTEM IS TO:

- ▸ promote diversification and integration of agricultural production systems;
- ▸ optimise use of on farm inputs, biological processes and local natural resources;
- ▸ minimise the use of external inputs, while increasing productivity and farmers income;
- ▸ disseminate information and redirect and increase support to agricultural research and technology development to promote sustainable land use systems;
- ▸ use tradition indigenous technologies, as well as strategic research, on biological processes that govern agricultural production.

WHAT TO DO - PLAN, DESIGN, COLLABORATE

As part of an integrated system design is one of the most important aspects in the improvement of the landscape's value. A range of actions assist in improving the design of new development and the landscape in rural areas and particularly on farm and include:

- ▸ a landscape and wildlife plan for the whole farm, to integrate nature conservation with the cropping programme;
- ▸ priority for the maintenance and enhancement of existing features;
- ▸ attention for non farmed and farmed areas, it must be remembered that much potential also lies in cropped land.
- ▸ consideration that mixed farms and rotations, which include a large range of crops, offer a wider range of options for wildlife.
- ▸ communication at all levels. Staff and contractors should be aware of areas of conservation value.;
- ▸ ensuring that new development draws on the best design in terms of layout, use of space and management of traffic;
- ▸ ensuring that new development should be complementary to existing buildings and in harmony with the surrounding countryside;
- ▸ avoiding irreversible losses of land and landscape at all costs.

When planning the environmental enhancement and management programme for the farm, existing habitats and landscape features should take a priority. At all costs damage to landscape features and wildlife areas should be avoided.

However, despite the fact that there are substantial savings that can be made through the attention to detail from an Integrated Crop Management system and that there are many grants for some specific conservation projects and developments, it must be remembered that conservation can cost money. Land taken out of production still incurs rent or mortgage charges. The overheads of the farm business become spread less widely than before. If trees are planted, the ultimate crop is unlikely to produce income during the lifetime of the forester or his successor but weeding and thinning must still be paid for.

In spite of these constrictions, farmers who care to devote a little throughout to a matter can to help sustain a varied population of wildlife. Any farming system can sympathetic to the environment. Many farmers have begun to understand that, for the sake of game and other species as well as general economic reasons, herbicides and fungicides should be used not as a matter of routine but only when essential. Time of application of chemicals can be chosen carefully to avoid undue damage to birds and many other species in the wildlife chain. Hedge bottoms can be allowed to become slightly wider and field work can be planned so that disturbances from noise and the human voice is arranged in such a way that game and other birds and mammals are encouraged towards shelter rather than away from it. Tranquillity is an essential part of sympathetic farm management. It does not exist in urban life and therefore should be all the more important to countrymen.

COMMUNITY INVOLVEMENT

It has been estimated that in 1994 some 590 million visits were made to the countryside. This level of public access inevitably gives rise to conflicting pressures. In some areas it is difficult to reconcile the privacy of ownership and conservation objectives with public access. Most local people value in both aesthetic and recreational terms, elements of their landscape, yet they are rarely involved in the decisions and processes that shape it. As land managers and owners, farmers clearly should be making decisions about how to best farm their land. But if responsibility is assigned to both farmers and the communities for landscape conserving activities, where local people are encouraged to become involved in local farming in an indirect way, then again more understanding would be created among different interests.

Indeed there are village specialists that help out on some of the LEAF Demonstration Farms. They come from all sectors and classes of the community and facilitate the integration of marginalised groups, so allowing their skills and knowledge to influence development priorities. Given the chance local people are able to monitor environmental change and so take action when required. (Pretty, 1994). Coupled with this it is of course the need for a better understanding of farming and the countryside by the public in the first instance and again this is where the LEAF Demonstration Farms have an active role to play.

There are thus many demands for the role of both the countryside and the landscape it needs to be:

▸   **an environmentally healthy countryside**

biodiversity is one of the key requirements of the Rio Summit and commitment from global governments. So much of our landscape features provide the habitats for a rich and varied wildlife ranging from birds to small insects.

► **a beautiful countryside**

visual appearance of the countryside is perhaps the most obvious impact for the countryside as it the most easily measured and most likely to cause concern in the publics eyes. This obviously includes the loss of habitats, such as hedgerows, trees, ancient woodland and grassland, the colour of the countryside, the introduction of oilseed rape and linseed ranging to the complete change in use of the countryside through gravel extraction and the creation of roads

► **a diverse countryside**

this indicates the importance of biodiversity and also the need for correct planning and control. Here the local topography and natural flow of the land is essential on the planning of the structure.

► **an accessible countryside**

there is a strong interest for the non rural community to visit the countryside but just as it is essential for farmers to be responsible in their roles as custodians of the countryside so it is important for those visiting the countryside to take a responsible attitude to care and concern.

► **a thriving countryside**

not only is it essential that there is a rich diversity in the countryside it is also essential that there is a financially thriving countryside.

As has been highlighted previously, Integrated Crop Management (ICM) is one practical and achievable solution for all these demands to be met by farmers. A whole farm system, ICM is not a prescriptive approach but is site specific, allowing flexibility in the system, which can take into account the topography, the importance of local history and design and the natural flow of the land.

ICM incorporates: site, rotations, crop protection, crop nutrition, pollution control, energy use, variety choice, organisation and planning and a positive and whole farm approach to landscape, wildlife and habitats. It brings into account the detail of a diverse variety choice, essential for better control of pests and diseases, cropping plan, a sound rotation is one of the most important contributing factors to better soil structure, better pest and disease control etc. ICM is a '*quiet revolution*', not a flash of dramatic change. It is a gradual process that builds on existing practices through the utilisation of the best of traditional methods and the best of modern technology. It is not only good common sense farming it is also involves commitment from all sides from the industry, from farm managers and farmers, and from farm staff and contractors.

At LEAF (Linking Environment And Farming), we are part of a pan European project that is demonstrating and developing Integrated Crop Management. We now have 17 LEAF Demonstration Farms, selected against stringent criteria, who are promoting Integrated Crop

Management to both farmers and non farmers. Furthermore we have developed both guidelines (LEAF 1994a) for Integrated Crop Management and the LEAF Environmental Audit (1994b). The latter is a unique document. Originally designed as a management tool to help farmers along the route of ICM it now also has potential as a customer assurance and assisting in securing a market, addressing the demands of 'traceability' and 'due diligence'.

Indeed now in its second year the LEAF Environmental Audit has assisted farmers in many ways. It has 'made them think', it has helped them identify areas where they are getting it right and give them credit for that and it has identified areas where risks could have led to environemtal destruction and pollution.

ENVIRONMENTAL AUDITING

Environmental auditing is basically the systematic examination of the interactions between any farm operation and its surrounding. This includes all emissions to air, land and water; legal constraints; the effects on the neighbouring community, landscape and ecology; and the public's perception of the farm in the local area.

The LEAF Environmental Audit aims to help safeguard the environment by:

▸ facilitating the management control of environmental protection;
▸ assessing compliance with farm's policies which would include meeting regulatory requirements.

There are three essential functions to starting a successful environmental audit:

COMMITMENT	-	from management, staff and/or contractors and advisers;
RESOURCES	-	environmental auditing can save money but resources can be required in management time;
LEADERSHIP	-	the audit involves everyone but responsibility does need to be allocated correctly.

OBJECTIVES OF THE LEAF ENVIRONMENTAL AUDIT

▸ Firstly define what you want to achieve;
Those carrying out the audit were able to identify areas that did have an impact on the environment. Many of which may have been identified in the past, but using LEAF Environmental Audit helped draw them together.
▸ The next step is assessing where changes can be made, by providing answers to the following questions:
What are we doing?
Can we do it better?
Can we do more?
Can we do it more cheaply?

Environmental auditing does not stop at compliance with legislation. Nor is it a 'green token' public relation exercise, although, done well it could bring considerable positive PR.

Rather it is a total strategic approach to the farm's activities.

The LEAF Environmental Audit is a practical, non-prescriptive way for farmers to look objectively at their business, assess strengths and weaknesses and identify cost saving operations and capital expenditure to minimise long term risk; it shows how the integration of the environment with food production can be profitable as well as desirable. The audit is a series of self assessment forms and provides a convenient and structured way which, when carried out on an annual basis, will monitor progress and help in determining priorities on the farm.

It addresses 8 principal areas:

Landscape features;	Conservation of energy;
Wildlife and habitats;	Pollution control;
Management of the soil;	Organisation and planning;
Crop protection;	Animal welfare.

As well as taking stock and giving credit to existing practices and identifying areas for future improvement, there are many long term benefits which can result from carrying out the audit.

These include:
- improving economic performance;
- enhancing environmental performance;
- meeting insurance requirements;
- gaining a marketing edge.
- ensuring environmental protection;
- meeting legislative requirements;
- addressing public concerns;

In this paper practical examples will be sighted highlighting the importance of ensuring that the decisions for landscape management are fully integrated into the whole farm philosophy and that the flow of land is paramount to the vision of the farm and the fabric of the countryside.

The starting point of the audit is the management policy statement which sets out the goals for the whole farm. The farmers are then encouraged to carry out a whole farm plan identifying areas of specific environmental value (eg from FWAG or ADAS). From there the series of self assessment sheets in the LEAF Environmental Audit assists farmers in looking at their farm objectively. Many of the questions are based on standard good agricultural practice but with the rapid advancement of technology there are other areas that can assist farmers in addressing and targeting situations where greater attention to detail is required. The audit makes people think and with increasing pressure on farmers to address both environmental and economic criteria it is a useful management aid.

From the analysis of the returns of the LEAF Environemtal Audit (1994) the farmers who carried out the audit recognised that not only are auditing and accountability where the future is going in terms of answering consumer demands, but also that Integrated Crop Management is the way forward for a profitable and environmentally responsible farming system.

There were strong indications that these farmers are 'getting it right' in terms of recognising the need for positive environmental management and responsible farming practices, and are taking the necessary action. Indeed, in a majority of cases, they have been doing so for many years.

Furthermore, there were some key areas identified by the LEAF Environmental Audit, which farmers wanted to develop and act upon to improve their farm management and achieve set targets.

More specifically, these included:

- the need to improve staff training and awareness;
- the implementation or updating of a conservation plan (for example a FWAG whole farm plan);
- familiarisation with MAFF codes of Practice;
- improved soil management, for example mapping areas where erosion, leaching or compaction is a problem;
- formulating an emergency plan;
- formulating a waste management plan;
- increasing staff involvement in farm policy and decision making;
- monitoring - both in terms of physical records, fuel and threshold levels for insects and disease incidence.

A PRACTICAL APPROACH

Landscape management is central to the whole farm policy and to illustrate this further to follow are a few examples of landscape management and whole farm policy on some of the LEAF Demonstration Farms.

At Cold Harbour Farm the landscape is very open and there are beautiful views across the Wolds. There is an absence of natural features on the farm and attention is payed to the hedges and their management. An active programme of hedge regeneration is being implemented on the farm with gapping and coppicing and in recent years over half a mile of new hedgerow has been planted. North south hedges are encouraged to grow larger with infrequent trimming. East west hedges are also allowed to thicken out at the base and trimmed to an A shape at 5-6 feet. Furthermore two tumuli have been grassed down and preserved and there are other Neolithic sites of particular historic interest on the farm.

On another, Ian Brown's farm management policy for Lee Moor Farm includes:
- Optimising the use of inputs and assets to achieve a profitable return, using appropriate technology in sympathy with the environment
- Developing areas of the farm unsuited to arable production into intensive conservation development;
- Encouraging public access for education and recreation.

His most visible influence on the farm however, is the transformation from a rather featureless arable holding to one that has an increasing number of interesting areas with surprises around almost every corner. This reflects Ian's belief that, as custodians of the countryside, farmers can do much to manage, enhance and extend the environmental character of their farms, while maintaining a profitable business.

In contrast Guiting Manor enjoys an abundance of natural features; miles of hedges, two streams that run the length of the estate, and hectares of scattered woodland, including an area of ancient woodland with a rich diversity of plants and wildlife. A management policy to maintain the condition and balance of these features has been reinforced entering a 10-year agreement under the Cotswolds Environmentally Sensitive Area scheme to restore and preserve hedges and walls, and limit fertiliser and pesticide use on old pastures. Indeed more than 30000 trees have been established in recent years many on four new plantations covering nine hectares in all under the Farm Woodland Scheme. Also on the estate is a strong commitment to the restoration of the village cottages and today the Trust lets its houses as far as possible to local countryfolk at affordable prices.

And a further example is Henry Cator of Rotac Farms partnership, who sets out the main aims of the estate as follows;
▸ To make trading profits while acting as custodians of the countryside for future generations;
▸ To preserve the traditional nature of the estate as a whole without sacrificing good husbandry or sound management;
▸ To act as the nucleus of a thriving rural village and community.

Conservation activities, sympathy for natural resources, consideration for the environment and a fully integrated cropping approach are all long standing principles on the farm. Management and renewal of the Estate's natural features is an ongoing activity - not something to engage in to collect grants or as a response to public opinion. Hedges have always been maintained and are cut every other year on a rotational basis to ensure there is no shortage of bird nesting sites. Tree planting takes up some 1.6 to 2 hectares a year, extending or supporting existing woodland in some cases, creating new plantations in others. An important Ramsar Site marshland, harbouring rare plant species and a great diversity of wild birds and mammals, falls within the estate's boundary. The Bure Marshes, a Site of Special Scientific Interest (SSSI) are run under a management agreement with English Nature. Other ancient grazing marshes have been reclaimed and ponds periodically dredged to prevent them silting up.

Buildings that are no longer practical for modern farming methods are kept in good condition until an alternative use can be found. Many now house a diverse selection of businesses which include an equestrian centre, an automotive engineer, boat storage and even a small brewery. Two cottages behind the brewery building which has previously housed sheep and animal feed milling equipment, have become the village pub or brewery tap.

A final example is Edward Darling who having restored the farming business to maximum profitability and restructuring the enterprise to separate farming and estate management activities, now includes priorities:
▸ Producing a quality product at optimum profitability;
▸ Enhancing the environment in which all involved in the farm live and work;
▸ Providing opportunities for other people, whether housing, employment or workplace.

One ambition is to see 20 people working on or from the farm again and to provide homes for a similar number. Indeed Mr Darling is well down the way to achieving this ambition.

Today Edward Darling continues the philosophy and combines sound land management with responsible conservation techniques. He has adopted an integrated approach focusing on woodland, hedgerow and grassland with the overriding aim of promoting diversity of habitat alongside economic crop management.

All the above farms are practising Integrated Crop Management. They are farmers who emphasise the importance of a farming business that is commercially viable at the same time as being environmentally sensitive.

CONCLUSION

In the management of the landscape one of the key facts to address is the flow of the land. The starting point is appropriate planning to the area and this is where the good house keeping and the discipline of an ICM system can help.

By identifying the existing features, areas of specific environmental sensitivity and areas for potential, the LEAF Environmental Audit together with a whole farm conservation plan is an excellent starting point for the stepping stone approach of Integrated Crop Management. This starts off by giving highest priority to those existing features that should first and foremost be managed positively.

And to follow is a ten-point plan to assist in planning for landscape management under ICM techniques:

- adopt a whole farm approach
- assess the flow of the land
- manage existing features
- develop plans in keeping with the local environment
- identify areas to improve natural flow of the land
- assess the cropping pattern in the lay of the land
- remember the importance of landscape goes well beyond the visual appearance of the countryside - the environment offer biological benefits and the cropping pattern can add to the biodiversity of the surroundings and less risk
- encourage public access across the landscape
- training for farm staff and the non farming community is essential to spell out the importance of environmental value
- PLAN.

REFERENCES

Baldock, D. (1992) Agriculture and Habitat Loss in Europe. WWF
LEAF, (1994a) A Practical Guide to Integrated Crop Management.
LEAF, (1994b) The LEAF Environmental Audit.
Quarrie, J. (1992) Earth Summit 1992. The United Nations Conference on Environment and Development.
Pretty, J.N. (1994) Regenerating Agriculture Policies and Practice for sustainability and self reliance.

Session 6
System Projects

Chairman and
Session Organiser   S OGILVY

RESEARCH INTO AND DEVELOPMENT OF INTEGRATED FARMING SYSTEMS FOR LESS-INTENSIVE ARABLE CROP PRODUCTION: PROGRESS 1989-1994

V.W.L. JORDAN, J.A. HUTCHEON, G.V. DONALDSON, D.P. FARMER

IACR-Long Ashton Research Station, Department of Agricultural Sciences, University of Bristol, Long Ashton, Bristol BS18 9AF, UK

ABSTRACT

The LIFE project is an interdisciplinary research study at IACR Long Ashton designed to address, exploit and integrate interactions of farming system components, holistically, and to provide the technology for economically viable, ecologically acceptable and environmentally benign production systems. The first 5-year cycle was completed at harvest 1994.

Over the 5-year period, adoption of less-intensive strategies based on integrated technology reduced overall yields of wheat and oilseed rape by up to 18%, and yields of barley and oats by 11%. Nevertheless, production costs were also reduced by 32% and overall profitability was maintained. Within this period, substantial reductions (kg ai ha^{-1}) in applied nitrogen (36%), herbicides (26%), fungicides (79%) and pesticides (78%) have been obtained over standard farm practices designed to reflect current arable crop production strategies. Data are presented on innovative strategies and decision making processes, and their implementation in two commercial "Demonstration Farms" in south-west England.

INTRODUCTION

The development and implementation of more environmentally benign, sustainable production systems is being increasingly recognised as the long-term objective for arable crop production systems. This will require a gradual and stepwise transition, taking on board new opportunities created by science and technology, in order to provide a more rational and balanced approach to economically sound agricultural production, aimed at harmony between agriculture and the environment. Most farmers are unlikely to change their practices radically in the short-term but are, nevertheless, seeking ways to reduce their unit cost of production. Increasingly, they are prepared to adopt more rational approaches for nutrient and pesticide use, and to exploit alternative measures that minimise risks of problems arising that would otherwise require treatment with chemicals.

There are many ways to reduce production costs, either selectively or holistically, as part of strategies to improve farm income. Selectively-focused component research in a number of areas has contributed to these by providing options for reductions in agrochemical use that minimise environmental contamination. These include integrated nutrient management, forecasting systems for pests and diseases, reliant upon a basic understanding of population dynamics and other risk factors; reduced doses of herbicides, based on knowledge of the effects and interactions of weed growth, weather and soil conditions on herbicide performance; and improved spray technology. New developments in mechanical weed control, either alone or in combination with low doses of herbicides (Caseley et al., 1993) may not only complement current weed control strategies, but also offer crop nutritional benefits from the nitrogen mineralised by mechanical intervention (Smith et al., 1994). Recent research, that encourages natural enemies of pests and antagonists of diseases, or substitutes the use of biological control agents or behaviour-controlling chemicals for persistent pesticides, should also offer future opportunities to reduce the agrochemical load on the environment.

Other options in the development of reduced input systems involve manipulation and integration of husbandry practices within crop management.

Crop rotation is a key component in reduced and integrated systems of production, with maximum use made of crops that contribute positively to soil fertility. Crop rotation also provides options for reduced use of fungicides by decreasing disease carry-over from crop to crop, and herbicide reductions by permitting selective control of troublesome grass weeds in broad-leaved crops in the rotation without use of persistent herbicides. However, these effects and their interactions need to be examined over full rotational sequences in order to exploit the cumulative benefits.

Whilst selective reductions in nutrients and agrochemicals will reduce input costs, profitability can only be maintained by full and optimal integration of these exogenous variables within the whole system of crop management to ensure reliable yield at reasonable cost with an acceptable margin of profit. The long-term, farm-scale Less-Intensive Farming and Environment (LIFE) research project at IACR-Long Ashton, investigates opportunities to combine and optimally exploit all the above techniques and methodologies within an integrated farming systems approach.
Research since 1989 (Jordan & Hutcheon, 1993; 1994) indicates that a less-intensive and integrated approach for arable crop production can maintain profitability by reducing the unit cost of production. There are also consistent indications of improved soil structure and quality, reduced agrochemical contamination and increases in soil flora and fauna, especially predators of key pests.

Based on the data generated from the LIFE project, prototype cropping systems, designed to be more environmentally benign than those currently in operation, have been formulated and implemented on two commercial farms in south-west England since 1992 in order to explore the feasibility and constraints of adopting such systems of production. This EU-funded Demonstration Project aims to demonstrate, to members of the farming industry, alternative methods and approaches that encourage farming practices which are compatible with environmental and natural resource protection; to provide on-site training in the principles and practices available for implementation; to appraise attitudes of members of the farming industry towards adoption; and to show that such systems are technically and economically viable.

MATERIALS AND METHODS

The LIFE Project

Established in 1989, this long-term, farm-scale experiment occupies a total of 23 ha. It comprises 20 field units (each of about 1 ha) within five fields, in order to compare four systems of production in fully-phased 5-course rotations. The four comparisons comprise a conventional rotation (CON) and an integrated rotation (IFS) each managed by standard farm practice (SFP), defined as that adopted by a technically competent farm manager and annually adjusted to reflect changes in conventional practice, and research-based lower input options (LI) (Jordan & Hutcheon, 1993). The crop rotations, husbandry practices and management decisions for the four systems of production have been well documented (Jordan & Hutcheon, 1994). Standard farm machinery is used throughout and a detailed diary of full husbandry records maintained. Comparative energy costs for machinery operations on the four production systems have been produced (Donaldson et al., 1994). Crop yields are determined by taking 16 measured combine-cuts from pre-determined reference areas across each field unit and quality parameters are measured. Production costs (variable costs) are calculated on the basis of IACR-Long Ashton Farm purchase costs for seed, basal fertiliser, nitrogen and other nutrients, fungicides, insecticides, molluscicides, plant growth regulators and desiccants. The values for grain output are based upon HGCA average market price for the UK during the first week of October each year, for October delivery.

Commercial "Pilot" Demonstration Farms

Trerule Demonstration Farm and Bake Farm at Trerulefoot, Cornwall are sited in a central part of a 600 ha arable enterprise at Trerulefoot, Cornwall. The farms are in an area of great landscape value, on free-draining silty loam soil, with heavy winter rain contributing greatly to the annual rainfall of 1060 - 1270mm. Underground springs, most of which are piped to the nearest open water-course, and pockets of clay create a management challenge on the farm. On the commercially farmed land, key crops are winter wheat for animal feed, winter barley for malting and winter oats for milling. Other crops (oilseed rape, linseed, peas, beans) provide a natural break for these cereals; 60 ha of woodland fall within the farms' boundaries. In addition to the 40 ha permanent pasture (for 300 ewes), there is 4 ha of rough ground on which 10 different habitats have been established - four ponds, wetland areas, hazel coppice, meadow, fir plantation and new woodland planting. Wildlife is promoted within the boundaries of Trerule and Bake Farms, and 150 wild flower species have been identified.

The Trerule Farm unit (32 ha) was therefore selected because it has a favourable farm infrastructure for exploitation of the principles of integrated production. It comprises six fields (average field size 5 ha), with established field boundaries of traditional raised banks accounting for 3% of the land area.
Thus, the following 6-course rotation was adopted for the integrated farming systems approach: winter wheat - winter oats - winter barley - setaside (natural regeneration) - "option crop" - spring crop (oilseed rape/linseed/peas) following a winter green cover.

Harnhill Manor Farm, near Cirencester, Gloucestershire, has been farmed by the Royal Agricultural College since 1987. The farm is situated on the edge of the Cotswold Limestone as it gives way to the alluvial soils of the Thames Valley and has a cropping area of 243 ha. Crops grown on the farm are mainly cereals (winter wheat, winter barley and spring wheat) and other combinable crops grown in rotation (winter beans, oilseed rape and linseed). Soil types are mainly Corn Brash and Forest Marble, with some overlying areas of clay loam and deeper alluvial soils. The farm is sited in a predominantly cereal growing arable area of the Cotswolds and is typical of the region. There are well structured hedgerows within and surrounding the farm, as well as managed woodland areas.

In contrast to Trerule Farm and the surrounding areas in Cornwall, many farms in the Cotswolds comprise large fields which are usually ploughed. As a consequence, soil erosion has been a problem in some years. In addition, one of the guidelines for Integrated Production states that *"the lateral dimension of an individual field should not exceed 100 m, otherwise fields need to be divided by annual or permanent vegetation to provide adequate ecological reservoirs"* (El Titi et al., 1993). Therefore, the approach adopted at Manor farm was to convert a large field (Driffield Bank - 30ha) typical of the area, into an integrated farming systems unit, by dividing the field into six manageable units. Headland and boundary strips ("raised-banks"- 4m wide) were established between each field unit in spring 1993, and sown with various grass and wild flower mixtures. A 14m strip was also prepared centrally in the 30 ha field for establishment of a tree line. The following 6-course rotation was adopted in Driffield Bank: winter wheat - winter barley - winter beans - winter wheat - setaside (natural regeneration)- winter oilseed rape.

With the exception of winter beans which, at Harnhill, are broadcast and ploughed in as the method of establishment, all other crops are established using minimum/non-inversion tillage, and managed according to the guidelines for integrated production. All farming operations and the economic evaluation (crop yields, production costs, gross/net margins) for both commercial demonstration farms are done by the farm managers.

RESULTS

The LIFE Project

At harvest 1994, the LIFE project had completed its first 5-year cycle, so that all crops in the rotation have been grown on each designated field unit. Set-aside was introduced in autumn 1992, converting the conventional rotation from a 4- to a 5-year rotation and substituting set-aside for winter beans in the integrated rotation. Over the 5-year period, the lower input options on the conventional rotation reduced yields of "first-wheats" by 8%, "second-wheats" (grown only in 1990-1992) by 10%, barley by 11% and oilseed rape by 1%. However, these lower input options resulted in savings in production costs of 40% for "first wheats", 32% for "second wheats", 26% for barley and 29% for oilseed rape. With the lower input options in the integrated rotation, wheat yields (all "first-wheats") were reduced by 18%, due mainly to the selection of inherently lower yielding, disease resistant, quality cultivars; oat yields were reduced by 11%, and oilseed rape yields by 18%, whereas the yield of beans was increased by 5% (Jordan & Hutcheon, 1994). Thus, in the systems comparison, although this resulted in an overall 10 and 15% yield reduction in the conventional and integrated rotation, savings in production costs were 33% and 35%, respectively. In terms of profitability over the 5-year period, standard farm practice on both rotations gave gross margins of £577 ha^{-1}, whereas the lower input options on the conventional rotation increased gross margin by £37 ha^{-1}; however, in the integrated rotation gross margin was reduced by £18 ha^{-1} (Table 1).

TABLE 1. Grain yields (t ha^{-1}), production costs and gross margins (£ ha^{-1}) from the LIFE project (1990-1994); 5-year means of the systems comparisons.

Crop	Yield (t ha^{-1})	Variable Costs (£ ha^{-1})	Gross Margin (£ ha^{-1})
Conventional rotation			
Standard Farm Practice	6.40	251.90	577.76
Lower Input Options	5.76	169.90	614.80
Integrated rotation			
Standard Farm Practice	5.92	230.28	576.60
Lower Input Options	5.05	148.96	558.45

Commercial "Pilot" Demonstration Farms

Trerule Farm, Cornwall:
In the 1993 harvest year, all crop yields, including those of crops grown conventionally on adjacent land (Bake Farm), were lower than the annual regional average for the previous 10 years (Table 2), due to climatically limiting variables. Nevertheless, responses and profitability from the "Pilot Farm" in this first transitional year, were most encouraging. By comparison with the 10-year conventional farm averages for the crops grown (GM = £621 ha^{-1}), the crops in the "Pilot Farm" grown under the guidelines for integrated production gave a farm average gross margin of £617 ha^{-1}, (Table 3).

TABLE 2. Economic Appraisal of Conventional Farm Practice at Trerulefoot (Previous 10-year Average) (GM = gross margin)

Crop	Production Costs (£ ha⁻¹)							Yield	GM
	Seed	Fert	Herb	Fung	Pest	Othr	Total	tha⁻¹	£ha⁻¹
W.Wheat	47.1	89.0	41.6	43.5	0.0	4.7	226	7.5	674
W.Barley	44.9	65.8	39.0	43.6	1.6	29.8	225	6.9	672
W.Oats	39.5	68.2	20.2	48.2	1.6	23.2	201	6.8	615
Linseed	39.5	38.2	58.6	0.0	0.0	0.0	136	1.5	567
Sp OSR	22.6	35.5	0.0	0.0	3.6	0.0	62	2.2	576

TABLE 3. Economic Appraisal of crops grown under IFS guidelines at Trerulefoot -

1993 harvest year

Crop	Production Costs (£ ha⁻¹)							Yield	GM**
	Seed	Fert	Herb	Fung	Pest	Othr	Total	tha⁻¹	£ha⁻¹
W.Wheat	52.7	49.2	33.3	21.4	6.0	0.0	164	5.2	497
W.Barley	57.6	49.7	57.3	17.9	0.0	0.0	183	5.6	619
W.Oats	49.1	49.7	45.4	21.4	0.0	0.0	166	5.0	569
Linseed	42.6	43.4	4.1	23.4	0.0	0.0	113	5.8	805
Sp OSR	53.6	32.3	10.2	0.0	0.0	0.0	96	1.7	595

1994 harvest year

Crop	Production Costs (£ ha⁻¹)							Yield	GM**
	Seed	Fert	Herb	Fung	Pest	Othr	Total	tha⁻¹	£ha⁻¹
W.Wheat	27.9	50.6	46.7	29.6	6.0	0.0	161	5.5	614
W.Barley	37.8	28.3	35.5	31.0	0.0	11.8	144	5.5	795
W.Oats	40.0	43.3	35.5	31.0	0.0	11.8	161	4.8	688
Linseed	46.7	33.0	9.9	10.1	0.0	0.0	100	5.2	638
Sp OSR	45.4	7.6	46.6	0.0	2.1	26.6	130	1.9	610

** Gross Margin includes area payment.

TABLE 4. Economic data from conventionally grown crops at Bake Farm, Trerulefoot - 1994 harvest year

Crop	Production Costs (£ ha⁻¹)	Yield (£ ha⁻¹)	GM** (£ ha⁻¹)
W.Wheat	175	6.6	672
W.Barley	189	5.4	699
W.Oats	134	5.9	652
Sp.OSR	125	2.6	691

** Gross Margin includes area payment.

In the 1994 harvest year, wheat yield was lower (< 1 t ha^{-1}) than conventionally grown wheat at Bake Farm, which resulted in a lower gross margin (Tables 3,4). This was attributed to the difference in the amount of applied nitrogen (99kg N ha^{-1} for IFS compared with 200kg N ha^{-1} for conventional). Therefore on this soil type, the decision-making process for N requirement, based on residual soil N, needs to be improved. Barley yields were similar to those conventionally grown and satisfied quality malting requirements (1.6% grain N). This resulted in a higher gross margin due to the reductions achieved in variable costs. The winter oats established well and were very competitive, therefore, no post-emergence herbicide was applied. In order to meet yield expectations and reduce the risk of crop lodging, only 50kg N ha^{-1} was applied (compared with 120kg N ha^{-1} on conventional oats). Whilst integrated oat yield was reduced and the growing costs lowered, a small (2%) loss in gross margin occurred. The spring oilseed rape, established after an overwinter green cover (forage rape), provided a reservoir for slugs and some damage occurred. Although a reasonable yield was achieved, the gross margin was eroded due to extra variable costs required in the establishment and treatment of the winter cover, prior to sowing the spring oilseed rape.

Driffield Bank, Harnhill:

Due to the late start for this farm conversion, spring-sown crops of wheat, barley, oats and beans were established in February 1993, to provide the correct crop entries for a winter-crop dominated integrated rotation to be sown in autumn 1993. Crops in the 6-course rotation at Harnhill for the 1993 cropping season were therefore: winter oilseed rape (established previously in autumn 1992) - spring wheat - spring barley - spring beans - spring wheat - spring oats (instead of "setaside" as derogation was not obtained in this preliminary year). All spring crops, except beans, were established using non-inversion tillage techniques and all crops were grown according to the guidelines for integrated production.

Whilst direct economic comparisons cannot be made between the spring-sown integrated crops (Table 6) and the previous year's averages for winter-crop dominated rotations (Table 5), the economic appraisal does provide an indication of the financial implications of such a transitional phase conversion. In addition, it does indicate the options for a spring-dominated cropping system managed under the guidelines for integrated production. However, caution should be taken in interpretation of these financial data, as the demonstration farm did not receive derogation for setaside due to the late start of the project. Therefore, two gross margin figures are provided, with gross margin adjustments for area payments given in parenthesis.

In the 1994 harvest year, both winter wheat crops were drilled under two different cultivation systems (Dutzi one-pass system or tined cultivation/ Accord drill) in order to compare crop establishment and yield. Initially, plant establishment was 48% lower with the Dutzi system than with the tined/Accord drill combination, but final yield was 16% higher in the areas sown using the Dutzi. In addition, whilst disease was notably less following establishment with the Dutzi system, weed infestations appeared greater, attributed partly to the weed

TABLE 5.   Economic Appraisal of Conventional Farm Practice at Harnhill
Manor Farm (Previous year's (1992) average)

Crop	Production Costs (£ ha^{-1})							Yield	GM*
	Seed	Fert	Herb	Fung	Pest	Othr	Total	tha^{-1}	£ha^{-1}
W.Wheat	47.7	74.8	42.8	31.0	2.2	4.1	202	6.0	485
W.Barley	31.5	67.4	67.0	40.8	0.0	2.2	209	6.0	456
WOSRape	24.5	80.0	50.0	21.1	13.9	0.5	190	3.0	568
Linseed	57.2	34.1	30.8	0.0	10.5	17.4	150	0.8	406
W.Beans	58.4	0.0	7.4	0.0	24.7	0.0	90	3.4	526

* Gross margins based on Sept 30, 1992 values for grain in store plus area
payments owing of £400 and £500 for oilseed rape and linseed respectively.

TABLE 6.   Economic Appraisal of crops grown under IFS guidelines at
Driffield Bank, Harnhill

1993 harvest year

Crop	Production Costs (£ ha^{-1})							Yield	GM
	Seed	Fert	Herb	Fung	Pest	Othr	Tot	tha^{-1}	£ha^{-1}
Sp.Wheat	84.8	37.6	36.0	18.5	0.0	0.0	177	4.6	265 (405)
Sp.Barley	56.9	36.4	53.1	18.3	0.0	0.0	165	4.5	390 (695)
WOSRape	27.4	59.8	36.3	0.0	20.2	0.0	144	1.8	117 (561)
Sp.Beans	117.9	18.6	4.2	27.8	0.0	0.0	167	4.0	232 (598)
Sp.Oats	68.6	34.1	17.9	9.5	0.0	0.0	130	4.7	447 (587)

(Figures in brackets include area payment)

1994 harvest year

Crop	Production Costs (£ ha^{-1})							Yield	GM**
	Seed	Fert	Herb	Fung	Pest	Othr	Total	tha^{-1}	£ha^{-1}
W.Wheat	84.8	47.1	47.5	10.0	0.0	0.0	188	6.5	639
	69.3	41.0	55.0	10.0	0.0	0.0	175	5.4	544
W.Barley	49.1	43.0	23.1	24.0	0.0	0.0	139	4.9	632
WOSRape	24.4	47.8	39.3	0.0	10.0	0.0	120	1.2	474
W.Beans	57.9	0.0	9.2	19.8	0.0	0.0	85	2.9	549

** Gross margins include area payment

transplanting ability of the Dutzi in areas where glyphosate was not used.
The integrated barley crop was sown in mid-October, and considered too late
by local farmers to achieve acceptable yield. However, despite the
relatively low yield (4.9 t ha^{-1}), malting quality was achieved. The winter
beans, grown at a cost of £85 ha^{-1} (seed, weed harrow, low dose herbicide

and fungicide), produced 2.9 t ha^{-1} with a very acceptable gross margin (Table 6). Winter oilseed rape, established using the Dutzi in late August, was slow to emerge. The crop reached cotyledon stage in September and hardly grew throughout the winter. Several factors may have contributed to this, such as poor seed/soil contact, cold temperatures, oat straw toxins and volunteers. Plant populations averaged 54 plants m^{-2}; the crop was attacked by pollen beetle at flowering and heavy rain occurred during pod-set. Thus, a low yield (1.2 t ha^{-1}) and gross margin were obtained (Table 6).

DISCUSSION

Data generated from research into less-intensive farming for environmental protection (the LIFE Project) over the past 5 years has shown, through an integrated farming systems approach, a positive trend in economics, agrochemical and pesticide reduction and enhancement of beneficial organisms and processes, and identified farming practices that can be selectively modified to provide quality production without economic loss.

With regard to the husbandry practices and decision-making processes adopted on the demonstration farms, the pest, disease and weed strategies gave adequate and satisfactory control, but the decision models for applied nitrogen were considered to be less reliable, especially for quality wheat production. These are being re-appraised and refined within the LIFE project, and appropriate modifications will be included in 1994/95.

There has been a mixed response to IFS practices and procedures from members of the farming industry. Those on marginal land and/or those with mixed farms favour integrated production. Others, on the more productive arable land, tend to have higher overheads and therefore consider that they need high yields on all crops each year. In addition, there is still much dependence on high yielding cash crops (wheat and barley), because growing lower-yielding combinable break crops (oats, beans, peas and oilseed rape) can decrease the rotational farming system gross margin, irrespective of environmental benefits. Although some farmers have already adopted a more integrated rotation and have moved partially towards an integrated approach, farm economics are still the most important factor, thus motivation for change remains dependent upon economic advantage.

The response generated from the Demonstration Farms and at other associated events has convinced many farmers, especially in marginal areas, that adoption of IFS farming methodologies is feasible and a practical proposition. This has led to "satellite groups" of farmers willing to implement IFS principles alongside conventionally grown crops in order to, collectively, achieve hands-on experience and understanding of less-intensive systems of production. Furthermore, the introduction of the Directive on Nitrate Vulnerable Zones (NVZ) has stimulated some farmers to adopt integrated methodologies as a way of complying with this regulation. Farmers with soil erosion problems are also undertaking practices demonstrated on the Pilot Farms, in order to reduce land loss and overland drainage problems. This, in turn, is a major cause of environmental concern because of of silting-up of river courses. Environmental legislation coupled with environmental incentives seem to be the factors likely to encourage farmers to adopt integrated farming systems approaches.

ACKNOWLEDGEMENTS

The authors wish to thank H & M Bond, A. Lister and staff at Trerulefoot; M.Limb, A.Norris and the staff at The Royal Agricultural College, Cirencester for developments on the demonstration farms. This research and development has been supported by the Ministry of Agriculture, Fisheries and Food and by the Commission of the European Communities (DGVI). IACR-Long Ashton receives grant-aided support from the Biotechnology and Biological Sciences Research Council of the United Kingdom.

REFERENCES

Caseley, J.C.; Wilson, B.J.; Watson, E; Arnold, G. (1993) Enhancement of mechanical weed control by sub-lethal doses of herbicide. *Proceedings. European Weed Research Society Symposium*, Braunsweig, pp 357-364

Donaldson, G.V.; Hutcheon, J.A.; Jordan, V.W.L.; Osborne, N.J. (1994) Evaluation of energy costs for machinery operations in the development of more environmentally benign farming systems. *Aspects of Applied Biology 40, Arable farming under CAP reform*, 87-92.

El Titi, A.; Boller, E.F.; Gendrier,J.P. (1993) Integrated Production: Principles and Technical Guidelines. *IOBC/WPRS Bulletin*, **16** (1) pp. 96.

Jordan, V.W.L.; Hutcheon, J.A. (1993) Less-intensive integrated farming systems for arable crop production and environmental protection. *Proceedings No.346, The Fertiliser Society, Peterborough, UK*, 32 pp.

Jordan, V.W.L.; Hutcheon, J.A. (1994) Economic viability of less-intensive farming systems designed to meet current and future policy requirements: 5 year summary of the LIFE project. *Aspects of Applied Biology 40, Arable Farming under CAP Reform*, 61-68.

Smith, S.P.; Iles, D.R.; Jordan, V.W.L. (1994) Nutritional implications of mechanical intervention for weed control in integrated farming systems. *Aspects of Applied Biology 40, Arable Farming under CAP Reform*, 403-406.

INTEGRATED FARMING SYSTEMS AND SUSTAINABLE AGRICULTURE IN FRANCE

P. VIAUX, C. RIEU

ITCF : Institut Technique des Céréales et des Fourrages, 8 Av, du Président Wilson F75116 PARIS - FRANCE

ABSTRACT

The concept of sustainable agriculture includes sociological, economical and agronomical aspects. In order to determine what the implications are for France, four trials on sustainable management for arable farms have been set up by ITCF and ACTA* in very contrasting regions. Conventional arable farming systems (CSF) are compared with integrated farming systems (IFS). CFS is the cropping system used by farmers in 1990 in the area where the experiments are situated. IFS represents a low input system and tries to minimise the environmental impact of the system. Trials are large-scale (15-75 ha) with large plots (2 to 5 ha) in order to measure the feasibility of the system and the economic and environmental parameters. After four years of the experiment, a significant reduction in the use of inputs and consequently input cost (25 to 37 %) was obtained especially for fertilisers, fungicides and insecticides. IFS strategies resulted in yield decrease (up to 32 % according to the crop) but the economic results were slightly better (with 1995 price conditions). However, many questions still need to be answered : intercrop management, minimum tillage techniques (ability to improve), weed control in an IFS context. It will be necessary to carry on the experiments for several more years in order to stabilise the system and verify the viability of the decisions making processes.
(*) **ACTA** : Association de Coordination Technique Agricole

INTRODUCTION

The concept of sustainable development was defined in 1987 by the Brundtland Commission in preparation for the Earth Summit which took place in June 1992 in Rio de Janeiro. The problem for the researcher is to put this concept into practice. For agriculture, we can say that the role of the farmers is to feed humanity, to preserve a safe environment (for the long term) and to use natural resources carefully. Then, there are three roles for sustainable agriculture : economical, ecological and social. Can agronomists be more positive and describe accurately cropping systems which are sustainable ? For the last ten years, many researchers have proposed the integrated farming system (IFS) as an answer to this question. A review of integrated farming experiments was produced by Holland and al. (1994).

Some results are available in this paper, but they are generally preliminary results. The long term implications of IFS or low input strategies on agronomy or environment are not known.

In this paper, the results obtained from three French sites after one complete rotation (3 or 4 years according to the experimental site), are presented.

MATERIALS AND METHODS

A four trial network was laid out in France during the cropping season 1990-1991. Three of these trials were set up by ITCF, the last one, not presented here, was set up by ACTA. During the first two years, this project was linked with an European network financed by the E.U. Research Program CAMAR.The project aims to provide economic and technical references for integrated arable farming systems (IFS).

The three trial locations were chosen according to their soil and climate characteristics (table 1).
The size of the experimental plots ranges from 2 to 5 ha. Every crop is present each year but there is no replication. At each experimental site, for each crop and for each devised system, the crop husbandry techniques were described before starting the experiment.

TABLE 1- Site details

	Boigneville	Saint-Hilaire	Montgaillard
Soil type	Loam	Loamy clay	Calcar. Clay hilly
Climate	Oceanic dry	Continental	Contrast
Annual rainfall	663 mm y^{-1}	794 mm y^{-1}	655 mm y^{-1}
Rotations	• Half deep soil :	- oilseed rape	- sunflower
	- peas	- winter wheat	- durum wheat
	- durum wheat	- spring barley	- winter peas
	- winter wheat		- durum wheat
	• Shallow stony soil :		
	- oilseed rape		
	- winter wheat		
	- spring barley		
Systems in comparison		Conventional	
		Integrated	

At each site, two main management systems are compared. A conventional farming system (CFS) where the most common practices of the local farmers are carried out. These crop management are not stable during the four years and are adapted each year to the farmers practices. For example, in Montgaillard, the variable costs in 1991 were 2669 FF./ha and have decreased to 1940 FF./ha in 1994.

The IFS is, as far as possible, close to the principles and technical guidelines defined by the IOBC/WPRS working group (El Titi *and al* ., 1993). The main aspects are, rotation as long as possible according to the soil and climate condition, minimum three course, shallow cultivation and a lower target yield, about 20 % less than the soil-climate potential. For winter wheat, cultivars resistant to diseases, are chosen, sown at a low density and a sowing date delayed by about one week. In regard to crop protection, all crops are managed in order to minimise disease development : less nitrogen, lower sowing density, etc... In addition, chemicals were only used when thresholds were reached. No growth regulators were used. During the intercrop period, mechanical weed control has been used to control weeds. On the crops, early treatment combined with low doses of herbicides was used.

Numerous observations are made : The crop establishment is subject to particular monitoring that is essential since the soil cultivation is different between CFS and IFS (deep cultivation with a plough in CFS and soil cultivation at 7-10 cm deep in the case of IFS). Emergence losses as well as the crop growth rates during early growth stages are recorded. Pests are especially observed during early growth stages : sitona (*Sitona lineatus*) and thrips (*Thrips angusticeps*) on winter or spring peas, wireworms (*Agriotes spp*) on sunflower (*Aphis fabae*) are especially observed and treated if necessary... The weed flora development is subject to accurate surveys and weeds are mapped in the plots. Diseases are monitored with direct observations on plants but also with the use of prediction models especially on winter wheat. Yields are measured on harvest areas that have been defined after a methodological study. Quality of the harvested products is obtained through sample analysis. The type of analysis is adjusted to suited each product. Economic aspects : all data necessary to do comprehensive calculations have been recorded, machinery costs included purchase costs, repair costs, fuel and lubricant consumption for each field operation according to soil conditions, in order to take into account the labour time.

For each purchased input (seeds, fertilisers, pesticides...) date of purchase and amounts applied have been registered. All these costs are processed to give a set of economic indicators such as gross margin, direct margin, net margin for each crop and each rotation. Additionally, with the help of an interactive computer program devised by ITCF (SIMU-GC), overall results for the farm : net income, balance sheet and cash etc, can be obtained. In addition, technical indicators such as working times, equipment wear and tear rate are analysed.

RESULTS

The results are a precise description of each system and of the decision making process for each crop in each site. This aspect is very important when cropping systems are studied because the main objective of the research is not really to compare the two systems but to improve each one. For that purpose, each year the gap between forecast and real results is analysed (Viaux, 1994). Another reason for this methodology is to facilitate the transfer of technology of the new system to farmers.

Soil tillage and crop establishment

In IFS, all the crops were established with non inversion tillage. After harvest, the crop residues were chopped and immediately incorporated in the soil in order to facilitate the decomposition of crop residues and also to provoke a flush weeds and volunteers. This was followed by an other cultivation two months later. This strategy has some advantages after several years. The trafficability and natural drainage are better and there is no soil compaction. But for spring crops (peas or spring barley), after a very wet winter, the sowing date was delayed. Nevertheless in Montgaillard, there was less erosion with the sunflower crop in may 1992 in the IFS treatment due to crop residues on the soil surface.

The losses at emergence are higher with IFS when compared with CFS. These losses can reach 50 % for a spring crop due to the bad position of seeds or, in some years, to pest development : thrips angusticeps on peas and wireworms (*agriotes* spp) on sunflower. After four years of trials, it seems that shallow cultivation can slightly increase some soil pests but no specific slug problems were observed.

Weed control

Weed control is probably the main problem to manage with IFS. This is partly due to the tillage technique. Herbicide costs can be reduced of 16 % at Montgaillard but in the two others sites they are higher in IFS. Since the first year of experiment, there have been some problems with annual weeds like cleavers (*Galium aparine*) in peas in Boigneville and Montgaillard and wild oat (*Avena fatua*) on wheat that could not be controlled correctly. After several years, the difficulties are increasing. Annual weeds like blackgrass (*Alopecurus*) are more significant in IFS compared with CFS in Saint-Hilaire. In Boigneville, there is a lot of Bromus (*Bromus sterilis*) on the edges of the plot. But the main problems are with perennial weeds : thistle (*Cirscium arverse*), couch grass (*Elymus repens*), convolvulus dock (*Rumex acetosella*), specifically in Montgaillard.

To improve the weed control, the intercrop period has been managed more carefully. Just after harvest, the soil is cultivated with a disc-harrow to incorporate crop residues and to favour weed and volunteer emergence. After about two months, a disc-harrow is used again in dry conditions to destroy these weeds and volunteers, 90 % of the emerged weeds can be destroyed by this method. For the spring crops and if the weather is dry, a third pass is done in November. Otherwise, glyphosate is used before drilling. At Boigneville this strategy used in 1994 allow a significant reduction of herbicide costs in IFS (539 FF./ha in 1992, 272 FF./ha in 1994).

Disease control

This is probably the most interesting result of this experiment. The holistic approach to disease control by combination of resistant cultivars, low nitrogen inputs, delayed sowing dates, etc...has allowed a strong reduction in fungicide inputs : 55 % in Montgaillard, 69 % in Boigneville, 87 % in Saint-Hilaire. Reductions are more important on cereals than on other crops. Generally speaking, the disease pressure is lower in Saint-Hilaire, that is why the highest reduction is observed on this site. Nevertheless, diseases are controlled in each case at the same level. Some diseases like eyespot are tolerated under 25 % of plants infected at stem elongation stage, some others are totally controlled like brown rust in the South West.

Figure 1 : Average percentage of reduction of variable costs with IFS compared to CFS (1991-1993 or 1991-1994)

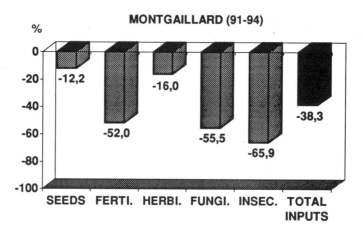

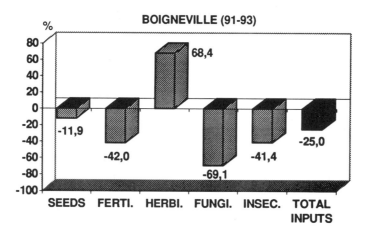

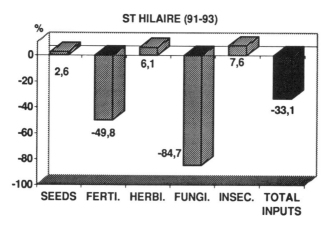

Fertilisation

The fertilisation is also largely reduced (by 42 to 52 % according to the site). All macro elements are included in this reduction : N, P, K. For example, in Montgaillard, the average N input in the rotation is 95 kg/ha in CFS and 48 Kg/ha in IFS, with a maximum of 222 kg/ha on durum wheat on CFS in 1994 versus 175 kg/ha the same year in IFS. On the same site 83 kg ha^{-1} yr^{-1} P$_2$0$_5$ is applied to CFS and 50 kg ha^{-1} yr^{-1} to IFS.

Nitrogen is generally reduced by about 20 to 30 % on cereal crops. The reduction could be higher on oil crops (no nitrogen on sunflower with IFS).

Total inputs

IFS can save a lot of inputs (Figure 1) when compared with CFS. On average for rotation, this reduction can reached 26 to 38 % according to the site. The target at the beginning of the experiment was 35 %. As seen above, the reduction in inputs is variable according to each input. The reductions are significant for fertilisation and fungicide, and very low for herbicides and seeds. The insecticide case is misleading because the reduction can be quite high (66 % at Montgaillard) but the input level to CFS is low in absolute terms (132 FF./ha).

YIELD AND ECONOMIC RESULTS

Whatever the crop and the site, the yields were lower in IFS when compared with CFS. For winter wheat, the gap ranged from 0.9 t ha^{-1} at Montgaillard to 2 t ha^{-1} at Boigneville and Saint-Hilaire. At Boigneville, the gap was higher than on the other sites, especially on spring crops, because of the poor plant establishment for the two first years.

In general terms, at all sites there has been an improvement in concerning the management of IFS and, in the last few years the gap between CFS and IFS is decreasing. For example, with yields of winter wheat at Boigneville the gap between IFS and CFS decreased from 34 % to 14.8 % (Table 2), when averaged over two rotations.

Table 2 : Yield of winter wheat at Boigneville. Average of two rotations (t ha^{-1})

	1991	1992	1993	1994
CFS	8.95	5.60	8.80	7.75
IFS	5.85	4.50	7.40	6.60
Δ (%)	- 34.6	- 19.6	- 15.9	- 14.8

These gaps in yields are nevertheless quite different from those observed in other European countries. The differences between IFS and CFS are generally lower than in France (about 3 % at Lautenbach in Germany, about 10 to 15 % at Long Ashton in U.K.). These differences are due to the absolute level of CFS. CFS is not really a high input system because the French farmers have already reduced their input in the regions where the trials have been set up for the last 10 years. The second reason is that nitrogen and fungicides which are the main factors for yields have been strongly reduced in the French IFS.

Despite this yield reduction, the economic results (Table 3) are really interesting. When 1995 prices are used, the gross margin are better at Montgaillard and at Saint-Hilaire for IFS than for CFS.

The net margins are better with IFS in every case. This is due to the mechanisation costs which are lower with IFS because of lower soil tillage costs. These net margins do not include labour costs. The time spent on IFS in the field is lower than in CFS : at Montgaillard, CFS needs 6 h ha^{-1} yr^{-1} of labour while IFS needs only 4.8 h ha^{-1} yr^{-1}. At Boigneville, CFS needs 8.1 h ha^{-1} yr^{-1} and IFS 4.8 h ha^{-1} yr^{-1}. These differences would theoretically increase the economical advantages of the IFS. Nevertheless, the time spent to observe the crops has not been measured and IFS needs more time for applying decision making processes.

Table 3 : Yields and economics results

Montgaillard (31)

Average 1991-1994

	CFS	IFS
Yield (Ton/ha)		
Durum Wheat	6,6	5,5
Winter Wheat	7	6,1
Winter Peas	4,4	4
Sunflower	2,4	2,4
Inputs (FF/ha)	2185	1348
Gross Margin (FF/ha)	4568	4870
Net Margin (FF/ha)	1355	1848

Boigneville (91)

Average 1991-1993

	CFS	IFS
Yield (Ton/ha)		
Durum Wheat	5,5	4,7
Winter Wheat	8,3	6,2
Spring Peas	5,7	3,5
Oil Seed Rape	2,3	1,5
Spring Barley	6,3	4,8
Inputs (FF/ha)	2118	1506
Gross Margin (FF/ha)	5205	3814
Net Margin (FF/ha)	815	942

St Hilaire (55)

Average 1991-1993

	CFS	IFS
Yield Ton/ha		
Winter Wheat	7,7	5,7
Oil Seed Rape	3,2	2,6
Winter Barley	6,4	4,7*
Inputs (FF/ha)	1979	1493
Gross Margin (FF/ha)	4679	4767
Net Margin (FF/ha)	512	736

* Spring barley in IFS

DISCUSSION

After four experimental years, it is not possible to conclude to make definite conclusion from these results. Firstly, because the IFS system is not really stabilised, eg. for weeds. Every year, some new problems arise or increase : bromus in Boigneville, rumex, thistle and many perennial weeds in Montgaillard or blackgrass in Saint-Hilaire. On the other hand, some improvements are seen, for example, an increase of the topsoil organic matter and more earthworms in IFS. Further information is needed on the environmental impact of these systems. Ceramic cups have been set up in Boigneville to measure nitrate leaching but it is too early to analyse these results. Nevertheless the economic results could be extrapolated to farmers.

The main difficulty in transferring the IFS technology to farmers is the variability of the results. At first, it is likely that the farmers, who adopted these techniques, will observe some variability in performance at the beginning , but as we have shown in this paper, this is mainly due to a lack of knowledge. Some other ITCF results (not published) show that low input systems do not result in an increase in variability of cereals if they are correctly managed.

An other unknown point is the time necessary to observe the crops before making a technical decision. We cannot measure this time in our experiments but it is certain that IFS needs more time than the other system.

This experiment highlights that there is a lack of technical references for optimising the integrated farming systems. We can give an unlimited list, for example we need more disease resistant cultivars of wheat with good baking quality even with low nitrogen inputs. There is a lack of knowledge concerning how to manage herbicides when shallow cultivation is used. We need some more efficient methods to appreciate thresholds in real fields. It is unacceptable for a farmer to spend too much time counting aphids in a wheat field when the price of an insecticide treatment is very low.

After this first period in which a complete rotation has been studied, the experiment is caring on but with a lot of changes. The rotation for the reference system is being simplified to be closer to the farmers practices. Generally, a two year crop rotation will be used : wheat/peas in Boigneville, oilseed rape/wheat in Saint-Hilaire and sunflower/wheat in Montgaillard. For IFS, there is little evolution except on soil tillage in Montgaillard . Because of the difficulties in controlling weeds, we have decided to introduce the plough one year out of four (this is also the case on the LIFE project in UK). In the next four years, we aim to increase the measurement of the environmental impact of the two systems.

Lastly, we are starting a new experiment with ACTA to improve our knowledge about the intercrop management with shallow cultivation and with or without catch crops. This experiment is partly financed by EU in the AIR III program.

REFERENCES

El Titi, A ; Boller, EF ; Gendrier, JP (1993) Integrated production : principles and technical guidelines *IOBC/WPRS Bulletin* ,**16,** 96 pp .

El Titi, A. (1990) Farming systems research at Lautenbach, Germany. *Swiss journal of Agricultural Research* , **29,** 237-247.

Farmer, DP ; Hutcheon, JA ; Jordan, VW (1994) Socio-economics of commercial implementation of integrated arable production systems under the reformed CAP. *Aspects of applied Biology* ,**40,** 81 -85.

Hani, F. (1990) Farming systems research at Ipsach, Switzerland - The « third way » project.
Swiss journal of Agricultural Science, **29,** 257-271.

Holland, JM. ; Frampton, GK. ; Çilgy, T. Wratten JD. (1994) Arable acronyms analysed. A review of integrated arable farming systems research in Western Europe. *Ann., appl. Biology,* **125,** 399-438.

Jordan, VWL. ; (1990) Long Ashton low input farming and environment (LA LIFE). *Swiss journal of Agricultural Research,* **29,** 389-390.

Jordan, VWL . (1992) Opportunities and constraints for integrated farming systems. *Proceedings of the 2nd ESA congress, Warwick University,* 318-323.

Vereijken, P (1990)   Research on integrated arable farming and organic mixed farming in the Netherlands. *Swiss journal of Agricultural Research,* **29,** 249-256.

Vereijken, P. ; Viaux, P. (1990) Vers une agriculture « intégrée » . *La Recherche,* **227,** 22-27.

Viaux, P. ; Roturier, C. ; Bouchet, C. (1993) Integrated arable farming systems in France. *Proceedings of Conference on Integrated Arable Farming Systems. Nitra,* 24-33.

Viaux, P. ; Fougeroux, A. ; Hilaire, A. ; Cavelier, A. ; Lescar, L. (1989) Planning of integrated farming systems experiments in France. *IOCB/WPRS bulletin. Wageningen The Netherlands,* **XII/5,** 16-20.

Viaux, P. ; Vazzana, C. ; El Titi, A. ; Jordan, VWL. ; Vereijken, P. ; Mikkelsen, C. (1993) Research into and development of integrated farming systems / European network - *short final report. CAMAR contract CT 90-00-10,* 13 pp.

EFFECTS OF REDUCING PESTICIDE INPUTS IN THE FIRST FOUR YEARS OF TALISMAN

P. BOWERMAN, J. E. B. YOUNG, S. K. COOK

ADAS Boxworth, Boxworth, Cambridgeshire, CB3 8NN

A. E. JONES

ADAS Drayton, Alcester Road, Stratford-upon-Avon, Warwickshire, CV37 9RQ

M. R. GREEN

ADAS High Mowthorpe, Duggleby, Malton, North Yorkshire, YO17 8BP

ABSTRACT

TALISMAN (Towards A Lower Input System Minimising Agrochemicals and Nitrogen) was started in autumn 1990 at three ADAS Research Centres. Standard and alternative six-year rotations are being tested under a Current Commercial Practice (CCP), applying full recommended rates of pesticides and nitrogen, and Low Input Approach (LIA) in which 50% of the nitrogen and a maximum of 50% of the pesticide amounts are used. The design allows the effect of reducing the rate of herbicide, fungicide, insecticide and nitrogen to be assessed separately. During the first four years the crops grown included most major combinable crops in rotations appropriate to the soil type of the centre concerned. Reduced rates of pesticides produced variable yield responses in cereals and the effects upon yields in other crops were small. Margins over pesticides and nitrogen costs in cereals increased most with reductions in herbicides at the full rates of nitrogen. Changes in margins with beans and linseed were small with reduced pesticide inputs.

INTRODUCTION

A major, multi-disciplinary project to investigate the environmental and economic effects of pesticide use in intensive cereal production was carried out at ADAS Boxworth in 1981 - 1988; this study became known as the Boxworth Project (Greig-Smith *et al.*, *1992*). The results from the Boxworth Project provided a basis for further research to be funded by the Ministry of Agriculture, Fisheries and Food (MAFF) including TALISMAN which commenced with sowing in autumn 1990 and will continue for at least six years.

TALISMAN, (Towards A Lower Input System Minimising Agrochemicals and Nitrogen) was designed to measure the economic, agronomic and, to a lesser extent, the environmental effects of adopting cropping systems which use lower levels of agrochemicals and nitrogen than conventional cropping systems. Crop inputs, yields and economic results from the first four years of TALISMAN are presented in this paper.

MATERIALS AND METHODS

The sites for TALISMAN are at ADAS Boxworth (well-structured clay), ADAS Drayton (heavy clay) and ADAS High Mowthorpe (silty clay loam). Standard and alternative six course rotations are being tested under a Current Commercial Practice (CCP) approach for pesticides and nitrogen and a Lower Input Approach (LIA) in which 50% of the nitrogen rate applied to CCP is used and a maximum of 50% of the pesticide rates. The standard rotations are typical for the individual sites and the alternative rotations are based predominantly on spring crops which have an inherently lower requirement for pesticides and nitrogen (Table 1).

TABLE 1. TALISMAN rotations at each site.

Year	Boxworth	Drayton	High Mowthorpe
Standard Rotations			
1	w. oilseed rape	w. oilseed rape	w. oilseed rape
2	w. wheat	w. wheat	w. wheat
3	w. wheat	w. wheat	w. wheat
4	w. beans	w. beans	w. beans
5	w. wheat	w. wheat	w. wheat
6	w. wheat	w. wheat	w. barley
Alternative Rotations			
1	linseed	s. beans	s. beans
2	w. wheat	triticale	w. wheat
3	s. wheat	triticale	s. barley
4	s. beans	s. oats	linseed
5	w. wheat	triticale	w. wheat
6	s. wheat	triticale	s. barley

Products used in CCP are those most widely used by farmers as indicated in the Pesticide Usage Survey Reports for Arable Crops (Davis *et al.*, 1992). The manufacturers' recommended label rates are used. During the experiment some recommendations, rates or active ingredients may be superseded and CCP evolves to reflect these changes. Nitrogen rates are determined using "Fertiplan", which is a fertiliser planning service based on previous cropping, soil type and yield prediction.

Wherever possible, the reduction in LIA pesticides is being achieved by omitting applications altogether. However, if it is estimated that the loss of crop value would be greater than 10% by withholding the agrochemical, then up to half of the rate applied to CCP can be used. A full rate application is allowed in very exceptional instances when there is already conclusive evidence that less than the full rate would result in a crisis. The cultivars, cultivations and sowing dates are the same in both CCP and LIA.

The experiment is designed to compare the individual effects of herbicides, fungicides and insecticides at CCP and LIA rates at normal and half rates of nitrogen in two rotations. TALISMAN has a split-plot design with rotation and nitrogen rate as main treatments and pesticides as sub-treatments. Combinations of the CCP and LIA rates of herbicides, fungicides and insecticides are represented in 5 sub-treatments so that the effect of reducing the rate of each pesticide can be assessed. The sub-treatments are: all pesticides applied at CCP rate; all pesticides applied at LIA rate; and three combinations of only herbicide, fungicide or insecticide at the LIA rate with the remaining two pesticide components at the CCP rate.

Main plots are 24m x 24m and divided into 5 equal sub-plots. There are three or four replicates at each site and an additional replicate without sub-treatments adjacent to a field boundary is used to monitor arthropods. The effects of pesticides on invertebrates and non-target soil micro-organisms are being studied in the SCARAB (Seeking Confirmation About Results At Boxworth) project (Bowerman, 1993).

Each rotation has two phases (except for the alternative rotation at ADAS Boxworth). Phase 1 started at the first year in the rotation and Phase 2 started at year 4.

RESULTS

The pesticide levels applied in the first four harvest years are shown in Table 2. The applications of a label recommended rate for herbicide, fungicide or insecticide (including molluscicide) has been taken as 1 pesticide unit, and half-rate applications as 0.5 unit.

TABLE 2. Pesticide units applied (mean per crop 1991-94).

Crop (No. in brackets)		CCP	LIA
First w. wheat	(9)	6.1	2.9
Second w. wheat	(5)	8.0	4.0
Spring wheat	(1)	4.0	2.5
Winter barley	(1)	3.0	1.5
Spring barley	(3)	4.0	2.3
Spring oats	(1)	2.0	1.0
Triticale	(2)	5.5	1.5
Oilseed rape	(6)	4.2	1.6
Linseed	(3)	1.7	1.0
Winter beans	(6)	3.0	1.0
Spring beans	(5)	3.0	1.0
TOTAL		44.5	20.3

TABLE 3. Mean yields of cereals and field beans (t ha^{-1} at 85% DM), and oilseed rape and linseed (t ha^{-1} at 91% DM) at full rates of nitrogen and pesticides, and differences of other treatments.

Crop (no. in brackets)	Nitrogen rate	Pesticide inputs				
		All high	Low herbicide	Low fungicide	Low insecticide	All low
			(a) ± 0.094		(b) ± 0.232	
First	100%	7.80	+ 0.05	-0.05	- 0.27	- 0.60
W. wheat (9)	50%	- 0.23	- 0.39	- 0.28	- 0.32	- 0.65
			(a) ± 0.085		(b) ± 0.181	
Second	100%	7.70	- 0.03	- 0.17	- 0.09	- 0.64
W. wheat (5)	50%	- 1.78	- 1.64	- 1.71	- 1.76	- 1.89
			(a) ± 0.199		(b) ± 0.252	
S. wheat (1)	100%	6.34	- 0.02	+ 0.05	+ 0.20	+ 0.12
	50%	+ 0.26	- 0.05	- 0.26	+ 0.30	- 0.14
			(a) ± 0.185		(b) ± 0.372	
W. barley (1)	100%	8.11	+ 0.03	- 0.01	- 0.40	- 0.70
	50%	- 1.68	- 2.85	- 2.32	- 1.63	- 1.93
			(a) ± 0.202		(b) ± 0.299	
S. barley (3)	100%	6.17	- 0.14	- 0.24	+ 0.09	- 0.14
	50%	- 0.98	- 0.85	- 0.68	- 0.11	- 0.99
			(a) ± 0.168		(b) ± 0.235	
S. oats (2)	100%	4.93	-0.08	-0.21	+0.02	-0.54
	50%	+0.01	-0.18	-0.32	-0.34	-0.08
			(a) ± 0.171		(b) ± 0.288	
Triticale (2)	100%	4.97	+ 0.09	- 0.12	+ 0.23	- 0.12
	50%	+ 0.03	+ 0.56	- 0.18	+ 0.51	+ 0.27
			(a) ± 0.081		(b) ± 0.108	
Oilseed rape (6)	100%	2.23	- 0.19	- 0.06	- 0.04	- 0.39
	50%	- 0.53	- 0.69	- 0.42	- 0.37	- 0.64
			(a) ± 0.032		(b) ± 0.046	
Linseed (3)	100%	2.03	- 0.09	+ 0.06	- 0.01	- 0.03
	50%	- 0.06	- 0.15	- 0.02	- 0.05	- 0.10
			(a) ± 0.075		(b) ± 0.078	
W. beans (6)	100%	4.32	+ 0.16	+ 0.04	- 0.11	+ 0.08
	50%	- 0.04	+ 0.01	+ 0.12	- 0.01	- 0.08
			(a) ± 0.110		(b) ± 0.127	
S. beans (5)	100%	3.47	- 0.12	0 .00	- 0.07	- 0.25
	50%	- 0.08	- 0.30	- 0.09	- 0.23	- 0.35

(a) SE for comparisons at same level of nitrogen.
(b) SE for comparisons between levels of nitrogen and interactions.

TABLE 4. Prices and area payments, 1994.

Crop	Price £/t	Area payment £/ha
Wheat	108	193.53
Barley	100	193.53
Oats	98	193.53
Triticale	95	193.53
Oilseed rape	175	436.54
Linseed	135	481.06
Beans	100	359.41

TABLE 5. Margins over costs of pesticides and nitrogen (£ ha^{-1}) for full rates of nitrogen and pesticides, including area payments, and differences of other treatments.

Crop	Nitrogen rate	Pesticide inputs				
		All high	Low herbicide	Low fungicide	Low insecticide	All low
First W. wheat	100%	885	+22	+14	-28	-25
	50%	+1	+2	+13	-8	-4
Second W. wheat	100%	799	+40	+16	+4	+14
	50%	-166	-110	-126	-150	-97
S. wheat	100%	755	+5	+13	+22	+29
	50%	+48	+22	0	+52	+21
W. barley	100%	895	+13	+17	-40	-42
	50%	-141	-248	-187	-136	-138
S. barley	100%	703	+5	-19	+20	+12
	50%	-79	-46	-44	-80	-54
S. oats	100%	608	0	- 12	+4	-33
	50%	+16	+4	-9	-18	+27
Triticale	100%	522	+31	+27	+34	+60
	50%	+18	+89	+37	+74	+112
Oilseed rape	100%	662	-6	+4	+8	-6
	50%	-62	-64	-28	-19	-19
Linseed	100%	701	-4	+7	-3	+5
	50%	+3	0	+8	+5	+7
W. beans	100%	734	+44	+24	-1	+49
	50%	+5	+29	+32	+10	+33
S. beans	100%	647	+9	+13	-1	+15
	50%	-8	-9	+3	-16	+6

Overall, the level of pesticides in LIA was reduced to 46% of that in CCP (Table 2); achieved mainly by reducing application rates rather than by omitting applications. The reductions with herbicides were less than those with either fungicides or insecticides indicating a greater availability and confidence in thresholds for disease and pest problems than for weeds. Crops grown in the standard rotations have had a greater requirement for nitrogen and pesticides than those in the alternative rotations.

Yields were significantly reduced in second winter wheat at full rate of nitrogen with the all low pesticide treatment compared to all the other combinations of pesticide input levels except low fungicide, and at the half rate of nitrogen with the same pesticide treatment but only compared with the low herbicide treatment (P = 0.05) (Table 3).

Based upon 1991-94 yields, mean margins for crop output less pesticide and nitrogen costs were calculated using commercial prices for inputs. The crop prices and area payments are shown in Table 4.

In the first and second winter wheat crops, which are the cereals on which most data is available, margins increased most with the reductions in herbicides at the full rates of nitrogen (Table 5). Reduction of pesticide inputs to beans and linseed gave increased or slightly reduced margins. Winter oilseed rape crops were redrilled with spring rape at ADAS Boxworth and ADAS Drayton in 1994 because both of the autumn crops were killed by pigeon and slug attacks.

Margins per year for the standard and alternative rotations at each site, presented in Table 6, were based upon the yields achieved with full rate of nitrogen (none for beans), and full rates of pesticides (except those for cereals and beans which were at the LIA rate for herbicides). It should be noted that no allowance has been made for the reduced area as a result of the set-aside requirement.

TABLE 6. Margins per year (£ ha^{-1}) of standard and alternative rotations at each site at 1994 price regimes.

Site	Standard rotation	Alternative rotation
Boxworth	684	656
Drayton	726	628
High Mowthorpe	977	775

DISCUSSION

Reductions in pesticide use have been achieved more frequently by the prudent use of reduced rates than the omission of applications. The effects on yields and margins of reducing nitrogen rates by 50% were greater than the effects of reductions in levels of pesticides used.

Cereals, in particular spring and winter wheat, have shown variable responses to reduced inputs. Yields of first winter wheats in 1992 at ADAS Boxworth and Drayton were greater with LIA nitrogen rates than with CCP rates. This was probably an effect of crops being unable to utilise the higher rate of nitrogen to fill grains because of the weather conditions, particularly a shortage of water; a phenomenon that occurred in other experiments in that year. One of the largest responses to reduced pesticide input to a cereal crop occurred at High Mowthorpe in the first winter wheat in 1991/92 where there was a late season infestation of grain aphid (*Sitobion avenae*) and rose grain aphid (*Metopolophium dirhodum*). At the CCP rate of nitrogen, an application of dimethoate to the CCP at the early dough stage (GS 83) resulted in yield increases in both phases of 7 and 8% compared with nil insecticide to the LIA. Smaller increases in yield were achieved with the same treatment at the LIA nitrogen rate.

The yield responses to reduced rates of pesticides in the other crops were usually small which indicate that it was possible to reduce the rates of application of herbicides, fungicides and insecticides by half, or to omit them at most sites.

Winter oilseed rape was vulnerable to the effects of reduced pesticide applications as it is a low yielding crop and problems can be damaging. Yield reductions appeared to be associated with poorer control of volunteer wheat at ADAS Boxworth (Clarke *et al.*, 1993; Cook *et al.*, 1995). In addition, the reduced rate of nitrogen on rape in LIA allowed greater competition from weeds.

Weed counts in successive crops have often shown higher weed numbers with treatments with reduced rates of nitrogen or herbicides or both. Weed seedbank numbers were higher at High Mowthorpe and Boxworth with reduced rate herbicide treatments after only three years of the rotations (Lawson, personal communication).

The margins of the standard rotations were greater than those of the alternative rotations at all of the sites. At Boxworth, the low margin of the standard rotation was mainly the effect of poor yields of rape in 1991 and 1994 and the first winter wheat crops in 1992.

TALISMAN is indicating that there are potential savings to be made with inputs to combinable crops, but large penalties can be incurred from omitting or reducing key inputs. The effects of reducing inputs to the various crops are discussed in more detail by Young *et al.*, 1994. The skill required to get the correct balance of inputs is likely to become greater as the pressure to reduce inputs increases. Rotations of combinable crops based upon those sown in the autumn produce higher margins than rotations which are based predominantly on spring sown crops.

ACKNOWLEDGEMENTS

The authors are grateful to many ADAS colleagues for their help. Financial support for this work from the Ministry of Agriculture, Fisheries and Food (MAFF) is gratefully acknowledged.

REFERENCES

Bowerman, P. (1993) Sequels to the Boxworth Project - studies of environmental, agronomic and economic effects of reduced crop inputs. *Journal of the Royal Agricultural Society of England,* **154**, pp. 45-60.

Clarke, J. H.; Bowerman, P.; Young, J.E.B.; Cook, S.K.; Jones, A.E.; Groves, S.J.; Green, M. (1993) Effect of recommended and reduced rate herbicides on weed number, yield and gross margin in TALISMAN: Report on the first two years. *Brighton Crop Protection Conference - Weeds,* **3**, pp. 1009-1014.

Cook, S.K.; Jones, A.E.; Green, M. (1995) A comparison of input levels in oilseed rape. *Proceedings of 9th International Rapeseed Congress, Cambridge,* (in press).

Davis, R.P.; Gowthwaite, D.G.; Thomas, M.R.; Bowen, H.M. (1992) *Pesticide usage survey report 108. Arable farm crops in England and Wales 1990.* London: MAFF, 89 pp..

Greig-Smith, P.; Frampton, G.K.; Hardy, A.R. (1992) *Pesticides, Cereal Farming and the Environment. The Boxworth Project.* London: HMSO.

Young, J.E.B.; Bowerman, P.; Cook, S.K.; Green, M.R.; Jones, A.E. (1994) The TALISMAN experiment - observations and implications for integrated pest and disease management of arable crops in the UK. *Brighton Crop Protection Conference - Pests and Diseases,* **1**, pp. 125-134.

THE EFFECTS OF REDUCED FERTILIZER AND HERBICIDE INPUT SYSTEMS ON THE YIELD AND PERFORMANCE OF CEREAL CROPS

D. L. EASSON

The Agricultural Research Institute of Northern Ireland, Hillsborough, BT26 6DR.

A. D. COURTNEY

Applied Plant Science Division, Department of Agriculture for Northern Ireland, Newforge Lane, Belfast BT9 5PX.

J. PICTON

Greenmount College of Agriculture and Horticulture, Antrim, BT41 4PU.

ABSTRACT

Results are presented from the first four years of a nine year project in Northern Ireland in which crops under six course arable and arable/grass rotations are treated with full rates, 50% rates or minimum rates of fertilizers and pesticides. Monitoring of the effects on yields, weeds, diseases, profitability, invertebrates and soil mineral status is being carried out. Soil fertility was high at the start and yields with full and half rate inputs with spring and winter barley were similar so that the reduced input plots had higher gross margins. With time the fertility of the reduced input plots has declined and the full rate fertilizer plots have become more profitable. However, the use of half rate herbicides have been sufficient in most cases to give adequate control of weeds. There is evidence of higher weed levels where crop competition has been poorer with reduced herbicide rates.

INTRODUCTION

The Reduced Input Systems of Cropping (RISC) experiment is a nine year project now completing its fifth season at two sites in Northern Ireland. The economic performance of full rate fertilizer and pesticide inputs is being compared with half rates for crops grown under six course arable and arable/grass rotations. The protocol being used is similar to that of the ADAS 'TALISMAN' project in which 'Current Farming Practice' levels of inputs are being compared with 50% rates (Bowerman, 1993), but additional 'Integrated' and 'Minimum' input treatments have been included in the present study. The arable/grass rotation includes four years arable cropping and two years ley and is an alternative rotation representative of Northern Ireland agriculture. Through monitoring changes in the weed burden, the invertebrate population and the soil mineral N content the project aims to investigate not only the economic aspects but also the sustainability of the systems and the environmental effects of adopting lower input strategies.

In this paper only the performance of the cereal crops will be considered with particular reference to the plots which received full rate N (CFP), 50% rate N (LIA) and no N (MIN) together with a) full rate herbicide, fungicide and insecticide, b) half-rate sprays, c) half-rate herbicide with full rates of other sprays, or d) no sprays.

MATERIALS AND METHODS

Replicated field experiments were laid out in 1991 at Hillsborough and 1992 at Greenmount in which two six course rotations are being compared. Each rotation is represented at two phases (Table 1). Rotation A represents a mixed arable system with a two year ley followed by potatoes, winter wheat, spring barley and with the early harvest of winter barley allowing a reseed into ley. In this rotation cattle slurry is normally applied prior to ploughing. In Rotation B oilseed rape and potatoes are the break crops which follow two years of cereals. The Hillsborough soil is a sandy clay loam which had been under ley for a number of years while the Greenmount soil is a clay loam which had been under arable cropping.

TABLE 1. *Rotations used in the RISC experiment at both Hillsborough and Greenmount sites*

Rotation	1991	1992	1993	1994	1995	1996	1997
Phase 1							
Rotation A	S Barley	W Barley	Grass	Grass	Potatoes	W. Wheat	S Barley
Rotation B	S Barley	Potatoes	W Wheat	W Barley	W OSR	W Wheat	S Barley
Phase 2							
Rotation A	Grass	Potatoes	W Wheat	S Barley	W Barley	Grass	Grass
Rotation B	S Barley	OSR	W Wheat	S Barley	Potatoes	W Wheat	W Barley

Within each rotation there are four levels of input, Current Farming Practice (CFP), Low Input Approach (LIA), Integrated Low Inputs (ILI) and Minimum Inputs (MIN) but the ILI treatment is not considered further in this paper. All input levels are represented in Phase 2, but only the CFP and LIA in Phase 1. The protocols for the CFP and LIA are based on surveys of current usage of pesticides and recommended fertilizer levels, with the use of 50% rates of these inputs in the LIA treatment. In the MIN treatment no fertilizer has been applied and pesticides have only been used to prevent crop failure. As well as agronomic data on yield and components of yield, weed biomass, levels of disease, soil mineral N and nitrogen uptake are monitored. Weekly monitoring of invertebrate species, mainly ground beetles, also takes place. Comparison of the economic performance of the treatments is made on the basis of gross margin (GM) analyses using costs and output values derived from current commercial values and relevant area payments under the full arable aid scheme.

In three of the four replicate blocks, the main 10 m by 20 m plots are divided into five 10 m by 4 m sub-plots to which reduced rates of herbicide, fungicide and insecticide are applied individually or in combination. The sub-plots therefore allow the effects of reduced rates of each of these components to be studied at both full and half rate nitrogen levels.

RESULTS

Spring and winter barley

The spring barley in the first year after grass ley at Hillsborough showed no yield response to fertilizer nitrogen or weed control and the yield of 6.4t ha^{-1} with CFP inputs was no higher than that with MIN inputs (Table 2). As a consequence the GM of £597 ha^{-1} with CFP inputs was £43 ha^{-1} less than with the use of MIN inputs (Table 3). The principal weeds present were knotgrass (*Polygonum aviculare*) and couch grass (*Elymus repens*). In the MIN plots to which metsulphuron methyl was applied at 1/8th rate the knotgrass dry weight at harvest was 60g m^{-2} compared with 28g m^{-2} with the full rate application on the CFP plots (Table 4).

TABLE 2. Spring barley grain yields (t ha^{-1})

Site	Nitrogen	Spray applications			
		Full-rate	Half-rate herbicide	All sprays half-rate	No sprays
a) 1991 Spring barley,		s.e.m. 0.35	6 d.f.		
Hillsborough	Full N	6.4	6.6	6.4	-
	Half N	6.3	6.3	6.3	-
	No N	-	-	-	6.5
b) 1994 Spring barley		s.e.m. 0.40	46 d.f.		
Hillsborough	Full N	6.4	6.4	6.2	
	Half N	5.0	5.0	5.0	
	No N	-	-	-	3.5
Greenmount	Full N	5.3	5.0	5.1	
	Half N	3.7	3.8	3.3	
	No N	-	-	-	2.9

When spring barley was grown in the same plots in 1994 at Hillsborough yields with the CFP treatments were very similar to those in 1991, but where half rate N had been used yields were much lower (Table 2). However the use of half rate metsulphuron methyl or all sprays at half rate had no significant effects on yields at either site. The GM was consistently lower at the lower N level, but was not significantly affected by the reduced rate spray treatments (Table 3). There was no evidence that weed competition had become any worse between 1991 and 1994 where half rate sprays had been used, but weed biomass at harveste tended to be greater at Greenmount (Table 4). On the plots at Hillsborough with no herbicide or fertilizer, broad leaved weed biomass was 100 g m^{-2} compared with 60 g m^{-2} in 1991.

In the 1991/92 season, winter barley with CFP inputs yielded 7.8 and 6.6 t ha^{-1} at Hillsborough and Greenmount respectively and yields were reduced by no more than 1.2 t ha^{-1} where the half rate fertilizer was used (Table 5). In most cases 50% rate inputs improved the GM in 1992, but the profitability of the plots with full rate N and half rate sprays was poorer

TABLE 3. Spring barley gross margins (£ ha^{-1})

Site	Nitrogen	Spray applications			
		Full-rate	Half-rate herbicide	All sprays half-rate	No sprays
a) 1991 Spring Barley					
Hillsborough	Full N	597	622	625	-
	Half N	604	637	637	-
	No N	-	-	-	639
b) 1994 Spring barley					
Hillsborough	Full N	856	868	848	-
	Half N	693	705	698	-
	No N	-	-	-	490
Greenmount	Full N	777	760	765	-
	Half N	539	569	507	-
	No N	-	-	-	396

TABLE 4. Broad leaved weed biomass in spring barley (g 0.5m^{-2})

Site	Nitrogen	Spray application			
		Full-rate	Half-rate herbicide	All sprays half-rate	No sprays
a) 1991 Spring barley			s.e.m. 3.02	6 d.f.	
Hillsborough	Full N	14.3	-	20.5	-
	Half N	9.0	-	17.9	-
	None	-	-	-	23.2
b) 1994 Spring barley after winter wheat			s.e.m. 3.06	46 d.f.	
Hillsborough	Full N	0.2	0.8	0.9	-
	Half N	0.8	3	3.6	-
	None	-	-	-	50.9
Greenmount	Full N	8.8	14.9	12.2	-
	Half N	3.1	5.7	10.8	-
	None	-	-	-	78.1

at Greenmount (Table 6). However, in the 1994 winter barley crops at both sites, although the CFP treatments had similar yields to 1992, the yield reduction with 50% N was consistently over 2 t ha^{-1} with the consequence that GMs were higher with full rate N (Table 6). The effects of reduced rate herbicide (isoproturon + trifluralin at Hillsborough and isoproturon + diflufenican at Greenmount) or other sprays on yields were small in comparison and generally non significant. At Hillsborough yield was reduced significantly with the combination of low N and half rate sprays (LIA treatment), or with no N or sprays (MIN). However at Greenmount, the results were more variable, but the MIN treatment was very low yielding. Weed competition was light at Hillsborough, but was significantly greater at Greenmount where half rate herbicide was used (Table 7). In 1994 all the low N plots had poor GMs (Table 6).

TABLE 5. Winter barley grain yields (t ha^{-1})

Site	Nitrogen	Spray applications			
		Full-rate	Half-rate herbicide	All sprays half-rate	No sprays
a) 1992 Winter Barley		s.e.m. 0.66	46 d.f.		
Hillsborough	Full N	7.8	8.1	7.4	-
	Half N	7.8	7.7	6.6	-
	No N	-	-	-	5.5
Greenmount	Full N	6.6	6.0	5.6	
	Half N	6.0	6.0	6.5	
	No N	-	-	-	4.4
b) 1994 Winter barley		s.e.m. 0.38	46 d.f.		
Hillsborough	Full N	7.7	7.2	7.0	
	Half N	5.1	5.3	5.0	
Greenmount	Full N	6.7	7.1	6.9	
	Half N	4.8	4.8	4.3	

TABLE 6. Winter barley gross margins (£ ha^{-1})

Site	Nitrogen	Spray applications			
		Full-rate	Half-rate herbicide	All sprays half-rate	No sprays
a) 1992 Winter Barley					
Hillsborough	Full N	742	803	764	-
	Half N	834	823	740	-
	No N	-	-	-	732
Greenmount	Full N	519	477	455	-
	Half N	511	542	634	-
	No N	-	-	-	539
b) 1994 Winter barley					
Hillsborough	Full N	969	857	855	
	Half N	619	637	657	
Greenmount	Full N	835	905	903	
	Half N	588	572	608	

TABLE 7. Broad leaved weed biomass in winter barley (g 0.5 m^{-2})

Site	Nitrogen	Spray application		
		Full-rate	Half-rate herbicide	All sprays half-rate
1994 Winter barley		s.e.m. 3.06	46 d.f.	
Hillsborough	Full N	0.0	0.0	0.9
	Half N	0.1	0.0	0.9
Greenmount	Full N	4.7	7.4	8.8
	Half N	9.2	20.6	13.6

Winter wheat

The winter wheat following oilseed rape in the autumn of 1992 yielded up to 7.5 t ha^{-1} at Hillsborough but yields at Greenmount were poorer (Table 8). The use of half rate herbicide (metsulphuron methyl) did not significantly increase the weed biomass at Hillsborough, but at Greenmount there was a significant increase with high N plots. Yields were not affected, however, and at the higher N level at both sites the higher GM were with the use of half rate herbicide, or half rate sprays (Table 9). At Greenmount, however, the combination of half rate sprays with half rate fertilizer was less profitable

Sowing of the winter wheat after potatoes was delayed until late January 1993 due to the late harvest of the potatoes and the wet autumn conditions. Establishment at both sites was very poor with a spring plant count of only 50 to 60 m^{-2}. This wheat did not mature until early

TABLE 8. Winter wheat grain yields (t ha^{-1})

Site	Nitrogen	Spray applications			
		Full-rate	Half-rate herbicide	All sprays half-rate	No sprays
a) 1993 Wheat after OSR		s.e.m. 0.45	46 d.f.		
Hillsborough	Full N	6.5	6.6	7.4	-
	Half N	6.0	5.8	6.2	-
	No N	-	-	-	4.0
Greenmount	Full N	5.8	6.5	6.1	-
	Half N	5.4	5.0	5.0	-
	No N	-	-	-	3.1
b) 1993 Wheat after potatoes		s.e.m. 0.45	46 d.f.		
Hillsborough	Full N	3.9	3.3	3.3	
	Half N	2.5	2.4	2.5	
Greenmount	Full N	3.6	3.5	3.0	
	Half N	4.4	4.0	4.0	

TABLE 9 Winter wheat gross margins (£ ha^{-1})

Site	Nitrogen	Spray applications			
		Full-rate	Half-rate herbicide	All sprays half-rate	No sprays
a) 1993 Wheat after OSR					
Hillsborough	Full N	725	747	864	-
	Half N	720	694	766	-
	No N	-	-	-	578
Greenmount	Full N	625	733	695	-
	Half N	631	587	602	-
	No N	-	-	-	466
b) 1993 Wheat after potatoes					
Hillsborough	Full N	360	306	322	
	Half N	236	245	264	
Greenmount	Full N	303	312	273	
	Half N	451	421	445	

TABLE 10 Broad leaved weed biomass in winter wheat (g 0.5 m^{-2})

Site	Nitrogen	Spray application			
		Full-rate	Half-rate herbicide	All sprays half-rate	No sprays
a) 1993 Wheat after OSR		s.e.m. 15.23		d.f. 64	
Hillsborough	Full N	4.8	2.1	1.2	-
	Half N	1.8	2.5	3.8	-
	None	-	-	-	7.9
Greenmount	Full N	6.1	40.0	26.3	-
	Half N	5.3	9.4	11.5	-
	None	-	-	-	41.0
b) 1993 Wheat after potatoes		s.e.m. 31.42		d.f. 64	
Hillsborough	Full N	15.8	48.8	73.4	
	Half N	28.9	109.8	80.8	
Greenmount	Full N	174.2	237.1	189.3	
	Half N	21.9	74.7	80.0	

October and yields of only 3.7 t ha^{-1} with CFP inputs and 2.8 t ha^{-1} with the LIA inputs were were achieved (Table 8). The poor stand of wheat was a poor competitor against weeds and relatively high weed biomasses were recorded at harvest. Weed biomass increased significantly where reduced rate herbicide was used (isoproturon + bromoxynil + ioxynil), and the weed problem was particularly severe with the high N plots at Greenmount (Table 10). A wide range of weed species were present including chickweed *(Stellaria media)*, fumitory *(Fumaria officinalis)*, hemp-nettle *(Galeopsis tetrahit)* and redshank *(Polygonum persicaria)*, and grass weeds also made a significant contribution. The yields, which were already low, were not reduced any further by the use of half rate herbicide alone, but the yields and GMs were lower when all sprays were applied at half rate.

DISCUSSION

In the first two seasons the application of 50% less N only led to slight reductions in yield due to the high initial fertility of the sites. The use of the half rate herbicides also did not lead to significant reductions in yield as weed control remained adequate. The GM of spring barley in 1991 and winter barley in 1992 therefore tended to be highest with the LIA treatment which included both half rate N and sprays. Even the minimum input treatments had a GM not far below the CFP treatments in some cases. In the 1993 and 1994 crops, however the pattern began to change and the yields of the lower input crops have been falling behind those of the CFP treatments. Weed problems were not particularly evident in early sown wheat after OSR, but the 50% N significantly reduced yields. The GMs were therefore higher with reduced sprays, but at the higher N level. The results from the late sown wheat after potatoes serve mainly as a reminder of how important crop competition is in suppressing weeds. The spring and winter barley crops in 1994 were high yielding only where the full N levels were applied and the continued use of half rate herbicide and other pesticides was sufficient to maintain these yields and thus to give similar or higher GM in most cases.

These results from Northern Ireland are similar to the inital results from the TALISMAN project (Clarke *et al*, 1993). While it would be unwise to extrapolate the results of the first three years into the longer term due to the likely build up of weed problems, and the steady decline in fertility with reduced fertilizer inputs, it is clear that the scope for reducing agrochemical inputs without significantly reducing profitability may be considerable. Data is being collected on the soil seed bank from samples taken at the start of the project and from samples to be taken at the mid-point and end of the project.

ACKNOWLEDGMENTS

The monitoring of the crops and recording of diseases, pests and weeds has been carried out by staff of the D.A.N.I. Science Service and their contribution is gratefully acknowledged, along with the assistance of staff of the Agricultural Economics and Biometrics Divisions in the interpretation of the results.

REFERENCES

Bowerman, P. (1993) Sequels to the Boxworth Project - studies of environmental, agronomic and economic effects of reduced crop inputs. *Journal of the Royal Agricultural Society of England* **154**, 45-60.

Clarke, J.H.; Bowerman, P.; Young, J.E.B. and Cook, S.K. (1993) Effect of recommended and reduced rate herbicides on weed number, yield and gross margin in TALISMAN: Report on the first two years. *Brighton Crop Protection - Weeds*, **3**, 1009-1014.

SCARAB: THE ENVIRONMENTAL IMPLICATIONS OF REDUCING PESTICIDE INPUTS

M.R. GREEN, S.E. OGILVY

ADAS High Mowthorpe, Duggleby, Malton, North Yorkshire, YO17 8BP.

G.K. FRAMPTON, T. ÇILGI

Department of Biology, University of Southampton, Biomedical Sciences Building, Southampton, SO16 7PX

S. JONES

School of Biological Sciences, University of Wales, Bangor, LL57 2UW

K. TARRANT, A. JONES

Central Science Laboratory, London Road, Slough, Berkshire, SL3 7HJ

ABSTRACT

> The SCARAB experiment is a major field-scale, long-term investigation of the ecological effects of two different pesticide regimes, Current Farm Practice (CFP) and Reduced Input Approach (RIA) on invertebrates, soil microbial biomass, earthworms, flora, crop pests and diseases. The pesticide regimes are compared on a total of 7 split fields, on three ADAS Research Centres, in six course rotations. Of the 139 pesticide units applied over 4 years, 55 herbicides, 55 fungicides and 29 insecticides were applied to CFP. RIA received only 32 herbicides, 29 fungicides and no insecticides. Adverse effects on some groups of invertebrates occurred following application of some broad-spectrum insecticides. Certain populations took longer than six months to recover. There has been a trend for soil biomass to increase in RIA. It has also been noted that total microbial biomass is closely linked to crop type and that, in general, levels are higher in cereal crops, compared with break crops. No overall treatment effects have yet been recorded on earthworm numbers under the two pesticide regimes, although earthworm populations vary widely between the sites according to soil type and organic matter content. In cereal crops, weed control has tended to be satisfactory when a reduced rate herbicide has been applied. Weed problems have occurred in RIA when lower rates of herbicide have been used in break or non-cereal crops. There have been no major problems from reduced pest and disease control.

INTRODUCTION

The SCARAB experiment, which is funded by MAFF, was specifically designed to pursue the results and hypotheses developed from the Boxworth Project (1981-1988) (Greig-Smith *et al.*, 1992). It was considered that the generality of environmental effects seen at Boxworth, and particularly those on invertebrate populations associated with intensive cereals, should be evaluated at other sites and tested in other crops.

The overall objective of SCARAB is to establish the broad ecological consequences of applying two different pesticide regimes to six-course arable crop rotations, which include cereals and break crops (Cooper, 1990). The comparison of systems involving lower pesticide inputs with conventional crop production is designed to identify which particular pesticide regimes are harmful to non-target species. SCARAB is focused on the ecology of key invertebrate species, primarily those of economic importance such as predators and parasites of crop pests.

More specific objectives are to monitor the effects of the two pesticide regimes on the numbers, taxonomic composition and trophic structure of arthropod faunas. Also, the project examines the effects of the two pesticide regimes on floral diversity and distribution, non-target soil micro-organisms, soil microbial biomass and earthworms.

MATERIALS AND METHOD

SCARAB is sited on three ADAS research centres: Drayton in Warwickshire, Gleadthorpe in Nottinghamshire and High Mowthorpe in North Yorkshire. On all three sites, baseline monitoring started in June 1990, with the first differential treatments applied in autumn 1990. Rotations at each site are typical of the locality and are shown in Table 1.

TABLE 1.   SCARAB rotations

| SITE | | |
Drayton	Gleadthorpe	High Mowthorpe
Winter wheat	Potatoes	Winter oilseed rape
Winter wheat	Spring wheat	Winter wheat
Grass ley	Winter barley	Spring barley
Grass ley	Sugar beet	Spring beans
Grass ley	Spring wheat	Winter wheat
Grass ley	Winter barley	Winter barley

Two pesticide regimes are compared at each site. CFP represents the pesticide use by a technically competent, financially-aware farmer in a farming situation comparable to the site. All pesticides are applied at label recommended rates. RIA is intended to contrast with

CFP in its intensity of inputs. RIA consists of minimal use of fungicides and herbicides, which are applied at half-rate or less. No insecticides are used on RIA unless a severe threat of crop loss is evident.

Treatments are applied to conventionally drilled or planted farm crops on a split field basis. Treatment areas range from 4 to 17 ha. Crop monitoring and assessments are done on fixed plot areas marked out in each half of each field. Each pair of plots is located on a common field boundary and extends 150 m into the centre of the crop. Plots are 84 m wide with a 36 m buffer zone between plots.

Summarised data for total pesticide use in CFP and RIA is shown in Table 2. A pesticide unit is defined as one application of a pesticide product at the recommended rate for a particular task, dependent on its target and its intensity, crop timing and environmental conditions.

TABLE 2. Total Pesticide units applied 1990-1994

Pesticide regime	CFP	RIA
Herbicides	55	32.5
Fungicides	55	29
Insecticides	29	-
Total	139	61.5

In the first four years of SCARAB, a total of 139 pesticide units at full label recommended rate were applied to CFP. During this time , RIA received 44 percent of the pesticide units applied to CFP and no insecticides.

Monitoring of crop development, weed distribution, pest and disease levels and assessment of yield were done by ADAS Science staff.

Invertebrates were sampled using pitfall traps and Dietrick suction samplers (D-Vacs). Pitfall traps were left open for seven days in every 14 day period, except between harvest and drilling, or during autumn cultivations (spring-sown crops). The traps were arranged in four transects parallel with the field boundary, one in the boundary and the other three at 10 m, 75 m and 150 m into the field. D-Vac samples were taken on average on 18 occasions each year, every fortnight during April to October. Catches of invertebrates were identified and analysed by researchers at Southampton University.

Soil microbial biomass was sampled at High Mowthorpe and Gleadthorpe only. Soil samples were taken following each pesticide application, and since 1992, before the pesticide application as well. Six samples (about 1 kg weight from 2 to 10 cm depth) were taken in an oblong grid pattern at 10 m intervals in each plot. The following chemical and microbiological parameters were measured by researchers at University of Wales, Bangor (Hancock *et al.*, 1993): pH and soil moisture content; total soil fungal biomass, total soil microbial biomass, organic carbon, soil organic nitrogen, microbial biomass nitrogen; vital fungal biomass; indirect (agar plate) counting of bacteria and fungi and soil mineralisation rates in carbon amended and non-amended soil.

The long-term effects of the two pesticide regimes on earthworm populations were monitored by scientists from the Central Science Laboratory (CSL) twice a year, when earthworm activity is expected to be the highest. Three samples were collected in each treatment area, using a 50 cm x 50 cm quadrat, dropped at random at 10-20 m intervals. Earthworms were hand sorted from soil samples dug down to plough layer and assessed for total numbers, biomass, species composition and age composition (Tarrant *et al.*, 1994).

The occurrence of gross short-term mortality was monitored by searching four 1 m x 100 m transects of the field surface in both the CFP and RIA areas. Samples of worms were collected from the top 50 mm of soil using a corer of diameter 150 mm. At least 15 cores were taken at 2 m intervals to provide a pooled sample of at least 5 g of earthworms which were frozen for residue analysis. The core depth was chosen to be consistent with that used for estimating exposure of earthworms to pesticides in risk assessment (EPPO/CoE, 1993).

Data from all assessments were subjected to analysis of variance where appropriate and significance tests. The design of the experiment in not orthodox on that it lacks true replication of treatments, which is inevitable given the large-scale plots necessary for this type of study (Greig-Smith *et al.*, 1992).

RESULTS

Arthropod monitoring

At the start of the SCARAB project, a baseline assessment was made of the taxonomic richness of arthropods at each site. A total of 258 taxonomic groups of arthropods was recorded in pitfall traps and 111 in D-vac samples. At all sites, pitfall trap samples were dominated by spiders (Araneae) and beetles (Coleoptera). The principle invertebrate families consisted of ground beetles (Carabidae), rove beetles (Staphylinidae), plaster beetles (Lathridiidae), blossom beetles (Nitidulidae) (Coleoptera) and money spiders (Linyphiidae) (Araneae). Invertebrates present in D-vac suction samples consisted mainly of Hemiptera, the lucerne-flea (*Sminthurus viridis*) (Colembola), thrips (Thysanoptera), flies (Diptera) and Coleoptera. The principle families of Coleoptera were Staphylinidae and Lathridiidae whilst the Hemiptera consisted mainly of aphids (Aphididae) and leafhoppers (Cicadellidae).

No irreversible and adverse long-term effects of full-rate pesticide use in CFP have been detected in pitfall trap catches of polyphagous predators (Çilgi & Frampton, 1994). However, winter use of chlorpyrifos did cause major reductions in several groups of Coleoptera which persisted for several months (Figure 1). Examination of the timing of two chlorpyrifos applications at Drayton indicate that they had similar effects on overall catches of Coleoptera (Figure 1), but differed markedly in their effects on individual species of Carabidae and Staphylinidae because different species were present at the time of each spray. Differences between CFP and RIA Collembola catches at Drayton persisted up to harvest 1994 in some taxa.

Figure 1. Pitfall trap catches of Coleoptera (all families grouped together) in SCARAB project Field 5 (Drayton)

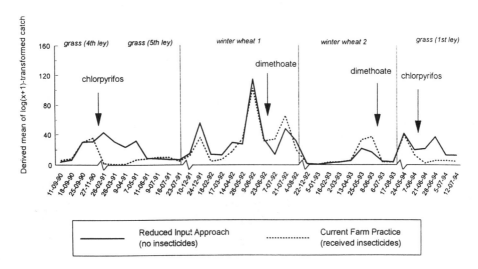

All pesticide effects detected so far have been attributed to insecticide use. No clear effects of fungicide or herbicide were apparent, but observed differences in arthropod populations between CFP and RIA not easily explained by pesticide use, could reflect subtle long-term indirect effects of herbicides and fungicides. Differences between the species composition of Collembola communities at High Mowthorpe in CFP and RIA could plausibly be explained by differences in weed communities which occurred between the CFP and RIA areas of fields because of reduced weed control in RIA.

Soil biomass dynamics

Although it represents a small fraction of soil organic matter (1-3% of soil organic carbon), soil microbial biomass is the major agent of chemical and biochemical transformations within soil ecosystems. Data from the first four years at two sites (High Mowthorpe and Gleadthorpe) have shown that the gross yearly biomass carbon levels appear to be strongly related to crop type and crop rotation. In three cereal crops at High Mowthorpe (winter barley and winter wheat), the pooled biomass data for Old Type field gave a value of 358 μg C g^{-1}. When break crops of spring beans and winter oilseed rape were grown in the same field, the pooled value was lower at 297 μg C g^{-1}. The same trend for lower biomass carbon levels in break crops was repeated at Gleadthorpe, despite the differences in soil type, microbial communities and crop rotation between the two sites. It is probable that the difference is due to increased root biomass in cereal crops. This increases the carbon flux via exudation from rootlets into the microbial soil component, which in turn, stimulates activity and resultant biomass size.

Periodic soil sampling around pesticide application events indicated that soil microbial biomass fluctuates more widely within the CFP management regime, than in RIA. Generally, the CFP regime has led to a reduction in microbial parameters, which has sometimes been statistically significantly different from RIA (data not presented). This trend can be seen in Old Type field North at High Mowthorpe from April 1990 to April 1995 (Figure 2).

Figure 2. % difference of CFP soil microbial biomass compared with RIA
Old Type, High Mowthorpe 1990-95

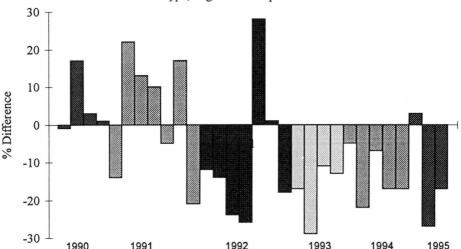

Earthworm populations

Earthworm monitoring did not start until spring 1993. Results since then have shown that there are large and statistically significant difference in earthworm biomass and numbers between the sites (Figure 3). However, there have been no consistent, ecologically significant differences between treatments at any of the sites. This is illustrated by the results at Drayton where biomass has been higher on different treatments in alternate seasons. At High Mowthorpe, although there have been large fluctuations in earthworm numbers and biomass over time, these changes have been similar on both treatment areas. At Gleadthorpe, earthworm numbers have continued to be very low.

Sampling after selected pesticide applications was done to determine any short-term effects on earthworms. No mortality was detected in earthworms, following applications of chlorpyrifos and propiconazole to grass. Residues of these pesticides in the soil were similar to the predicted values currently used in risk assessments. Residues of chlorpyrifos in earthworms were similar to predicted values, but propiconazole levels were higher than expected.

Figure 3. Changes in earthworm biomass at ADAS SCARAB sites 1993-1994

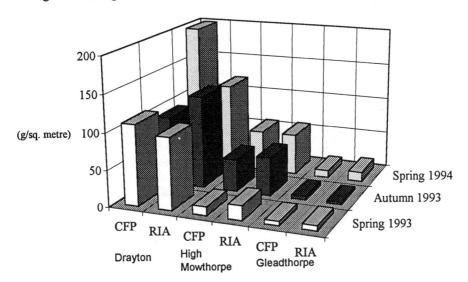

Pest, disease and flora monitoring

Pests

The major pests problems seen in SCARAB have been those normally associated with the range of crops grown at each site, and have followed normal cycles of development within years. Populations of grain aphids in winter wheat and winter barley crops at all three sites remained at low levels during autumn and spring, but rose above the threshold level of 66 % of tillers infested in late June/early July. Populations were well controlled by full-rate insecticides applied to CFP. Levels remained at threshold in RIA for periods of two to three weeks until natural decline through predation occurred . Pest levels in non-cereal break crops, oilseed rape and spring beans at High Mowthorpe and potatoes and sugar beet at Gleadthorpe, were in most cases below treatment thresholds, and did not contribute to significant reductions in yield and quality. High numbers of leatherjackets were found in the fifth year grass at Drayton. Generally, there has been no build up of pests through the rotations in the RIA treatment of any of the sites.

Diseases

Disease in cereals was common at all sites. *Septoria tritici*, *Septoria nodorum* and powdery mildew all developed to levels requiring treatment in most years. Single or two spray programmes were effective in controlling the diseases present in CFP, and in RIA, when applied at half-rate or less. Fungicides were applied to spring beans at High Mowthorpe in 1991 and at Gleadthorpe in 1993 to control chocolate spot and downy mildew. The half rate treatment in RIA was as effective as full-rate at High Mowthorpe, but much less effective at Gleadthorpe, where disease levels in RIA were almost three times greater than in CFP. Threshold levels of disease were notably absent in oilseed rape, sugar beet and grass. As a result few fungicides were applied to these crops. Because of the high risk of potato blight in potatoes grown at Gleadthorpe in 1991 and 1994, both CFP and RIA received a similar comprehensive fungicide programme.

Weeds

Application of reduced rate herbicide over four years has significantly increased weed numbers and weed seed return in RIA at High Mowthorpe and Gleadthorpe. Poor control of weeds in RIA in the first year of SCARAB at High Mowthorpe led to an increase in weeds/m^2 of 112 percent compared with CFP. This difference in weed numbers persisted in winter wheat in 1992 and in winter barley in 1993, where the high numbers of weed contributed to major problems in harvesting and drying, and affected grain quality. In winter oilseed rape in 1994, weed numbers in RIA were still double those of CFP, following an application of full-rate herbicide to the whole field. At Gleadthorpe, weed control at half-rate in RIA in cereals was better than at High Mowthorpe, as the weed spectrum was broad-leaved weeds, with no problem weeds such as cleavers and blackgrass. The rates of the herbicides used in potatoes at Gleadthorpe are limited by soil type. Reduced rate herbicide in potatoes gave acceptable levels of weed control. Problems were seen in sugar beet with previous crop volunteers of potatoes and oilseed rape, but these were adequately controlled at full and half-rates of the standard industry repeat low-dose programme.

The trend of increasing weed numbers in RIA has not been seen at Drayton. In the two years when both fields were in winter wheat, half-rate herbicide gave effective weed control. Following wheat, both fields were planted with perennial ryegrass and managed as a silage crop, which has proved to be very competitive against established and newly germinating weeds. Data on total numbers of weeds at all sites during the four years of SCARAB are shown in Figure 4.

Figure 4. Changes in total weed numbers 1991 - 1994

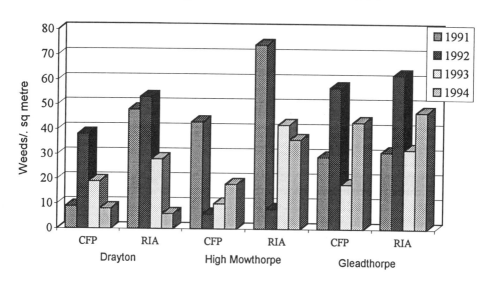

DISCUSSION

The first four years of monitoring in SCARAB have shown that the effects of adopting a CFP pesticide regime on the arable ecosystem are not as severe as those found in the intensive regime demonstrated in the Boxworth Project (1981-1988) (Greig-Smith *et al.*, 1992). The most serious adverse effects of insecticide use on polyphagous predators have been caused by the use of chlorpyrifos. In contrast, none of the dimethoate sprays used in SCARAB fields have had comparable lasting effects. The importance of life-cycles and dispersal abilities of arthropods, which determine their vulnerability to pesticide use, was demonstrated in the Boxworth project and has been endorsed by the results so far from SCARAB. They also show that temporal and spatial distribution patterns of species could be as important in determining patterns of exposure and responses to pesticide use. The effects of the two regimes on soil microbial biomass and earthworm populations has not been consistent at any of the sites, but there are indications that, as SCARAB progresses, numbers of both are tending to increase in RIA.

For most crops at most sites, the effect of reduced fungicide use in RIA has not led to significant differences in disease levels between RIA and CFP. Greater differences in disease development and control have been seen in winter wheat than in spring barley, oilseed rape and grass. Insecticides used on CFP have been very effective in controlling the target pests. Levels remained high in RIA areas, which received no insecticide treatment. Some aspects of quality, such as specific weight in cereals, may have been reduced as a result of feeding by high pest levels in RIA.

The efficacy of half-rate herbicide treatments in SCARAB has varied from very effective to inadequate. Weed control was generally poorer in the broad-leaved break crops such as oilseed rape and beans, where broad-leaved weeds predominated, and were often out of the range of the herbicides applied. Weed control in cereal crops was generally better and differences in weed number and diversity between CFP and RIA were small. The residual levels of weeds left after treatment on RIA, and to some degree, CFP, have led to a measurable increase in weed seed return.

ACKNOWLEDGEMENTS

Financial support for this work from the Ministry of Agriculture, Fisheries and Food is gratefully acknowledged. The authors thank all the staff involved in this Project for their assistance, support and advice.

REFERENCES

Çilgi, T & Frampton, G.K. (1994). Arthropod populations under current and reduced-input pesticide regimes: Results from the first four treatment years of the MAFF SCARAB project. *Brighton Crop Protection Conference - Pests and Diseases*, **6B-4**, 653-60.

Cooper, D.A. (1990). Development of an experimental programme to pursue the results of the Boxworth Project. *Brighton Crop Protection Conference-Pests and Diseases*, **1**, 153-62.

EPPO/CoE (1993) Decision-making scheme for the environmental risk assessment of plant protection products. Chapter 8. Earthworms. *EPPO Bulletin* **23**, 131-149.

Greig-Smith, P.W.; Frampton, G.K.; Hardy, A.R. (Eds) (1992). *Pesticides, cereal farming and the environment*, HMSO, London.

Hancock, M., Frampton, G.K., Çilgi, T., Jones, S.E and Johnson, D. B. (1993). Ecological Aspects of SCARAB and TALISMAN Studies. *Ecology and Integrated Farming Systems - Proceedings of the 13th Long Ashton International Symposium* **17,** 289-306.

Tarrant., S.A. Field., A. Jones., C. McCoy. (1994). Effects on earthworm populations of reducing pesticide use: Part of the Scarab Project. *Brighton Crop Protection Conference - Pests and Diseases*, **9C-2**, 1289-1294.

LINK INTEGRATED FARMING SYSTEMS: A CONSIDERED APPROACH TO C
PROTECTION

S.E. OGILVY, D.B. TURLEY

ADAS High Mowthorpe, Duggleby, Malton, N. Yorks YO17 8BP, UK

S.K. COOK

ADAS Boxworth, Boxworth, Cambridge CB3 8NN, UK

N.M. FISHER

SAC, Penicuik, Edinburgh EH26 0PH, UK

J. HOLLAND

The Game Conservancy Trust, Fordingbridge, Hants SP6 1EF, UK
Dept of Biology, University of Southampton, Southampton SO9 3TU, UK

R.D. PREW

IACR Rothamsted, Harpenden, Herts AL5 2JQ, UK

J. SPINK

ADAS Rosemaund, Preston Wynne, Hereford HR1 3PG, UK

ABSTRACT

Integrated farming requires a more considered approach to crop production
and protection, and seeks to integrate cropping sequences, husbandry
techniques, and disease resistant cultivars with more managed and efficient
agrochemical use and natural biological control. The LINK Integrated
Farming Systems project aims to develop practical and economically viable,
integrated arable systems, which are environmentally more acceptable than
conventional production systems. The first two years of the five year project
on six sites in the UK have shown that generally inputs can be reduced and
profitability maintained, but husbandry practices used to minimise leaching of
nutrients and to replace agrochemical inputs may increase management time
and result in higher operating costs. However, it will require three to five years
of the study to be completed before a full evaluation of the economic and
environmental effects can be made.

INTRODUCTION

Pressures on UK farmers to reduce inputs of agrochemicals have generally been less than in some European countries, where particular environmental problems associated with intensive pesticide use have had to be addressed. However, in the last decade, there has been increased pressure on farmers to reduce costs per unit of output to maintain profitability. There has also been strong pressure from the EU, the national government and consumers for farmers to become more concerned over environmental protection and food quality. Many farmers have already moved away from using high rates of inorganic fertilisers and from using prophylactic and insurance pesticides at full recommended rates, and are basing treatments on managed inputs, thresholds and appropriate rates. Although, reduced fertiliser and pesticide use will save costs, yields and profitability will be maintained only if husbandry practices are also modified to help limit leaching risk, pest, disease and weed problems.

Much previous and current agricultural research has been oriented towards single factors or problems, and short term studies. Alternative research methods are being adopted which look at farming systems over rotations to measure cumulative effects over five or six years, and to express the interactions between husbandry techniques, agrochemical inputs, pests, diseases, weeds and crop performance on large field areas over longer periods of time. Such integrated farming is being researched and encouraged in the UK and several other European countries (Vereijken & Royle, 1989; Jordan *et al.*, 1990; El Titi, 1992).

A large research project on integrated farming, the LINK Integrated Farming Systems (IFS) project, commenced in April 1992 as part of the LINK Programme "Technologies for Sustainable Farming Systems" (Wall, 1992; Prew, 1993). The project was set up on six sites to develop arable integrated farming systems which concentrate on practical feasibility and economic viability, but also take into account level of inputs and environmental impact. The integrated system is compared with local conventional practice at each site. This work is seen as a development of integrated farming in different geographical and climatic locations in the UK, over a wide range of soil types. ADAS, IACR Rothamsted, Scottish Agricultural College, The Game Conservancy Trust and Southampton University are collaborating in this project. Some of the results from the first two years of the project are presented in this paper.

METHODS

Sites and rotations

The project is being done at six sites with a wide range of soil types and geographical situations in the main arable areas of the UK. Four of the sites are on commercial farms, on the Manydown (MD) Estate near Basingstoke in Hampshire, on the Scott Abbott Arable Crop Station at Sacrewell (SW) in Cambridgeshire, on the Lower Hope (LH) Farms Estate near Hereford and on the Rosemains and Turniedykes farms at Pathhead (PH) in Midlothian. The other two are on ADAS Research Centres, at Boxworth (BW) in Cambridgeshire and High Mowthorpe (HM) in North Yorkshire. The integrated system is compared with a reference conventional system at each site on split or quartered fields, in five course rotations with all phases of the rotation present in each year (Table 1). Appropriate rotations were chosen for each site based on local practice. At least two fields are replicated at each site. Field-scale plots were chosen so that field operations could be carried out on a large enough scale to be commercially relevant and to minimise interference between treatments. Approximately 55 ha are devoted to this project at each site. Crop performance, environmental and economic impact assessments are recorded at each site.

TABLE 1. Crop rotations

Site	System	Rotational phase and crop				
		1	2	3	4	5
Sacrewell	IFS	W wheat	Set-aside	Peas	W wheat	Potatoes
	Conv	W wheat	Set-aside	Peas	W wheat	Potatoes
Boxworth	IFS	Linseed	W wheat	W beans	W wheat	W wheat
	Conv	WOSR	W wheat	W beans	W wheat	W wheat
H. Mowthorpe	IFS	W wheat	Set-aside	S beans	W wheat	Seed potatoes
	Conv	W wheat	Set-aside	WOSR	W wheat	Seed potatoes
Lower Hope	IFS	W wheat	Set-aside	S beans	W wheat	Potatoes
	Conv	W wheat	Set-aside	WOSR	W wheat	Potatoes
Manydown	IFS	W wheat	W wheat	S barley	Vining peas	WOSR
	Conv	W wheat	W wheat	S barley	Vining peas	WOSR
Pathhead	IFS	SOSR	W wheat	Set-aside	W wheat	S barley
	Conv	WOSR	W wheat	Set-aside	W wheat	W barley

System definitions

The integrated system is defined as a husbandry system which maximises profitability with a different balance of inputs to that used conventionally, and aims to achieve environmental benefits; whereas conventional practice is defined as crop husbandry which maximises profitability using external inputs applied within permitted limits to overcome constraints on production. Management controls have been built in so that treatment decisions are based on clear guidance to ensure a common approach wherever possible, and also to ensure a clear distinction between systems. However, the integrated approach does vary from site to site as a result of different soil and climatic conditions, different pest, disease and weed pressures and different environmental problems.

Integrated farming techniques

Crop protection

One of the main strategies in integrated farming is to increase the diversity of crop species in a rotation to prevent disease and pest carry-over from crop to crop. The aim is to have at least four different crops in a rotation and this has been achieved on five of the six LINK IFS sites. Disease resistant cultivars are used wherever possible, but yield and quality are also important considerations if crops are to remain financially viable. Alternative husbandry techniques such as mechanical weeding are used in conjunction with chemical control, and opportunities are taken in the integrated system to allow predators and parasites of pests to build up in sufficient numbers to help control crop pests. Time of establishment may be delayed to limit weed, pest and disease problems in the autumn. However, late establishment increases the risks of nutrients leaching from the soil profile between crops, so timing is chosen to address the most pressing environmental concern at an individual site. Pesticide use is optimised by basing decisions on thresholds, in-crop monitoring systems, crop mapping and patch treatments, trapping techniques and appropriate rates. The most specific pesticide is chosen where possible to minimise off-target effects so that species diversity is maintained.

Soil and nutrient management

Soil and nutrient management are key factors in an integrated farming system. Cultivations are planned on a rotational basis to minimise soil disturbance, retain soil structure and fertility, help minimise soil erosion, maximise crop performance and energy savings and encourage beneficial invertebrates such as earthworms and predatory beetles and spiders. However, it is recognised that cultivations are very soil specific and some soils do not respond well to repeated non-ploughing techniques, which can result in soil structure problems and increases in grass weeds. In addition, choice of crop can influence the degree of soil cultivation required, especially for crops like potatoes which require a deep tilth.

Nitrogen inputs are carefully calculated to balance with individual crop requirements, offtakes and existing soil residues, without leaving excess residues after harvest which could be lost by leaching or run off. Applications of other basal elements, such as phosphorus, potassium and magnesium, are calculated on a similar basis to nitrogen but dressings are usually applied on a rotational basis with the aim of maintaining soil fertility at an appropriate level rather than depleting soil reserves. Where practical, crop residues are returned to the soil to minimise loss of nutrients from the integrated system. Cover crops are used before all spring crops in the integrated system where weather conditions permit their establishment, and a green cover is maintained during the critical periods for leaching in the set-aside period, to retain soil nutrients in the top layer of the soil profile, to prevent soil erosion and the development of problem weeds.

RESULTS

The 1993 cropping year was the first treatment year following a baseline assessment year in 1992. The challenging weather conditions of this first year put the two farming systems to a rigorous test. A wet harvest in autumn 1992 resulted in delayed crop establishment, especially after potatoes, poor soil conditions, problems with slugs and difficulties in establishing effective autumn cover crops. Most crops eventually established quite well and gave acceptable yields. The wet summer and harvest of 1993 also caused numerous problems, notably very high blight risk conditions all season for the three potato crops and loss of quality in some milling wheat crops. The 1994 growing season was generally more favourable and most crops performed well. Individual problems with particular crops at some sites were apparant but these were not linked between sites.

Nitrogen use meaned across all crops in the rotation at each site is given in Figure 1. Overall, 15 and 16 percent less nitrogen was applied in the integrated system in 1993 and 1994 respectively compared with the conventional system. This was mainly a result of different nitrogen requirements for the crops on the two systems and differences in the residual levels of nitrogen in the soil.

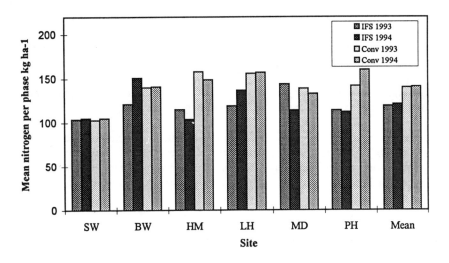

Fig. 1 - Mean nitrogen applied per phase kg ha -1

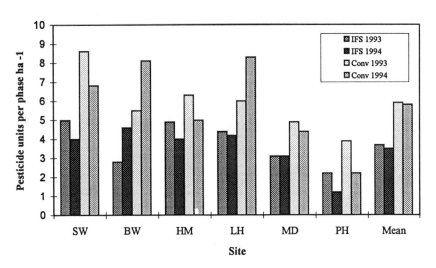

Fig. 2 - Mean pesticide units per phase ha-1

Pesticide use has been meaned on a similar basis to nitrogen. A pesticide unit in this situation is defined as one application of an active ingredient at the recommended rate for a particular task, dependent on the target and its intensity, crop, timing and environmental conditions. Overall, 37 and 39 percent fewer pesticide units were applied in the integrated system in 1993 and 1994 respectively (Figure 2). The lower inputs were achieved by growing a slightly different range of crops in the integrated system, including some spring-sown crops, altering cultivar choice for more disease-resistant cultivars, adopting delayed drilling in some cases to reduce the need for blackgrass control and autumn insecticide use, mechanical weeding especially in potatoes and some cereal crops, and using a more cautious, "wait and see" approach to treatment thresholds. There were fewer opportunities to replace pesticides with alternative techniques in the very high value, high risk, potato and vining pea crops in both cropping seasons. The more considered approach to crop protection in integrated systems does involve increased management time. This will be assessed as part of this project and will contribute to the full economic evaluation of integrated farming.

Over all the rotations, crop yields were not substantially affected by changing to an integrated system of production, when meaned over all sites in the first two treatment years (Figure 3). Wheat yields were generally lower on the integrated system compared with the conventional system because lower-yielding but higher-quality cultivars were chosen.

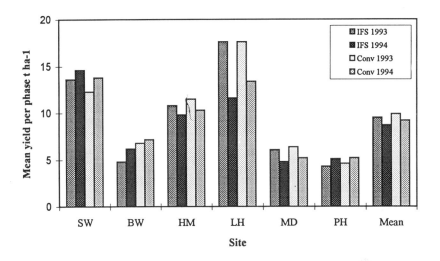

Fig. 3 - Mean yield per phase t ha -1

Financial returns for the two systems were meaned over each rotation for each site (Figure 4). Figures include the Arable Area Payments for eligible crops. It is recognised that the percentage of rotational set-aside to eligible crops is too high in these five-course rotations because of the design of the experiment. However, this will be taken into account when a full economic appraisal is undertaken at the end of the project. Gross output was generally lower on the integrated system, compared with the conventional system, but overall, the differences between the two systems were small. The three sites which grow potatoes, Sacrewell, High Mowthorpe and Lower Hope, had the highest mean gross outputs but also the highest costs. Variable costs were generally lower on the integrated system which reflected the reduced nitrogen and pesticide use. Overall, the two systems gave very similar gross margins, although there were individual site differences (Figure 4). In 1993, the integrated system was more profitable than conventional practice at Sacrewell, Lower Hope and Manydown and in 1994 was more profitable at Sacrewell, Boxworth and Pathhead. However, it will be three to five years before it is possible to make a reliable comparison of the profitability of the two systems.

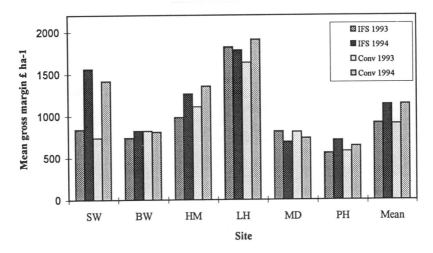

Fig. 4 - Mean gross margin per phase £ ha -1

Operating costs for all field operations and crop drying and handling were estimated to provide a measure of the efficiency of the integrated system when compared with conventional practice, but the data are not presented here. The initial estimate of operating costs reflected the higher costs on the potato sites and also the slightly higher overall costs of operating the integrated system. The implications of adopting an integrated system on labour and machinery requirements and work patterns are being investigated as part of this study and will be reported at the end of the project.

DISCUSSION

Research in Europe has shown that the more managed integrated approach to crop production, which combines natural regulatory components and husbandry practices into farming activities, with the aim of replacing purchased off-farm inputs, is feasible on a large scale and is a logical development for current agricultural practice (El Titi, 1992; Jordan & Hutcheon, 1994). Many successful integrated farming pilot schemes have put research into practice in Europe (Wijnands, 1992; Jordan et al., 1993). The main objectives of an integrated system of production are to address environmental concerns, to ensure inputs are managed efficiently and effectively, and to maintain or increase quality of produce and profitability. If successful, an integrated system will meet the most pressing concerns of industry, consumers and governments alike. The LINK IFS project seeks to develop the integrated system across a wide range of crops, soil types and locations in the UK, covering many environmental and cropping issues. Initial results from the first two years are encouraging, and the successful components will be built into future practice as the system continues to evolve. However, it is already apparant that there are many conflicts to be addressed in the development of an integrated system; for example techniques adopted to solve one environmental problem may exacerbate other problems. Further work is planned within the project to investigate in more depth the economic and environmental implications of altering farming practice, to ensure that practical, relevant and viable systems are available for the industry.

ACKNOWLEDGEMENTS

Funding from the project sponsors, the Ministry of Agriculture, Fisheries and Food, the Scottish Office Agriculture and Fisheries Department, the Home-Grown Cereals Authority (Cereals and Oilseeds), Zeneca Agrochemicals and the British Agrochemicals Association, is gratefully acknowledged. Thanks are also given to the host farmers, statisticians and all other colleagues involved in the project.

REFERENCES

El Titi, A. (1992) Integrated farming: an ecological farming approach in European agriculture. *Outlook on Agriculture*, **21**, 33-39.

Jordan, V.W.L.; Hutcheon, J.A.; Perks, D.A. (1990) Approaches to the development of low input farming systems. In: *Crop protection in organic and low input farming*, R. Unwin (Ed.), *BCPC Monograph No. 45*, Farnham: BCPC, pp. 9-18 .

Jordan, V.W.L.; Hutcheon, J.A.; Glen, D.M. (1993) *Studies in Technology Transfer of Integrated Farming Systems - Considerations and Principles for Development.* Bristol: AFRC, pp, 16.

Jordan, V.W.L.; Hutcheon, J.A. (1994) Economic viability of less-intensive farming systems designed to meet current and future policy requirements: 5-year summary of the LIFE project. *Aspects of Applied Biology*, **40**, *Arable farming under CAP reform*, pp.61-68.

Prew, R.D. (1993) Development of Integrated Arable Farming Systems for the UK. *Proceedings of the HGCA Cereals R & D Conference 1993*, London : HGCA, pp. 242-254.

Vereijken, P.; Royle, D.J. (1989) Current status of integrated farming systems research in Western Europe. *WPRS Bulletin*, **XII, 5**: 76pp.

Wall, C. (1992) A LINK collaborative research programme on Technologies for Sustainable Farming Systems. *Proceedings of the Brighton Crop Protection Conference - Pests and Diseases* 1992, **III**: 1107-1114.

Wijnands, F.G. (1992) Evaluation and introduction of integrated arable farming in practice. *Netherlands Journal of Agricultural Science*, **40**, *Research on integrated farming systems in the Netherlands*. 239-250.

THE ROLE OF NEW TECHNOLOGY IN PROMOTING SUSTAINABLE AGRICULTURAL DEVELOPMENT

J. TAIT, P. PITKIN

Scottish Natural Heritage, Research and Advisory Services Directorate, 2 Anderson Place, Edinburgh EH6 5NP.

ABSTRACT

Scottish Natural Heritage (SNH) has set up a project known as TIBRE (Targeted Inputs for a Better Rural Environment) to investigate how new technology (chemical, biological, IT and engineering) could be introduced into existing agricultural systems (both integrated and intensive) to improve their sustainability. The project is concentrating in the first phase on arable cropping. This paper explains how we have interpreted the concept of sustainable development, and the role of the TIBRE project in achieving greater sustainability.

THE CONCEPT OF SUSTAINABILITY

The concept of sustainability is often treated as self-explanatory and left undefined. While we are aware of the pitfalls in attempting to define it, we feel it is important to clarify our interpretation in the context of the TIBRE project (SNH, 1993).

We see the concept of sustainability as inherently systemic. It can only be applied in practice to the behaviour of a system of interacting variables, referring to the extent to which the system can continue to operate in its present form for the foreseeable future, i.e. its stability and long term viability.

We also see it as a concept that is related to the management of systems and which is irrelevant in the absence of any human interest in the system. For example, some natural ecosystems are relatively stable and, in the absence of external interference, will persist in their present form for the foreseeable future. Others are in transition states which, in the absence of external interference, will develop and change. However, the question of their sustainability only becomes relevant when people have an interest in exploiting the system for some purpose, for example to manage it for conservation purposes or to exploit its natural resources.

Many factors can affect the degree of sustainability of a managed system. At the subsystem level, the sustainability of natural resource use is the most obvious one. The Government's Strategy for Sustainable Development (Anon, 1994) identifies the following ways in which agricultural systems have become less sustainable as they have become more intensive: habitat loss; increased eutrophication of fresh and saline waters; pesticide contamination of land, water and air; loss of organic matter from soils; soil erosion; soil acidification; and contamination by chemicals, although many of these effects are reversible.

All managed systems have economic, political and social components, and these can often trigger a change in the stability and viability of a system before its natural components show signs of strain. In the case of agricultural systems, the pressures to develop integrated and organic cropping systems in the UK have been political rather than agronomic and have arisen from social perceptions and evidence from other countries that intensive agricultural systems are unsustainable. Despite numerous predictions that intensive arable agriculture would inevitably collapse into a degenerative spiral, there is as yet no evidence of this taking place in the UK. This has made it difficult to persuade arable farmers that their intensive cropping systems are agriculturally unsustainable and hence has made it more difficult to persuade them to take up integrated and organic approaches. (The economic sustainability of intensive systems is discussed below.)

In some intensive agricultural systems, the excessive use of pesticides has led to a breakdown of the supporting ecosystem to such an extent that the agricultural system was no longer sustainable. Examples are the growing of cotton in Texas (Curry & Cate, 1984), rice growing in the Philippines (Kenmore et al., 1987) and fruit growing in some parts of the UK (Solomon, 1987). In these cases, where lack of sustainability has been clearly demonstrated to them, farmers have shown a greater willingness to take up integrated cropping systems.

EXTERNAL SUPPORT FOR INTENSIVE ARABLE SYSTEMS

The agricultural revolution that brought about today's intensive farming systems began in the 1950s with the emergence of the modern agrochemical industry producing the inputs that enabled farmers to increase crop yields steadily over a forty year period. For example, cereal crop yields in the UK have increased by approximately half between 1971 and 1992 (MAFF, 1973; MAFF, 1992).

The core feedback loop of Figure 1, with the variables in bold type, illustrates some of the factors that were driving the industrial investment that, until recently, fuelled the growth of intensive systems. Increasing industry investment in research and development for new agrochemical products was fuelled by a combination of increasing crop yields and increasing crop prices. Increased investment in research and development led to more new products on the market which led in turn, through more intensive marketing by industry, to a greater level of uptake of these new products by farmers. The resulting higher crop yields, coupled to stable or increasing crop prices led to steadily increasing farm incomes which closed the feedback loop by encouraging yet higher levels of investment by industry. The main driving force for this positive feedback loop (which was regarded as a virtuous circle from the perspective of the agrochemical industry) was the level of UK Government, and later European Community, support for agricultural production. In addition to guaranteeing farmers a market for their crops at a favourable price, the Government policy set ever increasing standards for freedom from damage by pests and diseases which also encouraged the use of pesticides; a free advisory service was provided to encourage the uptake of new technology, reinforcing the marketing efforts of the industry; and publicly funded near-market research and development supported the development of new products by industry.

Less important, but still significant, drivers of investment are shown at the top of Figure 1. Regulation, which is often claimed to inhibit innovation, in this case acted as a stimulant, by withdrawing from the market out of date products that had outlived their patent protection, making way for a new generation of more expensive, patent-protected products. The emergence of resistance to pesticides among insect pests and diseases had a similar effect. (It was rare for this to occur within the patent protection period for a chemical.)

One output from this feedback loop since the 1960s was a continually increasing level of agricultural surpluses. This was seen as a cause for concern, but was not directly addressed before intensive farming systems had become firmly established.

During the 1980s farm incomes became less secure as the European Community, in order to reduce the cost of disposing of these surpluses and to remedy their distorting effect on world trade, took steps to avoid over-production. The resulting decline in government support for agricultural production has begun to convert this positive feedback loop from a virtuous to a vicious circle from the agrochemical industry perspective leading to a depressed market for new technology. This effect was compounded by the almost complete withdrawal of the free advisory service and a cessation of near market research and development work at public expense.

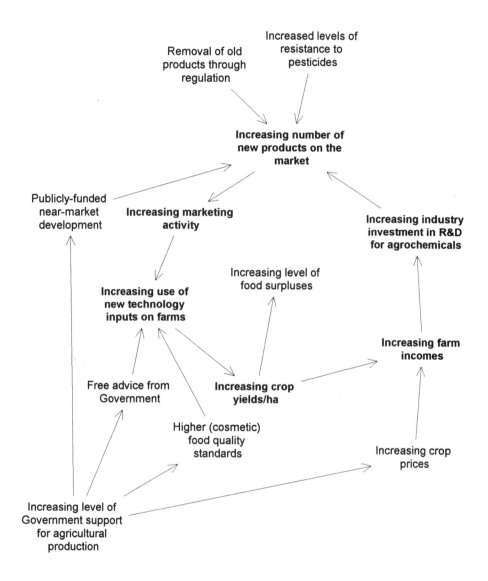

Figure 1. The system sustaining industrial investment in the development of new agrochemicals, in the 1960s and 1970s.

Thus, contrary to many predictions, it was the economic and political unsustainability of the system that finally applied the brakes to this bandwagon, rather than its ecological unsustainability.

ENVIRONMENTAL IMPACT AND SUSTAINABILITY OF FARM CROPPING SYSTEMS

So far, we have focused on the concept of agricultural sustainability and the factors driving agrochemical industry investment which is often seen as a threat to it. From a wider environmental perspective, in most parts of the world and under most agricultural systems, the greatest impact is caused by the basic act of farming the land. Compared to this, the differences in environmental impact between different types of farming system are generally small. However, this impact is not always negative in terms of the natural heritage. Some types of arable farming system, such as appropriately managed organic systems and traditional cropping in the crofting areas of Scotland, create habitats which it is SNH's responsibility to support.

In keeping with its remit "to secure the conservation and enhancement of, and to foster understanding and facilitate the enjoyment of, the natural heritage of Scotland", as stated in its founding legislation, the Natural Heritage (Scotland) Act 1991, SNH must take account of these environmental impacts and benefits. It has done this by attempting to influence the nature of agricultural systems, as outlined below, and also by supporting the creation and maintenance of wildlife habitats on farms.

This section considers the environmental opportunities and problems associated with organic, intensive and integrated farming systems. Table 1 gives our view of the sustainability, acceptability and environmental impact of these three types of system. The first three subsections make a comparison on the basis of the criteria in Table 1. The fourth subsection discusses the impact of policy-related issues on overall sustainability..

Organic cropping systems

Organic systems are widely regarded as the most sustainable form of agricultural system. In the sense that they attempt to operate in a self-contained manner without the aid of extraneous inputs, this is indeed the case (Table 1, criterion 1). Their impact on cropped and non-cropped areas of the farm is usually considerably less than that of more intensive systems and the crop rotations employed on organic farms can increase biodiversity on the farm, both in the soil and among and between the crops (criteria 2 and 3). Their external or wider environmental impact is also generally less than that of the other two types of system, although the amount of nutrient runoff from organic farms can be considerable (criterion 4).

However, organic systems are unlikely to be more than a partial answer to the problem of agricultural sustainability. As explained above, given the extent to which intensive systems have become embedded in the farming community, experience suggests that few farmers will be prepared to take them up unless encouragement is provided in the form of substantially greater public support or unless consumers are prepared to pay a high premium for organic produce (criteria 5 and 6).

Intensive farming systems

At the other end of the spectrum, intensive farming systems generally have the most negative environmental impact, on and off the farm (criteria 2-4), although probably less so in Scotland than in other parts of the UK where pesticide use is greater. Some aspects of intensive farming are not acceptable to many consumers but most are not sufficiently concerned about the source of the produce they buy to pay the present level of the organic premium (criterion 6). Most farmers have a very positive attitude to intensive farming systems (criterion 5) and, while they may be prepared to cut down on inputs to avoid the most obviously wasteful practices, they will not readily give up their allegiance to intensive farming systems (Carr & Tait, 1991).

As noted above, there is little evidence as yet that intensive farming systems are not agriculturally sustainable (criterion 1).

Table 1   Sustainability, environmental impact and acceptability of farming systems

| Criterion | Arable Farming Systems* | | |
	Organic	Integrated	Intensive
1. Agricultural sustainability	++	++	+/++
2. Impact on farmed areas	+/-	-	- -
3. Impact on non-farmed areas of the farm	+/-	-	- -
4. Off-farm environmental impact	o/-	--	- -
5. Acceptability to farmers	--	-	++
6. Acceptability to consumers	+	o/+	o/-

    *   + = favourable; o = neutral; - = unfavourable

Integrated farming systems

Integrated farming systems attempt to reduce the environmental impact and improve the sustainability of farming systems by more judicious use of inputs, combined with better diagnosis of crop problems, and crop rotations designed to reduce the need for inputs. There is no doubt that, through careful planning and management of an integrated system, the farmer can reduce inputs and maintain or improve yields (Conway & Pretty, 1991), leading to a greater likelihood of improved agricultural sustainability compared to intensive systems (criterion 1). However, generally, this is at the expense of a greater input of managerial skill and time, and often a higher level of risk and these factors reduce the acceptability of such systems to many farmers (criterion 5). As with organic systems, there is little evidence that most consumers are prepared to pay a significant premium for food produced by integrated systems (criterion 6).

The extent to which integrated systems will improve the environmental impact of the farming system will depend on the motivation of the farmer and the degree of skill with which they are implemented, but there is no doubt that they could have a significantly lower environmental impact than intensive systems (criteria 2-4).

Policy issues

The above analysis shows that integrated and organic approaches have considerable potential to reduce the environmental impact of current intensive farming systems, in line with Government policy to modify agricultural systems in order to improve their environmental sustainability. The UK Strategy for Sustainable Development states that environmental pressures from agriculture will be reduced by "... the reduction in levels of price support which will, in turn, reduce the incentives to intensive production ... which, from the point of view of the UK environment is the most important feature of the 1992 CAP reform." (Anon, 1994). The discussion of Figure 1 has already noted that these policy developments are affecting industry strategies.

The impact of these policies on farmer behaviour is more difficult to gauge precisely. As noted already, many farmers do not see the current levels of farming intensity as being unsustainable. This view is held most strongly by the largest and most intensive farmers who are likely to exert a greater impact on the environment. As discussed above, given the entrenched nature of intensive systems, these farmers are unlikely to be willing to change to integrated or organic systems without substantial financial incentives. This will be particularly so since commercial pressures and competitive forces are likely to give rise to price reductions in the currently available technologies from Western European and American companies and a flood of cheap products from Eastern Europe. Farm incomes have also risen rapidly since 1992 as a combined result of the Arable Area Payments Scheme and the ending of the link between the pound Sterling and the European Exchange Rate Mechanism. A strategy which could be adopted by some farmers in response to any future liberalisation of world trade in agricultural commodities may be a move to ever greater intensity of farming systems using the cheapest available technology which is likely to be the most environmentally damaging.

Looking at the agricultural system as a whole, it is reasonable to question whether the current policies attempting to deal with perceived overproduction of many crops are likely to remain in place in the long term. Factors which could lead to a change in this policy include major crop failures due to climate change or some other unforeseen factor, and opportunities to divert agricultural land to the production of non-food crops for industrial feedstock, leading to pressures to maximise production on the remaining land used for food production. In the latter case, both food and non-food crops are likely to receive high levels of inputs. A return to maximum production using existing technology is likely to lead to a significant increase in the environmental impact of agriculture.

A FLEXIBLE AND ROBUST APPROACH TO SUSTAINABLE DEVELOPMENT

It is unfortunate that this faltering in the pace of technological innovation for agriculture is taking place at a time when new opportunities for more environmentally sustainable technology are arising in the areas of biotechnology, chemical technology, information technology and engineering. The SNH TIBRE project is attempting to focus attention on this area with a view to encouraging the evolution of an overall strategy for sustainable agricultural development in the 21st Century which:

- takes account of the wide range of types of agricultural system that exist in the UK;
- is sensitive to the needs and attitudes of individual farmers;
- incorporates a range of approaches to sustainable development; and
- is robust in the face of policy changes.

SNH will continue to support initiatives leading to the creation and management of wildlife habitats and landscape features on agricultural land. However, on many farms, these areas are poorer and less diverse than they might be because of the impacts arising from farming activities on the cropped areas.

Recognising that most intensive farmers are not likely to take up integrated or organic approaches, the TIBRE project aims to encourage the uptake by all non-organic farmers of new, less environmentally damaging technological products and hence to lead to an improvement in the sustainability of intensive and integrated farming systems.

The fastest returns will arise from technology which is already on the market or will soon be available. However, the greatest long term benefits may arise from new technologies and products which are in an early stage of research and development and which may be being held back by companies because they do not see a viable market slot in competition with existing cheaper technology.

SNH have commissioned a series of studies to investigate the range of technological developments (chemical, biochemical, IT and engineering), in both categories, which would enable arable growers to farm in a more environmentally sustainable manner, and we shall report on these elsewhere.

For the short-term options SNH has consulted a group of farmers and others on the feasibility and acceptability of various options. In general there was strong support for the aims of TIBRE and an appreciation of the involvement of the farming community at this early stage in the initiative. There was a perceived need for information on new technology and its potential influence on the environment and considerable sensitivity to the role of public opinion in influencing the adoption of new technology in farming. There was also support for the setting up of farm level projects to demonstrate the managerial and economic feasibility of new technological developments.

Implementation of the longer term options will require the creation of clearly identifiable market opportunities to persuade industry to develop new technology which is more environmentally sustainable. This could be done through a range of policy initiatives operating in this case at the UK and EU levels. In both cases, SNH will be working with partners from Government departments, other agencies and industry, some of whom may eventually take over the lead in implementation of the TIBRE project.

The task of encouraging industry to bring forward new technological products which will enable intensive agricultural systems to become more sustainable seems daunting at first sight. However, the TIBRE project is in keeping with modern thinking on a wider front on constructive approaches to environmental improvement through technological innovation as evidenced by, for example, the UK Government LINK programme, the DoE/DTI Clean Technology Initiative, the European Commission Fourth Framework Programme and the Technology Foresight Initiative of the UK Office of Science and Technology. By adding our weight to this general trend we may be able to achieve a significant shift towards more sustainable development in the strategies of industry, farmers and others involved in land use.

There is no doubt that society has the scientific and technical competence to develop sustainable, intensive agricultural systems. Our social and political competencies are, however, less well developed. The flexibility and robusness of initiatives like TIBRE could have very significant benefits, in terms of both the natural environment and social cohesion. In order to achieve these benefits we will need to use our existing institutions and procedures more effectively and be more creative in developing new ones.

REFERENCES

Anon. (1994) *Sustainable Development: the UK Strategy* London: HMSO.
Carr, S; Tait, J. (1991) Differences in the attitudes of farmers and conservationists and their implications. *Journal of Environmental Management, 32,* 281-294.
Conway, G.R.; Pretty, J.N. (1991) *Unwelcome Harvest: Agriculture and Pollution.* London: Earthscan.
Curry, G.L.; Cate, J.R. (1984) Strategies for cotton - boll weevil management in Texas. In *Pest and Pathogen Control: Strategic, Tactical and Policy Models*, G.R. Conway (Ed), Chichester: Wiley, pp. 169-183.
Kenmore, P.E.; Litsinger, J.A.; Bandong, A.C.; Santiago, A.C.; Salac, M.M. (1987) Philippine rice farmers and insecticides: thirty years of growing dependency and new options for change. In *Management of Pests and Pesticides: Farmers' Perceptions and Practices*, J. Tait and B. Napompeth (Eds), Boulder: Westview, pp 98-108.
MAFF (1973) *Agricultural Statistics, United Kingdom, 1971.* Government Statistical Service, London: HMSO.
MAFF (1992) *The Digest of Agricultural Statistics, United Kingdom, 1991.* Government

Statistical Service, London: HMSO.

Scottish Natural Heritage (1993) *Sustainable Development and the Natural Heritage: the SNH Approach*. Publications Section, Scottish Natural Heritage, Battleby, Redgorton, Perth PH1 3EW, pp 26.

Solomon, M.G. (1987) Fruit and hops. In *Integrated Pest Management*, A.J. Burn, T.H. Coaker and P.C. Jepson (Eds), London: Academic Press, pp 329-360.

Session 7
Workshop Discussions

Chairman	M TALBOT
	Professor J B DENT
	C MACKIE
Session Organisers	D YOUNIE
	J HOOKER

WORKSHOP DISCUSSIONS

Topics

a: Research methodology in Sustainable Farming Systems

b: Approaches towards technical progress in Sustainable Farming Systems

c: Technology transfer in Sustainable Farming Systems

Session 8
Novel Chemistry

Chairman and
Session Organiser   G B STODDART

EXPLOITING CHEMICAL ECOLOGY FOR SUSTAINABLE PEST CONTROL

J.A. PICKETT, L.J. WADHAMS, C.M. WOODCOCK

IACR-Rothamsted, Harpenden, Hertfordshire, AL5 2JQ, U.K.

ABSTRACT

The study of chemical ecology, particularly involving pheromones and other semiochemicals that influence insect behaviour, promises methods of pest control as alternatives to the exclusive use of broad-spectrum toxicants. However, if the potential of semiochemicals in crop protection is to be realised, a greater understanding of insect/insect/plant interactions and insect chemical ecology generally is essential. Semiochemicals, when employed alone, often give ineffective or insufficiently robust pest control. Use of semiochemicals should therefore be combined with other approaches in integrated management strategies. The main components of such strategies are pest monitoring, to allow accurate timing of pesticide treatments, combined use of semiochemicals, host plant resistance and trap crops, to manipulate pest behaviour, and selective insecticides or biological control agents, to reduce pest populations. The objective is to draw together these approaches into a push-pull or stimulo-deterrent diversionary strategy (SDDS). In an SDDS, the harvestable crop is protected by host-masking agents, repellents, antifeedants or oviposition deterrents. At the same time, aggregative semiochemicals, including host plant attractants and sex pheromones, stimulate colonisation of pests on trap crops or entry into traps where pathogens can be deployed. Because the individual components of the SDDS are not in themselves highly efficient, they do not select for resistance as strongly as conventional toxicant pesticides, thereby making the SDDS intrinsically more sustainable.

Semiochemicals in pest control

The semiochemicals which have been used most successfully in pest control are the sex pheromones of Lepidoptera and the aggregation pheromones of Coleoptera (Howse *et al.*, 1995). Many commercially developed systems exist for using the sex pheromones of Lepidoptera in slow-release formulations to disrupt normal mate location. In the control of forest pests, aggregation pheromones of bark beetles are used in trap-out procedures. However, in dealing with the main pests of arable agriculture in Northern Europe, which principally comprise the aphids, alternative types of semiochemicals have to be employed, not only to control aphid pests directly, but also to reduce their transmission of plant virus diseases (Pickett *et al.*, 1994).

Antifeedants against aphid pests

In efforts directed towards control of aphids, strategies involving more sophisticated use of semiochemicals have been developed, including integration with population-reducing components such as biological agents. A number of antifeedant compounds, principally derived from plants, have been identified as potentially useful against aphid colonisation and feeding (Griffiths *et al.*, 1989). Some of these act sufficiently quickly to reduce virus transmission, even when the viruses are transmitted in the non-persistent or semi-persistent modes. In the laboratory, the drimane sesquiterpenoid antifeedant (-)-polygodial, extracted from the water-pepper plant, *Polygonum hydropiper*, reduced transmission of potato virus Y by the peach-potato aphid, *Myzus persicae*, by over 70%, thus demonstrating activity against a non-persistently transmitted virus disease which can be passed on to the plant after very limited contact with the aphid. In the field, the target was the bird-cherry-oat aphid, *Rhopalosiphum padi*, as a vector of the persistently transmitted barley yellow dwarf virus. Although polygodial can now be synthesised on a large scale using a method modified from Hollinshead *et al.* (1983), this procedure gives the racemic mixture and although the unnatural (+)-polygodial has similar antifeedant activity to the natural isomer (Asakawa *et al.*, 1988), a nature-identical material was considered more appropriate for field use. To this end, a quantity of *P. hydropiper* was cultivated and after liquefied carbon dioxide extraction of the plant material on an industrial scale, the product contained a high concentration of (-)-polygodial. This was applied to cereals at 50 g/ha on three occasions in the autumn and, compared with untreated plots, gave over a t/ha improved yield, equivalent to the yields obtained using cypermethrin, a broad-spectrum synthetic pyrethroid insecticide (Pickett *et al.*, 1987). The pesticide was employed at a similar application rate, but with only one treatment because of its longer half-life compared to the relatively unstable (-)-polygodial. However, such approaches, based on only one semiochemical type, were considered to be insufficiently robust for general farming practice and a direct population-reducing component was added.

Integrated use of the aphid alarm pheromone

When aphids are attacked, they release an alarm pheromone which causes other aphids in the area to disperse. The pheromone comprises the sesquiterpene hydrocarbon (E)-β-farnesene, and the synthetic product has been developed for use against aphid pests together with other agents causing direct reductions in population. Thus, in glasshouse trials using a hand-held electrostatic application system, the aphid alarm pheromone combined with spores of *Verticillium lecanii* gave a substantial improvement in control of the cotton aphid, *Aphis gossypii*, as compared to unsprayed chrysanthemums or the two treatments alone (Pickett *et al.*, 1986). In the field, a similar approach was adopted with a tractor-mounted electrostatic system, but with the contact pyrethroid permethrin as the population-reducing agent. Again, a highly significant improvement was obtained with the combined pheromone and pesticide treatment, as compared to the single treatments or unsprayed plots (Dawson *et al.*, 1990).

Integrated use of antifeedants

In addition to aphids, there are many coleopterous pests of arable agriculture in Northern Europe, but the drimane antifeedants, described above, are not as active against these insects. However, it was found that antifeedants in the clerodane class of diterpenoids, e.g. the ajugarins, were extremely effective against coleopterous pests, particularly in the family Chrysomelidae. Against adults of the mustard beetle, *Phaedon cochleariae*, a concentration of 0.00001% applied to leaves showed significant antifeedant activity, whereas with the diamondback moth, *Plutella xylostella*, a concentration of 0.01%, similar to the field rate of polygodial used against aphids, was necessary (Griffiths *et al.*, 1988). The ajugarins had virtually no activity against aphids. This high selectivity against Coleoptera, and particularly the Chrysomelids, is also demonstrated with the Colorado potato beetle, *Leptinotarsa decemlineata*. In simulated field trials, electrostatic spraying of an ajugarin protected the top parts of mustard plants, *Brassica nigra* (Griffiths *et al.*, 1991). Here, it was essential to combine use of the antifeedant with a population-controlling agent since, as the plants grew through the applied antifeedant, the insects would feed even more avidly on the growing tips. Thus, when the insect growth regulant teflubenzuron was simultaneously applied to the lower parts of the plants, the population of insects was reduced within 24 h to less than 1% and the top parts of the plants, where the flowers and seed would subsequently develop, were completely protected. Currently, this approach is being applied to oilseed rape, *Brassica napus*, with the insect growth regulant replaced by various species of fungal pathogen active against Chrysomelidae and other coleopterous pests (Pickett *et al.*, in press).

A new range of antifeedants is now being developed comprising the β-acids or lupulones (I) of the hop, *Humulus lupulus*. These compounds are produced as waste

I

products during hop extraction as they have no role in the brewing process. Nonetheless, they show strong antifeedant activity against a range of insect pests, including mites (Sopp *et al.*, 1990), and are being used to protect the very crop from which they are extracted (Jones *et al.*, submitted).

The push-pull or stimulo-deterrent diversionary strategy

It has been demonstrated that, for best use of semiochemicals, certainly against Northern European arable crop pests, more sophisticated regimes combining semiochemicals with population-reducing components should be adopted. Such approaches are encapsulated in the push-pull or stimulo-deterrent diversionary strategy (SDDS) (Pyke *et al.*, 1987; Miller and Cowles, 1990) (Figure 1), in which semiochemicals are deployed to "push" colonising insects away from the harvestable crop and also to attract predators or parasitoids into the area. At the same time, the pests are aggregated on a sacrificial or trap crop so that a selective control agent, e.g. a fungal pathogen, can be used directly to reduce the pest population.

"PUSH" (away from the crop)	"PULL" (into traps or trap crops)
Kairomone inhibition	Kairomones
Repellents, antifeedants, oviposition deterrents	Aggregation, sex and oviposition pheromones
	Visual cues
Attractants for parasitoids and predators	Selective control agents (e.g. pathogens)

Figure 1. The push-pull or stimulo-deterrent diversionary strategy (SDDS)

The model chosen for initial demonstration of the SDDS was the pea and bean weevil, *Sitona lineatus*, on field beans, *Vicia faba*. The "pull" component of the SDDS was the aggregation pheromone of *S. lineatus*, identified previously as a simple 1,3-diketone (II). The activity of this compound as an attractant was shown to be enhanced

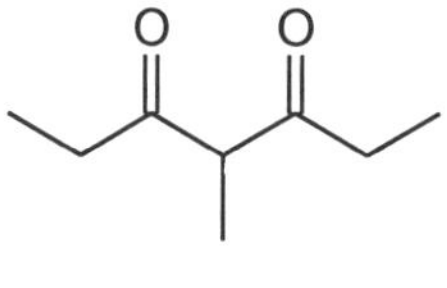

II

by plant components including (Z)-3-hexen-1-ol, (Z)-3-hexen-1-yl acetate and (R + S)-linalool. The identification of these compounds, particularly the aggregation pheromone, relied heavily on electrophysiological preparations from the antennae of *S. lineatus*, both the electroantennogram (EAG) and single-cell recordings (SCR), directly coupled with high resolution gas chromatography (GC) (Blight *et al.*, 1984, 1991). Although the plant components identified as synergising the attractiveness of the aggregation pheromone are ubiquitous in the plant kingdom, the insect nonetheless employs highly specific receptors for their detection. For example, the olfactory cells specifically responding to (Z)-3-hexen-1-ol are relatively insensitive to (Z)-3-hexen-1-yl acetate and (R + S)-linalool. These compounds together, or the aggregation pheromone alone, can be used as lures in yellow-coloured traps similar to those employed in cotton against the boll weevil, *Anthonomus grandis* (Blight *et al.*, 1991).

Traps baited with the aggregation pheromone should be commercially available for monitoring of *S. lineatus* within the near future.

The "push" component of the SDDS against *S. lineatus* involved use of a commercially available antifeedant based on an extract of the Indian neem tree, *Azadirachta indica*. Although there are many claims for the effectiveness of neem extracts in the general scientific literature, in arable agriculture such materials do not compare favourably with conventional pesticides. However, against *S. lineatus*, sufficient antifeedancy was observed to allow further investigation within the SDDS. Thus, in field trials comprising the "push" and "pull" components described, significantly fewer *S. lineatus* were found on the "push" plots and more on the "pull" plots relative to untreated. Although the "push" plots were insufficiently well protected for high input agriculture, these trials nonetheless demonstrated the principle of the SDDS (Smart *et al.*, 1994).

SDDS components of aphid control

The recent identification of aphid sex pheromones (Dawson *et al.*, 1987, 1990; Guldemond *et al.*, 1993) allowed the demonstration that these insects could make oriented flight to a distant source of semiochemical (Campbell *et al.*, 1990). This stimulated the search for plant-derived semiochemicals that might be involved in long-range selection of host plants.

The primary rhinaria on the fifth and sixth segments of the aphid antenna are implicated in detection of host plant chemicals. Single-cell recordings from olfactory nerve cells within these organs, made using electrolytically sharpened tungsten electrodes, allowed identification of a range of host plant attractants for these pests (Wadhams, 1990; Nottingham *et al.*, 1991). However, certain cells appeared to have no function in interactions with host plants, nor did they seem to be involved in insect/insect communication. Other groups working on larger insects have also noticed such apparently redundant cells. By investigating a number of non-host plants, it appeared that such cells did in fact have a role in detecting chemicals typical of plants upon which the aphid could not feed. Again using coupled GC-SCR, accomplished for the first time with aphids in connection with this work, it was possible to identify compounds from non-host plants to which the apparently redundant cells responded. Thus, the black bean aphid, *Aphis fabae*, which feeds on many plants but seldom on members of the Cruciferae (= Brassicaceae), detects specific isothiocyanates which are typical of these plants. It was shown that such compounds act as repellents for this aphid and also as masking agents for the normal attractancy of bean volatiles (Nottingham *et al.*, 1991). Similarly, when members of the Labiate family (= Lamiaceae) were investigated, other compounds having a similar role were identified, including the monoterpene oxidation product (-)-(1*R*,5*S*)-myrtenal, which again significantly reduced attractiveness of host plant volatiles (Hardie *et al.*, 1994a).

R. padi colonises *Prunus padus*, the bird-cherry, as its primary host for sexual reproduction in the autumn. However, in the spring, it must migrate to the summer or secondary host, cereal crops. One of the compounds shown to be highly active in GC-SCR work, using volatiles from the primary host, was methyl salicylate, which therefore became a candidate for repellent activity against the spring migratory morphs. In 1992, in field work conducted on barley in Sweden, over 50% reduction in population of *R. padi* was obtained using methyl salicylate, either released from an emulsifiable concentrate sprayed onto the crop or using slow-release vials (Pettersson *et al.*, 1994). Subsequently, high repellent activity was found with other species of cereal aphids, including the grain aphid, *Sitobion avenae*. Although *S. avenae* does not normally host-alternate, its employment of methyl salicylate as a repellent may indicate the role of this compound as a plant stress signal because of its relationship with the damage-inducible phenylalanine ammonia lyase pathway (Ward *et al.*, 1991). Thus, in 1993 and 1994, approximately 50% reduction of cereal aphid population was again achieved with release rates of 1-5 mg/plot/day, with plot sizes of 10 m^2 (unpublished results).

Parasitoid attraction in an aphid SDDS

Another important component of the SDDS against aphids is the attraction of parasitoids into the crop at an early stage of population development. This can be achieved by use of synomones, released by plants on feeding damage, which attract aphid parasitoids and stimulate foraging behaviour. Such synomones represent a learned response and it is not always possible to attract parasitoids into crops when they have been foraging on non-crops with non-pest aphids as a host source. However, GC-coupled electrophysiological studies, performed for the first time on braconid wasps such as the general aphid parasitoid *Praon volucre*, have shown that components of aphid sex pheromones, e.g. the nepetalactone isomer III, can act as potent attractants

III   IV

(kairomones) in an unlearned situation (Hardie *et al.*, 1993, 1994b). In field trials using potted plants, a four-fold increase in attack and egg-laying by *P. volucre* on *S. avenae* was obtained with this compound. For the more specific parasitoid *Aphidius ervi*, an important control agent for the pea aphid, *Acyrthosiphon pisum*, no attraction or increase in parasitism was observed in these trials. However, the sex pheromone of *A. pisum* comprises largely the nepetalactol isomer IV and when this compound was employed in field pot trials, there was a three-fold increase in parasitism (Pickett *et al.*, 1994). The exploitation of natural populations of aphid parasitoids in reducing aphid populations on a number of crops is now being investigated with combined Levy Board funding.

SDDS and sustainability

It has been shown that the SDDS comprises a number of components affecting different aspects of the behaviour and development of pests. Although each component, when compared to conventional broad-spectrum toxicants, is relatively ineffective, e.g. the plant-derived aphid repellents which reduce populations by only 50%, this has the advantage of not selecting efficiently for resistance and thus contributes to the sustainability of the SDDS.

Sustainability and transgenic crop plants

The defence chemistry of many modern crops is relatively inefficient, largely because this has been removed by long-term plant breeding programmes in the interests of high yield and nutritional value for human consumers. With recombinant DNA technology, it is now feasible to produce metabolites of value in crop protection within the parts of the plant that are not consumed (Hallahan *et al.*, 1992). Sustainability is again likely to be greater where the targets involve semiochemicals because, even if constitutively expressed, these agents would not select strongly for resistance as do the potent biological toxins, such as that from *Bacillus thuringiensis*, currently under commercial development. Two general strategies can be adopted: one is to modify existing secondary metabolism pathways for defence by altering the level of expression of endogenous genes. The alternative is to insert alien genes, preferably from other higher plants and particularly wild species, that have retained defences based on secondary metabolism, so as to augment an existing biosynthetic route. Such approaches are being adopted for oilseed rape (Pickett *et al.*, 1995), where the objective is to reduce the production of glucosinolate precursors for specific pest attractants and, at the same time, to produce highly attractive cultivars for use as trap crops. Such crops could also have industrial value in terms of the oil produced. With regard to the insertion of alien genes from higher plants, a number of terpenoids have been targeted, particularly components produced by plants in the *Nepeta* genus, including compounds III-IV (Hallahan *et al.*, 1995), which are related to the aphid sex pheromones. In the long term, it is intended to investigate modification and augmentation of phenolic production related to the phenylalanine ammonia lyase pathway, which would involve target semiochemicals including the methyl salicylate discussed earlier. Such work has a major world significance in that compounds, e.g. veratrole (1,2-dimethoxybenzene), found to be highly active to the rice pest *Nilaparvata lugens*, the brown planthopper, are also produced via an extension of this particular pathway (Cocking *et al.*, 1994).

ACKNOWLEDGEMENTS

This work was in part supported by the United Kingdom Ministry of Agriculture, Fisheries and Food.

IACR receives grant-aided support from the Biotechnology and Biological Sciences Research Council of the United Kingdom.

REFERENCES

Asakawa, Y.; Dawson, G.W.; Griffiths, D.C.; Lallemand, J-Y.; Ley, S.V.; Mori, K.; Mudd, A.; Pezechk-Leclaire, M.; Pickett, J.A.; Watanabe, H.; Woodcock, C.M.;

Zhang, Z-n. (1988) Activity of drimane antifeedants and related compounds against aphids, and comparative biological effects and chemical reactivity of (-)- and (+)- polygodial. *Journal of Chemical Ecology* 14, 1845-1855.

Blight, M.M.; Pickett, J.A.; Smith, M.C.; Wadhams, L.J. (1984) An aggregation pheromone of *Sitona lineatus*. *Naturwissenschaften* 71, S.480.

Blight, M.M.; Dawson, G.W.; Pickett, J.A.; Wadhams, L.J. (1991) The identification and biological activity of the aggregation pheromone of *Sitona lineatus*. *Aspects of Applied Biology* 27, 137-142.

Campbell, C.A.M.; Dawson, G.W.; Griffiths, D.C.; Pettersson, J.; Pickett, J.A.; Wadhams, L.J.; Woodcock, C.M. (1990) Sex attractant pheromone of damson-hop aphid *Phorodon humuli* (Homoptera, Aphididae). *Journal of Chemical Ecology*, 16, 3455-3465.

Cocking, E.C.; Blackhall, N.W.; Brar, D.S.; Davey, M.R.; Khush, G.S.; Ladha, J.K.; Pickett, J.A.; Power, J.B.; Shewry, P.R. (1994) Biotechnological approaches to rice genetic improvement. *Proceedings, Food Security in Asia, The Royal Society, London, 1st November, 1994*, pp. 23-26.

Dawson, G.W.; Griffiths, D.C.; Janes, N.F.; Mudd, A.; Pickett, J.A.; Wadhams, L.J.; Woodcock, C.M. (1987) Identification of an aphid sex pheromone. *Nature*, 325, 614-616.

Dawson, G.W.; Griffiths, D.C.; Merritt, L.A.; Mudd, A.; Pickett, J.A.; Wadhams, L.J.; Woodcock, C.M. (1990) Aphid semiochemicals - a review, and recent advances on the sex pheromone. *Journal of Chemical Ecology*, 16, 3019-3030.

Griffiths, D.C.; Hassanali, A.; Merritt, L.A.; Mudd, A.; Pickett, J.A.; Shah, S.J.; Smart, L.E.; Wadhams, L.J.; Woodcock, C.M. (1988) Highly active antifeedants against coleopteran pests. *Proceedings of the Brighton Crop Protection Conference - Pests and Diseases*, 1041-1046.

Griffiths, D.C.; Pickett, J.A.; Smart. L.E.; Woodcock, C.M. (1989) Use of insect antifeedants against aphid vectors of plant virus disease. *Pesticide Science* 27, 269-276.

Griffiths, D.C.; Maniar, S.P.; Merritt, L.A.; Mudd, A.; Pickett, J.A.; Pye, B.J.; Smart, L.E.; Wadhams, L.J. (1991) Laboratory evaluation of pest management strategies combining antifeedants with insect growth regulator insecticides. *Crop Protection*, 10, 145-151.

Guldemond, J.A.; Dixon, A.F.G.; Pickett, J.A.; Wadhams, L.J.; Woodcock, C.M. (1993) Specificity of sex pheromones, the role of host plant odour in the olfactory attraction of males, and mate recognition in the aphid *Cryptomyzus*. *Physiological Entomology*, 18, 137-143.

Hallahan, D.L.; Pickett, J.A.; Wadhams, L.J.; Wallsgrove, R.M.; Woodcock, C.M. (1992) Potential of secondary metabolites in genetic engineering of crops for resistance. In: *Plant Genetic Manipulation for Crop Protection*, A.M.R. Gatehouse, V.A. Hilder and D. Boulter (Eds.), Wallingford: C.A.B. International, pp. 215-248.

Hallahan, D.L.; West, J.M.; Wallsgrove, R.M.; Smiley, D.W.M.; Dawson, G.W.; Pickett, J.A.; Hamilton, J.G.C. (1995) Purification and characterization of an acyclic monoterpene primary alcohol:NADP$^+$ oxidoreductase from catmint (*Nepeta racemosa*). *Archives of Biochemistry and Biophysics* 318, 105-112.

Hardie, J.; Isaacs, R.; Nazzi, F.; Powell, W.; Wadhams, L.J.; Woodcock, C.M. (1993) Electroantennogram and olfactometer responses of aphid parasitoids to nepetalactone, a component of aphid sex pheromones. Behavioural ecology, augmentation and enhancement of aphidophaga. *Proceedings, 5th International Symposium of the Global IOBC Working Group: Ecology of Aphidophaga.*

Hardie, J.; Isaacs, R.; Pickett, J.A.; Wadhams, L.J.; Woodcock, C.M. (1994a) Methyl salicylate and (-)-(1R,5S)-myrtenal are plant-derived repellents for black bean aphid, *Aphis fabae* Scop. (Homoptera: Aphididae). *Journal of Chemical Ecology* **20**, 2847-2855.

Hardie, J.; Hick, A.J.; Höller, C.; Mann, J.; Merritt, L.A.; Nottingham, S.F.; Powell, W.; Wadhams, L.J.; Witthinrich, J.; Wright, A.F. (1994b) The responses of *Praon* spp. parasitoids to aphid sex pheromone components in the field. *Entomologia Experimentalis et Applicata*, **71**, 95-99.

Hollinshead, D.M.; Howell, S.C.; Ley, S.V.; Mahon, M.; Ratcliffe, N.M.; Worthington, P.A. (1983) The Diels-Alder route to drimane related sesquiterpenes; synthesis of cinnamolide, polygodial, isodrimeninol, drimenin and warburganal. *Journal of the Chemical Society, Perkin Transactions 1*, 1579-1589.

Howse, P.; Stevens, I.; Jones, O. (1995) *Insect Pheromones and Their Use in Pest Management*, London: Chapman & Hall, 256 pp.

Jones, G.; Campbell, C.A.M.; Pye, B.J.; Maniar, S.P., Mudd, A. Repellent and oviposition-deterring effects of hop beta-acids on the two-spotted spider mite *Tetranychus urticae*. *Pesticide Science* (submitted).

Miller, J.R.; Cowles, R.S. (1990) Stimulo-deterrent diversionary cropping: a concept and its possible application to onion maggot control. *Symposium on Semiochemicals and Pest Control*, Wageningen, October 1989. *Journal of Chemical Ecology*, **16**, 3197-3212.

Nottingham, S.F.; Hardie, J.; Dawson, G.W.; Hick, A.J.; Pickett, J.A.; Wadhams, L.J.; Woodcock, C.M. (1991) Behavioral and electrophysiological responses of aphids to host and nonhost plant volatiles. *Journal of Chemical Ecology* **17**, 1231-1242.

Pettersson, J.; Pickett, J.A.; Pye, B.J.; Quiroz, A.; Smart, L.E.; Wadhams, L.J.; Woodcock, C.M. (1994) Winter host component reduces colonization by the bird-cherry-oat aphid, *Rhopalosiphum padi* (L.) (Homoptera, Aphididae), and other aphids in cereal fields. *Journal of Chemical Ecology*, **20**, 2565-2574.

Pickett, J.A.; Cayley, G.R.; Dawson, G.W.; Griffiths, D.C.; Hockland, S.H.; Marples, B.; Plumb, R.T.; Woodcock, C.M. (1986) Use of the alarm pheromone and derivatives against aphid–mediated damage. *Abstracts 6th International Congress Pesticide Chemistry, IUPAC, Ottawa, 1986*, 2C–08.

Pickett, J.A.; Dawson, G.W.; Griffiths, D.C.; Hassanali, A.; Merritt, L.A.; Mudd, A.; Smith, M.C.; Wadhams, L.J.; Woodcock, C.M.; Zhang, Z-n. (1987) Development of plant–derived antifeedants for crop protection. In: *Pesticide Science and Biotechnology*, R. Greenhalgh and T.R. Roberts (Eds.), Blackwell Scientific Publications, pp. 125–128.

Pickett, J.A.; Wadhams, L.J.; Woodcock, C.M. (1994) Attempts to control aphid pests by integrated use of semiochemicals. *Brighton Crop Protection Conference – Pests and Diseases -1994*, 1239-1246.

Pickett, J.A.; Butt, T.M.; Doughty, K.J.; Wallsgrove, R.M.; Williams, I.H. Minimising pesticide input in oilseed rape by exploiting natural regulatory processes. *Proceedings of the 9th International Rapeseed Congress 1995, Cambridge, UK*, in press.

Pyke, B.; Rice, M.; Sabine, B.; Zalucki, M. (1987) The push-pull strategy - behavioural control of *Heliothis*. *Australian Cotton Grower*, May-July 1987, 7-9.

Smart, L.E.; Blight, M.M.; Pickett, J.A.; Pye, B.J. (1994) Development of field strategies incorporating semiochemicals for the control of the pea and bean weevil, *Sitona lineatus* L. *Crop Protection* 13, 127-135.

Sopp, P.I.; Palmer A.; Pickett, J.A. (1990) The effect of a plant-derived antifeedant on *Tetranychus urticae* and *Phytoseiulus persimilis*: "A first look". *SROP/WPRS Bulletin XIII/5 (1990)*, 198-201.

Wadhams, L.J. (1990) The use of coupled gas chromatography: electrophysiological techniques in the identification of insect pheromones. In: *Chromatography and Isolation of Insect Hormones and Pheromones*, A.R. McCaffery and I.D. Wilson (Eds), New York: Plenum, pp. 289-298.

Ward, E.R.; Uknes, S.J.; Williams, S.C.; Dincher, S.S.; Wiederhol, D.L.; Alexander, D.C.; Ahl-Goy, P.; Métraux, J.-P.; Ryals, J.A. (1991) Coordinate gene activity in response to agents that induce systemic acquired resistance. *Plant Cell* 3, 1085-1094.

THE EFFECT OF FOLIAR APPLIED POTASSIUM CHLORIDE ON *ERYSIPHE GRAMINIS* INFECTING WHEAT.

J.W. COOK, P.S. KETTLEWELL, D.W. PARRY

Crop and Environment Research Centre, Harper Adams Agricultural College, Newport, Shropshire. TF10 8NB

ABSTRACT

Potassium chloride fertiliser applied as a foliar spray in field and glasshouse experiments significantly reduced the percentage leaf area of wheat affected by powdery mildew (*Erysiphe graminis*) compared to an application of fertiliser to the soil. The response was dependent upon the concentration of the solution applied. The optimum concentration in glasshouse conditions was approximately 10% w/v. The foliar applied potassium chloride reduced the germination of powdery mildew spores on the leaf surface and also inhibited the establishment of infections. The inhibition of infection was associated with increases in the leaf water potential. Experiments using polyethylene glycol to produce biologically inert solutions of equivalent osmotic potential to potassium chloride revealed that both the inhibition of germination and the reduction of the disease symptoms may be due to the physico-chemical properties of the fertiliser rather than metabolic toxicity or nutritional effects on the host.

INTRODUCTION.

Work carried out in the late 1980's to investigate the response of cereal yields to foliar application of potassium chloride revealed that these applications could also reduce the leaf area affected by *Septoria* spp. and powdery mildew (Kettlewell *et al.*, 1990). It was decided that foliar applied potassium chloride may be a potentially inexpensive disease control agent. If the potassium chloride fertiliser applied to the crop could also reduce the severity of disease to below the threshold at which conventional fungicide control methods would be used it could potentially reduce the total resources used in crop production with economic and environmental benefits. The object of this work was to evaluate foliar applied potassium chloride for the control of powdery mildew and investigate the mode of activity.

FIELD WORK.

    Initially field experiments were carried out, on wheat,
to evaluate the effect of potassium chloride fertilisers on
a range of pathogens by sequential applications throughout
the season.   Potassium chloride was applied at different
timings throughout the growing season to identify pathogens
against which foliar applied potassium chloride was
effective.   Potassium chloride fertiliser was applied as
soil dressings or foliar sprays to investigate whether the
control was due to the nutritional properties of the
fertiliser or some factor intrinsic to the foliar
application.

    The field experiments were conducted at sites in
Shropshire during 1992 on cultivar Mercia and 1993 on
cultivars Riband and Apollo.   The sites had soil potassium
indices of one and zero respectively.   The experiments were
factorial designs with three factors which were growth
stages at application of 32, 39 and 51 each with two levels
which were potassium chloride applied as a foliar spray or a
solid powder.   Each level comprised 33.3 kg/ha potassium
chloride applied as a solid powder or a foliar spray of
15.136% w/v solution at 220 l/ha.   For comparative purposes
two treatments were included outside the factorial design.
These were standard practice which comprised 100 kg/ha soil
applied potassium chloride at growth stage 32 and standard
practice plus a three spray prophylactic fungicide
programme.   The experiments were assessed by visual
observation weekly after treatment for three weeks using
A.D.A.S. disease charts.

    There were no visible disease symptoms before
treatments were applied in either year.   The application
of potassium chloride as a foliar spray at growth stage 32
consistently reduced the leaf area affected by powdery
mildew compared to the solid fertiliser and standard
practice treatments for three weeks after treatment in both
years.   This is illustrated for 1992 (fig.1).   In both
years the leaf area affected by powdery mildew was not
significantly different between the foliar treatment and the
standard practice plus fungicides.   The leaf area of the
flag leaf affected by powdery mildew was very small in both
years .   There was no interaction between cultivar and
fertiliser application method.

GLASSHOUSE AND LABORATORY STUDIES.

    Having established the effectiveness of foliar applied
potassium chloride as an agent for the control of powdery
mildew on wheat, glasshouse and laboratory work was
conducted to investigate its mode of action.

Figure 1.   The effect of potassium chloride
applied to the foliage or soil at growth stage 32
on the percentage leaf area of winter wheat (cv.
Mercia) affected by powdery mildew in 1992.

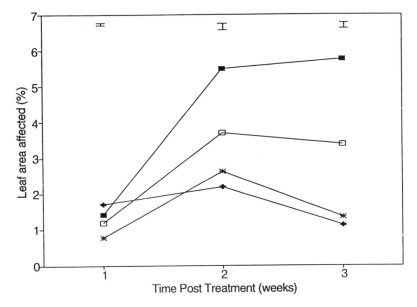

─■─ 33.3 kg/ha soli ─◆─ 220 l/ha 15.1% ─*─ Std. practice + ─▣─ Std. practice

Rate responses.

    Three initial experiments examined the effect of
different concentrations of potassium chloride solution
applied to the leaves of unvernalised winter wheat plants in
order to establish the optimum concentration for the control
of powdery mildew.

    The plants were sprayed with a precision pot sprayer at
a rate equivalent to 200 l/ha and inoculated by shaking pots
of artificially inoculated seedlings over the plants.

    In three experiments there was a linear decline in the
leaf area affected by mildew up to 10% w/v (fig.2).   Above
this concentration the leaf area affected increased.   This
increase probably reflected a decline in plant resistance to
the pathogen due to osmotic stress.

Figure 2. The effect of increasing the
concentration of potassium chloride solution on
percentage leaf area of wheat affected by powdery
mildew.

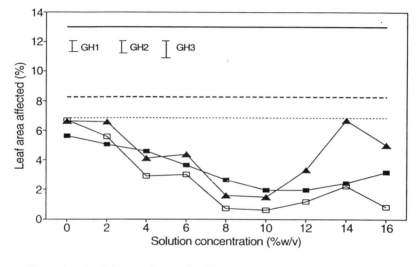

- Exp. 1 (cv.Apollo) — Exp. 2 (cv.Riband) — Exp. 3 (cv.Apollo)
- Control Exp. 1   — Control Exp. 2   --- Control Exp. 3

Timing Responses.

An experiment was conducted in which unvernalised wheat
plants (cv. Apollo) were treated with a foliar spray of
water at 200 l/ha or a 10% w/v potassium chloride solution
at 200 l/ha either as a foliar spray or a soil applied
solution. The treatments were applied at four timings which
were seven or three days before or after inoculation with *E.
graminis*).

The foliar applied potassium chloride resulted in a
lower percentage area of the wheat leaves being
affected by powdery mildew than soil applied potassium
chloride or water applied to the leaves. This indicated
that the control of powdery mildew achieved by foliar
applied potassium chloride was due to some intrinsic factor
concerning the solution and application to the leaf rather
than a property of the potassium chloride itself. There was
no significant difference between the percentage area of the
leaves affected by powdery mildew with regard to timing of
application. This indicated that the foliar applied
potassium chloride had both protective and curative
properties if applied within seven days of inoculation.

Spore germination *in vitro.*

In view of the linear relationship between percentage leaf area affected by powdery mildew and increasing concentration of the fertiliser solution it was proposed that potassium chloride solution might act on the spores of *E. graminis* by creating an osmotic potential and causing an eflux of water from the spores so reducing their viability rather than direct metabolic toxicity.

An experiment was carried out to examine the effect of solution osmotic potential on spore viability *in vitro.* Solutions of potassium chloride or polyethylene glycol with osmotic potentials of 0, 12.8, 25.6, 38.4, 51.3 and 64.1 bar were created. These were equivalent to potassium chloride solutions of 0,2,4,6,8, and 10% w/v. Polyethylene glycol was used as a control for the potassium chloride because it is a biologically inert osmotica. Spores were incubated in these solutions on glass slides. A minimum of thirty spores per slide were monitored at each assessment. The plasmolysis of spores and the production of germ tubes were recorded.

The spores did not collapse in distilled water but retained their integrity as described elsewhere (Corner, 1935; Manners and Housain, 1963). A proportion also retained the viability to produce germ tubes. All patterns in the response to osmotic potential were essentially evident after six hours. Increasing solution osmotic potential resulted in a negative linear response in spore germination (P=0.001) and a linear increase in plasmolysis (P=0.001). No consistent differences were detected between the effects of the two osmotica. Overall it appeared that the critical factor affecting spore germination was the osmotic potential of the solutions rather than the osmotica.

Spore Germination *in vitro.*

Since the wax cuticle interacts with surface solutions and spray deposition is uneven across the leaf it was difficult to draw direct comparisons between solutions of the same concentration on the leaf surface and *in vitro* results as the conditions are so different. Therefore an experiment was conducted *in vivo* examining the effect of solution osmotic potential on the germination of *E. graminis* spores.

The experiment was conducted using unvernalised wheat plants (cv. Apollo) with three fully expanded leaves. The plants were grown in a low potassium compost in a spore free propagator with an ambient air source. The plants were inoculated and sprayed with polyethylene glycol and potassium chloride solutions as detailed for the *in vitro* experiment at a rate equivalent to 200 l/ha. After twenty

four hours the upper fully expanded leaf of certain plants
were detached and cleared with a 3:1 mixture of acetic acid
and ethanol followed by rehydration with an alcohol series.
The leaves were stained with trypan blue and observed. The
remaining plants were visually assessed for percentage area
of the upper leaf affected by powdery mildew after two
weeks.

The area of the upper fully expanded leaf affected by
E. graminis declined in a linear manner (P=0.001) as the
osmotic potential of the solutions increased with no
significant difference between the two osmotica.
Microscopic examination revealed that the percentage
germination of the spores declined in a linear manner with
increasing osmotic potential of the solution applied
(P=0.001) with no difference between osmotica. However
virtually all spores which produced germ tubes continued to
develop and produce infection.

The means of percentage leaf area affected and spore
germination were expressed as percentages of their controls
and plotted together (fig.3).   A parallel regression was
carried out revealing that both data sets had similar
slopes.

FIGURE 3.   The reduction in spore germination
and percentage area affected by *E. graminis in
vivo* expressed as a percentage of control.

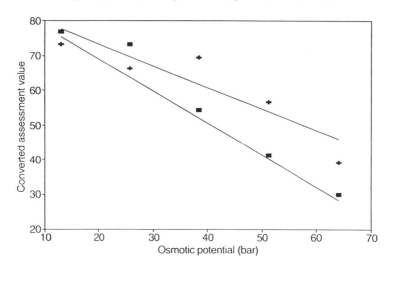

Leaf Water Potential.

Percentage germination was much higher than the equivalent percentage leaf area affected. This suggested that some other factor may have an influence upon leaf area affected by interfering with the establishment of the pathogen. Preliminary experiments had indicated that foliar applied potassium chloride was rapidly taken up by the leaves of wheat plants and that it also resulted in an increase in the leaf water potential. This elevation of the leaf water potential lasted over thirteen days and therefore such an effect could conceivably have been responsible for the control of powdery mildew observed when potassium chloride was applied seven days before inoculation.

An experiment was conducted to investigate the relationship between the osmotic potential of a foliar applied solution, the leaf water potential and the percentage leaf area affected. Unvernalised wheat (cv. Apollo) plants with four fully expanded leaves were treated with either polyethylene glycol or potassium chloride solutions and inoculated with spores from stock pots of the same wheat cultivar. The solutions had a range of osmotic potentials. Two plants per block received each treatment. One plant was destructively sampled after twenty four hours. The upper fully expanded leaf was detached and the leaf water potential determined using a pressure bomb. The other plant was visually assessed for the percentage area of the upper fully expanded leaf affected with powdery mildew after fourteen days. The leaf water potential was positively correlated with the osmotic potential of the solution applied. The leaf area affected was negatively correlated with the solution osmotic potential. This suggested that the leaf water potential increase could be responsible for the reduction in disease symptoms. Linear modelling revealed a possible correlation between leaf water potential and the leaf area affected after the removal of block effects. However the relationship was different with respect to the two osmotica. It was thought that the effect of solution osmotic potential on spore germination was obscuring the responses to changes in leaf water potential. Therefore the experiment was repeated using three plants per treatment per block. The upper fully expanded leaf of the third plant was detached, cleared and the percentage germination of the spores determined. A multiple regression model was then used to test the correlation of the leaf water potential and the percentage leaf area affected by powdery mildew after accounting for the differences in spore germination. This experiment confirmed the previous experient.

Microscopic examination of leaves four days after inoculation revealed that very few germinated spores developed haustoria on leaves treated with a foliar applied potassium chloride or polyethylene glycol of 64 bar osmotic potential. Development was normal on untreated leaves and those sprayed with water.

CONCLUSION.

The application of potassium chloride fertiliser as a foliar spray may be an effective method for the control of powdery mildew disease on wheat. The effect could be the result of the inhibition of spore germination on the leaf surface and partly by induced changes in the leaf physiology. The most likely change to affect pathogen establishment is a rise in the leaf water potential following application. The effect appears to be non nutritional and purely due to the physico-chemical properties of the solution. This suggests that other osmotically active fertiliser solutions could have a similar effect.

The control of powdery mildew by foliar applied potassium chloride was comparable with that provided by conventional fungicides. This suggests that with further development alternative fertiliser strategies using foliar potassium chloride could be developed leading to a reduction in fungicide usage. This would considerably reduce the costs of cereal production and the use of non renewable resources in agriculture leading to more sustainable crop production.

ACKNOWLEDGEMENTS

We thank The John Oldacre Foundation for providing financial support for this work.

REFERENCES

Corner, E.J.H. (1935). Observations on resistance to powdery mildews. *New Phytologist*, **34**, 180-200.

Kettlewell, P.S., Bayley, G.L., and Domleo, R., (1990). Evaluation of late season foliar application of potassium chloride for disease control in winter wheat. *Journal of Fertiliser Issues.* **7** (1), 17-23.

Manners, J.G. and Housain, S.M.M. (1963). Effects of temperature and humidity on conidial germination in *Erysiphe graminis. Transactions of the British Mycological Society.* **46**(2), 225-234.

OBSERVATIONS ON AZADIRACHTIN FOR THE MANAGEMENT OF CABBAGE
CATERPILLAR INFESTATIONS IN THE FIELD

A. JENNIFER MORDUE (LUNTZ), G. DAVIDSON

Department of Zoology, University of Aberdeen, Tillydrone Avenue, Aberdeen
AB9 2TN;

R.G. MCKINLAY, J. HUGHES
Scottish Agricultural College, West Mains Road, Edinburgh EH9 3JG.

ABSTRACT

Small scale field trials have been carried out involving an assessment of crop damage and
insect toxicity at the individual level. Azadirachtin and Neemazal-F were demonstrated to
give protection to brassicas against lepidopteran pests. No differences were found
between azadirachtin and Neemazal-F preparations at the same concentration and dose-
rate. Crop protection was demonstrated for *Mamestra brassicae*, *Pieris rapae* and
Plutella xylostella, however the caterpillars did not die immediately and remained on the
plants feeding very little or not at all. At the spray regime of 50 ppm at 400 l/ha
caterpillars died either before pupation or pupated after a long delay. Of those which did
pupate the majority died during diapause and before eclosion.

INTRODUCTION

Azadirachtin, from the neem tree *Azadirachta indica*, is a limonoid with marked antifeedant,
insect growth regulatory and reproductive effects against insects (Schmutterer, 1990; Mordue
(Luntz) & Blackwell, 1993). The feeding deterrency and growth disruptive effects of
azadirachtin have been well described for numerous species and stages of insects of many
orders (e.g. Mordue (Luntz) *et al.* 1985, 1986; Blaney *et al.*, 1990) and recent advances have
been made in the field using both commercial and semi-commercial preparations of neem
(Schmutterer, 1988; Locke & Lawson, 1990; Isman *et al.*, 1991).

Research into the mode of action and pest control potential of azadirachtin has been ongoing
for some 30 years. However, whilst the effects of azadirachtin have been well documented in
the laboratory, most field work has been carried out on a large scale using neem formulations
rather than pure compound. In an effort to bridge the gap between laboratory and fieldwork
small field experiments were set up to (i) compare the effectiveness of pure azadirachtin with a
neem based formulation, Neemazal-F, (Trifolio-M, GmbH Lahnau, Germany) (ii) to measure
crop damage at the individual plant level and (iii) to observe the feeding behaviour, growth and
development of individual insects from the treated crop. To achieve these aims natural cabbage
caterpillar infestations of brassicas were studied in (i) control ie. untreated, (ii) azadirachtin-,
(iii) Neemazal-F- and (iv) cypermethrin-treated plots, the latter being a standard treatment for
the control of lepidopteran pests of brassicas. Both crop damage and insect toxicity were
observed at the individual level.

Table 1 Site details and design of field experiments to assess the biological efficacy of the azadirachtin treatments in Midlothian, Scotland, 1991 and 1992.

Site details/ experimental design	1991	1992
Site details		
Name	Goshen, Musselburgh	Bush Estate, Penicuick
British National Grid Ref.	NT367729	NT244638
Soil series	Dreghorn	Darvel
Soil type	Freely drained brown calc. and brown forest soil	Freely drained brown calc. and brown forest soil
Height above sea level	20m	190m
Aspect	Level	Level
Experimental design		

Treatments	1. Untreated
	   Solvent (2% ethanol) +
	   wetter (0.1% Triton-X-100)
	   in water at 400 l/ha
	2. Cypermethrin
	   'Ambush C': ICI
	   Agrochem.
	   25g ai/ha in 400 l water
	3. Azadirachtin
	   50 ppm at 233 l/ha
	   (12g ai/ha)
	4. 250 ppm at 233 l/ha
	   (58g ai/ha)
	5. 500 ppm at 233 l/ha
	   (117g ai/ha)
	
	6. 750 ppm at 233 l/ha
	   (175g ai/ha)
		1. Untreated
		   Solvent (2% ethanol) +
		   water (0.1% Triton-X-100)
		   in water at 400 l/ha
		2. Cypermethrin
		   'Ambush C': ICI
		   Agrochem.
		   25g ai/ha in 400 l water
		3. Azadirachtin
		   50 ppm at 233 l/ha
		   (12g ai/ha)
		4. 50 ppm at 400 l/ha
		   (20g ai/ha)
		5. Neemazel-F
		   5% azadirachtin
		   50 ppm at 400 l/ha
		   (20g ai/ha)

Layout	Randomised block	Randomised block
Replicates	4 (except azadirachtin treatments which were replicated once owing to scarcity of material)	4
Crop and cultivar	Calabrese (cv Marathon)	Summer cabbage (cv Pedrillo)
Planting date	31 May	15 May
Row width	56cm	56cm
Plant spacing	28cm	30cm
Fertiliser	N 148 kg/ha P_2O_5 74 kg/ha K_2O 74 kg/ha	N 225 kg/ha P_2O_5 75 kg/ha K_2O 175 kg/ha
Plot size	2.24 x 2.8 m	2.4 x 1.68 m

MATERIALS AND METHODS

The biological efficacy of the azadirachtin treatments was evaluated by field experimentation carried out according to EPPO Guidelines (Anon., 1984) in Midlothian, Scotland during July and August, 1991 and 1992. Site and experimental details are given in Table 1.

Using an Azo portable field sprayer, all treatments were applied twice in each year with a two week interval between applications. In both years, heavy rain following the first application rendered all treatments ineffective; the results reported are for the second application only. Plant damage by caterpillars was assessed in a mid-group of 16 (1991) or 8 (1992) plants in every plot by counting the number of holes in all leaves per plant (Chalfant *et al.*, 1979; Workman *et al.*, 1980) and, in 1992, also by a leaf damage index scale of 0-1 varying from least to most damaged plant per sampling occasion.

During 1992, caterpillars of *Mamestra brassicae* (L.) (cabbage moth), *Pieris rapae* (L.) (small white butterfly) and *Plutella xylostella* (L.) (diamond-back moth) were collected from unassessed plants in untreated and treated plots one day after treatment application, kept individually in the laboratory at ambient room temperature (22°C) and humidity and fed on leaves from original host plants. The area of leaves eaten per day was recorded until mortality or pupation occurred. Survival and eclosion of those caterpillars which had pupated was recorded during 1993.

RESULTS

Field Experiments

In both 1991 and 1992, damage to the crop was high in untreated and low in cypermethrin-treated plots. Azadirachtin and Neemazal-F treatments were intermediate in effect and gave relatively good plant protection (Fig. 1A & B). During 1991, there were no apparent differences in effect between 50, 250, 500 and 750 ppm azadirachtin at a spray rate of 233 l water/ha and hence these data were combined (Fig. 1A). Fifty ppm azadirachtin at a spray rate of 233 l/ha gave significantly less protection to the plants than at 400 l/ha by day 4 post treatment (P = 0.05) although protection was not significantly different by day 10 and in both cases was still greater than in controls (P < 0.01 at day 4 and day 10) (Fig. 1B). At a spray rate of 400 l/ha 50 ppm azadirachtin gave the same protection as 50 ppm Neemazal-F (Fig. 1B). There was no evidence of phytotoxicity in any of the treatments.

Assessment of crop damage in 1992 by using both the number of holes per plant and a leaf damage index scale demonstrated that both assessment methods were consistent and reliable in that very similar results were obtained in both cases. The results shown in Fig. 1 related to the increase in the number of holes per plant post-treatment.

Observations on individual larvae

An assessment of caterpillar numbers both before and after spraying in 1991 demonstrated significant differences (P<0.01) between cypermethrin and azadirachtin/Neemazal-F treatments. Caterpillars were significantly reduced (P < 0.01) in number after cypermethrin treatment (Table 2), however no such drop in numbers was found in azadirachtin treatments when compared with controls (Table 2).

Figure 1

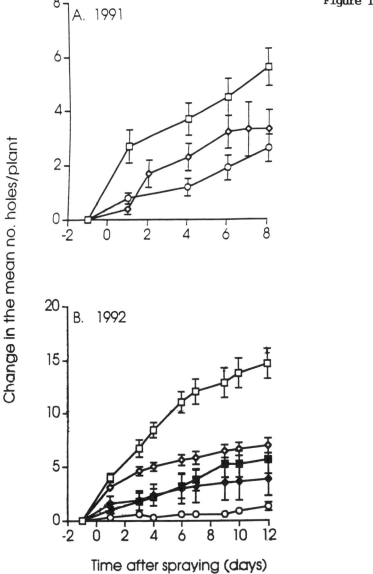

Fig. 1   The increase in damage to brassica crops, expressed as the number of holes per plant,
by natural infestations of cabbage caterpillars after treatment of the plants with
azadirachtin, Neemazal-F or cypermethrin (4 replicates of 16 (1991) or 8 (1992) plants).

A.   1991;   ☐ control; ◊ azadirachtin from 50-750 ppm (at spray regime of 233 l/ha);
O cypermethrin
B.   1992;   ☐ control; ◊ azadirachtin (at 50 ppm and 233 l/ha); ♦ azadirachtin (at 50 ppm
and 400 l/ha); ■ Neemazal-F (at 50 ppm and 400 l/ha); O cypermethrin

Table 2 The number of caterpillars observed per plant on day 8 after treatment of the crop with azadirachtin or cypermethrin on day 0; 1991 data (n=64 control and cypermethrin; n= 16 for azadirachtin)

Treatment (ppm)		Days post treatment		Significance (Student's t test)
		-1	+8	
Control		0.36 ± 0.08	0.41 ± 0.09	NSD
Azadirachtin	50	0.25 ± 0.11	0.30 ± 0.15	NSD
	250	0.12 ± 0.08	0.06 ± 0.06	NSD
	500	0.31 ± 0.15	0.19 ± 0.10	NSD
	750	0.31 ± 0.15	0.19 ± 0.10	NSD
Cypermethrin		0.17 ± 0.05	0.02 ± 0.02	P<0.01

Observations of feeding, growth and moulting were carried out during 1992 on insect larvae collected from crops treated one day previously. Overall numbers of insects available for collection was low, however by grouping together data from the azadirachtin treatments it was possible to gain some insight into the effects on cabbage caterpillar infestations in situations closely akin to that of the field. Experiments using larger numbers of reared insects are presently being carried out to confirm these preliminary results.

Mamestra brassicae
M. brassicae had been treated at the onset of the final instar. The larvae showed the classical effects of azadirachtin treatment; that is, a significantly reduced level of feeding and a delay in the time of pupation (Fig. 2). The average time of pupation was 25 days after spraying for controls and 30 days after spraying for azadirachtin-treated. Few mortalities occurred during the Vth instar and most individuals metamorphosed to diapausing pupae. However, whereas successful emergence occurred in May and June 1993 in 4 out of 5 controls, only 2 of 8 azadirachtin-treated *M. brassicae* emerged. Of these two, one appeared normal and the other died during eclosion in an unexpanded state. In both cases emergence was greatly delayed and occurred in August and September 1993 respectively. The remaining insects died as pharate pupae.

Pieris rapae
It was clear from five *P. rapae* larvae collected that treatment had also occurred towards the beginning of the final instar. The two control insects successfully pupated 20 days post-treatment after consuming an average of $2.9cm^2$ leaf per day, peaking at 9.5 cm^2 at mid-instar on day 8. The three azadirachtin-treated insects showed a strong feeding inhibition with an average of 0.23 ± 0.19 cm^2 leaf eaten per day throughout the period of the final instar. Two of the three treated insects died during the Vth instar whilst the third pupated after a long delay.

Plutella xyllostella
P. xyllostella larvae appeared to demonstrate a different pattern of response which was related to the production of a second summer generation of insects. Treated insects were nearing the end of the last larval instar at the time of treatment and the final stages of feeding and pupation were not affected (n=2). Controls (n=4) and one azadirachtin-treated insect emerged after 11 days, the other treated insect died in the pupal stage.

Figure 2

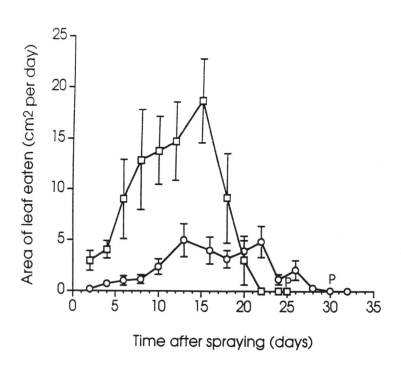

Fig. 2   Area of cabbage leaf eaten (cm²/day) by *Mamestra brassicae* Vth instar larvae taken
from crops sprayed with azadirachtin on day 0(1992). Insects were fed leaves from
their original host plant. (n = 5 for controls; n = 8 for azadirachtin-treated.
□ control; O azadirachtin.

DISCUSSION

In the field experiments, which covered a two year period, both pure azadirachtin and Neemazal-F preparations were shown to achieve good plant protection. At the recommended dose of 50 ppm and a spray regime of 400 l/ha such crop treatments achieved good protection which was almost as great as that achieved by cypermethrin treatment, a standard pyrethroid preparation for the control of lepidopteran pests. Effectiveness was lost in all cases, including cypermethrin, if rain immediately followed spraying supporting the argument for systemic treatment of the crop by azadirachtin. Such treatments have been shown to be effective in cabbage seedlings or leaves against *Pieris brassicae* (Osman & Port, 1990, Arpaia & van Loon, 1993) at doses of less than 10 ppm azadirachtin in the bathing medium (Arpaia & van Loon, 1993). Both azadirachtin and Neemazal-F gave similar results at the same dose, which reflects the importance of azadirachtin content in the overall activity of neem-based preparations (Isman *et al.*, 1990). Also the lack of any protective formulation did not affect the efficacy of azadirachtin under the summer conditions of Midlothian, Scotland, in 1991 and 1992 when conditions were cool and temperate.

Although insect numbers were low for the observational experiments it was still apparent that crop treatments with azadirachtin at 50 ppm were sufficient to give good control of cabbage caterpillars. Further detailed studies of different lepidopterous species in the field are required to reveal any differences in the sensitivity of different species to azadirachtin and the importance of the time of spraying in relation to the insect's life cycle.

Interestingly, in spite of good plant protection in azadirachtin and Neemazal-F treatments, caterpillar numbers remained high. Both the detailed observations of individuals and the assessment of insect numbers in the field emphasised the apparency of caterpillars on the crop for extended periods of time, although they were not actively feeding. Such insects showed the classical azadirachtin poisoning symptoms of reduced feeding, delayed moults and increased mortalities both prior to and during pupation. Such effects together with unsuccessful emergence both that season and the following year suggested a good potential control of future generations of insects. The apparency of the immobilised larvae on the crop must however be taken into account when assessing the acceptability of a neem-based product by farmers.

ACKNOWLEDGEMENTS

The gifts of pure azadirachtin from Professor E.D. Morgan, University of Keele and of Neemazal-F from Trifolio-M are gratefully acknowledged; also Mr J Cheyne for applying treatments in the field.

REFERENCES

Anon 19843. Guideline for the Biological Evaluation of Insecticides No. 83, 'Caterpillars on Leaf Brassicas'. European and Mediterranean Plant Protection Organisation, 1, rue Le Nôtre, Paris, 3 pages.
Arpaia, S. & J.J.A., van Loon, 1993. Effects of azadirachtin after systemic uptake into *Brassicae oleracea* on larvae of *Pieris brassicae*. *Entomologia experimentalis et applicata* **66** 39-45.

Blaney, W.M., M.S.J. Simmonds, S.V. Ley, J.C. Anderson & P.L. Toogood, 1990. Antifeedant effects of azadirachtin and structurally related compounds on lepidopterous larvae. *Entomologia experimentalis et applicata*, **24**, 58-63.

Chalfant, R.G., W.H. Denton, D.J. Schuster, R.B. Workman 1979. Management of Cabbage Caterpillars in Florida and Georgia by using Visual Damage Thresholds. *Journal of Economic Entomology*, **72**, 411-413.

Isman, M.B., O. Koul, A. Luczynski & J. Kaminski, 1990. Insecticidal and antifeedant bioactivities of neem oils and their relationship to azadirachtin content. *Journal of Agriculture and Food Chain*, **38**, 1406-1411.

Isman, M.B., O. Koul, J.T. Arnason, J. Stewart & G.S. Salloum (1991). Developing a neem-based insecticide for Canada. *Memoirs of Entomological Society of Canada*, **159**, 39-47.

Locke, J.C. and R.H. Lawson (Eds) 1990. Neem's potential in pest management programs. Proceedings USDA Neem Workshop. ARS-86.

Mordue (Luntz), A.J., P.K. Cottee, & K.A. Evans, 1985. Azadirachtin: its effects on gut motility, growth and moulting in *Locusta*. *Physiological Entomology*, **10**, 431-437.

Mordue (Luntz), A.J., K.A. Evans & M. Charlet, 1986. Azadirachtin, ecdysteroids and ecdysis in *Locusta migratoria*. *Comparative Biochemistry and Physiology*, **85**, 297-301.

Mordue (Luntz), A.J. & A. Blackwell, 1993. Azadirachtin: an update. *Journal of Insect Physiology* **39**, 903-924.

Osman, M.Z. & G.R. Port, 1990. Systemic action of neem seed substances against *Pieris brassicae*. *Entomologia experimentalis et applicata*, **54**, 297-300.

Schmutterer, H. 1988. Potential of azadirachtin-containing pesticides for integrated pest control in developing and industrialised countries. *Journal of Insect Physiology* **34**, 713-719.

Schmutterer, H. 1990. Properties and potential of natural pesticides from the neem tree, *Azadirachta indica*. *Annual Review of Entomology*, **35**, 271-297.

Workman, R.B., R.B. Chalfant, D.J. Schuster, 1980. Management of the Cabbage Looper and Diamondback Moth on Cabbage by using Two Damage Thresholds and Five Insecticide Treatments. Journal of Economic Entomology, **73**, 757-758.

THE USE OF SPRAY ADJUVANTS IN WINTER CEREALS IN SCOTLAND

K. P. DAWSON

CSC CropCare, Perth, Scotland

ABSTRACT

A number of field trials were carried out at a range of sites to examine the effect of adjuvants (a soya phospholipid/proprionic acid blend, latex, an alkoxylated amine, and two organosilicones) on a range of fungicides (fenpropimorph, flusilazole, tebuconazole, prochloraz, and tridemorph) in winter wheat in Scotland. The target diseases were *Pseudocercosporella* (eyespot), *Erisyphe graminis* (mildew), *Septoria nodorum* and *Fusarium nivale*. In addition a trial in winter barley on manganese uptake is reported. Significant improvements in disease control, trace element uptake and yield were achieved by a number of the product/adjuvant combinations. The most effective results were achieved with organosilicone surfactants. Reductions in fungicide dose rates were achieved with improved efficacy, which led to increased sustainability of production systems.

INTRODUCTION

Cool, wet weather and difficult disease patterns in the arable areas of Scotland provide problems for the grower in achieving adequate uptake and efficacy of systemic fungicides in cereals. Previous work in Scotland (Dawson & Ballingall, 1990 and Dawson 1992) has shown that by using adjuvants in an integrated barley production system improvements in efficacy and gross margin can be achieved. The aim of the trials programme reported here was to extend this work to the wheat crop and investigate the scope for improved efficacy and fungicidal dose reduction. Much previous work has been carried out in the greenhouse and/or with formulation composition of fungicides. As greenhouse studies may differ markedly from field conditions, substantive evidence from field trials is required using commercial programmes and more complex tank mixes.

METHODS AND MATERIALS

A series of replicated small plot field trials were conducted on wheat (*Triticum aestivum*) from 1990-1994 to examine the control of powdery mildew (*Erisyphe graminis*), eyespot (*Pseudocercosporella herpotrichoides*), *Septoria nodorum* and *Fusarium nivale* in winter wheat; trial sites were located in Mintlaw (Grampian Region), Aberdour (Fife Region), and Kelso and Berwick (Borders Region). A trial was also carried out in winter barley to examine the effect of adjuvants on manganese uptake.

The trials were a randomised block design with four replicates. The seed rate was 500 seeds/m^2 and the main variety used was cv Riband and the plot size was 2m x 12m. The plots were sprayed with an Azo small plot sprayer using a Lurmark flat fan 80° EO2-80 nozzle at 2.5 bar delivering 200l/ha of water, producing a medium quality spray (BCPC classification). Grain yields were taken with a small plot combine equipped with a load cell. Disease control given in the data tables was measured as a visual assessment, using twenty leaves or stems per plot for each treatment. The days after application (DAA) and crop growth stage are given in the data tables. The assessment of eyespot was additionally measured using an ELISA technique (an immunoassay technique for disease confirmation and quantification) (Smith et al,1990). Apart from relevant experimental treatments, trace elements were applied to the plots in response to soil and tissue analysis. Soil pH was in the range of 5.9-6.3. Manganese tissue analysis was by the nitroperchloric acid digestion followed by atomic adsorption, after prewashing foliage with distilled water. The Zadocks growth stage key was used in defining growth stages. In one trial some treatments received a wetting treatment prior to application to simulate a crop with a heavy dew on the leaves.

The fungicides used in the programme were fenpropimorph (750g active ingredient (ai) per litre), a formulated mixture of fenpropimorph and tridemorph (BAS 464;500 + 250g ai/l), prochloraz (400g ai/l), flusilazole (160g ai/l), tebuconazole (250g ai/l). These products contain an adequate adjuvant system in their formulation according to manufacturers. The adjuvants used were soya phospholipid and proprionic acid (750g/ha LI700; Loveland Industries) at 0.5% v/v of spray solution, synthetic latex (63g/ha Bond;Loveland Industries) at 0.08% v/v of spray solution. An alkoxylated amine (Arma blend;International Speciality Products) at 0.1% v/v of spray solution, and two organosilicone surfactants (Silwett L77 and Slippa, a blend of Silwett L77 and a linear alcohol;OSi Specialities) at 0.15% v/v of spray solution.

A. Stem base disease control

In the first trial the effect of an organosilicone adjuvant (Silwett L77) on the control of eyespot and *Fusarium* achieved with recommended and lower than recommended rates of prochloraz and flusilazole was investigated. The treatments were applied at first node and a standard fungicide (triadimenol) was applied to all treatments at flag leaf emergence and at full ear emergence, in order to remove the late season effects of foliar pathogens. The variety in the trial was Riband grown as a second wheat crop in the Borders.

B. Mildew control

A series of trials examined the effect of adjuvants on control of mildew in wheat with recommended and lower than recommended rates of fenpropimorph and fenpropimorph/tridemorph in combination. The treatments were applied at first node and a standard commercial fungicide (flusilazole) was applied to all treatments at flag leaf emergence and at ear fully emerged, in order to remove the late season effects of other foliar pathogens. In one trial (Table 3) plots were artificially wetted to the point of run off, prior to a spray application including synthetic latex, these are denoted as wet (W) and dry (D) in tables 2 and 3. The variety in all trials was Riband.

C. Septoria control

A trial in wheat examined the effect of adjuvants on *Septoria* control with tebuconazole at half the recommended rate at ear emergence. The predominant disease was *Septoria nodorum* and this was confirmed by an ELISA test. A standard programme of fungicides was applied to all treatments at Zadocks 31 and 37 and treatments were applied at ear emergence.

D. Manganese Uptake

The trial on winter barley examined the effect of an organosilicone and a phospholipid adjuvant on the uptake of manganese and grain yield.

ANOVA statistical analysis was conducted on the data and the values given in the tables which differ significantly from each other at a 95% confidence interval are denoted by a different letter.

RESULTS AND DISCUSSION

A. Stem base disease control

The control of stem base disease was examined in a second year wheat. The incidence of both eyespot and Fusarium were measured at several stages during the season by visual assessment (data are reported for GS75 assessments). The data for yield and the assessments are shown in Table 1. The table shows the data for eyespot control with both flusilazole and prochloraz, best control was achieved by the lower rate of each product in combination with Silwett L77. This result was confirmed by ELISA data. Prochloraz did not give similar control of eyespot to flusilazole. Only the combination of flusilazole and Silwett L77 gave significant control of stem base browning caused by *Fusarium*, which was a very visual effect in the field.

There was no lodging in any treatment due to the good seasonal conditions. In a normal Scottish harvest the effects of stem base *Fusarium* are a significant factor in lodging in many crops. In the absence of lodging yield differences between treatments showed few significant differences apart from comparison with the untreated control. The best treatments for *Fusarium* control was the lower rate of flusilazole with Silwett L77.

Table 1 Effect of fungicide and adjuvant combinations on control of Eyespot and *Fusarium* in winter wheat variety Riband (Borders 1990)

TREATMENT	Eyespot Control GS75 (%)	Fusarium Control GS75 (%)	Yield (t/ha)
UNTREATED	35a	44a	9.70a
(% mainstems infected)			
Flusilazole 200g/ha	45b	0a	10.65bc
Flusilazole 100g/ha	40ab	0a	10.35b
Prochloraz 400g/ha	55b	0a	10.96c
Prochloraz 200g/ha	40ab	0a	10.37b
Flusilazole 100g/ha + Silwett L77 0.15%(v/v)	60bc	56b	10.69bc
Prochloraz 200g/ha + Silwett L77 0.15%(v/v)	65c	0a	10.31b

B. Mildew Control

Control of mildew on the stem base in thick wheat crops is important if long term control is to be achieved. It also gives a useful measure of physical targetting of fungicide, as basipetal movement of fungicide down the plant post application is limited. Tables 2,3 and 4 show data from three trials which were carried out to examine the effect of adjuvants on mildew control in winter wheat. Adjuvant effect on fenpropimorph for leaf and stem mildew and yield increase was greatest from Silwett L77, and was significantly more effective than doubling the rate of fenpropimorph. The addition of soya phospholipid or synthetic latex applied either to wet or dry leaves gave similar increased mildew control to each other, but less than Silwett L77 and generally no better than doubling the rate of fenpropimorph. Artificially wetting the plots (W) prior to spray application also improved mildew control significantly on leaf 3 in the synthetic latex treatments 14 DAA (Table 2).

Table 2 Effect of fungicide and adjuvant combinations on severity of mildew in winter wheat variety Riband (Borders 1990)

TREATMENT	Mildew% Leaf 3 14DAA	Mildew% Leaf 3 23DAA	Mildew% Stem base 23DAA	Yield (t/ha)
UNTREATED	7.2a	9.0a	32.5a	8.51a
Fenpropimorph 375g/ha	2.5b	1.0b	18.5b	8.88ab
Fenpropimorph 187.5g/ha	3.5b	1.5b	21.5b	8.67a
Fenpropimorph 93.8g/ha	3.5b	2.5b	26.5ab	8.63a
Fenpropimorph 187.5g/ha + Silwett L77 (0.15%v/v)	0.2c	0.4c	0.5c	9.49c
Fenpropimorph 187.5g/ha + Soya phospholipid (0.5%v/v)	1.5b	1.0b	15.5b	9.01bc
Fenpropimorph 187.5g/ha + Synthetic Latex (0.08%v/v) D	1.0b	1.5b	15.0b	8.76ab
Fenpropimorph 187.5g/ha + Synthetic Latex (0.08%v/v) W	0.2c	1.0b	12.0b	9.21b

The use of synthetic latex on wet leaf (W) resulted in a significant yield increase over fenpropimorph alone in one trial. Addition of soya phospholipid to fenpropimorph gave variable results, significantly improving leaf mildew control in one of two trials, although no improvement in stem base mildew was achieved. However, despite this variable performance in mildew control, significant yield increases occurred. Grain yield generally followed the level of mildew control.

Table 3 Effect of fungicide and adjuvant combinations on severity of mildew in winter wheat variety Riband.(Aberdeenshire 1990)

TREATMENT	Mildew% Leaf 3 32DAA	Mildew% Stem Base 32DAA	Yield (t/ha)
UNTREATED	7.0a	20.5a	7.45a
Fenpropimorph 375g/ha	5.0a	12.5b	7.72a
Fenpropimorph 187.5g/ha	7.0a	15.0ab	7.65a
Fenpropimorph 187.5g/ha + Silwett L77 (0.15%v/v)	2.0b	1.5c	8.35b
Fenpropimorph 187.5g/ha + Soya phospholipid 0.5%(v/v)	5.0a	11.5b	8.15b
Fenpropimorph 187.5g/ha + Synthetic Latex (0.08%v/v) D	4.0ab	15.0ab	8.01ab
Fenpropimorph 187.5g/ha + Synthetic Latex (0.08%v/v) W	2.0b	7.5b	8.15b

A comparison of the two organosilicone surfactants in Table 4 shows the Silwett L77 blend to be superior in mildew control to Silwett L77 alone in this trial. The mildew control from the use of the organosilicone surfactants was more pronounced on the stem base, which should reduce reinfection.

Table 4 Effect of fungicide and adjuvant combinations on severity of mildew in winter wheat variety Riband (Fife 1993)

TREATMENT	Mildew% Leaf 3 32DAA	Mildew% Stem Base 32DAA	Yield (t/ha)
UNTREATED	5.0a	12.5a	9.51a
BAS464 (150 + 75g/ha)	1.5b	8.5a	10.28b
BAS464 (150 + 75g/ha) + Silwett L77 (0.15%v/v)	0.8b	1.5b	10.04ab
BAS464 (150 + 75g/ha) + Silwett L77 blend (0.15%v/v)	0.0c	0.0c	10.80c

C. Septoria Control

Further work was carried out to examine the effect of adjuvants added to tebuconazole on the control of *Septoria nodorum* at ear emergence. The data (table 5) show the disease control and yield benefits from a fungicide applied at ear emergence which was further enhanced by both adjuvants. The data show that the organosilicone surfactant has improved disease control in the lower leaf canopy, but has reduced control on the flag leaf (L1). It is believed that the fungicide has been washed down the plant at application and not redistributed into upper leaves, due to high soil moisture deficits. In contrast the alkoxylated amine improved disease control on the upper flag leaf and increased grain yield significantly. This affords the possibility of targetting sprays in the canopy using ELISA diagnostic techniques to determine the site of disease.

Table 5 Effect of fungicide and adjuvant combinations on control of *Septoria nodorum* in winter wheat variety Brigadier (Fife 1994)

TREATMENT	*%Septoria nodorum* Leaf 3 32DAA	*%Septoria nodorum* Leaf 1 32DAA	Yield (t/ha)
Untreated	65.4a	30.5a	12.95a
Tebuconazole (125g/ha)	18.5b	9.5c	13.62bc
Tebuconazole (125g/ha) +Silwett L77 (0.15%v/v)	5.5c	20.0b	13.50b
Tebuconazole (125g/ha) +Arma blend (0.1%v/v)	12.4b	2.5d	14.11c

D. Manganese Uptake in Winter barley

The addition of either a soya phospholipid adjuvant or an organosilicone surfactant increased both manganese uptake and grain yield in the barley variety Princess (Table 6).

Table 6 Effect of manganese and adjuvant combinations on uptake and grain yield of winter barley (Fife)

TREATMENT	Manganese (ppm) 21DAA	Yield (t/ha)
Manganese (620g/ha)	23a	10.64a
Manganese (620g/ha) + Soya phospholipid (0.5%v/v)	48b	11.02a
Manganese (620g/ha) + Silwett L77 blend (0.15%v/v)	59b	11.68b

The Silwett L77 blend was more effective than the soya phospholipid adjuvant in increasing yield on this manganese deficient soil. The variety used in this trial was Princess.

CONCLUSIONS

The evidence from the field trials data in this paper would suggest that adjuvants have an important part to play in increasing the efficiency of fungicides in winter wheat in Scotland, and increasing sustainability of production. The organosilicone surfactants,and in one trial, the alkoxylated tallow amine gave the best results with the fungicide and disease targets specified. There are possibilities for exploiting this potential either by maintaining efficacy and reducing fungicide dose or by maintaining the fungicide dose and increasing the effect. Both of these strategies may reduce cost of production per tonne and increase profitability, but only the former will reduce environmental load. It is clear from the data that fungicide/adjuvant/target interactions are specific and care must be taken in selecting the correct combinations for effective field use. A full understanding of product and adjuvant modes of action and disease epidemiology will be needed by the field adviser in order to exploit the potential of these management tools and improve the sustainability of inputs, from both economic and environmental standpoints.

REFERENCES

Dawson K.P. ; Ballingall M. (1990): *Proc Crop Protection in Northern Britain*: 267-274.

Dawson K.P. (1992): pp.587-593 *in* Foy,C.L.(Ed)*"Adjuvants for Agrichemicals"* CRC Press,Boca Raton.

Smith C.M.; Saunders D.W.; Allison D.A.; Johnson L.E.B.; Labit B.; Kendall S.J.; Hollomon D.W.; 1990: *Proc Brighton Crop Protection Conference*: 763-770.

The Synthesis And Biological Activity Of Analogues Of 1,7-Dioxaspiro-[5.5]Undecane. The Pheromone Of The Olive Fly *(Bactrocera oleae)*

DAVID R. KELLY, SCOTT J. HARRISON, SIMON JONES, ADAM WADDING

Dept. of Chemistry, University of Wales, Cardiff, P. O. Box 912, Cardiff, CF1 3TB

BASILIS E. MAZOMENOS AND DIMITRIS RAPTOPOULOS,

Institute of Biology, NCSR "Demokritos", P. O. Box 60228, Aghia Paraskevi 153 10, Greece

ABSTRACT

Eight analogues of the Olive fly pheromone (1) (Figure 1) have been prepared and their biological activity determined by electro-antennography (EAG) and laboratory bioassays. The nor-analogue (2) showed attraction comparable to the natural pheromone in laboratory bioassays, but low activity in the field.

INTRODUCTION

Olives are a major crop which generates 250 million days of employment a year in the EEC alone. Most olives are pressed for their oil and the rest are consumed as table olives. The premium product is Extra Virgin Olive Oil which is produced from the first pressing of intact olives. The total annual world production of olives and olive oil in the period 1988-1990, averaged 9.51Mt (million tons) and 1.79Mt respectively (FAO, 1991). On average 15% of the crop, worth £450M a year is lost to over thirty species of insect and £55M a year is spent on pest control (Claridge and Walton, 1992). Current control measures based on the use of conventional pesticides (eg. dimethoate) may have adverse effects on the olive ecosystem and undesirable residues accumulate in the fruit. Spain, Italy and Greece are in the course of developing IPM systems for olive production. Such systems offer the best prospects for economic and sustainable agricultural production. The monitoring of pests and natural enemies is central to the development and implemention of IPM systems. Pheromone (or other semiochemical) baited traps are highly effective for monitoring purposes. Moreover mass trapping or mating disruption using semiochemicals are viable alternatives to conventional pesticides.

The olive fly *Bactrocera (Dacus) oleae* (Gmel.) occurs throughout the countries of the Mediterranean basin and is the most serious pest of olive fruits. It has been estimated that *B. oleae* may account for 50-60% of total insect pest damage. The larvae bore within the fruit, causing premature fruit fall, decreased yield and quality of oil. The major component of the pheromone produced by virgin females was identified as 1,7-dioxaspiro[5.5]undecane (1) (Baker et al., 1980; Mazomenos et al., 1981; Garabaldi et al., 1983) and is usually referred to as olean or spiroketal. It was also isolated from male olive flies (Mazomenos and Pomonis, 1983). Spiroketal (1) is a strong attractant for males and causes females to aggregate near the pheromone source. Natural and synthetic (racemic) spiroketal (1) elicited a similar response from both male and female antennae in EAG experiments (Van der Pers et al., 1985). Haniotakis et al., (1989) reported that the orthoester; 1,5,7-trioxa[5.5]undecane (Figure 4a, X = O) was equally attractive to males. However due to its hydrolytic lability, this compound was only active for two weeks or less under laboratory or field conditions. The purpose of the current study was to determine the minimum structural elements of the spiroketal (1) which are responsible for eliciting attraction in order that more potent or cheaper attractants could be employed.

MATERIALS AND METHODS

Synthesis

The heteroannular ketals 1,7-dioxaspiro[5.5]undecane (1), and 1,6-dioxaspiro[4.4]nonane (3) were prepared by Claisen condensation of either δ−valerolactone or γ-butyrolactone, acid hydrolysis, decarboxylation and elimination of water to give the spiroketals (Erdmann, 1885; Baker et al., 1980). Treatment of a mixture of δ-valerolactone and γ-butyrolactone by the same method gave the three spiroketals (1) (2) (3) in a statistical mixture (25:50:25 respectively). These were separated by repeated flash column chromatography to give 1,7-dioxaspiro-[5.4]decane (2) in 99.2% purity, by capillary GC-MS. Oxaspiro[5.5]undecane (4) was prepared by addition of pentanyl-1,5-bis(magnesium bromide) to δ-valerolactone (Canonne et al., 1980) and acid catalysed dehydration (61%, 27% yield). Spiro[5.5]undecane (5) was prepared by Wittig reaction of cyclohexanone and 5-triphenylphosphonium-1-pentenyl bromide. The diene so formed (58% yield) underwent acid catalysed cyclisation (Zelinskii and Elagina, 1952) to give a mixture of alkenes (57% yield), which were hydrogenated (53% yield, Dixon and Naro, 1960). The "homoannular" ketals; 1,4-dioxaspiro[4.4]nonane (6), 1,5-dioxaspiro[5.4]decane (7), 1,4-dioxaspiro[4.5]decane (8) and 1,5-dioxaspiro[5.5]undecane (9), were prepared by acid catalysed ketalisation (Dainault and Eliel, 1973) and repeatedly distilled under vacuum until homogenous by GC (yields 25%, 33%, 24%, 58% respectively).

Electroantennogram studies

The insects used in laboratory based studies were derived from a colony established in NCSR "Demokritos", kept for many generations under laboratory conditions (25°C, 70% R.H.) in 14L:10D regime. EAG responses were obtained from antennae on excised *B. oleae* head (Van der Pers et al. 1984). Aliquots (10μL) containing 10mg of the compounds tested, were pipetted onto a filter paper strip (1cm x 2cm). The solvent was allowed to evaporate and the strip placed inside a Pasteur pipette. The stimulus was administered through the pipette by a "puff" of air. Stimulus duration was adjusted to 6L/min for 0.7sec, under a constant air current of 5L/min. The chemicals were randomly presented to the insects at 60 second intervals. The deflection of the control was subtracted from the deflection (-mV) produced by each analogue and expressed as percentages relative to the spiroketal (1).

Laboratory bioassays

Bioassays were conducted in a cage described by Mazomenos and Haniotakis (1981) using 100 sexually mature males, 5-10 days old, during the last three hours of the photophase. Samples of 10μl of the pheromone and pheromone analogues diluted in hexane (1μg/μL) were pipetted onto filter paper, hung about 3cm below the top of the cage in a weak airstream. The number of incoming insects was recorded over 5 minutes. Means comparisons were made with Tukey's multiple-range test (P=0.05).

FIGURE 1. Chemical structures of the sex pheromone of the female olive fly and the synthesised pheromone analogues.

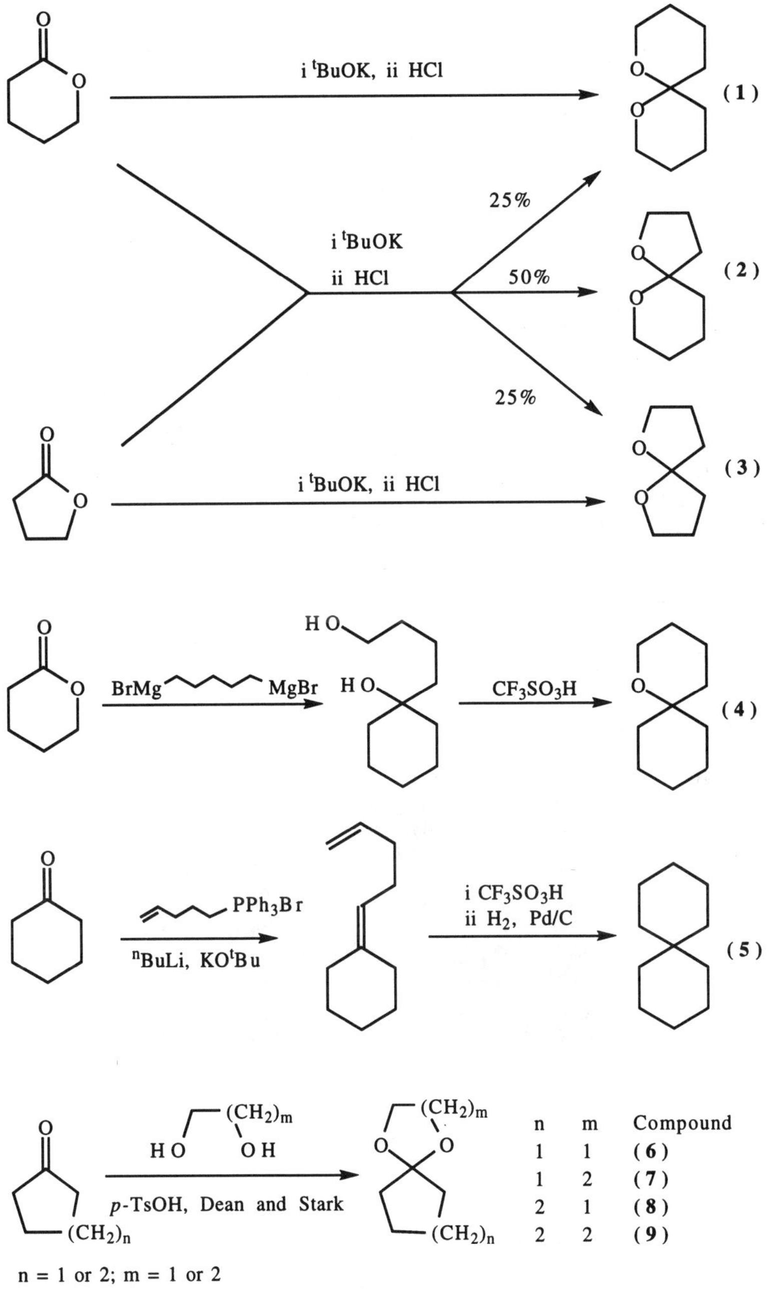

n	m	Compound
1	1	(**6**)
1	2	(**7**)
2	1	(**8**)
2	2	(**9**)

Field tests

During Sept. 15-Oct. 30, 1992 comparative male attraction field tests were conducted between the natural pheromone component (1) and the analogue (2) that showed promising attraction to males in laboratory bioassays. Five sticky plywood traps (20 x 20cm) for each treatment were baited with 1ml polyethylene vials containing, 20 and 50 mg of (1) or 2, 20, 200mg of (2). The traps were placed at a distance of about 50m in a completely randomised design in an abandoned olive orchard at Amaroussion, Attikis. Once a week, traps were serviced, rotated and the number of both sexes captured was recorded. Field experimental data were transformed to log(x+1) prior to statistical analysis. Mean comparisons were made with Tukey's multiple-range test (P=0.05).

RESULTS

Antennal response to the analogues

All analogues evoked similar responses in both male and female insects. The mean antennal responses of males are presented in Figure 2. The hydrocarbon analogues (4)(5) elicited comparable responses to the natural pheromone component (1) whereas all other analogues elicited lower responses ranging from 25-80% of the natural pheromone component (1). The weakest responses were to analogues with five membered rings and particularly those with two five membered rings (3)(6).

FIGURE 2. Relative EAG response (%) of *Bactrocera oleae* males to eight sex pheromone analogues, in comparison to 1,7-dioxaspiro[5.5]-undecane(1).

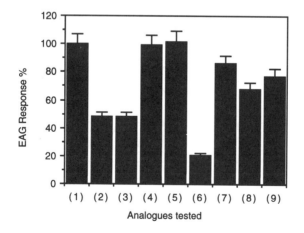

Laboratory bioassays

Male attraction to the natural pheromone component and the eight analogues is presented in Figure 3. The only compounds which caused attraction significantly different to the control (Tukey's multiple-range test, P=0.05) were the natural pheromone component (1) and the analogue (2). Moreover the latter elicited the same behavioural steps as the natural pheromone, with 70% of the attractivity.

Figure 3. Percent attraction of *Bactrocera oleae* males to eight sex pheromone analogues during bioassays conducted under laboratory conditions. Means followed by same letters are not significantly different, Duncan's multiple range test (P=0.05).

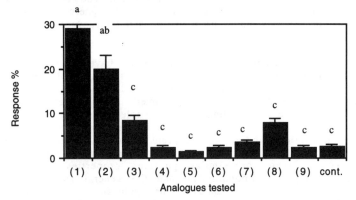

Field tests

The results from the field tests are shown in Table 1. The two concentrations of the natural pheromone component (1) attracted large numbers of males. The analogue (2) which elicited satisfactorily male attraction in the laboratory bioassays did not perform as well in the field. The total number of males captured by the 2mg and 20mg lures was not significantly different to the numbers of males caught in the unbaited control traps. At 200mg the analogue captured significantly more males than the controls, although this was only 8% of the total males caught by the 20mg natural pheromone traps. None of the analogue baited traps captured significantly more females than the controls.

TABLE 1. Number of *Bactrocera oleae* captured on sticky plywood traps baited with spiroketal (1) and analogue (2) at various concentrations (Sept. 15 - Oct. 30, 1992, Amaroussion, Attiki).

Baits	Traps Conc. (mg)	Total Males	Females	Insect/trap* Males	Females
1,7-dioxaspiro[5.5]undecane (1)	50	4832	180	268 a	10 a
""	20	4071	220	226 a	12 a
1,7-dioxaspiro[5.4]decane (2)	2	44	45	2.4 d	2.6 b
""	20	131	22	7.3 d	1.6 b
""	200	409	14	22.7 b	0.8 b
Control		32	20	1.7 d	1.1 b

* Means followed by same letters are not significantly different, Tukey's multiple range test (P=0.05)

DISCUSSION

Capillary GC analysis revealed that the analogue (2) contains 0.38% of the natural pheromone component (1). In order to show that this was not responsible for the observed attraction, a series of dilutions with concentrations of (1) equal to that contained in the sample of (2) were prepared and tested under laboratory conditions. In all cases the results were the same as the control. It has been reported (Mazomenos and Haniotakis, 1985; Haniotakis and Pitara, 1994) that the minimum threshold level for male orientation to the pheromone component (1) is 6ug, whereas the 10µ g samples used in the laboratory bioassay contained only 0.038µ g.

The majority of pheromone analogues which have been prepared thus far are fluoro-derivatives for binding studies (Wu et al., 1993) or stereoisomers prepared to confirm the stereochemistry of natural materials. The comparatively simple structure of the spiroketal (1) prompted us to undertake a more wide ranging study. Since we are dealing with a whole animal (or its antennae) the sequence of steps which transport the analogues from their source to the antennae and thereafter binding to odourant transport proteins, transport and binding to the receptor, signalling and finally clearance cannot be resolved (Prestwich 1993). Given the similar physical properties of the analogues (particularly volatility and lipophilicity), and the general purpose nature of the odourant binding system, it is unlikely that substantial discrimination of the analogues, occurs prior to the binding site. Comparison of EAG results and the behavioural assays enables the gross electrophysiological response of the antennae to be distinguished from the behavioural effects that the odourants elicit.

Figure 4a   Figure 4b

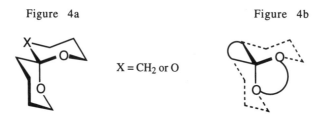

X = CH₂ or O

The first constraint on the structure of the binding site is that although the natural pheromone component (1) released by females is a racemic mixture (Fletcher et al. 1992), significantly more males are attracted by the (R)-enantiomer than the (S)-enantiomer (Haniotakis et al. 1986). Moreover the (S)-enantiomer is not an inhibitor and the orthoester; 1,5,7-trioxa[5.5]undecane (Figure 4a; X = O) has significant activity. This indicates that binding site can tolerate either an oxygen or a methylene group and that only two (rather than three) oxygens participate in binding. Figure 4a shows the lowest energy conformer (Deslongchamp 1983) of the (R)-enantiomer of the pheromone (1). This importance of the six membered rings for eliciting EAG responses is demonstrated by hydrocarbon analogues (4) and (5) and to a lesser extent the other six membered ring analogues eg (9). Moreover the compounds which have solely five membered rings (3), (6) gave the joint second lowest and lowest EAG responses respectively. In principle the "homoannular" ketals (6), (7), (8), (9) should be able to bind to the receptor in a crosswise fashion, with the oxygens located in the same positions as the natural pheromone component (1). This is illustrated in Figure 4b, in which the curved lines represent generalised carbon chains for the five and six membered ring analogues. Their lack of activity indicates that one or both of these regions are inaccessible. The spiroether (4) and the hydrocarbon (5) elicited EAG responses identical to those of the natural pheromone, but were amongst the least active in the laboratory bioassays. This indicates that the ring oxygens are essential for the behavourial response. Finally the "heteroannular ketals" (1), (2), (3) delineate the fine detail of the binding and the requisite signalling. Evidently both of the six membered rings are required to elicit the full behavioural response, although one six membered ring is sufficient to elicit reduced activity.

The work described illustrates the limitations of classical semiochemical techniques when applied to analogues. The EAG results gave reasonably good graded discrimination between the analogues but the most active analogue in the laboratory bioassay was ranked joint 7th out of the nine materials tested. Whereas the laboratory bioassay discriminated between the analogues so well that a structure activity relationship could not be deduced for the less active analogues. In the field bioassay only high concentrations of the analogue (2) were active. This is consistent with the laboratory bioassay because the $10\mu g$ of material presented as a lure in $0.12m^3$ space is equivalent to the amount of pheromone released by 100 females.

The results presented here indicate that the 1,7-dioxaspiro[5.4]decane (2) acts as a parapheromone, although it has no practical value due to its low level of activity. The lack of activity of the other analogues indicates that binding of the natural pheromone component (1) involves most, if not all of the bicyclic spiroketal structure.

REFERENCES

Baker, R., Herbert, R. H., Howse, P. E., Jones, O. T., Francke, W., and Reith, W. 1980. Identification and synthesis of the major sex pheromone of the olive fly, *Dacus oleae. J. Chem. Soc., Chem. Commun.* 52-54.

Canonne, P., Foscolos, G. B. and Belanger, D. 1980, One-step annelation. A convenient method for the synthesis of diols, spirolactones and spiroethers from lactones, *J. Org. Chem.*, **45**, 1828-1835.

Claridge, M. F. and Walton, M. P. 1992, The European Olive and its Pests-Management Strategies. In: P. T. Haskell (Ed), *Research collaboration in European IPM Systems*, BCPC Monograph No 52 pp, 23-30, 1992.

Dainault, R. A. and Eliel, E. L. 1973, 2-Cyclohexyloxyethanol, *Organic Syntheses Collective*, V, 303-310.

Deslongchamp, P. 1983, Stereoelectronic Effects in Organic Chemistry, Pergamon Press, Oxford, 6-10.

Dixon, J. A. and Naro, P. A. 1960, Synthesis of four spiro hydrocarbons, *J. Org. Chem.*, **25**, 2094-2097.

Erdmann, H. 1885. Umwandlung von lactonsauren in lactone und einwirkung von natriumathylat auf isocaprolacton *Ann* **228**, 176-186.

Fletcher, M. T., Jacobs, M. F., Kitching, W., Krohn, S., Drew, R. A. I., Haniotakis, G. E. and Francke, W. 1992, Absolute stereochemistry of the 1,7-dioxaspiro[5.5]-undecanols in fruit-fly species, *J. Chem. Soc. Chem. Commun.*, 1457-1459.

FAO 1991, Production Yearbook, Vol 44, Rome.

Gariboldi, P., Verota, L., and Fanelli, R. 1983. Studies on the chemical constitution and sex pheromone activity of volatile substances emitted by *Dacus oleae. Experientia.* **39**: 502-505

Haniotakis, G., Francke, W., Mori, K., Redlich, H. and Schurig V.1986. Sex specific activity of (R)-(-) and (S)-(+)-1,7-dioxaspiro[5.5]undecane, the major sex pheromone of *Dacus oleae, J. Chem. Ecol.*, **12**: 1559-1568.

Haniotakis, G. E., and Pittara, I. S. 1994. Response of *Bactrocera (Dacus) oleae* males (Diptera : Tephritidae) to pheromones as affected by concentration, insect age, time of day, and previous exposure. *Environ. Entomol.*. **23**: 726-731.

Haniotakis, G., Mavraganis, V., and Ragousis, V. 1989. 1,5,7-trioxaspiro[5.5]undecane a pheromone analogue with high biological activity for the olive fly *Dacus oleae. J. Chem. Ecol.* **15**: 1057-1065.

Haniotakis, G. E., Mazomenos, B. E., and Tumlinson, H. 1977. A sex attractant of the olive fruit fly, *Dacus oleae*, and its biological activity under laboratory conditions. *Ent. exp. & appl.* **21**: 81-87.

Kelly, D. R., and Mazomenos, B. E. 1992. The identification and synthesis of olive pest semiochemicals. In: P. T. Haskell (Ed), *Research collaboration in European IPM Systems*, BCPC Monograph No 52 pp, 23-30, 1992.

Mazomenos, B. E., and Haniotakis, G. E. 1981. A multicomponent sex pheromone of *Dacus oleae* (Gmelin), isolation and Bioassay. *J. Chem. Ecol.* **7**: 437-443.

Mazomenos, B. E., and Haniotakis, G. E. 1985. Male olive fruit fly attraction to synthetic sex pheromone components in laboratory and field tests. *J. Chem. Ecol.* **11**(3): 397-405.

Mazomenos, B. E., and Pomonis, J. G. 1983. Male olive fruit fly pheromone: Isolation, Identification and lab. bioassays. In: R. Cavalloro (Ed), *Fruit Flies of Economic Importance*. Proc CEC/IOBC Intern. Symp. Athens/Nov., 1982. A.A. Balkema, Rotterdam, pp, 96-103.

Prestwich, G. D. 1993. Chemical studies of pheromone receptors in insects, *Archives of Insect Biochemistry and Physiology*, **22**, 75-86.

Wu, W. Q., Bengtsson, M., Hansson, B. S., Liljefors, T., Lofstedt, C., Prestwich, G. D., Sun, W. C. and Svensson, M. 1993. Electrophysiological and behavioural responses of Turnip moth males, *Agrotis segetum* to fluorinated pheromone analogues, *J. Chem. Ecol.*, **19**, 143-157.

Van der Pers, J. N. C., Haniotakis, G. E., and King, B. M. 1984. Electroantennogram responses from olfactory receptors in *Dacus oleae. Entomologia Hellenica.* **2**: 47-53.

Zelinskii, N. D. and Elagina N. V. 1952. Synthesis of spiro[5.5]undecane, *Doklady Akad. Nauk S.S.S.R.*, **87**, 755-758 (*CA* 1954, **48**, 542c).

CONTROL OF FUNGAL PLANT PATHOGENS USING PUTRESCINE
ANALOGUES

D.R. WALTERS, N.D. HAVIS

Department of Plant Science, SAC, Auchincruive, Nr Ayr KA6 5HW, UK

D.J. ROBINS

Department of Chemistry, University of Glasgow, Glasgow G12 8QQ

ABSTRACT

The polyamines putrescine, spermidine and spermine are ubiquitous in nature
and are essential for cell division. It is now known that polyamine analogues
can perturb polyamine metabolism leading to powerful antiproliferative effects
in tumour cells. This paper reviews the results of a research programme
focused on the synthesis and evaluation of putrescine analogues as novel
fungicides. A number of aliphatic, alicyclic and cyclic diamines have been
shown to possess considerable fungicidal activity, but although many of these
compounds perturb polyamine metabolism in fungal cells, such changes are
not considered sufficient to account for the observed antifungal effects. It
seems likely that these putrescine analogues possess an additional mode of
action.

INTRODUCTION

The diamine putrescine, the triamine spermidine and the tetraamine spermine,
collectively known as polyamines, are ubiquitous in nature (Walters, 1995).
Spermidine and spermine have three and four net positive charges respectively and are
the most cationic small molecules in cells. They thus bind to negatively charged
macromolecules like DNA, and probably because of their interactions with nucleic
acids, they are essential for cell division (Walters, 1995). Spermidine appears to be the
major polyamine in fungi and in *Neurospora crassa*, for example, spermidine is an
essential metabolite (Pitkin & Davis, 1990).
    In higher plants, the first step in polyamine biosynthesis is the decarboxylation of
either ornithine or arginine, in reactions catalysed by the enzymes ornithine
decarboxylase (ODC; EC 4.1.1.17) and arginine decarboxylase (ADC; EC 4.1.1.19;
Fig. 1). Spermidine and spermine are formed by the subsequent addition of an
aminopropyl moiety onto putrescine and spermidine respectively. The aminopropyl
moiety results from the decarboxylation of S-adenosylmethionine by the enzyme S-
adenosylmethionine decarboxylase (AdoMetDC; EC 4.1.1.50). Most fungi appear to
synthesize polyamines using the ODC route only (Walters, 1995), prompting the
suggestion that specific inhibition of ODC should control fungal pathogens without
affecting the plant. Indeed, the ODC inhibitor α-difluoromethylornithine (DFMO) has
been shown to control powdery mildew on barley in glasshouse and field studies

(Walters, 1995). But polyamine metabolism can also be perturbed by polyamine analogues, leading for example, to depletion of intracellular polyamine levels and powerful antiproliferative effects in tumour cells (Porter & Sufrin, 1986). When the commercially-available analogue keto-putrescine was evaluated for fungicidal activity, it was found to reduce spermidine concentrations in the oat-stripe pathogen *Pyrenophora avenae,* and to control infections by several fungal pathogens on a range of crop plants (Foster & Walters, 1993). In the light of these results, it seemed prudent to examine the possibility that synthetic putrescine analogues might exhibit more powerful fungicidal activity.

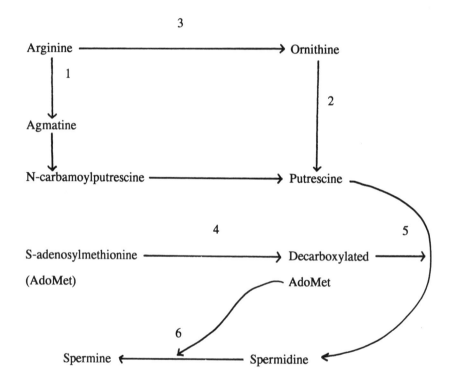

FIGURE 1. Pathways of biosynthesis of the major polyamines. 1 = arginine decarboxylase (ADC); 2 = ornithine decarboxylase (ODC); 3 = arginase; 4 = AdoMet decarboxylase (AdoMetDC); 5 = spermidine synthase; 6 = spermine synthase.

ALIPHATIC PUTRESCINE ANALOGUES AS NOVEL FUNGICIDES

In 1990, a programme of research was initiated, focused on the design, synthesis and evaluation of putrescine analogues as novel fungicides. During the first phase of this programme, a series of aliphatic diamines were produced and two analogues, (*E*)-

1,4-diaminobut-2-ene (E-BED) and (E)-(N,N,N',N'-tetraethyl)-1,4-diaminobut-2-ene (E-TED) (Fig. 2), were found to possess substantial activity against powdery mildew on barley, rust and chocolate spot on broad bean, powdery mildew on apple and late blight on potato, in glasshouse studies (Walters, 1995; Table 1). Interestingly, these analogues produced very different effects on polyamine metabolism. Despite leading to a substantial reduction in ODC and AdoMetDC activities in *P. avenae*, E-BED reduced spermidine concentration and actually increased putrescine concentration (Walters, 1995). On the other hand, E-TED reduced concentrations of both putrescine and spermidine in *P. avenae* in spite of increases in ODC and AdoMetDC activities (Walters, 1995). Since spermidine appears to be the major polyamine in fungi (Walters, 1995) and spermidine pools were not totally depleted following treatment with E-BED or E-TED, it seems unlikely that the main mode of action of these analogues is inhibition of polyamine biosynthesis. Recent work has shown that E-BED and E-TED both reduce appressorium formation by rust uredospores (Reitz *et al.*, 1995), and so it would be worth determining whether polyamine metabolism during germling development is particularly sensitive to polyamine analogues.

TABLE 1. Comparison of the fungicidal effects of (E)-1,4-diaminobut-2-ene (E-BED; 163 mg/litre) and (N,N,N',N'-tetraethyl)-1,4-diaminobut-2-ene (E-TED; 360 mg/litre).

Pathogen/host	% Disease control	
	E-BED	E-TED
Erysiphe graminis/barley	80	80
Uromyces viciae-fabae/broad bean	72	60
Botrytis fabae/broad bean	91	45
Podosphaera leucotricha/apple	75	75
Phytophthora infestans/potato	75	72

E-BED has also been shown to control powdery mildew on spring barley in field trials. For example, early season control of powdery mildew with 375 g/ha E-BED was as good as, if not superior, to that achieved with a mixture of 120 + 192 g/ha flutriafol + carbendazim (Havis *et al*, 1994; Table 2). Further, even though two sprays of flutriafol and carbendazim produced a greater reduction in mildew infection late in the season, two sprays of 375 g/ha E-BED produced an equivalent increase in grain yield compared to the untreated controls (Table 2). In this field study, E-BED was applied in 0.5 % Agral 90 (alkyl phenyl ethylene oxide; Zeneca). More recently, various formulations of E-BED were prepared and used in both glasshouse and field studies. Results to date indicate that although formulation altered the spectrum of activity of E-BED, it did not increase efficacy against powdery mildew in the field (Havis & Walters, unpublished results).

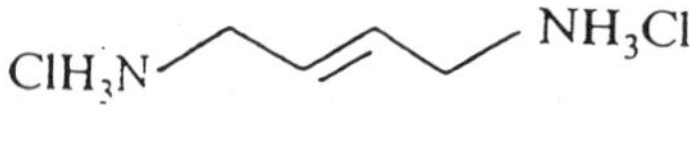

E-BED
(E)-1,4-diaminobut-2-ene

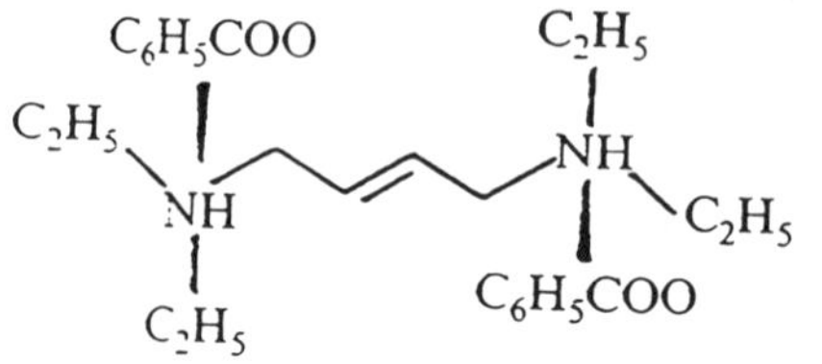

E-TED
(E)-(N,N,N',N'-tetraethyl)-1,4-diaminobut-2-ene

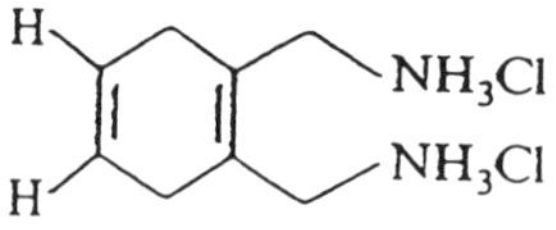

BAD
1,2-bis(aminomethyl)-4,5-dimethylcyclohexa-1,4-diene

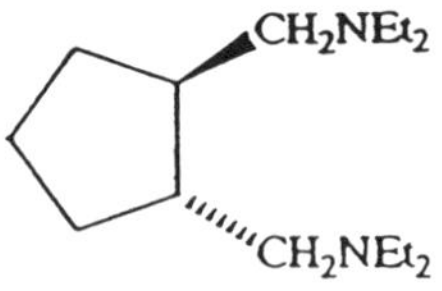

TBAMBH
trans-5,6-bis(aminomethyl)bicyclo[2.2.1]hept-2-ene

FIGURE 2. Chemical names and structures of four synthetic diamines

TABLE 2. Effects of E-BED (375 g/ha) and a flutriafol + carbendazim mixture (120 + 192 g/ha) on powdery mildew infection and grain yield of spring barley. Sprays were applied at GS 25 and at GS 59.

Treatment	Mean % powdery mildew infection				Grain yield (kg/plot)
	7 June	14 June	9 July	16 July	
Untreated	9.5	20.2	6.4	19.4	7.4
E-BED	3.4	10.5	3.5	10.5	9.2
Flutriafol + carbendazim	4.8	13.4	2.7	4.8	9.2
LSD ($P = 0.05$)	2.88	4.05	4.79	5.98	1.24

TABLE 3. Effect of alicyclic diamines on powdery mildew infection of barley seedlings

Compound	% reduction in mildew
BAD (211 mg/litre)	93
trans-BAD (211 mg/litre)	67
cis-BAD (211 mg/litre)	21
TCHMB (334 mg/litre)	52
TCCMB (243 mg/litre)	87
BACP (199 mg/litre)	74

Abbreviations

BAD: 1,2-*bis*(aminomethyl)-4,5-dimethylcyclohexa-1,4-diene
trans-BAD: trans-4,5-*bis*(aminomethyl)-1,2-dimethylcyclohex-1-ene
cis-BAD: cis-4,5-*bis*(aminomethyl)-1,2-dimethylcyclohex-1-ene
TCHMB: trans-N,N,N',N'-tetramethyl-1,2-cyclohexane *bis*(methaneamine)
TCCMB: trans-1,2-*bis*(dimethylaminomethyl)cyclobutane
BACP: 1,2-*bis*(aminomethyl)cyclopentene

ALICYLIC AND CYCLIC DIAMINES AS NOVEL FUNGICIDES

Following the synthesis and evaluation of a large number of aliphatic diamines, a series of alicyclic diamines were made and subsequently examined for fungicidal activity (Walters, 1995). Of these compounds, 1,2-*bis*(aminomethyl)-4,5-dimethylcyclohexa-1,4-diene (BAD; Fig. 2) was particularly active against powdery mildews, reducing barley powdery mildew infection by 93 % following post-inoculation treatment (Table 3). Results to date suggest however, that like E-BED and E-TED, the alicyclic diamines do not deplete intracellular polyamines sufficiently to account for the observed antifungal effects (Walters, 1995).

Interestingly, a number of cyclic diamines were synthesized and found to possess fungicidal properties and these analogues were also shown to perturb polyamine levels in fungal cells (Havis *et al.*, 1995). One of these compounds, trans-5,6-*bis*(aminomethyl)bicyclo[2.2.1]hept-2-ene (TBAMBH; Fig. 2), reduced concentrations of putrescine, spermidine and spermine in *P. avenae* by 84 %, 55 % and 64 % respectively, and although ODC activity was reduced, AdoMetDC activity was actually increased slightly (Table 4). Although in this case reductions in fungal growth could be attributed, in part, to reductions in polyamine levels, this is unlikely to be the sole mechanism of action.

TABLE 4. Effect of the cyclic diamine trans-5,6-*bis*(aminomethyl)bicyclo[2.2.1]hept-2-ene (TBAMBH; 240 mg/litre) on polyamine concentrations and enzyme activities in *Pyrenophora avenae*.

Treatment	Polyamine concentration, μmol g/f.wt		
	Putrescine	Spermidine	Spermine
Control	136 ± 9.9	129 ± 16.0	260 ± 12.6
TBAMBH	21 ± 4.0	57 ± 2.4	93 ± 8.4

Treatment	Enzyme activity, pmol CO_2 mg protein/h	
	ODC	AdoMetDC
Control	260 ± 20.7	54 ± 9.0
TBAMBH	170 ± 9.0	77 ± 21.4

Various mechanisms for the inhibition of cell growth by polyamine analogues have been proposed, including inhibition of polyamine biosynthesis by direct enzyme inhibition or regulation of polyamine biosynthetic enzymes (Porter & Sufrin, 1986). Results obtained with the different classes of putrescine analogue suggest that inhibition of fungal growth is not due to inhibition of enzymes of polyamine biosynthesis. It is known that a number of *bis*(benzyl)polyamine analogues inhibit growth of rat tumour cells partly by direct binding to DNA, with subsequent disruption of macromolecular function (Bitonti *et al.*, 1989). It would be useful to determine whether the putrescine analogues operate in a similar manner.

EFFECTS OF POLYAMINE BIOSYNTHESIS INHIBITORS ON MYCORRHIZAL FUNGI AND BIOCONTROL FUNGI

If compounds which perturb polyamine metabolism and/or function possess activity against phytopathogenic fungi, it is important to know whether these compounds have similar effects on mycorrhizal fungi. Given that mycorrhizal fungi are intimately associated with most plants and exert a profound influence on host physiology, any reduction in mycorrhizal infection may affect plant growth. In a series of experiments, Zarb (1995) and Zarb & Walters (1994a,b) found that different ectomycorrhizal fungi exhibit quite different responses when exposed to polyamine biosynthesis inhibitors. For example, mycelial growth of *Laccaria proxima* and *Hebeloma mesophaeum* was unaffected by exposure to polyamine inhibitors, while growth of *Paxillus involutus* and *Thelophora terrestris* was slightly reduced by inhibitor treatment. The lack of any effect of the polyamine inhibitors on growth of *L. proxima* and *H. mesophaeum* appeared to be due to the presence of ADC activity in these fungi (Zarb, 1995; Zarb & Walters, 1994a), and possibly to the compensatory formation of uncommon polyamines (Zarb & Walters, 1994c). It is important to remember that these data were obtained from *in vitro* experiments and it will be necessary to determine the effects of polyamine inhibitors on mycorrhizal infection and establishment on plant roots.

Interestingly, in preliminary experiments, a range of polyamine biosynthesis inhibitors was shown to have little effect on growth of various biocontrol fungi e.g. *Trichoderma harzianum* (Gemmell, McQuilken & Walters, unpublished results). Since biocontrol fungi can be used in Integrated Pest Management (IPM) systems, if fungicides with novel modes of action to be used in an IPM system, it will be important to know their effects on biocontrol fungi.

CONCLUSIONS

In the period 1961 - 1988, world food production increased by 3.67 % annually (De Waard *et al*, 1993), due in part to improvements in agricultural productivity. Given that plant diseases account for an estimated 12 % crop loss globally every year (Schwinn, 1992), it is likely that fungicide use was important in bringing about improvements in agricultural productivity. Fungicides will continue to play an important role in reducing crop losses due to disease. But the development of fungicide resistance, and the implementation of new criteria for pesticide registration based on the toxicological and environmental consequences of fungicide use, have led to a reduction in the number of active ingredients available (DeWaard *et al.*, 1993). There is clearly a need for fungicides with novel modes of action and it would seem prudent therefore, to further our knowledge of polyamine biochemistry in phytopathogenic fungi, and to determine the precise mode(s) of action of fungicidal putrescine analogues. But fungicides, including newer fungicides with novel modes of action, must be used wisely, as part of a more integrated approach to plant disease control. This does not mean using fungicides as a last resort when other strategies have failed, but using them as an integral part of sustainable disease control systems.

ACKNOWLEDGEMENTS

We are grateful to BTG for generous financial support and to Dr R A'Court for advice and guidance. SAC receives financial support from SOAFD.

REFERENCES

Bitonti, A.J.; Bush, T.L.; McCann, P.P. (1989) Regulation of polyamine biosynthesis in rat hepatoma (HTC) cells by a bisbenzyl polyamine analogue. Biochemical Journal **257**, 769-774.

DeWaard, M.A.; Geogopoulos, S.G.; Hollomon, D.W.; Ishii, H.; Leroux, P.; Ragsdale, N.N.; Schwinn, F.J. (1993) Chemical control of plant diseases: problems and prospects. Annual Review of Phytopathology, **31**, 403-422.

Foster, S.A.; Walters, D.R. (1993) Fungicidal activity of the putrescine analogue, keto-putrescine. Pesticide Science, **37**, 267-272.

Havis, N.D., Walters, D.R., Martin, W.P., Cook, F.M. & Robins, D.J. (1994). Evaluation of the putrescine analogue (*E*)-1,4-diaminobut-2-ene against powdery mildew on spring barley. Tests of Agrochemicals and Cultivars, **15**, 14-15.

Havis, N.D.; Walters, D.R.; Cook, F.M.; Robins, D.J. (1995) Synthesis and fungicidal activity of novel cyclic diamines. Pesticide Science (in press).

Porter, C.W.; Sufrin, J.R. (1986) Interference with polyamine biosynthesis and/or function by analogs of polyamine or methionine as a potential anticancer chemotherapeutic strategy. Anticancer Research, **6**, 525-542.

Reitz, M.; Walters, D.R.; Moerschbacher, B.; Robins, D.J. (1995) Effects of two synthetic putrescine analogues on germination and appressorium formation by uredospores of *Uromyces viciae-fabae* on artificial membranes. Letters in Applied Microbiology (in press).

Schwinn, F.J. (1992). Significance of fungal pathogens in crop production. Pesticide Outlook, **3**, 18-24.

Walters, D.R. (1995) Inhibition of polyamine biosynthesis in fungi. Mycological Research, **99**, 129-139.

Zarb, J.; Walters, D.R. (1994) The formation of cadaverine, aminopropylcadaverine and *N,N-bis*(3-aminopropyl)cadaverine in mycorrhizal and phytopathogenic fungi. Letters in Applied Microbiology, **19**, 277-280.

Session 9

Forecasting and Modelling

Chairman and
Session Organiser   Dr G EDWARDS-JONES

DECISION SUPPORT SYSTEMS FOR INTEGRATED CROP PROTECTION

J. D. Knight.

Imperial College Centre for Environmental Technology, Imperial College at Silwood Park, Ascot, Berks, SL5 7PY, United Kingdom.

ABSTRACT

The use of decision support systems in integrated crop protection systems is discussed. Present and future trends in the development of decision support systems in this area are addressed with appropriate examples where available. The possible reasons for the relatively limited number of such applications are presented in terms of the problems that face the development of these system. An attempt is made to estimate the benefits that such systems can make toward improving crop protection in terms of financial return and reduced environmental impact.

INTRODUCTION

Decision support systems have been developed for and used in agriculture over the past 12 years or so in an attempt to improve all aspects of decision making and therefore management. At the same time continuing research into integrated crop protection has resulted in extremely large amounts of information being produced. This information, or knowledge, may take many different forms, for example, raw data contained within databases, qualitative experience of experts, simulation models of crops and pests and research reports. By definition, an integrated approach requires that all different methods of pest control should be considered when making a management decision. Whilst it is possible to make decisions, albeit poor ones, in the absence of information, in order to make the best possible decision it is necessary to be able to both understand and have access to all relevant information. The transfer of the wealth of information from researchers to practitioners is often a weak point in the process of improving integrated pest management, but it is an area where decision support systems can work particularly well.

Decision support systems can be made up of either a single component such as a database, geographic information system, expert system or simulation model or a combination of these (Knight & Mumford, 1994). They are able to aid decision makers by providing management options for a given set of conditions, clarifying the problem by predicting the likely future development of it or by explaining it in terms of past events. Decision support systems can also allow users to examine the consequences of making different decisions and therefore allow a variety of options to be examined. The idea of combining information from databases with simulation models and expert systems is not new, but the ability to do this quickly and easily has only really come about in the past few years with the widespread availability of cheap and powerful computers and software. Decision support systems are reliant upon the development of their component parts i.e. databases, models and expert systems. The development of these components should only be undertaken when it is appropriate to the problem. It is important that before any decision support system is developed that the problem has been rigorously defined and the requirements of the decision maker are fully understood.

A number of reviews of the use of expert and knowledge-based systems in agriculture and resource management have been carried out (Edwards-Jones, 1992; 1993; Warwick *et al.*, 1993) which highlight the subject areas that the systems cover and the number and type that have been described in the literature over the last 10-15 years. There has been a gradual move away from expert systems in the true sense toward knowledge-based systems and now decision support systems. The objective of the systems have remained the same, making better crop protection decisions, but the method of achieving this has shifted from the computer providing the definitive answer to the computer providing the user with all relevant information and some interpretation of that information to aid in the decision making process.

PRESENT AND FUTURE TRENDS

Decision support systems are used in many areas of crop protection including diagnosis or identification, prescriptive control and strategic control. The use of decision support systems for pest identification has led to a simplification of the identification process. Instead of traditional paper keys that are linear and require the user to work through the entire key in a systematic fashion, the computerised keys allow the user to choose a question about a particular character. The answer is used by the computer to eliminate all non-matching taxa. The user then chooses the next character and so on until an identification can be provided. These programs also have the ability to cope with the possibility of variability within the data set by using error allowance levels. An example of this type of key is DELTA (DEscription Language for Taxonomy (Dallwitz & Paine, 1986)) which is used in a program called INTKEY for the identification of beetle larvae and also grass genera. The advantage of this approach is that should a feature be damaged or missing the user can elect to not to answer questions about it. In a traditional dichotomous key it would be much more difficult to complete an identification under these circumstances. Another approach to this problem is CABIKEY (Computer Aided Biological Identification KEY(White & Scott, 1994)),which has the added sophistication of allowing users to have characters presented as pictures rather than pictures being purely in addition to text.

A common use of decision support systems is to provide an estimate of the probable damage that will occur to a crop with a given level of pests. In this case users are prompted by the computer to enter information that is necessary to run a model of the insect population development. The model can be of any type but is most commonly a simulation or a regression model. Once the data is entered into the system the model is run and a projection of the future population and consequent damage is made. The output from the model can be used along with information about the economic injury level to make an appropriate decision on control. There are many examples of this type of approach, for example, Berry *et al* , (1991); Knight & Cammell (1994); Perry *et al*, (1990);Stone & Schaub (1990); Wilkin & Mumford (1994). This strategic use of decision support systems is particularly useful since to use a model of pest populations without some guidance would be confusing for many users, by embedding the model in a user friendly interface and leading the operator through the process of specifying the values to enter into the model this particular problem is overcome. However, by hiding the model behind such an interface the user is less well able to understand how the system works and it becomes a bit of a black box. Therefore, a compromise between ease of use and transparency needs to be made.

The use of models in decision support systems allows users to experiment with different control strategies to see which achieves their objectives within a specific budget. For example, Grain Pest Adviser is being developed (Wilkin & Mumford, 1994) to contain models of mites and insects that infest grain, cooling and insecticide degradation. The system displays a graphic representation of the pest populations both with and without the use of control options so the user can judge which method provides the level of protection that the user requires. The inclusion of information about the costs of carrying out any of these operations, for example, the price of grain, cost of insecticide and the cost of electricity to cool the grain, allows the user to get an economic analysis of that strategy so they can make a decision based on both the success and cost of the control method. Since there is a level of uncertainty associated with any of these decisions because of changes in grain prices in the future the user is able to change the price of grain to see what effect it has on the chosen option. The 'best' option under one set of conditions may be very sensitive to changes in grain prices and therefore an apparently less good, but more robust, solution may be chosen because it is less susceptible to changes in price.

In order to implement increasingly complex integrated pest management programmes decision support systems are becoming ever more sophisticated and often comprise a knowledge base, a database and a model of some description, however, they can still be useful if they only contain one of these components and complexity for complexities sake is not a good thing.

ADOPTION OF DECISION SUPPORT SYSTEMS

It is perhaps surprising that the uptake of these systems has not become more widespread as pest management becomes more complex and decision support systems can provide a means of handling this increased complexity easily and at reasonable cost. There are a number of possible reasons for this.

The first point to note is that many of the early systems were produced to explore the potential of using this approach for the solution of pest management problems so were never designed to be widely used in the field. They were simply intended to show that it was possible to extract information from experts and the literature and create a decision support system suitable for use in pest management. The result was that decision support systems were found to be suitable for handling the type of problems that occur in pest management. The next phase was the development of systems intended for application in the field by farmers and/or advisers. Whilst some systems were widely adopted, for example, EPIPRE in the Netherlands (Zadoks, 1981), a system that provided information and advice on the timing of pest control in cereal crops, other systems do not appear to have been adopted on a large scale. The reasons for this low level of uptake by the farmers is not very clear but is probably due to a combination of factors. Firstly, farmers tend, by nature, to be risk averse and therefore behave conservatively. They are therefore not likely to adopt new technology until they are sure that it will provide a benefit to them over their present system. Decision support systems were not only new in that they provided advice and information in a new way, but also relied on the farmer being able to use, and probably own, a computer. This was undoubtedly a major hurdle to be overcome in the 1980's. Today, however, most farms have a computer for record keeping and farmers obviously have the ability to use them. The problem of fear of the unknown should no longer be an obstacle to the adoption of this technology. The way in which the information is presented to the farmer is probably still very important. The majority of advice comes from trusted advisers with which the farmers have developed a working relationship. The concept of using advice from a computer is quite a significant step and one which many other people may also be hesitant in taking. One way around this problem is to provide decision support to the adviser, who may otherwise be unable to advise on a particular problem, by tailoring the decision support system for his use. The farmer receives his advice from the adviser who can act as a bridge between the new technology and the farmer. In time the farmer may feel that he is happy to trust the computer output and use it himself.

Other possible reasons for the low level of adoption are that the programs have been developed in isolation from the farmers and they do not address the problems that the farmers or advisers feel are the most important. The decision support systems have grown out of an existing research project and are not always relevant. This particular problem can be prevented by getting farmer or adviser input at an early stage to identify and define the most difficult management problems. If the problem is suitable for solution by a decision support system then one can be built using input from the farmer and adviser. In this way the farmers feel they have some ownership of the system and are more likely to use it since they will have greater understanding of what the system contains and how it works.

Some of the reasons for the low adoption of the systems could be due to the poor marketing of the systems many of which have been developed within the academic community where there is little or no experience of marketing. This could lead to products being incorrectly priced, for instance, if the product is given away free the user may perceive the product to be of no value and not worth using. Conversely, if the product is priced too highly then the farmer will be unwilling to pay for the system when it may be cheaper to get advice from an adviser without any capital investment. This situation is also changing as free advice is becoming harder to get and the number of government advisers is being reduced making it more difficult to get advice at the time that it is required.

In order to improve the adoption of decision support systems the following points should be addressed; greater effort should be made to make them relevant to the farmers needs, care should be taken that the target market has the ability, willingness and hardware to use the proposed system and the benefits of the system should be clearly demonstrated to the user.

POTENTIAL BENEFITS OF USING DECISION SUPPORT SYSTEMS.

One of the main advantages of the decision support system is its ability to handle complex situations much more easily and efficiently than humans can. To demonstrate the potential advantage of using a decision support system over manual decision making a hypothetical example will be used.

The basis of most crop protection decision making is founded on the concept of the economic injury level as defined by Stern *et al* (1959). The objective of pest management is to reduce or maintain pest levels below the economic threshold and the concept is therefore valid for both preventative tactics and curative ones. The cornerstone of the economic injury level is knowing the damage function for the pest and crop in question and using this to calculate the economic injury level using the value of the crop and the cost of any control measures that may be used. If damage only occurs at a specific point in the crop life cycle then a damage function can easily be determined by a few relatively simple experiments. However, if the pest can attack the crop during many different developmental stages it is possible that the plant will be able to sustain more or less damage depending on when it is attacked. An example of this is the response of cotton (*Gossypium hirsutum*) to attack by the bollworm (*Helicoverpa zea*). During the early stages of development yields of cotton can actually be higher when the plant is attacked by bollworm than when it is not attacked because the plant over-compensates for the loss of any reproductive organs. At later stages of plant development the same level of injury will result in a much reduced yield since the plant will be unable to compensate for any losses. This particular problem means that for the economic injury level to be estimated with any degree of accuracy different damage functions should be used at different stages of development. The idea of using a different regression function for each growth stage of the cotton plant has been recommended by Stone & Pedigo (1972), with the assumption that the effect of insect injury on plant yield does not vary within the growth stage which is probably not true since the response will change gradually rather than at one specific point in time A refinement of this approach has been explored by Ring *et al*. (1993) by using a response surface for the damage function which illustrates that the economic injury level is not a static thing but is dynamic even when the value of the crop and control is constant. If the variability in crop price and the cost of control is added in, the variability in the economic injury level is even greater. It therefore becomes difficult to calculate the economic injury level manually and some sort of decision support system can be of great use.

Once the damage threshold has been determined in order to calculate the economic injury level the farmer has to know the cost of the control method. In the simplest terms the choice will be made by considering the efficacy of a particular treatment and the cost. The economic injury level can then be assessed and any action taken. Where there is a large range of control products the economic injury level may vary quite largely depending on the price of the particular control. The problem of which control method to choose and the consequent calculation of the economic injury level becomes even more complex if the environmental costs of the control is added into the equation. Attempts have recently been made to estimate the environmental costs associated with using a range of insecticides (Higley & Wintersteen, 1992) with a view to incorporating these into the calculation of the economic injury level.

In this example the crop is assumed to be worth $500/ha and the cost of control is $20/ha. Three scenarios are examined; firstly, using just one regression equation, secondly, using three regression equations for different growth stages (1100, 1500 and 1700 dd), and thirdly, using a response surface. The damage relationships are not real but could be representative of the sort of responses that can occur. The relationships are shown in Figure 1.

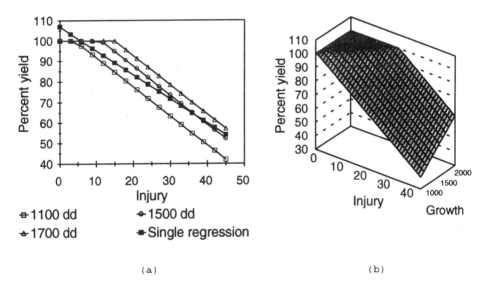

(a)   (b)

Figure 1. Damage relationships for a hypothetical crop: (a) a single estimate for all growth stages; separate estimates for 1100, 1500 and 1700 day degrees and (b) a response surface for all combinations of plant age and injury level.

It is assumed that the farmer checks his fields on three occasions (growth stages 1200, 1600 and 1800 day degrees) and finds injury levels of 7%, 15%, and 8% respectively. The farmer then has to make a decision to spray or not. The recommendations on whether to spray and the financial consequences of the actions are shown in Table 1.

It can be seen from the results that the use of the different methods can result in large differences in the financial returns to the farmer. The less sophisticated systems costing the grower more money. When the farmer also has to make a decision on which pesticide to use the economic threshold could change yet again depending on the cost of the method. The inclusion of the environmental costs into the calculation of the economic injury level increases the complexity of the decision still further. The above example is relatively simple but serves to demonstrate the point. Crops with variation in their responses to pest attack at different growth stages would be well suited to this approach.

Table 1. Comparisons of the economic injury levels and consequences of using them for a single regression estimate, three separate regressions and a response surface.

Method of calculating Economic Injury Level		Economic Injury level	Cost or benefit of method over response surface
Single regression	1200 DD	9.3	0.0
	1600 DD	9.3	33.5
	1800 DD	9.3	48.5
Three regressions	1200 DD	7.0	9.5
	1600 DD	14.2	9.0
	1800 DD	17.7	0.0
Response surface	1200 DD	8.9	-
	1600 DD	16.0	-
	1800 DD	19.5	-

CONCLUSIONS

Decision support systems are extremely useful and deserve to be used more widely than at present. For models requiring input, decision support systems can assist the user by reducing the need to have an extensive knowledge of the required values and the inner workings of the model. Similarly, the system can aid the user in interpreting results and suggesting decision options according to a wide range of scenarios. Decision support systems that include expert systems, geographic information systems simulation models and databases allow greater flexibility as they can be modified to take account of geographically specific parameters, assuming the required data is available. The process of specifying a system often helps to clarify a decision problem to both experts and farmers who are consulted during the procedure.

The increasing numbers of farm computers, the decline in traditional advisory services and improved user interfaces should help to improve the uptake of any new systems. It is interesting to note that systems for improving pest forecasting have been successfully deployed in Africa when it might be imagined that it would be more difficult than in developed countries. It is possible that the recipients in these countries have fewer preconceived ideas about what computers can and cannot do for them.

The ability of decision support systems to provide and interpret information for users is going to be of increasing importance as pest control becomes more and more complex with the balancing of a variety of control methods rather than just the traditional chemical approach. In addition, changes in crop protection legislation such as the potential need to include the environmental impact of any control decisions in the costs of control will also make the farmers task ever more complex and difficult.

REFERENCES

Berry, J.S.; Kemp, W.P.; Onsager, J.A. (1991) Integration of simulation models and an expert system for management of rangeland grasshoppers. *AI Applications, 5*, 1-13.

Dallwitz, M.J.; Paine, T.A. (1986) User's guide to the DELTA system; A general system for processing taxonomic descriptions. 3rd edition. *Report, CSIRO Division of Entomology, 13*, 1-106.

Edwards-Jones, G. (1992) Knowledge-based systems for pest management: An applications-based review. *Pesticide Science, 36*, 143-153.

Edwards-Jones, G. (1993) Knowledge-based systems for pest management: Theory and practice. *Crop Protection, 12*, 565-578.

Higley, L.G.; Wintersteen, W.K. (1992) A new approach to environmental risk assessment of pesticides as a basis for incorporating environmental costs into economic injury levels. *American Entomologist, 38*, 34-39.

Knight, J. D. & Cammell, M. E. (1994) A Decision Support System for Forecasting Infestations of the Black Bean Aphid, *Aphis fabae*, Scop. on spring sown field beans, *Vicia faba*. *Computers and Electronics in Agriculture, 10* 269-279.

Knight, J.D. & Mumford, J.D. (1994) Decision support systems in crop protection. *Outlook on Agriculture, 23*, 281-285

Perry, C.D.; Thomas, D.L.; Smith, M.C.; McClendon, R.W. (1990) Expert system coupling of SOYGRO and DRAINMOD. *Transactions of the American Society of Agricultural Engineers, 33*, 991-997.

Ring, D.R.; Benedict, J.H.; Landivar, J.A.; Eddleman, B.R. (1993) Economic injury levels and development and application of response surfaces relating insect injury, normalized yield, and plant physiological age. *Environmental Entomology, 22*, 273-282.

Stern, V.M.; Smith, R.F.; van den Bosch, R.; Hagen, K.S. (1959) The integrated control concept. *Hilgardia, 29*, 81-101.

Stone, N.D.; Pedigo, L.P. (1972) Development and economic-injury level of the green cloverworm on soybeans in Iowa. *Journal of Economic Entomology, 65*, 197-201.

Stone, N.D.; Schaub, L.P. (1990) A hybrid expert system/simulation model for the analysis of pest management strategies. *AI Applications, 4*, 17-26.

Warwick, C.J.; Mumford, J.D.; Norton, G.A. (1993) Environmental management expert systems. *Journal of Environmental Management, 39*, 251-270.

White, I.M.; Scott, P.R. (1994) Computerized information resources for pest identification: A review. In: *The identification and characterization of pest organisms*, D.L. Hawksworth (Ed.) Wallingford, CAB International, pp.129-137.

Wilkin, D.R.; Mumford, J.D. (1994) Decision support systems for integrated management of stored commodities. *Proc. 6th Int. Working Conference on Stored-Product Protection*, Canberra, Australia, April 1994. (in press)

Zadoks, J.C. (1981) EPIPRE: A disease and pest management system for winter wheat developed in the Netherlands. *EPPO Bulletin, 11*, 365-396.

THE USE OF MODELS IN INTEGRATED CROP PROTECTION

C.J. DOYLE

The Scottish Agricultural College, Auchincruive, Ayr KA6 5HW

ABSTRACT

An examination is made of the role that mathematical modelling can have in developing integrated strategies, involving reduced dependence on chemicals, for controlling weeds, pests and diseases in crops. From a detailed survey of the application of models to weed control, it is concluded that modelling has fundamentally been concerned with answering issues that contribute comparatively little to our understanding of 'sustainable' systems of crop protection. There has been a strong concentration of effort on simulating the impacts of herbicides and on identifying 'threshold' levels for spraying. Issues such as herbicide resistance, weed-crop interference and integrated weed management systems have received scant attention. Nonetheless, given the complexity of the management systems involved in Integrated Crop Protection, mathematical modelling would seem to be a potentially valuable tool. In the case of some forms of control, such as biological and genetic, it is also arguable that modelling the consequences is imperative, given the potential risks involved.

INTRODUCTION

'Integrated Crop Protection' is not a new idea and, since the term was first coined, it has come to mean a variety of things. In its simplest form it is used to describe a pest, disease or weed control strategy in which a variety of biological, chemical and physical control measures are combined to give stable long-term protection to the crop. More recently, the term has been used to describe more biologically-oriented control strategies that have arisen following problems with solely using chemicals (Lockhart et al., 1990; Swanton & Weise, 1991; Zimdahl, 1993). Equally, its underlying aims have been open to a variety of interpretations. Thus, when considered in the context of monitoring and forecasting pest, disease and weed infestations, the objective has often been seen to be pure profit maximisation (Cammell & Way, 1987). However, elsewhere Integrated Crop Protection has been seen as being concerned with the reduction of all inputs, as a means to safeguarding natural resources and minimising the impacts on the environment (Burn, 1987).

Nevertheless, if Integrated Crop Protection is to be seen as playing a pivotal role in the evolution towards 'sustainable' farming systems, then it seems more appropriate to accept the narrower definition of Burn et al. (1987), namely that it is the planned integration of a range of techniques to minimise the use of pesticides, herbicides and fungicides on the environment. As such, its aims are twofold:

i)   to steer the use of chemicals away from from prophylactic towards selective use, while still achieving an economically justifiable outcome; and

ii)   to minimise the use of chemical inputs and maximise the use of physical and biological controls for weeds, diseases and pests, so as to reduce environmental damage.

Thus, fundamental to devising an Integrated Crop Protection strategy is a sound biological understanding of the pest, disease or weed problem and the efficacy of different methods of control.   On this depends the choice of the subsequent control strategy. Equally important is an understanding of the benefits and risks involved for the grower, as well as for the rest of society. On the surface, it should be possible to gain useful insights into these issues using mathematical models, where a 'mathematical model' is defined as either an equation or a set of equations that represents the behaviour of a system (France & Thornley, 1984). Thus, models of weed infestation, population growth and control have served as a framework for organising biological information on weeds and for developing weed control strategies (Mortimer et al., 1980; Doyle, 1991). In particular, they have helped to identify information gaps, set research priorities and suggest control strategies (Maxwell et al., 1988). Furthermore, their value has arguably extended beyond being simply useful research tools. Thus, several key questions in weed control cannot be answered using conventional field trials, because of the constraints of cost, time or complexity (Doyle, 1989). As such, models have come to serve as experimental 'test-beds'.

However, a review of mathematical modelling work in weed control (Doyle, 1991) suggests that it is concerned with addressing a range of questions, which are tangential to the main thrust of Integrated Crop Protection. Thus, there has been a very marked concentration on studying the mechanisms for *controlling* rather than *preventing* weed infestation and within the range of control mechanisms to focus on chemical control. Virtually, without exception, all the published models concerned with the population biology of weeds have confined themselves to projecting what happens, given an initial level of weed infestation. In contrast, the mechanisms by which weeds are spread and dispersed, an understanding of which is central to preventing infestation and certainly to biological and genetic control, has been seriously neglected by mathematical modellers. Equally, in response to the rising on-farm expenditure on chemical control, modelling efforts have concentrated on evaluating the critical weed infestation levels above which chemical control is justified.

At the risk of some simplification, it would appear that mathematical models of weed, pest and disease control have largely directed their attention to the scientific question 'what' rather than the practical question 'how'. Thus, weed management models have primarily addressed three questions (Mortimer, 1987; Doyle, 1991):

i)   *what* is the relationship between the level of weed infestation and the crop losses;

ii)   *what* is the level of any control measure required to contain the infestation or totally eradicate the weed; and

iii)   *what* is the level of weed or pest infestation above which control measures are justified.

However, in respect of Integrated Crop Protection, these questions are subordinate to the more central issues of:

i)   *how* is it possible to promote the more selective use of herbicides, while ensuring economically acceptable levels of weed control;

ii)   *how* is it possible to minimise the environmental impacts of herbicides through the use of biological and physical control techniques; and

iii)   *how* are the economic risks to growers of switching to non-chemical controls to be minimised.

As shown by Tait (1987), this 'mis-match' between the aspirations of Integrated Crop Protection programmes and the main areas of enquiry conducted using mathematical models is not unique to weed control, but is evident in respect of pest control.

A ROLE FOR MATHEMATICAL MODELS ?

This apparent inapplicability of much past modelling work to the study of Integrated Crop Protection has led to widespread disillusion with so-called 'hard' quantitative models (Tait, 1987). This has been compounded by the fact that many crop disease, weed and pest problems are developing very rapidly and, by the time that models have been developed and potential solutions generated, the solutions have often ceased to be relevant. For mathematical modelling to make a contribution to sustainable systems of crop management, there is a need:

i).   to re-focus the areas of enquiry, so as to address more directly the issues raised by Integrated Crop Protection; and

ii)   to move away from modelling specific applied pest or weed problems to studying more strategic issues in relation to chemical resistance and the mechanism of biological controls.

To illustrate the type of scientific issues that need to be addressed and to assess the contribution that mathematical modelling can make to these, the case of Integrated Weed Management is considered.

In terms of future research on Integrated Weed Management systems, it is generally agreed (Swanton & Weise, 1991; Gressel, 1992; Wyse, 1994) that the key objectives are to identify:

i)   how to lower dependence on herbicides;

ii)   how to prevent or delay the development of herbicide-resistant strains of weeds;

iii)   how to use crop-weed interference techniques; and

iv)   how to integrate several weed control techniques, including selective
        herbicides.

<u>Lowering chemical dependence</u>

One of the primary aims of sustainable farming systems is to lower the use of chemical inputs. For a considerable period now, mathematical models have been used to examine ways of reducing the frequency and application rates of herbicides (Doyle, 1989; 1991). In these models, attention has mainly focused on modelling low weed infestation densities with the aim of identifying the minimum or 'threshold' density that justifies expenditure on weed control Specifically, the threshold density occurs where the cost of chemical control is equal to the net benefit in terms of enhanced crop yields. In estimating, the threshold levels for weed species, several existing models have taken account of the benefits in subsequent years (Doyle *et al.*, 1986; Mortimer, 1987; Doyle, 1989). This is important because chemical applications may affect not only the present weed population, but also indirectly future populations by preventing a build-up of seed in the soil. Clearly, this is directly relevant to a study of *sustainable* cropping systems.

Nevertheless, the primary motive behind these models has been to improve the *cost-effectiveness* of using herbicides, rather than reducing adverse 'environmental' impacts (Doyle, 1991). Nonetheless, they provide an existing body of knowledge, which can be applied to studying the opportunities for reducing the level of herbicide usage. Nor is it difficult conceptually to see how environmental impacts could be incorporated into such models by treating losses of biodiversity or water pollution as 'costs' associated with herbicide use. The inclusion of such environmental costs would then merely modify the threshold density at which herbicide use would be justified.

However, 'threshold' models have come under attack in recent years on four counts. First, they are dependent on experimental evidence regarding weed-crop competition. In many instances, the experiments are conducted at weed densities that are of limited relevance to the determination of economic thresholds (Dent *et al.*, 1989). Second, virtually all the threshold models developed have assumed that the weeds are uniformly distributed across the field. However, many weed species exhibit a marked tendency to cluster, leaving large areas of a field relatively free of infestation. Compared with a field in which the weeds are uniformly distributed, the crop yield loss will be less (Dent *et al.*, 1989; Brain & Cousens, 1990) and the consequent threshold density will tend to be higher.

The third criticism of threshold models is linked to the existence of uncertainty (Auld & Tisdell, 1987). In weed control, there are three principal sources that may modify the perceived optimal threshold density for spraying: i) the potential weed density; ii) the form of the crop loss function; and iii) the form of the herbicide dose-response function. A major factor in decision-making about whether to use a herbicide is the size of the weed population. Where a pre-emergence herbicide is to be used, then there must be some uncertainty about this. In the second place, although the general form of the crop loss function may be known, its precise shape varies with location and agronomic factors (Reader, 1985; Cousens *et al.*, 1988). Thus, the economic threshold for spraying will vary accordingly. Finally, the efficacy of a given herbicide in controlling weed infestation is sensitive to site and management practices (Zimdahl, 1993). Not only do these factors mean that the economic threshold density for a weed is subject to uncertainties, but the very existence of uncertainty is known to modify grower behaviour (Auld & Tisdell, 1987; Doyle, 1987; Pannell, 1990). If farmers are risk averse, then they are more likely to use herbicides in a prophylactic way and to apply them annually as a security against weed invasion The consequence of all this is that specific weed threshold densities become less relevant.

The other major conceptual problem with threshold models is that, in practice, treating the damaging external environmental effects of herbicides as a cost is not really workable. Apart from the problem of whether environmental damage, such as loss of plant and species diversity can be measured in 'economic' terms, the resultant threshold densities may be unacceptable. Basically, the effect of increasing the overall costs of applying chemical control is to increase the threshold weed density at which spraying is justified. It is conceivable that the inclusion of environmental costs raises the threshold to a level at which significant crop losses occur and which the grower would not be prepared to tolerate. Thus, in the absence of alternative means of controlling the weeds, the credibility of the predicted thresholds is subject to attack.

Accordingly, even to be useful research tools, threshold models will need to be adapted to take account of the four problems identified. This will require more information on: i) crop yield responses to low levels of weed infestation, ii) the patchiness of weed distributions; iii) the apparent variability of yield and dose responses between sites; and iv) the effect of uncertainty on the behaviour of growers themselves. In addition, if they are to be used to explore 'sustainable' weed control strategies, they will need to extended to include consideration of non-chemical means of control.

Managing herbicide resistance

However, the search to lower dependence on chemicals should not be allowed to mask the underlying problems. First, as Gressel (1992) and Wyse (1994) have observed, if crop producers had not had a fixation with weed-free fields and weed science had not concentrated so single-mindedly on chemical control, then the development of weed species with herbicide resistance would not now exist and the constant search for new and more powerful herbicides would not be necessary. For a long time it was assumed that, because weeds have relatively long life cycles and the same chemical is not used repetitively on the same land, weeds would not develop resistance to herbicides in the same way as insects have to insecticides. However, over 100 cases of herbicide resistance have now been reported in one or more of 15 herbicide chemical families (Zimdahl, 1993). As a result, understanding the evolution and dynamics of herbicide resistance in weed populations has now become a major issue in weed science. Arguably the complexity of the the biological processes which influence herbicide resistance dictates a research approach that focuses on the interactions between life history processes and population genetics. Mathematical models can serve such a function and can provide a tool for evaluating management tactics.

However, the vast majority of mathematical models developed to study weed control strategies are fundamentally empirical. That is to say they can describe the response of the weed population to a given herbicide dose, but they cannot explain the mechanism by which reduction in weed numbers occurs. To do this, it is necessary to develop *eco-physiological* models of weed growth and development which simulate key physiological processes, such as photosynthesis and partition of photosynthate. Only in the last two or three years have such models begun to emerge for a limited range of weed species (Kropff & Spitters, 1992; Cousens *et al.*, 1992; Weaver *et al.*, 1993). The primary constraint has been, and continues to be, shortage of detailed experimental data on weed physiology.

Therefore, studying herbicide resistance by means of mathematical models has remained more an aspiration than a practical reality. Nonetheless, Maxwell *et al.* (1990) did make a serious attempt to address the issue using a simulation model, based on gene flows. Basically, the flow of genes is seen as directly altering the proportion of herbicide-resistant and non-resistant alleles in the weed population. Herbicide-resistant genes are introduced into the population both by immigration of pollen and seed and by genetic drift within the existing population. Attempts to manage herbicide resistance then involves two distinct strategies, namely i) the use of alternative herbicides to remove 'resistant' plants and ii) the manipulation of the non-resistant type gene to increase its incidence in the population. The authors conclude from the modelling exercise that the latter strategy may be more cost effective.

Modelling crop-weed interference

However, a reduction in herbicide requirements is only the first step towards 'sustainable' systems of crop protection. The need is to develop non-chemical methods, which may be used in conjunction with low levels of herbicides to control weeds. Crop interference with weeds is one of the primary non-chemical methods of weed control (Wyse, 1994). Most of the cultural practices adopted by growers, as part of their production systems, are designed to create an environment that allows the crop to interfere with weeds to the greatest extent possible. However, the complexities of crop-weed interference mean that mechanistic models, such as the so-called *eco-physiological* models referred to previously, are required. The reason for this is that it is through the mechanisms of competition for water, nutrients and light that the crop interferes with weed development. Most of the weed control models surveyed in Doyle (1991) purely rely on empirical relationships based on plant densities to determine competitive effects between the weed and crop. Typical of such models is the one by Firbank & Watkinson (1985) in which the density of both weeds and crop plants (D), together with the corresponding yields per individual plant (Y), are presumed to be subject to density-dependent mortality, which can be represented by a non-linear reciprocal model of the type:

$$D_1 = D_0 [1 + \alpha_0 (D_1 + \alpha_1 D_2)]^{-1}$$
$$Y_1 = Y_0 [1 + \alpha_2 (D_1 + \alpha_3 D_2)]^{-1}$$

(1)

where α_0 to α_3 are constants and D_0 and Y_0 denote the initial plant densities and the mean yield of isolated plants respectively. This model depicts the density (D_1) and yield (Y_1) of the crop as a function of the densities of both the weed and crop species. If the weed density is very low, then the crop yield is projected to be a linear function of the crop density. However, at higher densities, the total crop yield becomes constant, that is to say independent of the density of the crop. The argument for this is that, as weed competition becomes more severe, the combined biomass yield of weeds and the crop becomes restricted by the total availability of resources in the given habitat. While this model 'describes' biological reality, it is clearly not capable of explaining the mechanism by which the weed interferes with the crop, through competition for light, water and nutrients.

At the same time, it is rare for a crop to be invaded by a single weed species. As such it is better to envisage a weed population as comprising a multi-species assemblage (van Groenendael, 1988). Although this concept has little place in the modelling of weed-crop competition, it is central to neighbourhood models used for studying ecological competition among plants (Pacala & Silander, 1985; 1987). The basic idea is that the performance of an individual plant can be determined from the number, distance and type of neighbours. For each individual plant, it is possible to identify a neighbourhood area within which there is interference from neighbours and outside which such effects are negligible. However, from a practical standpoint, the use of neighbour models in weed management for situations involving more than a few species is likely to prove computationally expensive (Doyle, 1991). Although Swinton & King (1994) have tried to overcome this, it is arguable that they have achieved it only at the expense of losing all the 'mechanistic' properties of the model.

Modelling Integrated Weed Management

While the search for more sustainable farming systems focusses attention on alternatives to chemical control, Integrated Weed Management is fundamentally concerned with studying how a variety of control methods, including herbicides, may be combined to produce acceptable levels of weed control (Swanton & Weise, 1991). Defined as such, Integrated Weed Management has been studied only to a limited extent using mathematical models, despite the obvious potential (Pandey & Hardaker, 1995). Some early attempts, including studies by Cousens et al. (1986) and Doyle et al. (1986), did investigate the implications of cultivation techniques and straw-burning on economic thresholds for herbicide spraying for winter wheat. However, they were not primarily concerned with evaluating the possibilities for integrating different control techniques. Only in the last few years have models of weed systems, which explore combined control techniques, been developed. In many cases, the impetus has come from an interest in assessing the economic risks and benefits associated with biological and genetic control methods (Pandey & Medd, 1990; Volker & Boyle, 1994). However, there have been one or two attempts to use models to explore the interactions between cultivation practices and herbicide usage as a way or reducing dependence on chemical controls in arable systems (Frank et al., 1992; Rasmussen, 1992).

What is strikingly absent is any attempt to use mathematical models to explore the social, physical and economic risks of non-chemical weed control techniques. As Pannell (1990) demonstrated using a model, risk considerations may lead to on-farm practices deviating from weed management recommendations. Thus, models can help to reveal inconsistences between recommended practices and the needs and circumstances of growers. This is especially the case, where integrated control methods are being proposed, often involving complex management decisions in the field. For biological and genetic control techniques, there are additional ecological and environmental risks and Gibson (1994) has shown how useful models may be in assessing these, before the technique is introduced into practice. However, in terms of Integrated Weed Management, these are all uncharted areas as far as modelling is concerned.

CONCLUSIONS

Thus, from a general survey of modelling work centred on weed control, it is evident that, although there is a general acceptance of the potential for using models to assist in the development of Integrated Crop Protection, the reality falls far short of the potential. Issues critical to the development of more sustainable crop protection systems, such as an understanding of the mechanisms for delaying herbicide resistance, the use of crop interference techniques and of integrated control systems, have not formed the basis of mathematical models. The reasons for the gap are fourfold. First, models of weed, pest and disease control in crops have concentrated strongly on simulating the impacts of chemical control and within that on identifying the threshold level of infestation for spraying. Thus, non-chemical control methods have received scant attention, as has the notion of combining different control methods. Second, the criteria used to determine the 'optimal' method of control has been 'cost effectiveness'. Other criteria, like minimising envionmental damage have been largely ignored. This has encouraged the selection of chemical methods of control. Third, models are only as good as the data on which they are based and, while a large body of data exists on herbicide response, much less data exists on the impact of other forms of control. There is also an acute shortage of information relating to the physiological processes involved in the growth and development of weeds. In consequence, currently it is often not possible to develop mechanistic models, which are needed if some of the alternative methods of control are to be explored. Fourth, the need for Integrated Crop Protection programmes often arises from the fact that control of more than one species of pest or weed is required. However, the techniques so far developed for simulating multi-species competition and control are compuitionally expensive, limiting their use. Nonetheless, given the complexity of the management systems involved in Integrated Crop Protection, mathematical modelling would seem to be a potentially valuable tool. Moreover, in the case of some forms of control, such as biological and genetic, it is also arguable that modelling the consequences prior to practical application is imperative, given the potential ecological and environmental risks involved.

ACKNOWLEDGEMENTS

The comments of Gareth Edwards-Jones on a first draft of the paper are gratefully acknowledged.

REFERENCES

Auld, B.A.; Tisdell, C.A. (1987) Economic threshold and response to uncertainty in weed control. Agricultural Systems, **25**, 219-227.

Burn, A.J. (1987) Cereal crops. In: Integrated Pest Management, A.J. Burn, T.H. Coaker and P.C. Jepson (Eds), London: Academic Press, 209-256.

Burn, A.J; Coaker,T.H.; Jepson, P.C. (1987) Preface. In: Integrated Pest Management, A.J. Burn, T.H. Coaker and P.C. Jepson (Eds), London: Academic Press, vii-viii.

Brain, P.; Cousens, R. (1990). The effect of weed distribution on prediction of yield loss. Journal of Applied Ecology, **27**, 735-742.

Cammell, M.E.; Way, M.J. (1987) Forecasting and monitoring. In: Integrated Pest Management, A.J. Burn, T.H. Coaker and P.C. Jepson (Eds), London: Academic Press, 1-26.

Cousens, R.; Doyle, C.J.; Wilson, B.J.; Cussans, G.W. (1986) Modelling the economics of controlling Avena fatua in winter wheat. Pesticide Science, 12, 1-12.

Cousens, R.; Firbank, L.G.; Mortimer, A.M.; Smith, R.G.R. (1988) Variability in the relationship between crop yield and weed density for winter wheat and Bromus sterilis. Journal of Applied Ecology, 25, 1033-1044.

Cousens, R.D.; Johnson, M.P; Weaver, S.E.; Martin, T.D.; Blair, A.M. (1992) Comparative rates of emergence and leaf appearance of wild oats (Avena fatua), winter barley (Hordeum sativum) and winter wheat (Triticum aestivum). Journal of Agricultural Science, Cambridge, 118, 149-156.

Dent, J.B; Fawcett, R.H.; Thornton,P.K. (1989) Economics of crop protection in Europe with reference to weed control. British Crop Protection Conference Weeds 1989, 917-926.

Doyle, C.J. (1987) Economic considerations in the production and utilization of herbage. In: Ecosystems of the World 17B: Managed Grasslands-Analytical Studies, R.W. Snaydon (Ed), Amsterdam: Elsevier, 217-226.

Doyle, C.J. (1989) Modelling as an aid to weed control management. British Crop Protection Conference -Weeds 1989, 937-942.

Doyle, C.J. (1991) Mathematical models in weed management. Crop Protection, 10, 432-444.

Doyle, C.J.; Cousens, R.; Moss, S.R. (1986) A model of the economics of controlling Aloopecurus myosuroides Huds. in winter wheat. Crop Protection, 5, 143-150.

Firbank, L.G.; Watkinson, A.R. (1985) On an analysis of competition within two-species mixtures of plants. Journal of Applied Ecology, 22, 503-517.

France, J; Thornley, J.H.M. (1984) Mathematical Models in Agriculture, London: Butterworths, 350pp.

Frank, J.R.; Schwartz, P.H.; Potts, W.E. (1992) Modelling the effects of weed interference periods and insects on Bell peppers (Capsicum annuum). Weed Science, 40, 308-312.

Gibson, G. (1994) THe nature of systems modelling and its use as a tool for policy analysis. In: Systems Modelling and the Rural Economy: A Study into the Feasibility of Using Mathematical Models as a Decision Aid to Policy Makers, G. Gibson, J. Crawford, A. Sibbald, R. Aspinall and C. Doyle. Report prepared for Scottish Office Agriculture and Fisheries Department, Edinburgh: Scottish Agricultural Statistics Service, 27-58.

Gressel, J. (1992) Addressing real weed science needs with innovation. Weed Technology, 6, 509-525.

Kropff, M.J.; Spitters, C.J.T. (1992) An eco-physiological model for interspecific competition, applied to the influence of Chenopodium album L. on sugar beet I: Model description and parameterization. Weed Research, 32, 437-450.

Lockhart, J.A.R; Samuel, A.; Greaves, M.P. (1990) The evolution of weed control in British agriculture. In: Weed Control Handbook: Principles, R.J. Hance and K. Holly (Eds), Oxford: Blackwell Scientific Publications, 1-42.

Maxwell, B.D.; Wilson, M.V.; Radosevich, S.R. (1988). Population modelling aproach for evaluating Leafy Spurge (Euphorbia esula) development and control. Weed Technology, 2, 132-138.

Maxwell, B.D.; Roush, M.L.; Radosevich, S.R. (1990) Predicting the evolution and dynamics of herbicide resistance in weef populations. Weed Technology, 4, 2-13.

Mortimer, A.M. (1987) The population ecology of weeds - implications for integrated management, forecasting and conservation. British Crop Protection Conference Weeds 1987, 935-944.

Mortimer, A.M.; McMahon, D.J.; Manlove, R.J.; Putwain, P.D. (1980). The prediction of weed infestations and cost of differing control strategies. British Crop Protection Conference - Weeds 1980, 415-422.

Pacala, S.W.; Silander, J.A. (1985) Neighbourhood models of plant population dynamics I: Single-species models of annuals. American Naturalist, 125, 385-411.

Pacala, S.W.; Silander, J.A. (1987) Neighbourhood interference among velvet leaf (Abutilon theophrastis) and pigweed (Amaranthus retoflexus) communities. Oikos, 48, 217-224.

Pandey, S.; Medd, R.W. (1990) Integration of seed and plant kill tactics for control of wild oats: An economic evaluation. Agricultural Systems, 34, 65-76.

Pandey, S.; Hardaker, J.B. (1995) The role of modelling in the quest for sustainable farming systems. Agricultural systems, 47, 439-450.

Pannell, D. (1990) Responses to risk in weed control decisions under expected profit maximisation. Journal of Agricultural Economics, 41, 391-403.

Rasmussen, J. (1992) Testing harrows for mechanical control of annual weeds in agricultural crops. Weed Research, 32, 267-274.

Reader, R.J. (1985) Temporal variations in recruitment and mortality for the pasture weed Hieracium floribundum. Implications for a model of population dynamics. Journal of Applied Ecology, 22, 175-183.

Swanton, C.J.; Weise, S.F. (1991). Integrated weed management: The rationale and approach. Weed Technology, 5, 657-663.

Swinton, S.M.; King.R.P. (1994) A bioeconomic model of weed management in corn and soybean. Agricultural Systems, 44, 313-315.

Tait, E.J. (1987) Planning an integrated pest management system. In: Integrated Pest Management, A.J. Burn, T.H. Coaker and P.C. Jepson (Eds), London: Academic Press, 189-208.

van Groenendael, J.M. (1988) Patchy distribution of weeds and some implications for modelling population dynamics: A short literature review. Weed Research, 28, 437-441.

Volker, K.; Boyle, C. (1994) Bean rust as a model system to evaluate efficiency of teliospore induction, especially the potential mycoherbicide Puccinia punctiformis. Weed Research, 34, 275-281.

Weaver, S.E.; Kropff, M.J.; Cousens, R. (1993) A simulation model of Avena fatua L. (wild-oat) growth and development. Annals of Applied Biology, 122, 537-554.

Wyse, D.L. (1994). New technologies and approaches for weed management in sustainable agriculture systems. Weed Technology, 8, 403-407.

Zimdahl, R.L. (1993). Fundamentals of Weed Science, London: Academic Press, 450pp.

FORECASTING ATTACKS BY INSECT PESTS OF HORTICULTURAL FIELD CROPS

R.H. COLLIER

Horticulture Research International, Kirton, Boston, Lincs PE20 1NN

S. FINCH, K. PHELPS

Horticulture Research International, Wellesbourne, Warwicks CV35 9EF

ABSTRACT

The timing of attack by pest insects can vary greatly both from region to region and from year to year. A simulation method, based on rates of insect development, has been developed for forecasting the timing of insect attacks. The method is based on using a fixed number of individuals from one generation to the next; and simulates the timing of events in the life-cycle of the pests rather than the population dynamics of the insects. Forecasts produced for the cabbage root fly, the carrot fly, the bronzed-blossom beetle and the large narcissus fly have been validated using pest monitoring data. Forecasts can be generated on a regional basis from standard meteorological data, or on a local basis from air and soil temperatures collected by participating growers.

INTRODUCTION

Insect pests of horticultural crops are often controlled by spraying insecticide onto established crops. Such sprays are used against root-feeding insects because the activity of insecticides applied at drilling or planting has diminished by the time later generations of the pest are active. Such sprays are also the most feasible way of controlling foliar pests such as aphids and caterpillars. Since the majority of currently-recommended insecticides are of relatively short persistence, treatments are most effective if they are targeted to coincide with periods of peak pest activity. Unfortunately, the timing of such peaks can vary considerably from region to region and from year to year. Although it is possible to monitor the activity of many pest species using insect traps, routine monitoring is laborious and often requires specialist knowledge. For a few pests, such as the large narcissus fly (*Merodon equestris*), an effective monitoring technique has not yet been developed.

An alternative is to use weather data to forecast the timing of pest attacks. Forecasting systems have been developed for many insects. For example, Finch (1989) cites references to ten separate models for forecasting the timing of attack by four pest species of *Delia* (Diptera, Anthomyiidae). Many of these forecasts have been based on day-degrees (e.g. Eckenrode & Chapman, 1972). However, day-degree forecasts have severe limitations, as their accuracy is based on the assumption that the relationship between the rate of insect development and temperature is strictly linear (Baker, 1980). In addition, day-degree forecasts can be used only to predict the start and/or the peak of activity of the population. They cannot readily predict the spread of activity nor can they cope easily with insect populations which have polymodal patterns of activity. For example, the cabbage root fly (*Delia radicum*) can occur as one of two developmental biotypes, that emerge either 'early' (in April-May) or 'late' (in June-July) in the season (Finch & Collier, 1983; Finch *et al.*, 1988). Within a particular locality, the population of cabbage root flies may consist primarily of one biotype or be a mixture of both. Finally, further problems in the use of day-degree models occur when attempting to interpret the overall effects of periods of insect dormancy, either diapause or aestivation, induced by changes in temperature or photoperiod. Since there is considerable variation between individuals in their rates of development, it is usual for only a proportion of the population to respond at any one time to a particular environmental cue. This is true of cabbage root fly populations during both aestivation and diapause (Collier & Finch, 1983; Finch & Collier, 1985).

At Horticulture Research International, a simulation method, based on rates of insect development, has been produced for forecasting the timing of attack by a number of pest insects (Phelps *et al.*, 1993). Variability between insects in their rate of development is also incorporated. The simulation method has been used to develop forecasts for the cabbage root fly, carrot fly (*Psila rosae*), bronzed-blossom beetle (*Meligethes* spp.) and large narcissus fly. The forecasts produced are now being validated with growers. Similar forecasts for certain pest aphids and caterpillars are under development. The biological basis, validation and practical uses of such forecasts are discussed in this paper.

THE MODEL

The forecasts for cabbage root fly, carrot fly, bronzed-blossom beetle and large narcissus fly were developed using a Monte Carlo simulation method (Phelps *et al.*, 1993). The method uses a fixed number of individuals (usually 500, to obtain repeatable simulations) from one generation to the next and simulates the timing of events rather than the population dynamics of the insects. To develop each model, individuals at each stage of development (egg, larva, pupa, adult) were reared in cooling incubators at a range of constant temperatures between 6 and 30°C to determine the relationship between rate of insect development and temperature. Linear or non-linear (Gompertz) curves were fitted to these data to provide equations which could be incorporated into the model. In addition, variability was incorporated using the 'same-shape property' (Sharpe *et al.*, 1977; Shaffer, 1983). This implies that the coefficient of variation of the rate of insect development is constant at all temperatures. Account was also taken of periods of dormancy (aestivation and diapause) and of activity thresholds which might affect the outcome of the forecasts.

Ideally the forecasts should be run using daily maximum and minimum air temperatures and maximum and minimum soil temperatures at a depth of 6-10 cm. However, maximum and minimum soil temperatures are not available from standard agro meteorological stations in the UK. Therefore, the program uses several equations to estimate soil maximum and minimum temperatures from the air maximum, air minimum and 10 cm soil temperatures recorded daily at 09.00 h GMT. If soil maximum and minimum data are available then the forecasts could be run equally-well using these.

The forecasts are validated using appropriate insect monitoring data. Cabbage root flies are monitored using yellow water traps (Finch & Skinner, 1974) or by sampling eggs (Finch *et al.*, 1975), carrot flies using orange sticky traps (Collier *et al.*, 1990a), bronzed-blossom beetles using yellow sticky traps (Finch *et al.*, 1990a) and large narcissus fly by the emergence and subsequent egg-laying activity of insects maintained in field cages (Finch *et al.* 1990b).

At present, pest forecasts are based on Meteorological Office data collected from the network of weather stations, some of which are not particularly close to areas of commercial vegetable/flower production. The actual forecasts are projected forwards using weather data from a previous, warm, year. In a few instances forecasts have also been generated using growers' own weather data. During the last four years, and as part of the validation process, forecasts of pest activity, based on weather data from 38 weather stations throughout the UK, have been made available to Horticultural Development Council levy payers. Forecasts have been sent to growers each week for several weeks before and during the period of pest activity.

Cabbage root fly

The cabbage root fly forecast (Collier *et al.*, 1991) was developed originally for timing the application of mid-season insecticide treatments to long-season brassica crops such as swedes. Other uses of the forecast include warnings of the likely onset of third generation attack to Brussels sprout buttons and to autumn-sown crops of oilseed rape. At present, most insecticide treatments to leafy brassicas are applied prophylactically, before or soon after transplanting, and treatment against subsequent generations of this fly is usually unecessary. However, the forecast could be used to indicate 'windows' where treatments would not be required. Figure 1 shows comparisons of observed and forecast cabbage root fly activity at Wellesbourne, Warwicks in 1985 and Feltwell, Norfolk in 1989.

Local variations in cabbage root fly activity include the co-existence in certain regions of the two developmental biotypes, with diapause of different durations (Finch & Collier, 1983; Collier *et al.*, 1989). Late-emerging flies emerge several weeks later than early-emerging flies so that their generations alternate. Similar damage to brassicas is caused by the closely-related turnip fly (*Delia floralis*) in Scotland and in some areas of south-west Lancashire (Finch *et al.*, 1986). The presence of the two cabbage root fly biotypes and turnip fly in areas of south-west Lancashire means that there is continuous root fly pressure to brassica crops throughout the summer. The cabbage root fly model produces forecasts for populations containing specified proportions of the two biotypes. A turnip fly forecast has not yet been developed.

The cabbage root fly model has also been used to indicate what might happen as a result of global warming (Collier *et al.*, 1990b) and to predict cabbage root fly phenology in Spain, where calabrese production for the UK market has been affected severely by larval damage to the florets.

Figure 1. Comparisons of observed and forecast cabbage root fly activity. Forecasts of egg-laying are compared with the numbers of females captured in water traps.

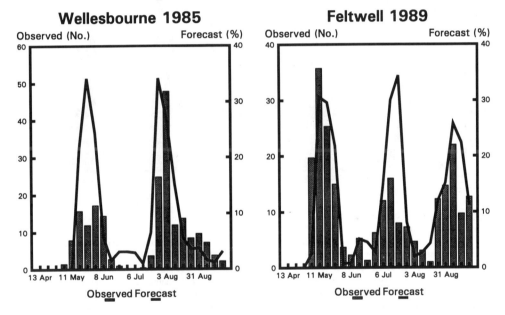

Carrot fly

The carrot fly forecast (Collier *et al.*, 1992) was developed to target mid-season insecticide treatments against attacks by second and third generation carrot fly. The carrot fly overwinters either as a pupa in diapause or remains in the larval stage, forming a non-diapause pupa in the spring. In the spring, adults emerge earlier and over a shorter period from insects that overwinter as larvae than from those that overwinter as pupae. The model can produce forecasts for populations containing specified proportions of diapausing pupae provided such information is known. If not, it is usually assumed that 50% of the population have overwintered in diapause. Figure 2 shows comparisons of observed and forecast carrot fly activity at Wellesbourne in 1986 and Ely, Cambridgeshire in 1989.

Carrot fly is a relatively non-mobile pest and the timing of activity in a particular crop can also be influenced by factors such as drilling date and the proximity of previous infestations. The program will produce forecasts for crops drilled on specific dates. Some mid-season insecticide treatments probably work mainly as larvicides whilst others, such as the pyrethroids, cypermethrin and the newer product lambda-cyhalothrin, mainly kill adults (Dufault, 1994). Therefore, the various insecticides may need to be applied at different times in the life-cycle of the pest.

The carrot fly forecast could also be used to predict the onset of damage in carrots so that they can be lifted and put into cold storage (Jonsson, 1992). This is a technique used widely in Northern Europe and North America.

Figure 2.   Comparisons of observed and forecast carrot fly activity. Forecasts of carrot fly egg-laying are compared with the numbers of flies captured on sticky traps.

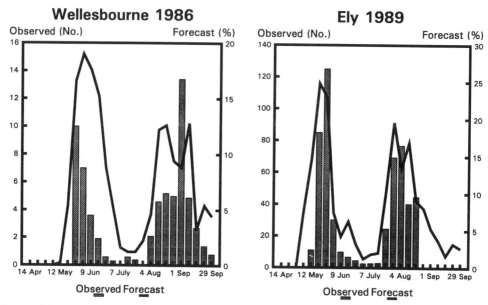

Bronzed-blossom beetle

Feeding by adult bronzed-blossom beetles in mid-summer, prior to hibernation, damages the curds or florets of cauliflower and calabrese (Finch et al., 1990a) so that spray treatments are sometimes necessary. The forecast is used to predict the emergence of adult beetles from pupae within the previous host crop; usually oilseed rape. However, beetle infestations are not inevitable and seem to depend both on the proximity of oilseed rape crops and on the occurrence of warm, humid conditions during the main period of beetle migration. Figure 3 shows comparisons of observed and forecast bronzed-blossom beetle activity at Wellesbourne and at Stockbridge House, Cawood, Yorkshire in 1990.

Large narcissus fly

An effective monitoring technique has not yet been developed for the large narcissus fly. Apart from general field observations, growers have no other way of determining when the fly is active. Following the withdrawal of aldrin in 1989, alternative control measures based on less-persistent insecticides and cultural techniques, are now being developed.

One possibility is that insecticides could be used to kill adult flies prior to egg-laying, which would require an accurate forecast of adult emergence. Alternatively, cultural techniques could be aimed at the prevention of egg-laying by, for example, charring or flailing narcissus foliage (S.Tones, personal communication), or lifting the crop early to avoid invasion of narcissus bulbs by newly-hatched larvae. Premature destruction of foliage, or the lifting of bulbs early, may reduce subsequently both bulb and flower yields. Forecasts would be required to indicate the latest date by which cultural operations could achieve adequate pest control without reducing yield.

Figure 3.   Comparisons of observed and forecast bronzed-blossom beetle activity. Forecasts of the summer migration of bronzed-blossom beetles are compared with the numbers of beetles captured on sticky traps, both in edible brassicas and oil seed rape.

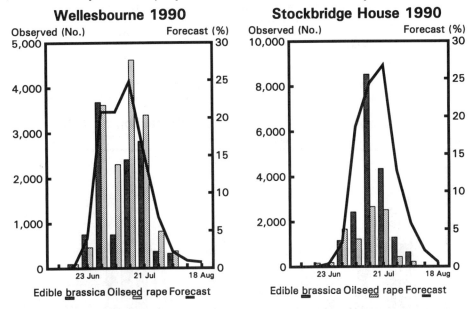

Adult large narcissus flies have a critical activity threshold, close to 20°C. Flight activity, and mating, is prevented at lower temperatures. Therefore, even if females have completed their 2-3 day pre-oviposition period, they remain hidden in crevices in the soil and are unable to lay until temperatures are high enough for mating. Once a female has mated, egg-laying can occur at lower temperatures (Collier & Finch, 1992). This threshold for mating has been incorporated into the forecast, since it determines the timing of egg-laying. In addition, reduced fly activity during periods of cold weather appears to reduce considerably the efficacy of insecticides applied against adult flies (S.Tones, personal communication). Figure 4 shows comparisons of observed and forecast large narcissus fly activity at Wellesbourne and Starcross, Devon in 1990.

FORECAST VALIDATION

Forecasts have been validated against as many sets of insect monitoring data as possible. The timing of pest activity may vary by 3-5 weeks between years and both the cabbage root fly and carrot fly may have 'partial' third generations in very warm years. The activity of cabbage root fly and carrot fly have been monitored at Wellesbourne for over 10 years. Earliest second generation cabbage root fly activity at Wellesbourne was recorded on 12 July in 1989 and latest activity on 5 August in 1986. Similarly, earliest second generation carrot fly activity was recorded on 21 July in both 1989 and 1990 and the latest on 26 August in 1986. Extensive monitoring data, to validate the pest forecasts, have also been collected over a number of years from the major areas of intensive vegetable production.

Use of the models has indicated that monitoring data must consist of >100 insects per generation if estimates of the timing of pest attacks are to be accurate to within one week (Collier & Phelps, 1994). As the forecasts provide an indication only of the timing of pest activity and not of the severity of attack, forecast data are generally expressed as percentages. When the times to 10% and 50% activity have been predicted and compared with the monitoring data, the majority of the pest forecasts have been accurate to within one week.

Forecasts have been produced using a network of rather widely-dispersed weather stations. However, there are obviously local differences in climate and in the degree of shelter which might affect the timing of pest activity in a particular field. With very mobile insects such as the cabbage root fly, there may be little point in recording temperatures in individual fields since the infesting population will have experienced the climate of the previous weeks, or months, in a different, unknown location. An intensive study in the Vale of Evesham (Finch & Skinner, unpublished data) indicated that there was very little difference in the timing of cabbage root fly activity from crop to crop (Figure 5). Although timing of activity may vary little within a region, intensive sampling in south-west Lancashire showed that the relative proportions of the two cabbage root fly biotypes varied considerably over relatively short distances (Finch *et al.*, 1986).

Figure 4.   Comparisons of observed and forecast large narcissus fly activity. Forecasts of large narcissus fly emergence are compared with the emergence of flies into large field cages.

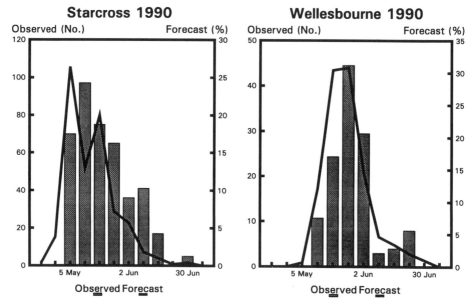

There are indications that the timing of carrot fly activity within a region may be more variable than that of the cabbage root fly but this variation may be attributable to differences in drilling date and to distance from sources of previous infestation. Both factors are now being investigated. With both species of fly, adults can be trapped close to their emergence sites during the pre-oviposition period, which during cold springs may last for several weeks (Collier & Finch, 1985). Therefore, in certain situations, data collected at emergence sites should not be used to discredit the accuracy of the egg-laying forecasts, as the two sets of information relate to different, not comparable, phases in the life-cycle of the pest.

The 20°C activity threshold of the large narcissus fly presents practical problems for validating forecasts of emergence since, on cool days, inactive insects are very difficult to find even in heavily-infested plots of narcissus grown within field cages.

FUTURE DEVELOPMENTS

The use of less-persistent insecticides, and pressure from consumers and retailers to reduce the number of insecticide treatments applied to crops means that growers need to target insecticide treatments more accurately. During a recent survey (Parker & Phelps, 1994), 78% of brassica growers indicated that they would find systems for forecasting pest attacks useful. In future, such forecasts could be produced by continuing to use Meteorological Office data to which could be added refinements to allow corrections for altitude and coastal effects. Alternatively, if growers were to obtain weather stations they could easily run the pest forecast models on their own computers. The brassica growers' survey (Parker & Phelps, 1994) showed that two thirds of growers already own a computer and a similar number make use of weather data collected either by themselves or by the Meteorological Office. Therefore, the option to run their own forecasts may already be feasible for many growers.

The next logical step, after determining the timing of pest activity, is to develop treatment thresholds to determine which of the various treatments are actually necessary. However, this will require a considerable amount of basic research, to produce robust systems that will enable final crop damage to be forecast accurately from the numbers of insects monitored during the early stages of crop infestation.

Figure 5.   The times of 50% first and second generation cabbage root fly egg-laying activity in the Vale of Evesham in 1980.

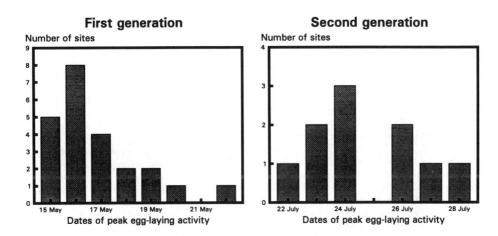

ACKNOWLEDGEMENTS

We thank MAFF for supporting this work as part of Project HH0103JFV and the HDC as part of Projects FV13, FV44, FV127 and BOF1. We thank also participating growers, Miss M. Elliott and Mrs W. Riggall of HRI, and Dr J. Blood Smyth, Mr B. Emmett, Mr S. Tones and Dr W. Parker of ADAS for their help in collecting the monitoring data; and the Meteorological Office for providing the weather data.

REFERENCES

Baker, C.R.B. (1980). Some problems in using meteorological data to forecast the timing of insect life cycles. *EPPO Bulletin* **10**, 83-91.

Collier, R.H. & Finch, S. (1983). Effects of intensity and duration of low temperatures in regulating diapause development of the cabbage root fly (*Delia radicum*). *Entomologia experimentalis et applicata* **34**, 193-200.

Collier, R. H. & Finch, S. (1985). Accumulated temperatures for predicting the time of emergence in the spring of the cabbage root fly, *Delia Radicum* (L.) (Diptera: Anthomyiidae). *Bulletin of Entomological Research* **75**, 395-404.

Collier, R.H., Finch, S. & Anderson, M. (1989). Laboratory studies on late-emergence in the cabbage root fly (*Delia radicum*). *Entomologia experimentalis et applicata* **50**, 233-240.

Collier, R.H., Finch, S., Emmett, B.J. & Blood-Smyth, J. (1990a). Monitoring populations of the carrot fly, *Psila rosae*, with sticky and water traps. *Annals of Applied Biology*, **116**, 441-446.

Collier, R.H., Finch, S., Phelps, K. & Thompson, A.R. (1990b). Possible impact of global warming on cabbage root fly (*Delia radicum*) activity in the UK. *Annals of Applied Biology*, **118**, 261-271.

Collier, R.H.,. Finch, S. & Phelps, K. (1991). A simulation model for forecasting the timing of attacks of *Delia radicum* on cruciferous crops. *EPPO Bulletin* **21**, 419-424.

Collier, R.H. & Finch, S. (1992). The effects of temperature on development of the large narcissus fly (*Merodon equestris*). *Annals of Applied Biology*, **120**, 383-390.

Collier, R.H., Finch, S. & Phelps, K. (1992). The feasibility of using models to forecast carrot fly attacks in commercial carrot crops. *Integrated Control in Field Vegetable Crops. IOBC/WPRS Bulletin XV/4*, 69-76.

Collier, R.H. & Phelps, K. (1994). Carrot fly monitoring as an effective tool for pest management: how many flies have to be trapped? *Aspects of Applied Biology* **37**, 259-263.

Dufault, C.P. (1994). The influence of spray placement, rate of application and rainfall after treatment on the effectiveness of four insecticides for the control of *Psila rosae* (F.) on carrots. *Proceedings of the Entomological Society of Ontario* **125**, 67-79.

Eckenrode, C.K. & Chapman, R.K. (1972). Seasonal adult cabbage maggot populations in the field in relation to thermal unit accumulations. *Annals of the Entomological Society of America* **65**, 151-156.

Finch, S. & Skinner, G. (1974). Some factors affecting the efficiency of water traps for capturing cabbage root flies. *Annals of Applied Biology* **77**, 213-226.

Finch, S., Skinner, G. & Freeman, G. H. (1975). The distribution and analysis of cabbage root fly egg populations. *Annals of Applied Biology* **79**, 1-18.

Finch, S. & Collier, R. H. (1983). Emergence of flies from populations of overwintering cabbage root fly pupae. *Ecological Entomology* **8**, 29-36.

Finch, S. & Collier, R.H. (1985). Laboratory studies on aestivation in the cabbage root fly (*Delia radicum*). *Entomologia experimentalis et applicata* **38**, 137-14.

Finch, S., Collier, R. H. & Skinner, G. (1986). Local population differences in emergence of cabbage root flies from south-west Lancashire: implicatons for pest forecasting and population divergence. *Ecological Entomology* **11**, 139-145.

Finch, S., Bromand, B., Brunel, E., Bues, M., Collier, R.H., Dunne, R., Foster, G., Freuler, J., Hommes, M., Van Keymeulen, M., Mowat, D.J., Pelerents, C., Skinner, G., Stadler, E. & Theunissen, J. (1988). Emergence of cabbage root flies from puparia collected throughout northern Europe. In: *Progress on Pest Management in Field Vegetables*. Eds. R. Cavalloro and C. Pelerents, P.P. Rotondo - D.G. XIII - Luxembourg No. EUR 10514. Balkema, Rotterdam, pp. 33-36.

Finch, S. (1989). Ecological considerations in the management of *Delia* pest species in vegetable crops. *Annual Review of Entomology* **34**, 117-137.

Finch, S., Collier, R.H. & Elliott, M.S. (1990a). Seasonal variations in the timing of attacks of bronzed-blossom beetles (*Meligethes aeneus/Meligethes viridescens*) on horticultural brassicas. *Proceedings 1990 Brighton Crop Protection Conference - Pests and Diseases*, 349-354.

Finch, S., Collier, R.H. & Elliott, M.S. (1990b). Biological studies associated with forecasting the timing of attacks by the large narcissus fly, *Merodon equestris*. *Proceedings 1990 Brighton Crop Protection Conference - Pests and Diseases*, 111-116.

Jonsson, B. (1992). Forecasting the timing of damage by the carrot fly. *Integrated Control in Field Vegetable Crops IOBC/WPRS Bulletin XV/4*, 43-48.

Phelps, K., Collier, R.H., Reader, R.J. & Finch, S. (1993). Monte Carlo simulation method for forecasting the timing of pest insect attacks. *Crop Protection* **12**, 335-342.

Parker, C. & Phelps, K. (1994). Putting the byte into brassicas. *Grower*, October 13 1994.

Shaffer, P.L. (1983). Prediction of variation in development period of insects and mites reared at constant temperatures. *Environmental Entomology* **12**, 1012-1019.

Sharpe, P.J.H., Curry, G.L., DeMichele, D.W. & Cole, C.L. (1977). Distribution model of organisms development times. *Journal of Theoretical Biology* **66**, 21-38.

GROWING MALTING BARLEY WITHOUT THE USE OF PESTICIDES

I.A. RASMUSSEN

Danish Institute of Plant and Soil Science, Dept. of Weed Control and Pesticide Ecology, Flakkebjerg, DK-4200 Slagelse, Denmark

K. KRISTENSEN

DIPS, Dept. of Biometry and Informatics, Thorvaldsensvej 40, DK-1871 Frederiksberg

S. STETTER

DIPS, Dept. of Plant Pathology and Pest Management, Lottenborgvej 2, DK-2800 Lyngby

ABSTRACT

A decision support system for growing malting barley without the use of pesticides is being developed. The system, which is based on a Bayesian network, estimates the probability distribution of yield and quality. To counter the disadvantages of not being able to use pesticides, varieties chosen are not only suitable for malting but also possess resistance against fungal diseases. Weeds are recommended controlled by harrowing. A submodel for fungal diseases estimates yield with or without use of pesticides, based on the area of a triangle similar to the area under the disease progress curve (AUDPC). Other submodels are being developed. Economic return at different prices is calculated from experimental results. The results indicate that growing malting barley without the use of pesticides can result in satisfactory quality and a yield loss that may be countered by less expenses or better prices.

INTRODUCTION

Research concerning mechanical weed control and disease resistant crop varieties has given the potential for developing a growing strategy entirely without the use of pesticides, and to find out whether this causes loss of quality or economic return. Malting barley for beer production was chosen for the work for several reasons. First, mechanical weed control can be undertaken succesfully in spring barley. Second, several spring barley varieties which possess resistance against the most important fungal pathogens are available. Third, there was at the time the project was initiated a potential market for malting barley grown without the use of pesticides. Fourth, a better price is paid for malting barley than for fodder barley - on the condition that certain quality demands are met. The present project concerning "Production of beer from Danish malting barley grown without the use of pesti-cides" was initiated in collaboration between the Danish Institute of Plant and Soil Science (DIPS) and the Carlsberg Breweries.

For the farmer to grow malting barley without the use of pesticides, it is an advantage if a decision support system is available. The farmer must know which important questions to take into consideration before choosing whether or not to attempt using the pesticide-free strategy. He has to know something about the potential risks and gains. The growing strategy must be planned before the growing season and carried through. And finally, during the growing season, there must be some way to assess the development of the yield and quality of the crop - is the crop being quenched by weeds, will an attack by pests or diseases result in an unacceptable reduction in yield or quality?

The work of the project aims at
a) determining the influence of single factors (weed control, fungal diseases etc.) on yield and quality,
b) setting up growing strategies that optimize these factors and evaluate the interaction,
c) developing a computer-based decision support system,
d) assessing the quality of the malting barley, the malt and the beer produced in the pesticide-free growing strategies,
e) developing methods for analyzing pesticide residues in malting barley and malt,
f) determining the level of contamination by microorganisms in malting barley produced with or without the use of pesticides,
g) calculating the economic consequences, risks and gains, for the farmer and the industry.
In this paper the focus is on the decision support system.

THE DECISION SUPPORT SYSTEM

The decision support system is based on a Bayesian network. A Bayesian network is useful in situations where causal relationships are a natural way of relating concepts, particularly when uncertainty is part of the relations (Jensen et al., 1990). The network can be presented as a set of nodes connected by directed links. The nodes represent the concepts, the links represent the causal relationships. The latter are given as conditional probabilities of the states of the influenced nodes, given the combination of the causal nodes. For further details see Rasmussen et al. (1990) and Kristensen & Rasmussen (1993).

The decision support system for growing malting barley without the use of pesticides consists of a main model and four submodels. The main model relates the basic input factors such as soil type, nitrogen fertilizer, variety, seed rate, sowing time etc. and estimates the distribution of expected yield and quality, assuming average weather and no attacks of pests or pathogens, no weeds, and optimal harvest and post-harvest treatments. The main model is represented in Figure 1. Although the model is rather simple it should be able to take into account the most important factors affecting the yield and quality.

In order to easily adapt new varieties into the model, the effect of varieties is described through different characteristics of the variety. In the main model 5 characteristics are used: germination speed, tillering, kernels per ear, size of kernels and protein content. Each of those characteristics is described by a distribution, e.g. the level of the potential protein content of varieties A, B and C may be described by the figures in Table 1. Other parts of the model estimate the available nitrogen in the soil. The final probability distribution of the protein content is estimated from the combination of the nitrogen potential, the variety specific modification and their interaction.

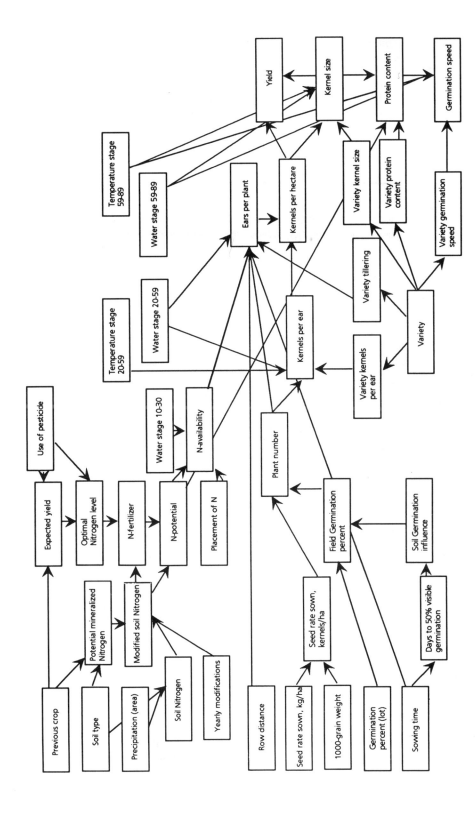

FIGURE 1. Main model of the decision support system for malting barley grown without the use of pesticides. Boxes represent nodes, arrows represent causal relationships, see text.

TABLE 1. An example of a probability distribution of potential protein content level for varieties A, B and C.

Variety	Potential protein content level		
	Low	Medium	High
A	50	30	20
B	25	50	25
C	10	30	60

The four submodels each predict the change in yield and quality caused by one of the following: fungal disease, aphid attacks, weeds/weed control and harvest/post-harvest treatments.

FIGURE 2. Submodel for fungal diseases. Legend as Figure 1.

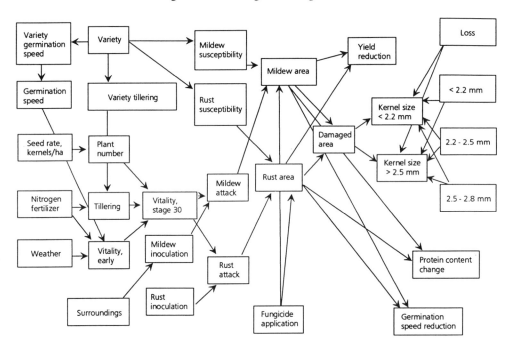

Submodel for fungal diseases

The model which estimates the effect of attacks of fungal diseases on yield and quality is shown in Figure 2. At the present time, only the effect on yield and kernel size is incorporated into the model. One part of the model is planned to estimate the time when attack by fungal pathogens, mildew (*Erysiphe graminis*) or rust (*Puccinia spp.*), may be expected to occur. The other part of the model estimates the reduction of yield and kernel size given the specified variety and a known, or estimated, time of attack by a fungal pathogen. The latter part of the model will be described in detail as an example.

This model is based on results from one year of experiments. When the effect of a single fungal attack was considered, the estimate of the factor the yield is reduced by was approximated as

$$f = e^{\Theta A}$$

where

    $0 \leq f \leq 1$, when $f = 1$ there is no yield reduction, as f decreases the yield reduction grows,

    $\Theta < 0$ is a constant specific to each fungal disease (mildew, rust)

    $A = 0.5 * D * p$ is the area of a triangle assumed to be proportional to the severity of the attack,

    D is the number of days the fungal attack affects the crop and

    p is the susceptibility of the variety measured as pecentage leaf coverage by the fungal attack.

The variety susceptibility measured as percentage leaf coverage (p) was taken from the most recent Danish annual publication of cereal varieties (Rasmussen et al., 1994) where susceptibility to fungal diseases, measured as percentage leaf coverage, are observed for the varieties every year at a number of different locations.

FIGURE 3. Area (A) of the triangles described in the text for mildew attack of two barley varieties (1 & 2) with different susceptibilities (p) and different times of visible attack (d_b) with and without fungicide use.

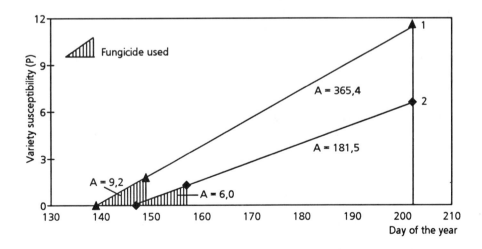

435

The triangle, A, was defined by the variety susceptibility, p, the day of visible attack, d_b (> 0.01 % coverage) and the day the crop reaches stage 89, d_{89}, so that the number of
days the fungal attack affects the crop, $D = d_{89} - d_b$. If the crop was sprayed with fungicide, the fungal attack was expected to affect the crop only for 10 days, and the triangle was then defined as the left part of the original triangle that was cut off by a vertical line after 10 days, see figure 3.

To be able to estimate the yield loss at the time of attack, the day stage 89 would be reached must be estimated. For the experiments this could be done by
$$S = -93.70 + 0.90336 * d$$
where d is the day of the year and S is the growth stage the crop is expected to be in at that time.

When both mildew and rust attacks were observed the combined effect was calculated by multiplying the reduction factors:
$$f = f_{mildew} * f_{rust}$$
where f_{mildew} and f_{rust} each are estimated as shown above.

In the model, the uncertainties for each of the factors: 1) the variety susceptibility, which varies between locations, 2) the area of the triangle, and 3) the effect of the fungus on yield (Θ) are taken into account. The values are categorized at each step and the probability distributions are propagated through the system. When a fungal pathogen attack becomes visible in the crop, expected yield and quality can be estimated in the two situations: fungicide use or no fungicide use.

ECONOMIC CALCULATIONS

An experiment with growing systems for malting barley without the use of pesticides was carried out at six different locations in 1993 and 1994 (Rasmussen 1995). Results from 9 of these experiments were used for economic calculations. The two treatments considered were treated the same for variety (Alexis), number of viable seed sown per square meter (300), nitrogen fertilizer (80 kg N per hectare); but differed in pesticide treatment: A (not treated with any pesticides, weed-harrowed), E (treated with pesticides against weeds, fungal diseases and insect pests according to need as adviced by PC-Plant Protection). Expenses for plant protection were calculated from the actual pesticides used, the number of times spraying was carried out and the number of times weed harrowing was conducted. The price received for the malting barley was calculated for the proportion of the yield that did not pass the 2.2 mm sieve, for the fodder yield all the harvest was included. Economic return was calculated at 3 prices: fodder barley, malt barley, and malt barley grown without the use of pesticides at ordinary price and at a price 20 percent above the ordinary price. Net economic return is seen in Figure 4. In four cases, the economic return of malting barley grown without the use of pesticides, given a price 20% greater than ordinary malting barley, exceeds that of malting barley grown with the use of pesticides, in three cases the economic return is almost the same, and in two cases the economic return of ordinary malting barley exceeds that grown without pesticides.

FIGURE 4. Net economic return, Danish kroner per hectare (dkr/ha) from 2 growing systems, with or without pesticide use. For each system and location, actual expenses for plant protection are subtracted. Fodder yield is everything harvested, malt barley yield is the proportion that did not pass the 2.2 mm sieve.

With pesticide use			Without pesticide use		
+	fodder yield,	95 dkr/hkg	o	malt barley yield, 115 dkr/hkg
x	malt barley yield, 115 dkr/hkg		*	malt barley yield, 138 dkr/hkg

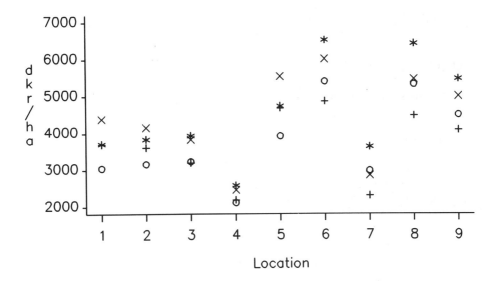

DISCUSSION

The decision support system is based on a Bayesian network. Other bases for decision support systems may be regression analyses, systems of partial differential equations (Hansen et al., 1990) or knowledge based expert systems (Dindorp, 1992; King et al., 1991). The main advantage of Bayesian networks over those systems is that a Bayesian network can supply the user with information on the uncertainty of the predicted outcome. The network propagates this uncertainty through the network in a logical way and as further information or assumptions are given, the final uncertainty is decreased. The main drawback of a Bayesian network is that all information has to be categorized into a limited number of states. In order to let the network be managable it might be necessary either to simplify the models (or parts of the models) or to restrict the number of states for certain variables to rather few states. We think that the benefits of the information on uncertainty is very important and counter the disadvantages.

The triangle used to predict yield loss at fungal attack is calculated somewhat similarly to the area under the disease progress curve (AUDPC) (Steffenson & Webster, 1992), which has sometimes been mentioned as a means of estimating yield loss (Wright & Gaunt, 1992). The area of the triangle was preferred to the AUDPC for two reasons: it permits an early estimate of the effect of fungal attack and requires only one direct measurement (time of attack) combined with published data, as opposed to AUDPC which requires repeated measurements during the season. Whether the estimates of yield loss are accurate remains to be validated.

It is still too early to conclude whether the decision support system for growing malting barley without the use of pesticides will be succesful, as the system is not finished and has not been validated by experimental work. However, the results from the experiments indicate that growing malting barley without the use of pesticides can be carried out. The yield can be of the required quality, and the expected yield reduction may be countered by less expenses and/or a greater price.

ACKNOWLEDGEMENTS

The project was funded partly by the Directorate for Agricultural Development.

REFERENCES

Dindorp, U. (1992) Agricultural applications of knowledge based systems concepts. *Tidsskrift for Planteavls Specialserie*, S **2201**, 97 pp.

Hansen, S.; Jensen, H.E.; Nielsen, N.E.; Svendsen, H. (1990) DAISY - Soil Plant Atmosphere System Model. *NPO-forskning fra Miljøstyrelsen*, **A10**, 272 pp.

Jensen, F.V.; Olesen, K.G.; Andersen, S.T. (1990) An algebra of bayesian belief universes for knowledge-based systems. *Networks*, **20**, 637-659.

King, J.P.; Broner, I.; Croissant, R.L.; Basham, C.W. (1991) Malting barley water and nutrient management knowledge-based system. *Transactions of the ASAE*, **34**, 2622-2630.

Kristensen, K.; Rasmussen, I.A. (1993) Models to assist the choice of strategies for growing malting barley without the use of pesticides. *Workshop on Computer-based DSS on Crop Protection, SP-report*, **1993** (7), 39-46.

Lancashire, P.D.; Bleiholder, H.; van den Boom, T.; Langelüddeke, P.; Stauss, R.; Weber, E.; Witzenberger, A. (1991) A uniform decimal code for growth stages of crops and weeds. *Annals of Applied Biology*, **119**, 561-601.

Rasmussen, L.K.; Thysen, I.; Pedersen, K.M. (1990) An application of causal probabilistic networks to examine reproduction of dairy cows. *Fællesberetning*, **SF1**, 59-68.

Rasmussen, J.; Nielsen, B.S.; Pedersen, J.B.; Olsen, C.C.; Welling, B. (1994) *Kornsorter*. Statens Planteavlsforsøg & Landsudvalget for Planteavl. 32 pp.

Rasmussen, I.A. (1995) Integrated production - an example concerning malting barley grown without the use of pesticides. *SP-rapport*, **1995** (3), 255-268.

Steffenson, B.J.; Webster, R.K. (1992) Quantitative resistance to *Pyrenophora teres f. teres* in barley. *Phytopathology*, **82**, 407-411.

Wright, A.C.; Gaunt, R.E. (1992) Disease-yield relationship in barley. I. Yield, dry matter accumulation and yield-loss models. *Plant Pathology*, **41**, 676-687.

THE ROLE OF SYNOPTIC MODELS IN THE DEVELOPMENT OF CROP PROTECTION FOR SUSTAINABLE CROP PRODUCTION SYSTEMS.

N.McRoberts, G.N. Foster*

Plant Science Department and *Environmental Sciences Department, SAC, Auchincruive, Ayr KA6 5HW

D.H.K. Davies

Crop Systems Department, SAC Bush Estate, Penicuik, Midlothian, EH26 0PH

K.A.Evans, R.G. McKinlay

Crop Science & Technology Department, SAC, West Mains Road, Edinburgh EH9 3JG

& S.J. Wale

Crop Biology Department, SAC, King Street Aberdeen, AB9 1UD

ABSTRACT

Crop protection practices in sustainable production must be more closely tied to the requirements of individual crops. Currently in intensive production, crop protection takes the form a number of reactive steps, which are generally triggered either by rather arbitrary threshold values for pest (invertebrate, disease, or weed) damage, or by crop growth stages, in which case no account is taken of the actual risk to the crop in any individual year. More efficient and sustainable crop protection systems can be developed when chemical and cultural control measures are seen as components of the overall production method and not as add-in solutions to the occurrence of individual pests. Our approach to developing these new systems has similarities to recent developments in several tropical production systems. Initially the crop production system is modelled by multivariate methods to identify advisory domains. The domains are described by different sets of management and pest variables and are therefore associated with different levels and types of risk. Farms (or individual fields) can be classified by the type of domain to which they belong and the crop production system designed from this starting point. Variation in occurrence of individual pests, or combinations of pests, may be modelled further to develop advisory aids through the application of generalised linear or probability-based models. Crop monitoring is essential both in the obtaining the initial data to describe the domains and in operating successful sustainable crop protection within them. Particularly in the case of chemical control measures, significant reductions in use without loss of reliability of yield, will be possible if pesticides are applied only when required and the applications are timed for maximum efficiency. These objectives will not be achieved if the farmer or advisor has no idea of the pest status of the crop throughout the season.

INTRODUCTION

The approach adopted in this paper follows from the view expressed by Vereijken (1992) that the development of sustainable agricultural systems depends on increased localisation of markets and production systems. This philosophy, when applied to crop protection, leads to the idea that in order to obtain sustainability each crop (field) must be treated individually, or at least its general features must be characterised, and crop protection measures applied specifically in response to locally important problems which arise. One way by which this objective might be achieved is through the development of synoptyic models for the crop production system which can be used strategically to improve the efficiency of extension work and applied research. A number of different types of model which can be classed as synoptic will be introduced and one of these approaches will be described in more detail using data from the COIRE (Crop Optimisation by Integrated Risk Evaluation) project.

SYNOPTIC MODELS IN CROP PROTECTION

Qualitative analysis

In the most general sense a synoptic model is one which provides an overview or summary of a system. One such model is shown in Figure 1, which is redrawn from Vereijken's (1992) paper and is a representation (causal graph) of the interacting factors in the world agricultural market. Although this type of model cannot be used directly for quantitative analysis of the system it may identify interactions which need to be studied in more detail.

One argument put forward by Vereijken (1992) was that the system represented in Figure 1 is essentially unstable. While this assertion cannot be tested directly with the model in its original format it can be explored in a second type of synoptic model developed in community ecology (May, 1974) and system analysis (Taber, 1991). An analysis of Vereijken's model has been conducted (Figure 2). In this approach the interacting factors are represented as a square matrix, in which each row represents the effects of one of the factors on each of the others in turn. A Markov chain process is used to examine the stability of the system over time by multiplying the matrix by a vector of initial conditions (given as cycle 0 in Figure 2) to produce a rectangular output matrix, as shown in Figure 2. The models briefly introduced so far may be used for qualitative analysis of systems. However, for more detailed examination of individual systems a quantitative approach to synoptic modelling is required.

Quantitative analysis

The term synoptic was introduced to crop protection by Stynes (1980) to describe a synecological approach to modelling crop losses. Stynes's method attempted to capture the complex interactions between the crop, its environment and production constraints (pests, weeds, diseases, and poor management practices) in simple regression models following initial data reduction by multivariate analysis. Related techniques have subsequently been developed by Savary, Zadoks and co-workers, and their application demonstrated in a number of tropical

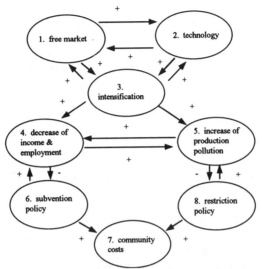

Figure 1. A causal graph model of interacting factors in intensive agricultural production. Redrawn from Vereijken (1992).

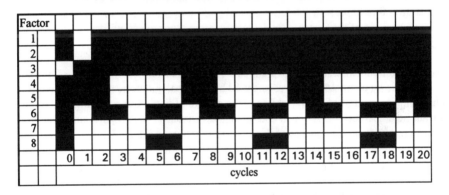

Figure 2 The predicted behaviour of the system represented in Figure 1. A solid cell in a any column indicates that the factor represented by the corresponding row will increase in the cycle represented by the column, while empty cells indicate that the factor will reduce or be unaffected for that cycle. The first three factors (free market, technology and intensification continually increase after the first cycle, while factors such as income and subvention policy oscillate on cycles of different periods.

cropping systems (Savary, 1991; Savary *et al.*, 1994). Recent interest in the application of GIS technology to crop protection has added a further class of related modelling techniques which may be termed synoptic (Nelson *et al.*, 1994). Irrespective of the analytical details, all of the quantitative synoptic modelling methods involve three general steps:/

1. Collection of multivariate data by surveys of real crops.
2. Data reduction.
3. Extraction of categories for the crops in the data set.

This group of modelling techniques share one further characteristic which is of interest in the context of developing sustainable production systems. In all of the methods variation between crops at different locations is taken into account and examined in detail. However, while these methods provide the potential to characterise the production constraints on the local scale required for efficient crop protection, they also provide a summary of the production system at a larger spatial scale; *e.g.* at a regional or national scale depending on the extent of the complete survey programme.

Example Of A Possible Methodology: Autumn-Sown Wheat In Scotland

The general aims and methodology of the COIRE project have been described previously (McRoberts *et al.*, 1994). For each of 50 fields, chosen to represent the arable area of Scotland, approximately 300 items of data were collected, including information on surrounding land use, pest, weed and disease populations, soil characteristics, and husbandry practices. In the COIRE project, which will end in 1996, data collection will be repeated for three full growing seasons for both wheat and autumn sown oilseed rape, and synoptic models for the crops will be developed from these sets of data. However, for simplicity, the current example will illustrate the approach using data collected at one survey only, immediately before harvest in 1994.

Field Characteristics

Data for 20 field characteristics were recorded on a presence/absence basis. The data matrix of fields by characteristics was analysed by principal components analysis to obtain a graphical representation of the variability in the sample of fields and a ranking of the important field characteristics which determine the variability. The separation of the fields in the first two principal components of this analysis is shown in Figure 3.

Weed data and Disease data

The number of visible weedy patches and the species composition of the patches were recorded in each field. A principal components analysis of the correlation matrix of these data was conducted to examine inter-field variation, as in the case of the field characteristics data, and is summarised in Figure 4 . The severity of 11 types of fungal infection was recorded in each of the 50 fields. The data were analysed in a similar manner as the other two sets of data. The separation of the fields and the association of the disease variables with the first two principal axes are shown in Figure 5.

Comparison of the groupings of fields suggested by the independent analyses

Overall agreement between the principal components for the three independent analyses was conducted by pair-wise canonical correlation analysis (CCA) (Digby & Kempton, 1987) Results from the CCAs are shown in Table 1.

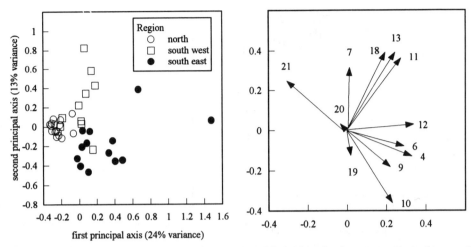

Figure 3. *The separation of wheat fields (left-hand figure) in the first two principal axes of an analysis of 20 field characteristic variables, and the association between the variables and the principal axes (right-hand figure, only a sub-set of the variables are shown for clarity).* 4, fresh water; 6, salt marsh; 7, moor land; 9, farm buildings; 10, shelter belt; 11, fallow land; 12, urban area; 13, waste ground; 18, uncultivated strip; 20, water course; 21, crop growth stage.

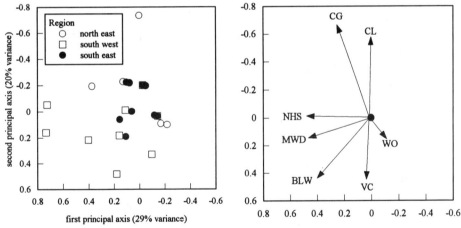

Figure 4. The separation of wheat fields in the first two principal axes of an analysis of their pre-harvest weed populations (left-hand figure) and the association between the weed variables and the principal axes (right-hand figure). NHS, number of weed hot-spots; MWD, mayweed, BLW mixed broad-leaved weeds; VC, volunteer cereal; WO, wild oat; CL, cleavers; CG couch grass.

The canonical correlations can be interpreted in the same way as standard correlation coefficients   and are the highest correlations which can be obtained between principal axes of the separate sets of data.

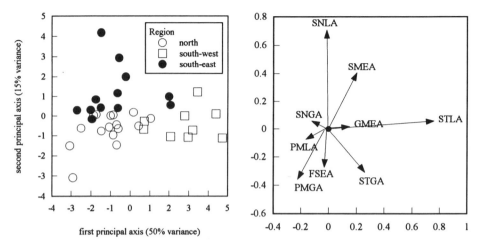

Figure 5. The separation of wheat fields in the first two principal axes of an analysis of their pre-harvest disease populations (left-hand figure) and the association between the disease variables and the principal axes (right-hand figure). SN; Septoria nodorum; ST, Septoria tritici; PM powdery mildew; FS Fusarium spp.; GM, grey mould; EA; % ear area infected; LA; % leaf area infected; GA, % glume area infected.

TABLE 1. Canonical correlations between the principal axes of analyses of site characteristics, weeds and diseases for fields of autumn-sown wheat.

sites and weeds data		sites and diseases data	
Canonical axis		Canonical axis	
1	2	1	2
0.6	0.3	0.7	0.6

Summarising the model of the crop system

Synoptic modelling should lead to a series of practical conclusions. The conclusions will vary depending on the aims of the modelling exercise (*e.g.* crop loss prediction, risk assessment, etc.). The following list of conclusions is illustrative of the type of information which one might expect to gather from synoptic modelling as described above. It is stressed that the example analyses and the conclusions drawn from it are intended only as examples of the methodology.

1. Two or three advisory domains can be recognised in the current analysis. The first domain consists of arable production in a mixed farming background (predominantly in the south west), with a variety of surrounding land uses. Cropping in this domain tends to be associated with weed infestations and diseases which attack the ears and glumes of the plant. The second domain (which may be split into two geographical areas) contains

crops in predominantly arable areas where weed control is apparently efficient, but where crops are more likely to be attacked by common leaf diseases.

2. Advisory domains in the cropping system should be established at a geographical scale no larger than the regions represented in the original data, since there is a clear regional variation at this scale irrespective of the type of data analysed. The agreement between the three sets of data probably results from the underlying correlation of all three types of data to broad climatic differences between the regions. Given the intra-regional variation expressed in some the data, further analysis should be conducted to determine whether a smaller geographical scale should be adopted for the advisory domains.

3. Increased productivity might be expected in the south west region by the adoption of improved weed control, particularly for broad-leaved weeds and volunteer cereals. Farmers may be asked to consider adoption of additional cultural methods to control these weeds if required, or alerted to the possibility of reducing herbicide doses by applying herbicides at an early stage of weed growth.

4. Farmers in the south west should be encouraged to consider early maturing varieties to reduce the effects of fungal diseases which damage the leaves and ears of the crop during wet weather (*Septoria tritici*, grey mould).

5. Crops in the south east and north east were more commonly attacked by leaf infecting diseases such as powdery mildew. The occurrence of these diseases is at least partly related to the density of wheat growing in these regions, and farmers should be encouraged to consider using mosaics of varieties from different resistance groups, within and between farms. If control of common leaf diseases is likely to be a priority in the south and north east, farmers should be encouraged to adopt assessment-based reduced dose spray programmes such as those discussed by Wale (this volume). A practical extension priority in this area would be to demonstrate the management, economic, and environmental benefits to the farmer of adopting this approach over a critical-point threshold approach.

DISCUSSION

Modelling methods which allow a detailed analysis of cropping systems and a hypothetical set of recommendations arising from one of these procedures have been presented. While the qualitative analyses may be useful in identifying areas for research, their application to examination of specific cropping systems is more limited, and Murdoch (1975) has pointed out several difficulties which may arise in attempting to apply analyses from theoretical ecology to crop protection problems.

The quantitative synoptic analyses presented here produce recommendations which could form the basis of general policies for crop protection in the development of sustainable systems. In addition, the comprehensive surveys conducted during the modelling process provide a snap-shot of the system which can be used as a reference as more sustainable production methods are adopted. The importance of monitoring as part of the development and practice of sustainable crop protection is stressed. The cost of making accurate assessments of crops is an inevitable part of making crop production more sustainable. It should not be taken for granted that farmers will respond positively or uniformly to suggestions about improving the efficiency of crop protection (Zadoks, 1989) or increasing sustainability (Fujisaka, 1994). Farmer's responses to advice can be included as a separate set of variables in the synoptic model (Savary *et al.*, 1994), and the rapidity with which new practices are taken up by farmers can then be related to the features of the advisory domain.

ACKNOWLEDGEMENTS

The authors thank the field staff involved in the COIRE project for collecting the data. SAC receives financial support from SOAFD, who fund the COIRE project.

REFERENCES

Digby, P.G.N. ; Kempton, R.A. (1987) *Multivariate Analysis of Ecological Communities*. Chapman & Hall, London, 206 pp.

Fujisaka, S. (1994) Learning from six reasons why farmers do not adopt innovations intended to improve sustainablility of upland agriculture. *Agricultural Systems*, **45**, 409-425

May, R.M. (1974) *Stability and Complexity in Model Ecosystems*. Princeton University Press, Princeton, 265 pp.

McRoberts, N. ; Foster, G.N. ; Duffus, C. ; Evans, K.A. ; McKinlay, R.G. ; Davies, K. ; Wale, S.J. (1994). COIRE: Crop Optimisation by Integrated Risk Evaluation. A new approach to crop protection in arable crops. In: *Sampling to Make Decisions* P. Brain, S.H. Hockland, P.D. Lancashire, L.C. Sim (Eds) *Aspects of Applied Biology*, **37**, 207-215.

Murdoch, W.W. (1975) Diversity, complexity, stability and pest control. *Journal of Applied Ecology*, **112**, 795-802.

Nelson, M.R. ; Felix-Gastelum, R. ; Orum, T.V. ; Stowell, L.J. ; Myers, D.E. (1994) Geographical information systems and geostatistics in the design and validation or regional plant virus management programs. *Phytopathology*, **84**, 898-905.

Savary, S. (1991). *Approches de la pathologie des cultures tropicales*. Karthala/ORSTOM , Paris, 288 pp.

Savary, S. ; Elazegui, F.A. ; Moody, K. ; Litsinger, J.A. ; Teng, P.S. (1994) Characterization of rice cropping practices and multiple pest systems in the Philippines. Agricultural Systems **45**, 385-408.

Stynes, B.A. (1980) Synoptic methodologies for crop loss assessment. In: *Crop Loss Assessment. Miscellaneous Publications of the Agricultural Experiment Station of the University of Minnesota*, **7**, 166-175.

Taber, W.R. (1991) Knowledge processing with fuzzy cognitive maps. *Expert Systems with Applications*, **2**, 83-87.

Vereijken, P (1992) A methodic way to more sustainable farming systems. *Netherlands Journal of Agricultural Science*, **40**, 209-233.

Wale. S.J. (1995) An integrated disease risk assessment system for winter barley. In : Integrated Crop Protection: Towards Sustainability? BCPC Publications.

Zadoks, J.C. (1989) EPIPRE, a computer-based decision support system for pest and disease control in wheat: its development and implementation in Europe. *Plant Disease Epidemiology*, **2**, 3-29.

TOWARDS AN INTEGRATED DISEASE RISK ASSESSMENT SYSTEM FOR WINTER BARLEY

S.J. WALE, FIONA MURRAY

Crop Biology Department, SAC Aberdeen, 581 King Street, Aberdeen AB9 1UD

ABSTRACT

A prototype Integrated Disease Risk (IDR) system for winter barley is outlined. It is based on a system first developed for winter wheat diseases. Four factors are integrated in equations for each of the main foliar diseases of winter barley. Values for each factor are determined from tables and inserted into IDR equations. The resulting score for each disease relates directly to the fungicide dose. Preliminary evaluation of the system has shown it to be user friendly and to result in acceptable levels of disease control.

INTRODUCTION

Deciding whether or not to apply a fungicide to winter barley and if so what dose to apply is a complex decision. Many factors influence the decision making process. The majority of factors related to such decisions have been listed by Paveley & Lockley (1993) and Wale (1994). However, the most important are probably inoculum, weather conditions, the disease resistance ratings of the variety grown and the crop sensitivity (sensitivity to disease induced yield loss) at a particular growth stage. Paveley (1993) has integrated these factors for wheat by developing equations for each foliar disease to permit decisions on timing and dose to be made. The decision support system (DSS) Paveley has developed is called Integrated Disease Risk (IDR) Strategy. Values are inserted for each factor in the equation and a score determined. From that score the dose to be applied, which ranges from nothing to full, is determined.

This paper describes some of the background to the development of a parallel system for winter barley. *Rhynchosporium secalis* (leaf blotch) will be used to illustrate how the system is being developed.

FACTORS IN INTEGRATED DISEASE RISK

Diseases

Winter barley can be infected by a range of foliar diseases. The most important are mildew (*Erysiphe graminis*), Rhynchosporium leaf blotch (*Rhynchosporium secalis*), brown rust (*Puccinia hordei*) and net blotch (*Pyrenophora teres*). Knowledge of the biology of each pathogen provides information as to when disease is likely to develop and when control is likely to be required. Thus for *R. secalis*, most yield loss occurs when the two last formed leaves become infected (Chiarappa, 1971). The component of yield most affected is seed weight. Clearly, prevention of infection of the top two leaves is crucial to minimise yield loss. However, since the most effective fungicides provide only 60 to 80% control when the inoculum potential is high, fungicide use earlier in crop development to delay epidemic development is vital to long term control.

The rate of epidemic development for polycyclic diseases is driven by the initial inoculum and other factors that influence the epidemic. Of these other factors infection frequency, latent period and sporulation are largely determined by the resistance of the variety. IDR incorporates a value for host resistance in the equation for each disease. The remaining factor that affects epidemic development is spore dispersal efficiency. This is driven by weather factors which are also accounted for in the IDR equation.

Inoculum

If a DSS is to be used on farm it must involve a straightforward assessment of inoculum. Measuring spore production is impractical. Assessment of leaf tissue area infected by a particular disease is widely used as a measure of disease in research but on a farm level could result in large variation from the actual infection due to the subjectivity of the assessment method. Provided that correct identification is possible, the most unambiguous way to record disease is on a presence or absence basis. On the premise that the greater the degree of infection, the higher up the plant disease will occur, measuring presence or absence on a critical leaf layer is an objective way to measure inoculum provided the diseases concerned do not have long latent periods.

In order to determine a value for the inoculum factor in each IDR equation the user is asked to identify one of four levels of infection on a critical leaf layer. A value is assigned according to the level, as shown for *R. secalis* in Table 1. The critical leaf layer is the third top fully expanded leaf at the time of assessment until ear emergence when the second top leaf is considered.

TABLE 1. Values for different inoculum levels of *R. secalis*

Value	Level of infection
0	No leaf blotch on critical leaf layer and no obvious Rhynchosporium on plants
1	<10% leaves on critical leaf layer with at least one lesion or no infection on critical leaf layer but some leaf blotch detected on plants
2	10-25% leaves on critical leaf layer with at least one lesion of leaf blotch
3	>25% leaves on critical layer with leaf blotch and also detected on leaves above

Weather

Until the advent of on-farm weather stations and sufficient knowledge relating disease epidemiology to crop micro-climate, the integration of complex weather criteria into a DSS is unlikely. The prototype IDR for winter barley uses simplified criteria based on basic weather parameters that can be measured easily on-farm.

Few attempts to link climatic conditions to infection and development of *R. secalis* have been made. One of the earliest (Ryan & Clare, 1975) determined the relationship of the period of leaf wetness and temperature with infection. This relationship has been adapted in the UK to identify 'Rhynchosporium risk periods' using the duration of high humidity in met screens 2m above the ground following rainfall to determine the length of leaf wetness and the average temperature over the period of leaf wetness (Polley & Clarkson, 1978). In studies on spring barley Wale (1983) found that the best relationship between Rhynchosporium risk periods and the disease progress came from accumulating consecutive periods. On farm, most growers can only record daily rainfall and daily maximum and minimum temperatures. Whilst a much greater level of sophistication is possible, the relationship between weather and epidemiology of most barley diseases remains unclear.

For *R. secalis*, the prototype IDR system utilises the criteria used in PC-Plant Protection, a Danish DSS (Murali, 1991). This simply utilises the number of days in the previous 14 when there was rainfall of 1mm or more. Values for the weather factor in the IDR equation for Rhynchosporium are determined as shown in Table 2.

TABLE 2. Values for the *R. secalis* weather factor

Value	Weather conditions in last 14 days
1	Unfavourable - 1 day or less with 1.0 mm or more of rain
2	Average - not falling into favourable or unfavourable categories
3	Favourable - 5 or more days with 1.0 mm or more of rain

Varietal disease resistance

Disease resistance ratings for recommended cereal varieties are published annually (e.g. Anon.,1994 ; Anon., 1995). These ratings relate to the average infection on plots not receiving any fungicide in variety trials around the UK. They are a mean of three seasons trials. For certain diseases account is also taken of polytunnel tests where specific races of pathogen are screened against the same varieties as grown in variety field trials. Disease resistance ratings are an indicator of disease risk. They describe in a single figure the likely severity of infection when conditions favour disease development and compatible races of the pathogen are present.

Whilst the resistance rating for each disease is a crucial part of a risk assessment system, potential errors should be borne in mind. Firstly, the rating is a three year average. If there has been a change in the pattern of races over the seasons and virulence has increased, by using a three year average, the extent of the change in susceptibility may not be apparent from the rating. Secondly, for certain obligate pathogens, such as *Erisyphe graminis*, where resistance is controlled by a few major genes, the development of a new virulent race able to overcome resistance genes may invalidate the resistance rating. Thirdly, races are known to vary in different localities often in relation to the varieties grown in that region. This is not apparent from a single figure for resistance for the UK. Indications of changes in the virulence of pathogens in localities or over time may be indicated in pen notes for varieties in recommended lists.

Where a highly susceptible variety is grown the assumption that disease will automatically develop should be avoided. Without sufficient inoculum and suitable climatic conditions, even a susceptible variety will remain uninfected. This confirms the importance of integrating all major factors influencing disease in a risk assessment. In the prototype IDR the published resistance ratings are used to determine values which are inserted into the IDR equation for each disease (Table 3).

TABLE 3. Values for variety disease resistance ratings

Value	Disease resistance rating
0	9
1	8
2	7
3	6
4	5
5	4
6	3 or 2 or 1

Crop sensitivity

This factor is perhaps the most understudied component of this prototype system. For each disease the extent of potential yield loss will vary at different stages of growth. In winter barley, it is generally accepted that the greatest and most consistent yield response to fungicide occurs around the first node stage - GS 31 (Zadoks et al, 1974). A further important, but less consistent timing, is at flag leaf emergence (GS 37-49).

Responses to treatment at different times of fungicide application have been evaluated in different ways. For example, in a series of trials in north-east Scotland, Wale (1987) used programmes of fungicides with different combinations of four timings: early spring (end of tillering to pseudostem erect (GS 30), first node (GS 31), flag leaf emerged (GS 39-49) and ear emergence to end of flowering (GS 51-69). By subtraction it was possible to determine the response to each timing and an average response determined over the series of trials. The average responses and percent of occasions when the responses were cost effective are shown in Table 4.

Mildew and Rhynchosporium were the predominant diseases in this series of trials. For other diseases the response to timings are likely to be different. For example *Puccinia hordei* is likely to cause greater yield loss after ear emergence than other pathogens as it is capable of infecting awns.

TABLE 4. Yield response (t/ha) for different timings of fungicide application in winter barley and percentage of occasions cost effective

Timing	Yield response (% occasions cost effective)	
	Sown before 21 Sept	Sown after 21 Sept
Early spring	0.31 (80)	-0.36 (0)
	All sowing dates	
GS 31	0.65 (100)	
GS 39-49	0.69 (85)	
GS 51-69	-0.04 (29)	

Other trials have utilised 'wave' designs where single fungicide applications targeted at a single disease are applied at progressively later times. In such trials the optimum timing for yield response usually coincided with the optimum timing for disease control. The drawback of these experiments is that in practice more than one disease is usually present and the optimum time for one may not be optimum for the other.

In the prototype IDR system, for each disease, values are ascribed according to the potential relative yield loss at each timing. For Rhynchosporium the values are shown in Table 5.

TABLE 5. Values for *R. secalis* crop sensitivity

Value	Crop growth stage
0.75	up to GS 30
2.0	GS 31 to GS 37
3.0	GS 39 to GS 49
1.0	GS 51 onwards

IDR EQUATIONS

In developing IDR equations for each disease, the format developed by Paveley (1993) has been adopted. The equations have been formulated using knowledge obtained from field and trial experience. The equations are open to modification as more experience is gained. The equation for Rhynchosporium is:

IDR score = $(A + 2B + C) \times D$

Values are inserted for inoculum (A), weather (B), variety disease resistance (C) and crop sensitivity (D).

The scores for each disease are then translated into a fungicide dose. The relevent doses for Rhynchosporium are shown in table 6. Doses relate to a specific broad spectrum mixture of a triazole and a morpholine fungicide. The table uses fungicide increments of a quarter dose but apart from a quarter dose as the minimum, there is no reason why dose should not be continuous in relation to the IDR score. In constructing the table, accommodation is made for two considerations. Firstly that control of *R. secalis* is unlikely to be complete and secondly that the dose selected is appropriate to provide sufficient disease control for the most cost effective yield response. A higher dose might give a greater degree of disease control but less profitably.

TABLE 6. Fungicides doses relating to scores
derived from the IDR equation for *R. secalis*

Score	Fungicide dose
0 - 8.0	No fungicide
8.1 - 16.0	1/4 dose
16.1 - 24.0	1/2 dose
24.1 - 34.0	3/4 dose
> 34.1	Full dose

The decision to apply a fungicide will relate to when a previous application was made. In this prototype IDR it is assumed that no fungicide is required within three weeks of an earlier application

FUNGICIDE CHOICE

The activity of fungicides against foliar diseases of barley varies greatly. Relative performance at doses below the full recommended dose is scantily understood. Part of a current Home-grown Cereals Authority project on Appropriate Fungicide Doses for Winter Barley is determining relative dose response curves for many of the commonly used fungicides. This information is vital if IDR scores are to be translated in dose recommendations for a range of fungicide options. Ultimately the fungicide cost needs to be included to determine the most cost effective treatment.

RESULTS FROM A FIELD TRIAL

In a winter barley fungicide trial with the variety Pastoral, four treatments were compared in a randomised block design with three replicates. A 'full dose' programme was compared to an untreated control and programmes of fungicides applied at the same timings as the full dose programme but with fungicide dose decisions based on IDR or PC Plant Protection. The same triazole and morpholine fungicide mixture was used throughout. Rhynchosporium was present early in the spring and developed rapidly in cool, wet conditions.

It was the primary disease throughout the spring and summer. Mildew and net blotch were present but at low levels and spraying for these diseases was never triggered in the IDR equations. The trial is due to be harvested in August 1995 but disease assessments are shown in Table 7. The IDR treatments kept Rhynchosporium levels close to that of the full dose programme, but this was achieved using a half dose equivalent less fungicide.

TABLE 7. Comparison of fungicide programmes in which fungicide dose was determined using IDR or PC Plant Protection with a full dose programme and an untreated control. Variety Pastoral. Tillycorthie, Grampian. 1994/5.

Date & GS	22/3/95 GS 25		29/4/95 GS 32			24/5/95 GS 49			16/6/95 GS 69
Treat	%I	Dose appl.	%S	%I	Dose appl.	%S	%I	Dose appl.	%S
UT	88	0	15.9	98	0	34.1	100	0	47.3
Full	88	0.5	2.0	44	1.0	1.5	29	1.0	2.7
IDR	88	0.25	5.3	62	0.75	3.7	78	1.0	2.3
PC	88	0	18.0	95	0.75	8.4	82	0.9	9.1
S.E.D.			1.57			1.88			4.61

All assessments relate to the third top leaf at each growth stage
UT = untreated control, Full = Full doses applied at GS 32 and GS 49 and a half dose at GS 25. IDR = Integrated disease risk strategy, PC = PC Plant Protection.
%I = % incidence, %S = % leaf area infected. Sen. = Senescent

Experience using IDR equations for determining dose has been encouraging. The system has been relatively straightforward to use and the same conclusions on the appropriate dose to use have been reached by different assessors. Several seasons of trials are required to test the system, identify any weaknesses and, if required, modify the equations. There is no reason why additional factors to the four used could not be incorporated into the equation if more precision is required. However, it is unlikely that precision in determining the fungicide dose at any particular timing will be feasible unless detailed monitoring of the crop, disease and weather are possible. This IDR system is not designed to be precise but rather to give guidance and to present a more rational approach to determining dose. Inevitably, a small degree of insurance is built into the system to cover for unforeseen eventualities.

Rational approaches, such as IDR, to determining whether a fungicide is necessary and if so what dose should be applied will be crucial if farmers are to remain competitive when cereal prices within the EU fall to that of world prices. A system such as IDR also lends itself to incorporation into computerised DSS's which are likely to become important tools in the farm office in the future.

ACKNOWLEDGEMENTS

The authors would like to thank staff of the Crop Biology Department of SAC-Aberdeen for help with the field trial. IDR for winter barley is currently being evaluated as part of a Home-grown Cereals Authority project on Appropriate Fungicide Doses.

REFERENCES

Anon. (1994) SAC Cereal Recommended list for 1995. 16pp.

Anon. (1995) Cereal Variety Handbook. National Institute of Agricultural Botany. 160pp.

Chiarappa, L. (Ed.) (1971) Rhynchosporium secalis on Hordeum spp. *Crop Loss Assessment Methods no. 3. F.A.O. Manual on the evaluation and prevention of losses by pests, diseases and weeds.* 2pp.

Murali, N.S. (1991) An information system for plant protection: 1. Development and testing of the system. *Colloquium on European data bases in plant protection, Strasbourg. Annales ANNP.* **2**, 143-148.

Paveley, N.D. (1993) Integrated Disease Risk (IDR) - a quantitative index system for wheat disease control decision support. *Proceedings of the Workshop on Computer-based DSS on Crop Protection. Parma, Italy.* 47-56.

Paveley, N.D.; Lockley, K.D. (1993) Appropriate fungicide doses for winter wheat - balancing inputs against the risk of disease induced yield loss. *Proceedings of the Cereals R&D Conference, Robinson Hall, Cambridge.* Home-Grown Cereals Authority, London. 177-197.

Polley, R.W.; Clarkson, J.D. (1978) Forecasting cereal disease epidemics. In: *Plant Disease Epidemiology*, P.R. Scott & A. Bainbridge (Eds.). Blackwell, Oxford, pp 141-150.

Ryan, C.C.; Clare, B.G. (1975) Effects of light, temperature and period of leaf surface wetness on infection of barley by *Rhynchosporium secalis. Physiological Plant Pathology,* **6**, 93-103.

Wale, S.J. (1983) Forecasting epidemics of Rhynchosporium secalis on spring barley in north-east Scotland. *4th International Congress of Plant Pathology, Melbourne.* Abstract no. 882.

Wale, S.J. (1987) Effect of fungicide timing on yield of winter barley in northern Scotland. *Proceedings Crop Protection in Northern Britain 1987.* pp 61-66.

Wale, S.J. (1994) Reduced fungicide doses for cereals - a practical perspective on their use. *Brighton Crop Protection Conference - Pests and Diseases - 1994.* pp 695-702.

Zadoks, J.C.; Chang, T.T.; Konzak, C.F. (1974) A decimal code for the growth stages of cereals. *Weed Research* **14**, 415-421.

Session 10

Integrated Weed/Pest/Pathogen Management: Implications for Wildlife, Education and Training

Chairmann and
Session Organiser   Dr R G McKINLAY

THE IMPLICATIONS OF INTEGRATED CROP PROTECTION APPROACHES FOR EDUCATION AND TRAINING

M.J. JEGER

Department of Phytopathology, Wageningen Agricultural University, P.O. Box 8025, 6700 EE Wageningen, The Netherlands

ABSTRACT

Integrated crop protection represents an approach to the control of weeds, pathogens and invertebrate pests that places the emphasis on the crop and cropping system rather than the individual disciplines that still dominate approaches to pest management. Broadly the challenge of integrated crop protection is two-fold: to synthesise available knowledge from a range of sources (agronomy, crop ecology, pest disciplines, control technologies, socio-economics) in order to achieve tactical, strategic or policy goals (which may be economic, environmental or health-related); and to identify key processes and interactions that require further analysis and research. This paper deals with the first of these challenges and considers how training and educational courses can best be geared for the intending practitioner (consultant, extension worker, researcher), the industry (seed, agrochemical and biotechnology companies) and policy-makers. Examples of present curricula at Wageningen Agricultural University (undergraduate, M.Sc. and short advanced courses) and developments involving linkages between Universities of the North and South will be outlined. Examples will be given from both Europe and tropical countries of where some progress has been made towards integrated crop protection and where training and education has been a key factor in its adoption. As with integrated pest management (IPM) there is a danger that integrated crop protection becomes an arcane excercise for academic researchers, aid donors, and environ- mentalists that bears little relationship with the concerns, realities and vocabulary of the farmer. Ultimately the success of integrated crop protection as a key component of sustainable crop production will depend not on ideology or idealised argument, but on implementatation in practice. That is the challenge for education and training.

INTRODUCTION

Crop protection plays a key role in the practice of agriculture throughout the world, across a wide range of farming systems and socio- economic circumstances, and across all agro-ecological zones. There are virtually no circumstances in which consideration need not be given to the ' crop losses caused by weeds, pests and pathogens. A key issue in the current debate on sustainable agriculture is the extent to which crop protection is part of the problem or can contribute to evolving solutions. It is important to keep in mind the different rates of adoption of new agricultural practices in the developed (mostly temperate) and less developed (mostly tropical) countries; and that the key concepts and practice of sustainability should not be predicated solely on the perhaps myopic view of developed intensive agriculture, which is arguably looking to step back from its current position rather than move forward in a sustainable way to a more productive agriculture - the concern of the developing countries. This paper aims to outline the concept of integrated crop protection, examine its historical roots in integrated pest management, consider the types of educational and training courses that are currently available and their adequacy, briefly review the role of integrated crop protection in sustainable agriculture and finally posit the kinds of new courses that will be required to service these trends.

IPM AND CROP PROTECTION

For several decades now, integrated pest management (IPM), has been promoted as the key approach to crop protection. One major aid donor regards IPM as 'the concept of the future for achieving environmentally compatible agriculture' (Deutsche Gesellschaft für Technische Zusammen- arbeit, 1992) and includes IPM as a major theme in all agricultural projects, including those involving training. Other reports are more circumspect, noting that there has been insufficient implementation to demonstrate adequately the effectiveness of IPM as a component of sustainable agriculture (National Research Council, 1992). One problem with evaluating IPM is that, despite several formal definitions, the term continues to have different meanings to different individuals with different agendas, as noted perhaps acidly by Geissbühler (1981) in reviewing the agrochemical industry's approach to IPM, which has long purported to take a serious approach (Sechser, 1981). Any approach to IPM must of course take into account the practices and views of the agricultural industry in its widest sense. The food retail industry has major interests in crop protection practices involving agrochemicals and IPM can be seen as a 'half way' house between conventional agricultural methods, using pesticides, and organic production (Spiegel, 1992).

Initially there is little doubt that the impetus for IPM came from the foresight of the scientific community who were able to influence and persuade funding bodies that change in crop protection practices was required. Paradoxically in evaluation of IPM programmes the criteria used and the conclusions drawn, tend to emphasise the economic considerations (Thompson et al., 1980; Linder et al., 1983; Trumble & Alvarado-Rodriguez, 1993) rather than the concerns of the early proponents, which were largely related to technical efficiency or environmental concerns, and the reduction in pesticide use. Increased revenue to farmers and growers is seen as an important perhaps necessary condition for the successful implementation of IPM. In the developing countries where the impetus for IPM programmes comes more from external aid donors rather than from intrinsic economic imperatives there has equally been concern over the lack of implementation of IPM research (NRI, 1992a, b), despite the success of some programmes (Matthews, 1991) in reducing pesticide usage on specific crops such as cotton. Croft's (1981) prediction - that possibilities for reducing the amounts of pesticides applied, when adjustments between reductions in active ingredient and increases in area treated are made, will be few - has broadly been correct. The influence of policy, however, on reducing pesticide usage in various countries, notable Scandinavia and The Netherlands, has been reviewed by Pettersson (1992, 1994). In The Netherlands the government launched a Multy-Year Crop Protection Plan in 1990, in order to halve the use of pesticides by the year 2000, curtail pesticide emissions to the environment, and implement stricter requirements for pesticide registration. As well as providing legislation, research funding and farmer incentives, the plan has included special extension and education programmes as policy instruments.

INTEGRATED CROP PROTECTION AND PRODUCTION ECOLOGY

One problem with IPM is the contrast presented between research and implementation. Is there such a thing as IPM research rather than research on its component disciplines, e.g. nematology (Roberts, 1993), or is the whole essence of IPM in implementation and farming practice? Certainly much of the research, and the conventional outputs of research such as scientific publications, remain solidly discipline (effectively pest taxa)-based. Also, it is clear that the very term IPM may tend to consolidate this trend. In practice farmers do not manage pests, they protect crops (and the investments made in producing them) by whatever means they have at their disposal. This suggests that a subtle change in emphasis to integrated crop protection is called for, in which the emphasis shifts to the crop and the cropping system - not as a change to hide a failed concept but as a more realistic approach that bears a closer relationship with agricultural practice, offers new research opportunities, and identifies needs for education and training. In fact this approach, often stemming from plant pathology (the least taxonomically-oriented of the traditional plant protection disciplines), has been espoused on several occasions (Wiese, 1982; Teng, 1987; Pfender, 1989). Of course traditional accounts of pest management in specific crops continue to be published (e.g. Grayson et al., 1990) but accounts which focus more on the crop or cropping system, or broader policy issues, have been made (Rola & Pingall, 1993).

A logical and consistent approach to the place of crop protection in sustainable agriculture has been proposed by production ecologists at Wageningen (Rabbinge et al., 1994). As well as defining issues at different hierarchical levels, often according to policy, strategy and tactics (Conway, 1984), crop growth and production are considered under categories that are successively growth defining, growth limiting, and growth reducing. Within this schema crop protection lies firmly in the growth reducing area. This analysis clarifies many of the concepts discussed previously by Wiese (1982) and provides a strong conceptual framework for the place of crop protection in sustainable agriculture (Chadwick, 1993). This context for integrated crop protection then sets an agenda for alleviating constraints caused by the whole range of biotic (also abiotic) factors. As well as helping to define a research agenda for integrated crop protection, there are also clear implications for the role of education and training in the implementation of such research and the practice of integrated crop protection. In considering education and training it is important to recognise the full range of beneficiaries, including the farmers, practitioners and the agricultural industry in its widest sense. It is also important to recognise there are differing views of education and training, that impinge particularly on extension, depending on the view taken of whether the 'Technology transfer' or 'Farmer participation' model is more appropriate. An extreme parody, but nevertheless present in accounts of the virtues of traditional farming (Thurston, 1990), is that it is the researcher or professional who has most to learn from the farmer. This contrast has been most sharply posed in developing countries and aid programmes but is by no means absent in the developed world.

EDUCATION AND TRAINING IN INTEGRATED CROP PROTECTION

Under this heading can be included all formal instruction ranging from short courses on aspects of crop protection, undergraduate and M.Sc. courses in crop protection or its component disciplines/technologies, to research training for Ph.D. degrees, especially where the latter includes formal course work. In practice emphasis will be placed upon the first two types of course and to illustrate the issues involved reference will be made to courses held at or run by the Wageningen Agricultural University. This paper is not an inventory of such courses, but rather considers the types of issues that determine the courses that exist or are likely to develop in the future. Any viewpoint expressed is solely that of the author.

The WAU is uniquely placed at the heart of agricultural research and training in The Netherlands. The Crop Protection Departments of the University (Phytopathology, Entomology, Nematology and Virology), the Research Institute for Plant Protection (IPO-DLO) and the Plant Protection Service (PD), collaborate extensively, and with the International Agricultural Centre (IAC) form the Wageningen Crop Protection Centre (WCPC) for developing countries. The IAC, together with other research/training institutes in the Netherlands, hosts an International Course on Integrated Pest Management for 15 weeks each year. The course is now in its 24th year. The programme for the course is summarised in Table 1.

TABLE 1. International Course on Integrated Pest Management, Programme, 1995

Subject	% course content
Introduction in Plant Production & Protection	5
Mycology	6
Bacteriology	3
Entomology	8
Virology	6
Nematology	8
Weed Science	8
Development of IPM	20
Pesticide Management	10
Presentation Techniques	5
IPM Extension	3
Case study Development of IPM	4
Library	14

Although much of the course is developed to subject matters based on pest taxa (40%) the remainder is allocated to generic topics, including IPM development, pesticide management, extension, and cropping system case studies. The participants in the course include researchers, practitioners, lecturers and private sector representatives. Refresher courses for alumni are regularly held in different regions, e.g. Latin America and S.E. Asia, and in general every effort is made to engender a sense of continuing involvement. A rather different short course is an International Postgraduate Course on 'Modern Crop Protection: Developments and Prospectives' (Zadoks, 1993). This is an intensive 1-week course which broadly addresses three topics: (1) Crop Protection Chemicals; (2) Biotechnology in Crop Protection; and (3) Integrated and Biological Control. In some cases again, however, these topics are sub-divided along conventional taxonomic lines, and presented by discipline specialists.

The Master's degree at Wageningen is generally for overseas students whether from Europe or the tropics. Crop Protection features in several Master's courses, e.g. in Ecological Agriculture, and is a major component of a Masters Course in Crop Science, with specialisation in crop breeding, crop production, crop protection and protected cultivation. Crop protection features in each of these specialisations. Dutch students normally graduate with the Ir. degree, broadly of the same level. As well as traditional courses offered by the individual Departments there is a wide range of courses available which are interdisciplinary in nature and address crop protection issues (Table 2).

TABLE 2. Range of courses in crop protection available in undergraduate curricula at Wageningen Agricultural University

Level	Course
Introductory	Orientation on crop protection Crop protection
Intermediate	Disease and pest development Phytopharmacy Analysis of a problem in crop protection Crop protection and society Biotechnology in crop protection Plant protection in the tropics Excursions (Netherlands and foreign country) Simulation and systems management in crop protection
Advanced	Crop loss Integrated control: plant diseases : insects Research training Practical training

A more recent development has arisen with NATURA (a Network of European Agricultural (Tropical and Subtropical Oriented) Universities and Scientific Complexes Related with Agricultural Development) founded in October 1988. A major initiative within NATURA has been the European Community Training Programme for Agricultural Universities in the South (which carries the acronym NECTAR). The aim of NECTAR through cooperation between NATURA members and Southern Universities is to establish and implement new courses and curricula in eight thematic fields, within agricultural development. The complete list of thematic fields in given in Table 3. These courses/curricula may be restricted to short courses or form the basis for M.Sc. curricula. One of the thematic fields is Sustainable Crop Protection which is co-ordinated by Wageningen Agricultural University. Five modules are planned within this thematic field: biological control, integrated pest management, decision tools for crop protection, recent developments in epidemiology, and weed management. Other than in the last module a purely pest taxonomic approach has been avoided and in all modules crops and cropping systems relevant to the southern partners are stressed. About six European Universities have been involved thus far in curriculum development for sustainable crop protection, which it is planned will finish in 1997. Implementation of the curricula is currently being discussed with the Universities of Zimbabwe, Harare, Cotonou, and Benin. A weakness of the early curricula development was the lack of involvement of Southern Universities in the early planning; this has subsequently been addressed.

TABLE 3. Thematic fields for curriculum development supported by the NECTAR programme

Theme	Co-ordinating University
Food and nutrition sciences	Royal Veterinary and Agricultural University, Copenhagen
Sustainable crop protection	Wageningen Agricultural University
Biomolecular sciences in sustainable agriculture	Athens Agricultural University
Rural environment and development interventions	Universidade de Tras os Montes e Alto Douro, Vila Real
Agricultural economics and policy reforms	Hohenheim Universität & Université Catholique de Louvain
Agricultural economics and rural development	University of Reading
Development operations, preparation and follow up	Agropolis, Montpellier
Water	Universita degli studi di Firenze

THE IMPORTANCE OF TRAINING AND EDUCATION FOR IMPLEMENTATION OF INTEGRATED PEST MANAGEMENT AND CROP PROTECTION

In the early fervour for integrated pest management approaches there was little attention given to requirements for formal training and education. Pimentel (1982) for example fails to mention these elements as being a constraint to IPM or its implementation. The report by the Natural Research Council (1992), under 'educational constraints', only places emphasis on inadequate extension, lack of trained personnel to interface between researchers and the farmer, and the general lack of information about IPM. Education and training must involve more than this, even if the technology transfer model of implementation is accepted. Linder et al. (1983) in their economic evaluation of an IPM programme, while recognising the need for continuing education, state little more than the efforts of the extension services to educate the farmer have been positive, and certainly provide no evidence to support this view.

At a recent BCPC symposium, several speakers addressed the issue of training especially for developing countries (Cox, 1994; Croxton, 1994; Ledru et al., 1995; Marshed-Kharusy, 1994), although the perception of training ranges from the technology transfer model to one in which it is the scientists and planners who actually need the training, not the farmers. In what may be the largest exercise in IPM adoption, on rice in

S.E. Asia, the very concept and contrast of the 'trainers' and the 'trainees' is considered as something to be avoided. Participants learn to obtain personal ownership of the knowledge that will help them to improve their crop protection practices (Chapter discussions in Chadwick, 1993).

Training in the safe use of pesticides might be considered desirable whatever the views on use of pesticides *per se*, but in some ways such training can be seen as effectively part of a pesticide subsidy which most aid donors now exclude from funding. The corollary of this is that where storage and disposal of previously-supplied agrochemicals presents a major hazard then there is a direct responsibility of the donor to provide assistance, especially through appropriate training (Cox, 1994).

In 1978 several chapters of the Proceedings of the IXth International Plant Protection Congress (Kommedahl, 1979) were devoted to education and training in Plant Protection; this trend continued at the XIIIth Congress in 1995 at The Hague, with major sessions held on the transfer of knowledge in crop protection (Anon., 1995), especially in relation to institutional constraints and the use of computer models in education and extension. The latter topic will be discussed at the end of this paper.

ROLE OF CROP PROTECTION IN SUSTAINABLE AGRICULTURE

A whole body of literature is now burgeoning on sustainable agriculture, which includes much of relevance for the development of integrated crop protection. It is only possible to do justice to a few of the issues within the confines of this paper. Much of the debate on sustainable agriculture concerns the need to conduct on-farm research as opposed to the perceived artificiality of the experimental station (Anderson, 1992). An added bonus is the involvement of the farmer in the design, implementation and dissemination of research. By definition, such research must include the interdisciplinarity lacking in IPM programmes (Pimentel, 1982). It should be recognised, however, that 'agro-ecosystems' approaches do not necessarily involve on-farm research, especially those where a regional network of sites is involved (Peterson et al., 1993). Nevertheless if research on integrated crop protection as a component of sustainable agriculture is increasingly to be done on-farm, with an increased participation of farmers, then changes in the way researchers are trained will be inevitable. Although it has been argued that sustainable agriculture must derive from applied ecology (Thomas & Kevan, 1993) it would be a delusion to believe that the issues relating to sustainability are entirely scientific (Levin, 1993). This is a particularly apposite point in relation to crop protection with a wide range of lobby and pressure groups competing with the supposed objective data provided by the scientific community. On the other hand the scientific community must better appreciate its limitations in determining policy and decision making (Miller, 1993) and accept that agricultural and environmental questions are, at this level, trans-science. It is simply no longer acceptable to train crop protection practitioners who can deal only with hard facts and cannot cope with the wider rationale and arguments relating to crop protection; these cannot, with profit, be dismissed.

There is of course a danger that the development of ideas on sustainable agriculture will carry all the vices (and some virtues) of traditional disciplines. One danger is the development of a specialised vocabulary, often justified as necessary for precision, but in reality operating to distance the practitioner from his/her subject matter. The theory of IPM rapidly developed its own distinctive vocabulary far removed from the farmers perceptions of pest problems (Norton, 1980). How many farmers, for example, would recognise that 'Pest Management is a multistage decision process in a stochastic and observable system' (Schoemaker, 1981); or in the context of agricultural systems, that 'Sensitivity in systems shows a negative relationship with the degree of internal articulations among different farming activities' (Viglizzo, 1994). However useful such language may be for a small circle of initiates, it serves little purpose for the implementation and practice of integrated crop protection and sustainable agriculture. A severe challenge in education and training is how to avoid such jargon in getting over the key concepts and elements of integrated crop protection.

NEW REQUIREMENTS IN EDUCATION AND TRAINING

Other than the requirement for education and training in on-farm research, farmer participation, and possibly a better appreciation of trans-science issues (but in a systems context), undoubtedly an element that will increase in importance is the use of different types of models ranging from biophysical simulations to econometric analysis, and to expert systems and decision tools combining both formal and informal elements.

TABLE 4. Papers presented in a session on Use of Computer Models in Education and Extension, XIIIth International Plant Protection Congress, The Hague (Anon., 1995).

Paper title	Country affiliation
A global survey of computer users in agricultural extension	Israel
Communication technologies for information provision and advisory communication in Dutch agriculture	Netherlands
IPM training approach used in Vietnam	Vietnam
Information technology in support in NGO based extension and education in Ghana	Ghana
Information technology supporting capacity building for rice systems research in National Agricultural Research Centres in Asia	Netherlands
Targeting your software to your audience	USA
Genetic software tools for pest management	Australia
A decision support system for integrated crop management of greenhouse crops	Canada

The use of systems analysis and simulation modelling has a long history in IPM and crop protection and increasingly in its relation to sustainability (Rabbinge et al., 1989). Pest management models are increasingly being formulated as crop management models (Hearn & de Roza, 1985). Many models remain that are solely concerned with pest taxa (e.g. Gonzalez-Andujar & Fernandez-Quintanilla, 1993; Doyle, 1991; Duncan, 1991) or control practices (e.g. Gallant & Moore, 1993; Russell & Layton, 1992) and this can also apply to the less formal expert systems or decision models (Cook & Royle, 1984; Luo Yong & Zadoks, 1992; Rossing et al., 1994a, b; Shtienberg et al., 1990; Travis & Latin, 1991; Teng & Yong, 1993). Some decision-models are broader in scope within crop protection (Edwards-Jones, 1993) and in their place at farm level (Milham, 1994). The field of decision-making has developed rapidly since the early 1980's (Austin, 1982). At the XIIIth Interaction Plant Protection Congress held in The Hague in July 1995 a session was devoted to the use of computer models in education and extension. The list of presentations (Table 4) gives an indication of the wide range of applications that have been tested.

Not only will computer modelling be linked more closely to on-farm research, but increasingly much of the exposure to integrated crop protection will come through courses involving computer exercises rather than traditional experimentation. This may be anathema to some but, as in other areas of life, computer education has become inevitable in dealing with the full range of complexities inherent in agriculture in general, and crop protection in particular.

REFERENCES

Anderson, M.D. (1992) Reasons for new interest in on-farm research. *Biological Agriculture and Horticulture*, **3**, 235-250.

Anon. (1995) *Abstracts XIIIth International Plant Protection Congress, The Hague, The Netherlands, 2-7 July 1995. European Journal of Plant Pathology* (unnumbered).

Austin, R.B. (Ed.) (1982) *Decision Making in the Practise of Crop Protection. BCPC Monograph no. 25*, Croydon: BCPC Publications, 238 pp.

Chadwick, D.J. (Ed.) (1993) *Crop Protection and Sustainable Agriculture*, Chichester: John Wiley & Sons, 285 pp.

Cook, R.J.; Royle, D.J. (1984) Computer aided cereal disease management: problems and prospects. *1984 British Crop Protection Conference - Pests and Diseases*, 699-705.

Conway, G.R. (Ed.) (1984) *Pest and Pathogen Control: Strategic, Tactical, and Policy Models*, Chichester: John Wiley & Sons, 488 pp.

Cox, J.R. (1994) Requirements for the safe and effective management of pesticides in less-developed countries. In: *Crop Protection in the Developing World*, R. Black & A. Sweetmore (Eds), *BCPC Monograph No. 61*, Farnham: BCPC Publications, pp. 21-27.

Croft, B.A. (1981) Use of crop protection chemicals for integrated pest control. *Philosophical Transactions of the Royal Society, London, B*, **295**, 125-141.

Croxton, S. (1994) Crop protection for subsistance cultivators: what are the answers? In: *Crop Protection in the Developing World*, R. Black & A. Sweetmore (Eds), *BCPC Monograph No. 61*, Farnham, BCPC Publications, pp. 59-66.

Deutsche Gesellschaft für Technische Zusammenarbeit (1992) *Integrated Pest Management: The Most Promising Concept for Environmentally Compatible Agriculture*, Eschborn: GTZ, 24 pp.

Doyle, C.J. (1991) Mathematical models in weed management. *Crop Protection*, **10**, 432-444.

Duncan, L.W. (1991) Current options for nematode management. *Annual Review of Phytopathology*, **29**, 469-490.

Edwards-Jones, G. (1993) Knowledge-based systems for crop protection: theory and practice. *Crop Protection*, **12**, 565-578.

Gallant, J.C.; Moore, I.D. (1993) Modelling the fate of agricultural pesticides in Australia. *Agricultural Systems*, **43**, 185-197.

Geissbühler, H. (1981) The agrochemical industry's approach to integrated pest control. *Philosophical Transactions of the Royal Society, London, B*, **295**, 111-123.

Gonzalez-Andujar, J.L.; Fernandez-Quintanilla, C. (1993) Strategies for the control of *Avena sterilis* in winter wheat production systems in central Spain. *Crop Protection*, **12**, 617-623.

Grayson, B.T.; Green, M.B.; Copping, L.G. (Eds) (1990) *Pest Management in Rice*, London: Elsevier Applied Science, 536 pp.

Hearn, A.B., Da Roza, G.D. (1985) A simple model for crop management applications for cotton (*Gossypium hirsutum* L.). *Field Crops Research*, **12**, 49-69.

Kommedahl, T. (Ed.) (1979) *Proceedings of Symposia, IX International Congress of Plant Protection, Washington, D.C., USA, August 5-11, 1979*, Vol. 1, 411 pp.

Ledru, X.; Thomas, M., Prêtot, C. (1994) Promoting soft and effective use of pesticides in the developing world: the need for integrated and co-ordinated approach. In: *Crop Protection in the Developing World*, R. Black & A. Sweetmore (Eds), *BCPC Monograph No. 61*, Farnham, BCPC Publications, pp. 69-78.

Levin, S.A. (1993) Science and sustainability. *Ecological Applications*, **3**, 545-546.

Linder, D.K.; Wetzstein, M.E.; Musser, W.N.; Douce, G.K. (1983) An economic evaluation of the Georgia Extension Service integrated pest management programs for cotton. University of Georgia. College of Agriculture, Experiment Stations, *Research Bulletin*, **293**, 32 pp.

Luo Yong; Zadoks, J.C. (1992) A decision model for variety mixtures to control yellow rust on winter wheat. *Agricultural Systems* **38**, 17-33.

Marshed-Kharusy, M.N. (1994) Plant protection in the developing world: problems and needs. Lessons from Zanzibar. In: *Crop Protection in the Developing World*, R. Black & A. Sweetmore (Eds), *BCPC Monograph No. 61*, Farnham, BCPC Publications, pp. 3-10.

Matthews, G. (1991) Comment. Cotton growing and IPM in China and Egypt. *Crop Protection*, **10**, 33-34.

Milham, N. (1994) On incorporating ecological thresholds in farm-level economic models of resource management. *Journal of Environmental Management*, **41**, 157-165.

Miller, A. (1993) The role of analytical science in natural resource decision-making. *Environmental Management*, **17**, 563-574.

National Research Council (1992) *Toward Sustainability. An Addendum on Integrated Pest Management as a Component of Sustainability Research*, Washington, D.C.: National Academy Press, 35 pp.

National Resources Institute (1992a) *A Synopsis of Integrated Pest Management in Developing Countries in the Tropics*. Chatham: NRI, 20 pp.

National Resources Institute (1992b) *Integrated Pest Management in Developing Countries: Experience and Prospects*, Chatham: NRI, 77 pp.

Norton, G.A. (1980) The role of forecasting in crop protection decision making: an economic viewpoint. *EPPO Bulletin*, **10**, 269-274.

Petersson, G.A.; Westefall, D.G.; Cole, C.V. (1993) Agroecosystem approach to soil and crop management research. *Soil Science Society of America Journal*, **57**, 1354-1360.

Pettersson, O. (1992) Pesticides, valuations and politics. *Journal of Agricultural and Environmental Ethics*, 103-106.

Petterson, O. (1994) Reduced pesticide use in Scandinavian agriculture. *Critical Reviews in Plant Sciences*, **13**, 43-55.

Pfender, W.F. (1989) Cultural control of plant pathogens in IPM.

Proceedings National Integrated Pest Management Symposium/Workshop, Las Vegas, Nevada, April 25-28, 1989, pp. 58-66.

Pimentel, D. (1982) Prospectives of integrated pest management. *Crop Protection*, **1**, 5-26.

Rabbinge, R.; Ward, S.A.; Van Laar, H.H. (Eds) (1989) *Simulation and Systems Management in Crop Protection*, Wageningen: Pudoc, 420 pp.

Rabbinge, R.; Rossing, W.A.H.; Van der Werf, W. (1994) Systems approaches in pest management: the role of production ecology. *Proceedings Plant Protection in the Tropics*, Eds A. Rajan & Y. Ibrahim, Malaysian Plant Protection Society, pp. 25-46.

Roberts, P.A. (1993) The future of nematology: integration of new and improved management strategies. *Journal of Nematology*, **25**, 383-394.

Rola, A.C.; Pingall, P.L. (1993) *Pesticides, Rice Productivity and Farmers' Health: An Economic Assessment*, Manila: International Rice Research Institute, 100 pp.

Rossing, W.A.H.; Daamen, R.A.; Jansen, M.J.W. (1994a) Uncertainty analysis applied to supervised control of aphids and brown rust in winter wheat. Part 1. Quantification of uncertainty in cost-benefit calculations. *Agricultural Systems*, **44**, 419-448.

Rossing, W.A.H.; Daamen, R.A.; Jansen, M.J.W. (1994b) Uncertainty analysis applied to supervised control of aphids and brown rust in winter wheat. Part 2. Relative importance of different components of uncertainty. *Agricultural Systems*, **44.**, 449-460.

Russell, M.H.; Layton, R.J. (1992) Models and modeling in a regulatory setting: considerations, aplications, and problems. *Weed Technology*, **6**, 613-616.

Schoemaker, C.A. (1981) Applications of dynamic programming and other optimization methods in pest management. *IEEE Transactions on Automatic Control*, **26**, 1125-1132.

Sechser, B. (1981) An approach to integrated pest management from the chemical industry. *Acta Phytopathologica Scientiarum Hungariae*, **16**, 239-243.

Shtienberg, A.; Dinoor, A., Marani, A. (1990) Wheat Disease Control Advisory, a decision support from management of foliar diseases of wheat in Israel. *Canadian Journal of Plant Pathology*, **12**, 195-203.

Spriegal, G. (1992) Integrated pest management and modern agriculture. *Journal of Biological Education*, **26**, 178-182.

Teng, P.S. (Ed.) (1987) *Crop Loss Assessment and Pest Management*, St. Paul, Minnesota: APS Press, 270 pp.

Teng, P.S.; Yong, X.B. (1993) Biological impact and risk assessment in plant pathology. *Annual Review of Phytopathology*, **31**, 495-521.

Thomas, V.G.; Kevan, P.G. (1993) Basic principles of agroecology and sustainable agriculture. *Journal of Agricultural and Environmental Ethics*, 1-19.

Thompson, P.; How, R.B.; White, G.B. (1980) An economic evaluation of grower savings in a pest management program. *HortScience*, **15**, 639-640.

Thurston, H.D. (1990) Plant disease management practices of traditional farmers. *Plant Disease*, **74**, 96-102.

Travis, J.W.; Latin, R.X. (1991) Development, implementation, and adoption of expert systems in plant pathology. *Annual Review of Phytopathology*, **29**, 343-360.

Trumble, J.T.; Alvarado-Rodriquez, B. (1993) Development and economic evaluation of an IPM program for fresh market tomato production in Mexico. *Agriculture, Ecosystems and Environment*, **43**, 267-284.

Viglizzo, E.F. (1994) The response of low-input agricultural systems to environmental variability. A theoretical approach. *Agricultural Systems*, **44**, 1-17.

Wiese, M.V. (1982) Crop management by comprehensive appraisal of yield
    determining variables. *Annual Review of Phytopathology*, **20**, 419-432.
Zadoks, J.C. (Ed.) (1993) *Modern Crop Protection: Developments and
    Perspectives*, Wageningen: Wageningen Press, 309 pp.

THE INFLUENCE OF PESTICIDES AND PEST MANAGEMENT STRATEGIES ON WILDLIFE

V. GUTSCHE

Biologische Bundesanstalt für Land- und Forstwirtschaft in Berlin und Braunschweig

Institut für Folgenabschätzung im Pflanzenschutz Kleinmachnow

Stahnsdorfer Damm 81

D - 14532 Kleinmachnow

(Federal Biological Research Centre in Agriculture and Forestry

Institute for Technology Assessment in Plant Protection)

ABSTRACT

On base of 12 chemical-physical parameters and 10 ecotoxical and toxical parameters the model SYNOPS 1.0 to assess and compare the risk potentials of active ingredients of pesticides was developed. The ranking of the 6 compared substances is visualized by risk graphs.

A strong integrated pest management (IPM) can reduce the entry of pesticides. The main IPM-elements are: usage of thresholds, dosing of pesticides according to situation, use of resistant and tolerant cultivars, usage of forecasting models, use of biological methods etc..

Edges, linking biotopes and set-aside programmes do also have positive effects on wildlife.

In two production experiments the effect of intensity of production factors (fertiliser, pesticides) and different farm management systems (organic-traditional), respectivly, were investigated.

In the intensity-experiment only a slight negative influence on wildlife was observed, in the system-experiment (survey) a some bigger difference could be shown.

INTRODUCTION

For centuries man's agricultural activities have created a landscape which forms an essential basis for the diversity in biotopes and species reaching a height in the 19th century. However, intensification of agriculture over the last decades (fertilisation, pest management, specialisation and concentration in production) is regarded to have primarily caused the rapid decline in plant and animal species. For this reason, ways must be found to connect better nature and landscape conservation with crop and animal production without losing sight of the prime aim of agriculturists - the production of foodstuff. Such new strategies have to be elaborated based on scientific knowledge of the effect of crop management systems on wildlife.

In the following greater attention is given to the effect of plant protection. First the action of individual pesticides is investigated with the help of a model for the synoptic assessment of the risk potential of active ingredients of pesticides.

The development of mathematical models which show the effect of pesticides in ecosystems is still at the beginning and becomes very difficult with increasing complexity of the ecological system. It seems to be hardly possible to elaborate models which quantitatively forecast the effect of whole strategies within a crop management system. We can only make general qualitative statements and produce individual quantitative results from comparative investigations.

Therefore, the second part of the present work attempts to compare the effect of components of integrated pest management, the role of edge and linking biotopes and the effect of land set-aside with the demands of nature conservation and to provide quantitative evidence for it. This is based on two long-term experiments and surveys at different agricultural sites of Germany comparing various intensities as well as forms of crop production (organic - conventional)(BARTELS and KAMPMANN (ed.), 1994; KÖNIG et al., 1989).

THE COMPARATIVE ASSESSMENT OF THE RISK POTENTIAL OF ACTIVE INGREDIENTS OF PESTICIDES

As far as plant protection in Germany is concerned the amendment of the Plant Protection Act of 1986 has enhanced the attention to aspects of environment protection for the registration of pesticides. For instance, a large number of chemical and physical and ecotoxicological data with respect to the environmental fate of the chemicals have to be presented. Ecotoxicological risk assessments are mostly based on monospecies tests. In this connection the amount of aquatic (empiric) data is bigger than that of terrestrial ones. Investigations with the help of model ecosystems (mezocosmos) are still at the beginning and show a wide range of methodological difficulties (SCHLOSSER, 1994).

The available data form the basis of a model (SYNOPS 1.0) to assess and compare the risk potentials of active ingredients. To assess as many active ingredients as possible a compromise between the size of the assessment model and the available data had to be made. The necessary input data are:

chemical physical parameters

DT50	soil	(disappearance time 50)
DT90	soil	(disappearance time 90)
DT50	water	
DT90	water	
DT50	photolysis	
DT50	hydrolysis	
DT50	volatilisation	
VP	vapour pressure	
Kp	partition coefficient solid-water	
Koc	partition coefficient organic carbon water	
lPow	logarithmic partition coefficient n-octanol-water	
SOL	solubility	

ecotoxicological and toxicological parameters

LC50	earthworm(lethal concentration 50)
NOEC	earthworm(no observed effect concentration)
BCF	bioconcentration factor earthworm
LC50	Daphnia
NOEC	Daphnia
LC50	fishes
NOEC	fishes
BCF	fishes
NOEL	mammals (no observed effect level)
NOEL	birds

The assessment model consists of 8 steps. They will be below explained. A presentation of the mathematical formulas is renounced for reasons of space.

Sphere of application and load

Starting point for the determination of the risk potential of each active ingredient is its sphere of application. It is mentioned in the registration and we understand by it the number of all possible applications against certain harmful organisms (group of harmful organisms) on a certain crop at a certain growth stage of this crop. In case of repeated applications against a harmful organism the maximum number given in the registration is assumed. On applying a pesticide not only the target (for instance the leaves of a crop) is loaded, but also other compartments (plant, soil, surface water). To calculate the load of the compartments a simple table is used containing the percentage distribution of the applied volume to plants, soil and drift. An example is shown in the following table:

Distribution of active ingredient after application

	from EC	to EC	plant (%)	Soll (%)	soilwater (%)
sugar beet	0	0	0	99,9	0,1
sugar beet	1	12	14,9	85	0,1
sugar beet	13	19	34,9	65	0,1
sugar beet	20	33	64,9	35	0,1
sugar beet	34	49	94,9	5	0,1
hop	0	0	0	96	4
hop	1	39	51	45	4
hop	40	59	71	25	4
hop	60	99	86	10	4
winter wheat	0	0	0	99,9	0,1
winter wheat	1	29	14,9	85	0,1
winter wheat	30	39	54,9	45	0,1
winter wheat	40	99	89,9	10	0,1

Indices and environmental exposure

Indices of the environmental exposure are the short-term predicted environmental concentration (sPEC) and the long-term predicted environmental concentration (lPEC) of the substance in the compartments soil, surface water and air. The short-term predicted environmental concentration reflects the amount of substance entering the above-mentioned compartments immediately after application. As a consequence of chemical-physical reactions, but mainly microbial processes, the active ingredient decomposes with time. How long how much of the substance is still present in the environment is calculated with the long-term predicted environmental concentration.

A proportion of the substance deposits to soil or sediment particles of the surface water in a way that it cannot decompose. It is also represented by an index (Cm).

Indices of the biological risk

The biological risk of the substance is calculated as an acute biological risk (abr) and a chronic biological risk (cbr) for earthworms, aquatic invertebrates and fish. The acute biological risk is the ratio of the short-term predicted environmental concentration to the lethal dose 50 of an active ingredient, i. e. the dose at which 50% of the load organisms die. The chronic biological risk is the ratio of the long-term predicted environmental concentration to the no observed effect level (NOEL) of the active ingredient, NOEL is the dose at which no effect on the organism under investigation could be observed.
To take into account the risk that a substance can accumulate in fatty tissue of an organism, the so-called bioconcentration factor (BCF) is determined for earthworms and fish as another index of the biological risk. The accumulation risk is compared with the NOEL of the substance for birds and mammals. The established ratio is called food chain risk (fcr). It expresses the risk that a substance also affects more highly-developed animal species.

Visualisation of the risk potentials

It is a general problem of technology assessment, which comprises also risk assessment of active ingredients, to visualise complex situations in such a way that they are understandable at first glance.
Finally SYNOPS procedure results in a ranking of the compared active ingredients. The ranking is illustrated with the help of a risk graph.

This risk graph is a circle divided into six sectors. Each sector is in turn divided into segments corresponding to the individual indices. The lower 3 sectors characterise the presence of the substance in the environment. Left the environmental exposure in water, right in soil and in the middle in the air. The upper 3 sectors represent the biological risk. Left to aquatic life, right to soil life and in the middle accumulation and food chain risk.

<u>Example assessment of the risk potentials of six active ingredients</u>

The following active ingredients were assessed: the insecticides methamidophos, endosulfan and lindane, the fungicides mancozeb and carbendazim and the herbicide isoproturon.
Methamidophos, mancozeb and isoproturon are according to the Industrieverband- Agrar e.V. (IVA) (Federation of Agricultural Industries) the best-selling active ingredients of their respective groups in Germany in 1993. Endosulfan and lindane are two "old" active ingredients which are not registered any more or only with restrictions.
To enable an overall assessment of their risk potentials, the registration situation of 1990 is assumed. Consequently various groups of active ingredients as well as "old" and "new" active ingredients are compared. The number of potential applications varied between 7 and 11 with the exception of mancozeb. Because of the high number of repeated treatments 26 possible applications are assumed.

Ranking of the compared active ingredients

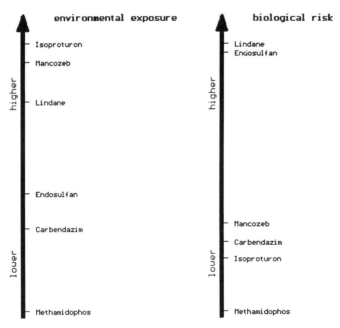

472

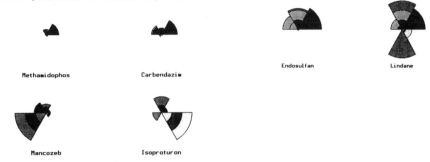

risk-potentials of active ingredients

Methamidophos

Carbendazim

Endosulfan

Lindane

Mancozeb

Isoproturon

The figure shows risk ranking and risk potentials. A difference is made between environmental exposure and biological risk. No doubt the more important group is the biological risk because the mere presence of a substance in the environment does not provide a risk. The risk graphs of the active ingredients demonstrate the ranking procedure. Large areas correspond to a high risk potential, small ones to a low one.
The "old" active ingredient lindane has both the highest total and the highest biological risk. Although isoproturon ranks second in the total risk, its biological risk is far lower. Its risk graph reveals that the relatively high environmental exposure (lower segments) does not change to a high biological risk (upper segments). The fungicidal active ingredient mancozeb shows a similar assessment. The situation is dramatically reversed for endosulfan. Despite a low environmental exposure it involves a high biological risk.
The new active ingredients reach only 50% of the biological risk potential of the "old" ones, whereas methamidophos, mancozeb and isoproturon, the best-selling ones, have the lowest biological risk potentials.

POSSIBILITIES TO REDUCE THE ENTRY OF PESTICIDES BY MODERN PEST MANAGEMENT

The observation of the principles of integrated pest management is theoretically part of good conventional farming. But in practice there is especially in arable farming a considerable deficit. In the following we explain several elements of integrated pest management which lead to a reduction of the entry of chemical-synthetic pesticides into the environment and thus contribute to the conservation of wildlife.

Use of thresholds

Germany has known thresholds for more than a hundred harmful organisms (including weeds). They are partially based on careful examination of the infestation-damage ratio, partially on simple calculations and experience (FREIER et al., 1994). From a critical point of view the use of fixed thresholds is doubtful because infestation and infection, yield formation and economic efficiency of pest management are subject to strong dynamics (LAUENSTEIN, 1991). Therefore scientists have the task to provide variable thresholds depending on the situation for the most important harmful organisms.
Initial considerations are concerned with beneficial thresholds for pest enemies in agroecosystems. Beneficial thresholds represent the minimum density of beneficial organisms to control pest populations (FREIER, 1993). They are important especially when pest pressure is low.

Dosing of pesticides according to situation

The reduction of dose rates, especially of fungicides and herbicides, is of growing importance in German pest management. Numerous experiments have been carried out by the Plant Protection Services of the Lands and the results are used for advising agriculturists. A reduction of the dose rate by 25-50% is not unusual. For fungicides there are investigations to make the dose rate dependent on infection pressure. It is planned to use epidemic models for the calculation of the infection pressure.

Use of forecasting models for pests and diseases

Model forecasting procedures for important harmful organisms should be possible to reduce prophylactic chemical treatments. At present a model project supported by the Bundesministerium für Ernährung, Landwirtschaft und Forsten (Ministry for Food, Agriculture and Forestry) tests nation-wide 11 model-based forecasting procedures and introduces them into practice. The below figure shows as an example the result for potato blight. It reveals a considerable potential to economy on numbers of spray applications.

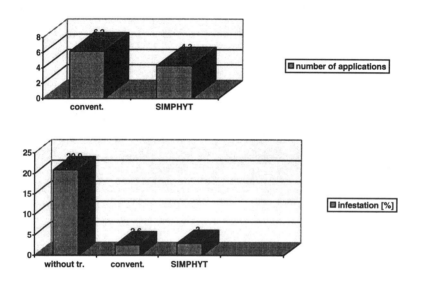

Use of modern technology

Feedback/Recovery sprayers (tunnel devices) can economise to 1/3 of the dose rate in fruit and wine growing (GANZELMEIER et al., 1993). This leads to a considerable reduction of soil load through pesticides and airborne drift and is especially important for the conservation of the environment.

Use of resistant and tolerant species

Classical breeding as well as genetic engineering are more and more aimed at disease resistance. In addition research is concerned with the induction of defence mechanisms inherent in plants by specific pathogens, microbial metabolic products or saprophytes (induced resistance). Resistance or increased tolerance to plant diseases means automatically renunciation or strong reduction of fungicide applications.

Use of biological measures

This is above all the use of microorganisms and of predators and parasitoids against harmful organisms or the disturbance of their behaviour by pheromones and repellents. Biocontrol is already of great importance for glasshouse crops. For outdoor crops there are only a few examples (*Trichogramma evanescens* against the corn borer, virus preparations against the codling moth, *Bacillus thuringiensis* against butterfly larvae and the Colorado beetle).

THE ROLE OF EDGE AND LINKING BIOTOPES FOR NATURE CONSERVATION

Consolidation and extension of plots have produced monotonous agricultural landscapes in many parts of Germany. This is a very serious problem in the East German Lands, where arable fields of more than 100 ha are not unusual. This increase has a decisive influence on the area of edge biotopes (hedges, arable field margins, boundary strips). An example of changing of hedge structure in a agricultural landscape between 1877 and 1979 is given in the following figure (KNAUER, 1988).

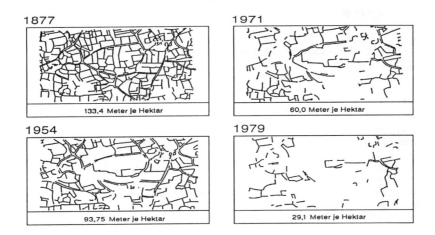

1877 — 133,4 Meter je Hektar
1971 — 60,0 Meter je Hektar
1954 — 93,75 Meter je Hektar
1979 — 29,1 Meter je Hektar

Edge biotopes are of eminent importance for the maintenance of the natural species diversity in agricultural landscapes. A compromise has to be made between economic field size and biotope linking. In this connection those calculations on the basis of the data collection of the Kuratorium für Technik und Bauen in der Landwirtschaft (Committee for Technology and Construction in Agriculture) (DOHNE, 1989; HEISSENHUBER and HOFMANN, 1992) are of interest which say that the amount of labour increases dramatically for fields smaller than 5 ha, but does not decrease significantly for fields larger than 10 ha.

All Lands of Germany have schemes facilitating the maintenance and creation of edge and linking biotopes (PLANKL, 1995).
The following facts have been gleaned from various scientific investigations regarding the role of edge and linking biotopes for nature conservation.

Hedges
(GLÜCK and KREISEL, 1989; NACHTIGALL, 1994; STECHMANN and ZWÖLFER, 1988; ZWÖLFER
et al., 1984)

- An increasing number of woody plants is increasingly ecologically important for
  animals.
  Hedges are an important habitat for birds. A study in Hessen Bundesland (Hesse Land)
  recorded 60 species within 2 years.
- Hedges are an important habitat for beneficial organisms. Out of 55 species 83% were
  entomophagous ones.
- Hedges are linking elements especially for soil-living animals with high mobility (e.g.
  Carabidae).
- They form a greater reservoir for beneficial arthropods than for pests.

Boundary strips
(KOTKA, 1984; MOLTHAN, 1990; NACHTIGALL, 1994; SCHWENNINGER, 1987; WELLING, 1990)

- They are an important source of food for carabid species.
- Margin width, number of flowers and number of syrphid flies form a positive correlation.
- Their flowering plants are a source of food for a large number of beneficial insects.
- Boundary strips 0.6 m wide had 18 species of syrphid flies, field margins 1.5 m wide had
  28.
- Minimum width of strips should be 1-2 m, otherwise they have insufficient effect.
- Aphid infestation on a field edge adjacent to a field margin of 3-4 m was found to be 50%
  lower than in the field.
- Migration of carabids up to 200 m into the field has been reported.
- Herbaceous boundary strips showed a larger number of species (carabids) than
  gramineous ones.

Arable fields margins
(KLINGAUF, 1988; NACHTIGALL, 1994; SCHUMACHER, 1984; WELLING et al., 1988)

- Field margins of arable fields are of special importance for the maintenance of arable
  weeds.
- 160 arable weed species have been found on uncultivated margins.
- The number of arable weeds showed only a slight increase in species diversity even
  after a 5-year experiment (Rheinland-Pfalz Bundesland (Rhineland-Palatinate Land)),
  because the whole region had been intensively cultivated for many years. The seed
  potential was presumed to be exhausted.

THE EFFECT OF LAND SET-ASIDE

The primary aim of the European set-aside scheme (VO 2293/92 EEC) is an easing of
the market. The scheme allows one-year rotational fallows and five-year permanent fallows.
Both options have secondary effects on nature conservation.
Rotational fallows are distinguished by naturally regenerated, sown and bare fallows. They
have different effects on wildlife. Bare fallows have to be considered to be inappropriate
from an ecological point of view and because of the risk of erosion and stronger nitrate
leaching. They have been forbidden in Germany.

Naturally regenerated fallow
(BURTH, 1993; ELSEN and GÜNTHER, 1992; MAYKUHS, 1989; OBERGRUBER, 1990)

- A large number of species in arable weed communities is often reported.

- The contrary is also known. Many plots in Brandenburg Land (fertilised sand soil) were nearly completely covered with *Erigeron canadensis*.
- The prehistory of the fallow (nutrients, seed bank) and water content of soil are decisive for the formation of a floristic species diversity.
- In case of a species-rich flora, rotational fallows are of similar importance for carabids and syrphids to flower- rich field margins.
- Naturally regenerated fallows can raise problems with soil-borne harmful organisms (nematodes, cereal diseases), but also with snails and mice in subsequent crops. The same is true of the increase in the seed potential of weeds for subsequent crops.

Sown rotational fallow
(BAUER and ENGELS, 1992; RUPPERT, 1993; NACHTIGALL (1994))

- They are used to prevent above all nitrate leaching and an increase in the weed seed bank.
- The frequent mixture of annual flowering plants with a high proportion of *Phacelia tanacetifolia* is favourable for bees, but not optimal for syrphid flies and Hymenoptera.
- However, a mixture of indigenous weeds and weak competitive grasses (rapid soil cover) seems to be very favourable for various beneficial arthropods.

Permanent fallows
(HARRACH and STEINRÜCKEN, 1990; NACHTIGALL, 1994; SCHUMACHER, 1990)

- Permanent fallows precede scrub and woodland consequently reducing species diversity.
- Only in the first two to three years can species-rich pioneer communities be observed.
- Permanent fallows are negatively assessed in regions with extensive farming because they evolute from species-rich arable weed communities into grass and ruderal communities.

COMPARATIVE INVESTIGATION OF THE EFFECT OF DIFFERENT FARM MANAGEMENT SYSTEMS:

The results of a comprehensive production experiment and a production survey are shown. The experiment was carried out on loess soil in the Ostbraunschweiger Hügelland (East Brunswick Hills) examining four types of intensive crop production (intensity trial). The survey was made at a loess site in the Kölner Bucht (Cologne Embayment) comprising two farms with different farm management systems (organic - conventional) (farm management trial).

Intensity trial (1987-1989)
(BARTELS and KAMPMANN (ed.), 1994)

Rotation:	Sugar beet, winter wheat, winter barley.
Intensity degree I_0:	Without pesticides, little application of N fertiliser.
Intensity degree I_1:	Suboptimal application of pesticides and N fertiliser.
Intensity degree I_2:	Application of pesticides and N fertiliser according to recommendations of integrated crop production. Optimisation of natural yield with minimum production costs.
Intensity degree I_3:	Employment of all admissible production factors. Maximisation of natural yield.

The effect of the various intensities on soil flora and fauna was studied. The following indicator groups were used:

Algae:	cyanobacteria, eukaryotic algae;
Fungi:	oomycetes, zygomycetes, basido-, asco- and deuteromycetes;
Fauna:	nematodes, earth worms, soil mites, collembolans.
Microbial activity:	dehydrogenase activity.

The most important results are summarised as follows:
1. Cultivation and varying conditions over the three trial years had a stronger influence on the indicators than the intensity of chemical production factors.
2. The intensity of production had hardly any influence on the number of species.
3. Degree 3 led to a significant reduction of the abundance of algae and collembolans. Dehydrogenase activity was slightly reduced. The same is true of the earth worm abundance (compare the following figure).

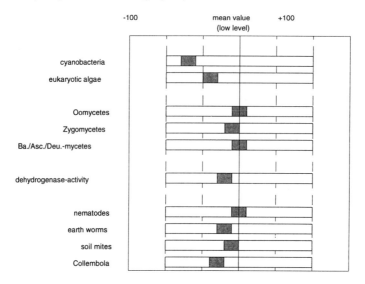

Altogether a slight negative influence was observed, and algae and collembolans recovered rapidly during the vegetative period because of their fast multiplication. But it has to be taken into account that the experiment was carried out in a region with many years' intensive agricultural management and drastic changes in the agroecosystem took place a long time ago. It can be assumed that the soil biocenosis consists now of euryoecious species which are hardly sensitive to disturbances.

Farm management trial (1982-1986)
(KÖNIG et al., 1989)

Organic farm	Conventional farm
- 60 ha more than 10 years organic farming mixed cattle breeding - livestock number 0.5-0.6 per ha	- 100 ha - crop including form wheat, barley and sugar beet
8-year rotation	3-year rotation
Fertilisation with stable manure and rock meal	Mineral nitrogen, phosphate, potassium, magnesium and lime fertiliser
As pesticides only siliceous and herb extracts	Chemical-synthetic pesticides - herbicides (partially pre-emergence) - insecticides against aphids - fungicides (partially repeated)

Cultivation of soil showed hardly any difference. The indicators under investigation were:

1. soil physics, humus and nutrient contents;
2. microflora and soil fauna: microbial activity, earth worms, enchytraeids, collembolans, soil mites, diplopods and wood lice;
3. arable weeds;
4. arable fauna: Carabidae, Staphylinidae, Catopidae, Chrysomelidae, Silphidae, Nematocera, Brachycera, Syrphidae, Heteroptera, Homoptera, Arenae, Lycosidae, Opiliones, Diplopoda, Chilopoda, Isopoda.

The following conclusions can be drawn from the survey with respect to the indicator groups:

Soil physics, humus and nutrient contents
- Conventionally cultivated soils have a stronger tendency to drying up; other physical soil parameters did not show any difference.
- Organically cultivated soils show a tendency to acidification.
- Humus content and quality, substance content and C/N ratio did not differ.

Microflora and soil fauna
- Only earth worms are facilitated in organically cultivated soil.
- The other parameters are by far strongly influenced by natural site factors. An influence of the farm management system could not be proven.

Arable weed vegetation
- Organically cultivated soil produced in general the whole species spectrum of arable weed communities.
- The percentage of weed cover was substantially higher on organic soil.
- Conventional soil produced only fragmented communities which indicated an accumulation of nitrogen.
- Organic soil also exhibits in the long run an increase in nitrogen indicators (general eutrophication of agriculture).

Arable fauna
- Faunistic data are in close correlation with the species diversity of weeds.
- In 8 of 13 arthropod groups under investigation, the mean number of species is far higher on organically cultivated plots than on conventional plots.
- Mean catch number (or activity abundance) was higher on organic plots for 11 groups of animals and on conventional plots for three groups.
- Ground beetles (Carabidae) and spiders (Arenae) were significantly more numerous on organic plots. The other 8 animal groups did not reveal any difference.

REFERENCES

Bartels, G.; Kampmann, T. (ed.) (1994) *Mitteilungen aus der Biol. Bundesanstalt für Land- und Forstwirtschaft, Berlin-Dahlem, Heft 295*, 405 pp.

Bauer, M.; Engels, W. (1992) *Apidologie*, **23**, 340-342

Burth, U. (1993) persönl. Mitt.

Dohne, E. (1989) *Schriftenreihe agrarspectrum*, **15**, 57-75

Eigen, TH.; Günther, H., (1992) *Z. Pfl. krankh. Pfl.Schutz, Sonderheft XIII*, 49-60

Freier, B. et.al. (1994) *Nachrichtenbl. Deut. Pflanzenschutzd.*, **46** (8), 170-175

Ganzelmeier, M. (1993) *Pflanzenschutz-Praxis 3/93*, 14-15

Glück, E.; Kreisel, A. (1988) *Berichte Akademie für Naturschutz u. Landschaftspflege, 10/86*, 64-83

Harrach, T., Steinrücken, U. (1990) in: *Hess. Min. f. Landw., Forsten und Naturschutz: Flächenstillegungen in der Landwirtschaft*, 21-22

Heissenhuber, A.; Hoffmann, H. (1992) in: *Bayerisches Staatsministerium für Landesentwicklung und Umweltfragen: Vorstudien Materialien 84*, 55-116

Klingauf, F. (1988) *Mitteilungen aus der Biol. Bundesanstalt für Land- und Forstwirtschaft, Heft 247*, 7-15

Knauer, M. (1988) *Mitteilungen aus der Biol. Bundesanstalt für Land- und Forstwirtschaft, Heft 247*, 147-161

König, K. et.al. (1989) *Schriftenreihe der Landesanstalt für Ökologie, Landschaftsentwicklung und Forstplanung Nordrhein-Westfalen, Band 11*, 286 pp.

Kotka, CH. (1984) Typische Feldraine und ihre Entomofauna im Hessischen Ried., *Dipl.-Arb., Univ. Aachen* , 99 pp.

Lauenstein, L. (1991) *Gesunde Pflanzen*, **43**, 346-550

Magkuhs, F. (1989) *Gesunde Pflanze*, **41**, 210-214

Molthan, J. (1990) *Mitt. Dtsch. Ges. Allg. Angew. Ent. 7*, 368-379

Nachtigall, G. (1994) *Mitteilungen aus der Biol. Bundesanstalt für Land- u. Forstwirtschaft, Heft 294*, 98 pp.

Obergruber, H. et al. (1990) in: *Hess. Min. f. Landw., Forsten und Naturschutz: Flächenstillegungen in der Landwirtschaft*, 28-29

Plankl, R. (1995) Institut für Studienforschung der Bundesforschungsanstalt für Landwirtschaft Braunschweig/Völkenrode: *Arbeitsbericht 1* , 129 pp.

Schlosser, I. (1994) in: *Deutsche Forschungsgemeinschaft - Ökotoxikologie von Pflanzenschutzmitteln - Sachstandsbericht, VHC Verlagsgesellschaft Weinheim* , 291-301

Ruppert, V. (1993) *Agrarökologie*, **8**, 149 pp.

Schumacher, W. (1984) *LÖLF-Mitt.*, **9** (1), 14-20

Schumacher, W. (1990) in: *Hess. Min. f. Landw., Forsten und Naturschutz: Flächenstillegungen in der Landwirtschaft*, 64-66

Schwenninger, H. R. (1987) *Mitteilung Dtsch. Ge. Allg. Angew. Ent.*, **6**, 364-370

Stechmann, D.-H.; Zwölfer, H. (1988) Schriftenreihe BML, Reihe A: Angew. Wiss., **365**, 31 - 55

Welling, M. (ed.) (1988) *Mitt. Biol. Bundesanst. Land- und Forstwirtschaft, Heft 247*, 165 pp.

Welling, M. (1990) Förderung von Nutzinsekten, insbesondere Carabidae, durch Feldraine und herbizidfreie Ackerränder und Auswirkungen auf den Blattlausbefall in Winterweizen, *Diss. Univ. Mainz* , 160 pp.

Zwölfer, H. et.al. (1984) *Berichte Akademie für Naturschutz und Landschaftspflege, Beiheft 3, Teil 2* , 155 pp.

Symposium Overview

Chairman and
Session Organiser   Dr R G MCKINLAY

CROP PROTECTION IN SUSTAINABLE FARMING SYSTEMS

D ATKINSON AND R G McKINLAY

SAC West Mains Road, Edinburgh, EH9 3JG

ABSTRACT

Using the contributions presented at the conference on "Integrated Crop Protection: Towards Sustainability?" as a basis, current major issues are identified. These would seem to be the operational description of sustainability, the potential contribution of basic ecological understanding to sustainable crop protection, the importance of early diagnosis of problems, the optimal balance between food production and other land uses and the appropriate form of experimentation needed to develop sustainable agricultural systems. Key questions in relation to these areas are identified.

INTRODUCTION

The title of the conference, "Integrated Crop Protection: Towards Sustainability?" poses a series of questions. What is crop protection? Can it ever be integrated? What do we understand by sustainability? Do we believe that complete sustainability is achievable? How important do we believe sustainability, as it is being defined, to be relative to the need to feed an increasing world population? Although the title poses questions, it also states certainties; the critical importance of crop protection for any food production system, the importance of how we protect our crops today for future generations and the belief that current studies will produce answers to the above questions. To provide a balanced perspective on what has been achieved to date and where the future should lead us, it is helpful to visit some of the above questions and certainties.

SUSTAINABILITY

A definition of sustainability is implicit or explicit in all of the papers presented at the conference. The Bruntland (1985) definition of sustainability, "development which meets the needs of the present generation without compromising the ability of future generations to meet their own needs" emphasises intergenerational transfers but is vague in relation to what is being transferred. Consequently, the definition requires further expansion before it becomes actionable. Tait and Pitkin (1995) draw from the generalised concept of sustainability, a number of aspects which they feel require to be emphasised; the ability of the system to continue to operate in its present form, the importance of a human interest in the system and the interaction between the desire to manage or exploit and sustainability. They also develop the concept of degrees of sustainability, i.e. sustainability not as an absolute concept but with the potential for systems to be more or less sustainable.

Atkinson *et al* (1994) emphasised the breadth of vision which must be applied to require a reappraisal of appropriate methods, so as to give full weight to scale-related factors and the interactions between species.

INTEGRATED CROP PROTECTION: TOWARDS SUSTAINABILITY?

The integration of the various crop protection and production aims seems likely to result in systems based on sound ecological principles which optimise resource use. This will arise from an enhancement of the role of native organisms and natural processes in regulating the extent of weeds, pests and diseases and consequently from the more targeted, and less extensive, use of crop protection materials. Such systems, in the context of the definitions given earlier, will thus be more sustainable. The title of the conference asks an explicit question. On the basis of the information presented here, it is possible to answer it with an unequivocal "yes" but also to identify some of the further research required if we are to go still further towards sustainability.

REFERENCES

ATKINSON D (1990). Land: agricultural resource or wildlife reserve? Reorganisation in the food factory. Aberdeen University Review 183, 218-225.

ATKINSON D, CHALMERS N, COOPER D, CORCORAN K, CRABTREE R, DENT B, WATSON CA (1994). The sustainability of lowland management systems pp 9-17. Scottish Sustainable Systems Project. Scottish Office Agriculture and Fisheries Department.

BRUNTLAND (1985). Mandate for change: Key Issues, Strategy and Workplan.

DOYLE C J (1995). The use of models in integrated crop protection. ibid.

FOX R T V (1995). Detection technology for plant pathogens. ibid.

FRY G L A (1995). Landscape ecological principles and sustainable agriculture. ibid.

GROSSBARD E, ATKINSON D (1985). The herbicide glyphosate. Butterworth UK.

HELENIUS J (1995). Regional - crop rotations for ecological pest management (EPM) at landscape level. ibid

KNIGHT J D (1995). Decision support systems for integrated crop protection. ibid.

MCQUILKEN M P (1995). Promoting natural biological control of soil-borne plant pathogens. ibid.

OGILVY S E, TURLEY D B, COOK S K, FISHER N M, HOLLAND J, PREW R D, SPINK J (1995). LINK integrated farming systems: a considered approach to crop protection. ibid.

sustainability, even within the context of European lowland agricultural systems. They suggested the importance of considering land quality, natural heritage, the rural population, energy and rural infrastructure in any discussion of sustainability. All of these elements are equally important when assessing crop protection in sustainable systems.

A more practical starting point to a pragmatic discussion of sustainability might begin by asking the questions:- Why are our conventional agricultural systems not considered to be sustainable? What is wrong with today's farming? Most answers to these questions identify issues which relate to changed priorities in relation to the use of non-renewable resources, principally fossil fuels and a changed view on the acceptable balance between food production and environmental and wildlife impacts. Essentially the opportunity costs of current farming systems appear to be felt to be too high, so that a cost benefit analysis of farming today would seem to conclude that costs substantially outweigh benefits. This change in public attitude (Atkinson, 1990) to farming, which has occurred over a short time scale, suggests that what is today considered sustainable may not be considered so in the next decade as public perceptions of priorities change. Flexibility in approach seems likely to be a key element in the development of sustainable systems. In this context, both understanding and anticipation will be important to the future planning of agricultural systems.

ECOLOGICAL BASIS OF SUSTAINABILITY

Biological control, landscape management and forecasting all depend upon, and contribute to, our understanding of the basic ecological processes which control the range of interactions between the organisms which are managed within farming systems. Fry (1995) and Helenius (1995) both identify that decisions related to agricultural sustainability cannot be taken at field and farm levels alone, but that meta-population dynamics, and consequently between-population rather than within-population processes will have a major effect on the persistence of both pests and beneficial species. Helenius (1995) identifies the aim of adjusting spatial and temporal scales of rotations so as to drive pests, usually the species of early successions, into local extinction. He also identifies that in ephemeral habitats, such as arable fields, the role of specialist natural enemies to pests and diseases is likely to be restricted. Other papers within the conference volume provide case histories in relation to these basic theories.

Practically, the most attractive elements of conventional crop protection with chemicals are simplicity, predictability and the absence of the need for a full understanding of basic processes. For a well translocated broad spectrum herbicide, such as glyphosate (Grossbard and Atkinson, 1985), the most important understanding required for successful weed control relates to how to apply the chemical to the target species, rather than knowledge of the ecology or physiology of the species. In systems dominated by chemical control, knowledge of the interactions between species and the autecology of individual species is of limited importance. Many studies of biological control have adopted a similar conceptual approach and emphasised the bringing together of host and pathogen, or predator and prey (Wilson *et al* 1995). This is appropriate for specific relatively aggressive predators, pathogens and a targeted approach, but is less valuable in more complex situations. Here the introduced organism must first establish itself within a niche before it can begin to interact with its target and other species. In most real farming situations interactions between species will be important. Complex interactions are also important where the time scale extends over more than a single season. In situations such as these there is need for a true understanding of the ecology of the system, i.e. there is no substitute for knowledge. In analysing the problems of promoting biological control through the use of soil borne plant pathogens McQuilken (1995) documents the importance of this approach and hence the need for holistic crop protection to be based upon a clear understanding of the ecology of crop plants, weeds, pests, predators, pathogens and naturally occurring micro-organisms.

All crop protection systems, even those using targeted chemical control, involve the use of some knowledge of the ecology of the crop and the other organisms found within it. Without the use of information technology, the amount of information which can be used in decision-making is limited. The inability to process information has, in the past, limited our ability to design crop protection systems which more fully use the full range of available ecological information. Doyle (1995) indicates that mathematical models have now developed to a stage where they should be able to aid decision-making in relation to the management of herbicide resistance, weed-crop interactions and integrated weed management systems. In addition, complex decision support systems (Knight, 1995) seem likely to become more common, as part of crop protection practice, in the near future.

In the section above on "sustainability", we suggested that conventional agriculture has become less acceptable to the population in general, and hence unsustainable, at least in part, because the current balance between food production and biodiversity was felt to be inequitable, especially in relation to wildlife and conservation. Practice would seem to indicate that good farming and successful biodiverse management can be achieved on the same land area. Many of the recent adverse interactions between farming and conservation have occurred because of a lack of understanding of the complex interactions between species which may control the balances which exist within a biodiverse assemblage, i.e. field margins, hedge rows and species rich grass land. The ecological understanding, identified above, as necessary for soundly based biological control and decision support systems is also needed if wildlife is to become a normal component of sustainable agricultural systems (Atkinson *et al*, 1994).

ANTICIPATION AND SUSTAINABILITY

A high proportion of the agrochemical which is applied to all of our crops is wasted. Some is wasted because it is applied to give protection against diseases or pest outbreaks which do not occur, or to control weeds which are absent from a particular soil area. With hindsight, a significant amount of chemical could have been saved by making application only when pest or disease outbreaks develop to levels with economic significance or to areas from which weeds emerge. The balance of risk and benefit, however, normally indicates the wisdom of an appropriate insurance strategy. Reducing the level of insurance applications requires a greater confidence in the identification of the presence of the pest/pathogen/weed at a stage when it is most susceptible to treatment. It also requires a clear view of the likely development of the pathogen from an initial low level under the particular field conditions found at the site for treatment. The development of the decision support systems discussed above will aid the second of these aspects. The first is currently being aided through improved diagnostic techniques (Fox, 1995). Developments in this area suggest that this approach is likely to become more common.

Anticipation also involves the development of new technologies and modifications to existing technologies so as to meet changed needs. The papers in this volume indicate the continued development of chemical control agents, together with that of more novel strategies, such as a stimulo-deterrent diversionary strategy, where semio-chemicals are used to modify behaviour and, as a consequence, give protection (Pickett *et al*, 1995). The availability of such materials is essential to the success of the TIBRE (Targeted Inputs for a Better Rural Environment) approach detailed by Tait and Pitkin (1995). Here the sustainability of systems is enhanced, not by a reduction in input but, primarily, by using inputs with reduced environmental side effects.

FOOD PRODUCTION AND SUSTAINABILITY

The evolution of the sustainability concept, from absolute to relative, and the role of public acceptability in defining priorities for resource use is leading to a redefinition of key targets for sustainable systems. Perspectives on what are acceptable risks due to crop protection and the methods of crop protection permissible within sustainable agricultural systems seem likely to vary with time. The implications of these changes must be that flexibility in approach in public attitudes, and the development of methods firmly based upon ecological principles, are likely to be the best ways ahead. Only in this way will it be possible to increase either the intensity of agricultural production or the area of land given over to food production without the loss of, recently gained, benefits to biodiversity. Where biodiversity is being achieved in parallel with viable food production, this is likely to represent a sustainable system, as defined by Atkinson *et al* (1994).

EXPERIMENTATION ON SUSTAINABLE FARMING SYSTEMS

Sustainable systems of crop protection seem likely to be based upon a fuller understanding of ecological principles and especially upon knowledge of inter-specific interactions and the effects of scale-related factors, i.e. metapopulations. In any situation where either interactions or scale are important, attempts to develop a blue print of systems from the sum information derived from a series of targeted studies, i.e. effects of single bio-control agents, is likely to be unsuccessful. Only studies including the monitoring of whole systems will provide information based upon the full range of complexities. This volume includes information on five system experiments currently in progress in northern Europe. These experiments can involve substantial areas of land, i.e. up to 75 hectares per plot in the ACTA experiments (Viaux and Rieu 1995) and many sites, six in the LINK IFS project (Ogilvy *et al* 1995). The size of such experiments, and hence their cost, means that considerable care is required to ensure that designs deliver answers to substantial questions and provide information which is generally, and widely, applicable. The changes in the way in which research is currently funded in the United Kingdom and some other European countries, with increased emphasis on shorter term projects, clearly makes the maintenance of such experiments difficult. Field experiments, such as those detailed above, are expensive to establish. In the absence of committed funding, such experiments will either be ended prematurely, thus failing to obtain the best return on establishment and monitoring costs, or will simply not be done. Changes in the objectives of crop protection research clearly

PICKETT J A, WADHAM L J, WOODCOCK C M (1995). Exploiting chemical ecology for sustainable pest control. ibid.

TAIT J, PITKIN P (1995). The role of new technology in promoting sustainable agricultural development. ibid.

VIAUX P, RIEU C (1995). Integrated farming systems and sustainable agriculture in France. ibid.

WILSON M J, HUGHES L A, GLEN D M (1995). Developing strategies for the nematode *Phasmarhabditis hermaphrodita* as a biological control agent for slugs in integrated crop management systems. ibid.